Neglected Plant Foods of South Asia

Tariq Ismail • Saeed Akhtar
Claudia E. Lazarte
Editors

Neglected Plant Foods of South Asia

Exploring and Valorizing Nature to Feed Hunger

Springer

Editors
Tariq Ismail
Department of Food Science
and Technology
Bahauddin Zakariya University
Multan, Punjab, Pakistan

Saeed Akhtar
Faculty of Food Science and Nutrition
Bahauddin Zakariya University
Multan, Punjab, Pakistan

Claudia E. Lazarte
Department of Food Technology
Engineering and Nutrition
Lund University
Lund, Sweden

ISBN 978-3-031-37076-2 ISBN 978-3-031-37077-9 (eBook)
https://doi.org/10.1007/978-3-031-37077-9

This Springer imprint is published by the registered company Springer Nature Switzerland AG
The registered company address is: Gewerbestrasse 11, 6330 Cham, Switzerland

Paper in this product is recyclable.

To the small farmers of South Asia for bringing us food, peace, and happiness

Preface

The global burden of hunger and food insecurity in a preview of the growing world population and the rapidly changing climate calls for immediate measures to upgrade staple crops' over-reliance on world food systems. The agriculture sector in South Asia and other parts of the world has experienced an intensive increase in crop production to increase food supplies. Contrarily this one fits for all solution has posed substantial damage to the agrobiodiversity and regional climate in particular, and agroecosystem in general. Neglected or underutilized plant species inherit the potential to survive under an array of biotic and abiotic stresses for which conventional or mainstream crops are hardly resilient. At large, neglected food plant species are rich in caloric energy, essential amino acids carrying proteins, dietary fibers, minerals, vitamins, and a diverse range of phytochemicals of human health significance.

South Asia is considered a center of global agrobiodiversity and has a wide range of nutrients dense, climate-resilient, locally adaptable, profitable, wild, semi-domesticated, and domesticated plant species. Food and Agriculture Organization has declared many neglected and underutilized plant species of South Asia as Future Smart Foods that can better fight against various forms of malnutrition in the region. Since neglected food plant species have never been a priority in our modern food production system, they often fail to cross the boundaries and borders of the farms and native regions of cultivation. According to the World Bank statistics, 57–81% of the people of South Asian countries belong to rural areas indicating a larger share of people in this region rely on their farm resources to ensure sustainability. Genetic resources of neglected or underutilized plant species if better conserved and sagaciously exploited, can guarantee food and nutrition security of the poor and under-privileged rural communities. Strong commitments and outcome oriented roles of governments and other stakeholders are mandated to implement neglected, under-utilized plant species operational framework that includes but is not limited to the selection of resilient and nutrients dense crop species, market assessment and establishing value chains, planning initiatives aimed at developing domestic and export market for neglected plants species, and development of the policies to mainstreaming of neglected underutilized plant species in modern agriculture system.

Neglected plant foods of South Asia – *Exploring and valorizing nature to feed hunger* comprehensively addresses all critical aspects of neglected, underutilized plant foods in the context of food security and climate change in South Asia. Chapters included in this book cover the topics of biodiversity, agrobiodiversity, climate change, essential criteria to include neglected plant species in food security planning, neglected plant foods of South Asia with particular emphasis on India, Pakistan, Bangladesh, Sri Lanka and Nepal. Nutrient rich, neglected foods such as millet, quinoa, amaranth, moringa, date palm, taro, jackfruit and fenugreek are presented in detail, highlighting their great potential to be included in the agro-food value chains to achieve sustainable and resilient food systems for zero hunger and better nutrition. Written by a team of experts in agrobiodiversity, food security, climate change, postharvest horticulture, crop production, food technologies, nutritional and social sciences, this book is an essential read for researchers and experts from all domains of food systems.

We would like to appreciate support of doctoral scholars of food value-addition research group at Faculty of Food Science & Nutrition, Bahauddin Zakariya University, in organizing and formatting contents of this book.

Multan, Punjab, Pakistan — Tariq Ismail
Saeed Akhtar
Lund, Sweden — Claudia E. Lazarte

Contents

Editors' Biography

Tariq Ismail is Associate Professor at Department of Food Science & Technology, Bahauddin Zakariya University, Multan, Pakistan. Dr. Tariq received his doctoral degree in Food Sciences from Pakistan and did his post doctorate from Department of Food Technology, Engineering and Nutrition, Lund University, Sweden. Salient research interests of Dr. Tariq include food value addition, food waste valorization, food safety and nutrition aspects of food security. He is leading a research group on exploration and value addition of indigenous food plant species of Pakistan. During the last couple of years, he has worked as research investigator on national and international projects emphasizing value addition of indigenous vegetables of Pakistan to develop cost effective, nutrient dense and safer food formulations as means to mitigate hidden hunger and other chronic health ailments. He is supervising master and doctoral research in food value addition and development of functional foods for the dietary management of nutritional inadequacies and metabolic disorders. Dr. Tariq has more than 60 scientific publications and book chapters on his record with high ranked international journals and publishers. He is on the panel of more than 15 peer-reviewed scientific journals with editorial and reviewer assignments. Dr. Tariq has a diverse experience of working with industry, public sector research and academia organizations in various capacities. He is a lifetime member of Pakistan Society of Food Scientists and Technologist and Pakistan Agricultural Scientists Forum.

Saeed Akhtar is an experienced educationist with a broad-based administrative and research profile. He earned his doctoral degree in Food Sciences from Pakistan and later completed postdoctoral training at Oregon State University, USA. He laid the foundation of a new era in the field of food and nutritional sciences in Pakistan. He is the Founder Dean of the Faculty of Food Science & Nutrition at Bahauddin Zakariya University Multan, Pakistan which has evolved as a premier national-level educational entity imparting knowledge in this field of specialization. His tireless efforts for raising awareness of the role of good food as an indispensable ingredient for human health among the people of the southern part of the Pakistani province of Punjab have been greatly valued at the societal level. Dr. Saeed has a wonderful research profile with around 125 international publications and book chapter contributions. He has won a multitude of research grants at national and international levels and has supervised more than 100 doctoral and master research theses on food safety, food security, food waste management, and food value addition. Recognizing his pedagogical skills and the knack for teaching, the Higher Education Commission of Pakistan conferred upon him the Best University Teacher Award for the years 2016 and 2021–22. Similarly, Pakistan Council for Science & Technology declared him one of the most productive scientists of Pakistan during 2013–2017. He is a member of the leading national and international organizations operating for the promotion of food technologies, nutritional sciences, microbiology, and the environmental protection. He believes that a rational application of food technologies can provide an optimum solution for food shortage, hunger, nutritional deficiencies, and food security in Pakistan.

Claudia E. Lazarte is an Associate Senior Lecturer at the Department of Food Technology, Engineering and Nutrition, Lund University, Sweden. Claudia received her PhD. from Lund University, thereafter she worked at the Food Agriculture Organization of the United Nations in development research projects investigating dietary diversity in developing countries particularly Tajikistan and Nigeria. In 2017 she returned to work in academia at Lund University, where her research interests are focused on nutritional and anti-nutritional compounds in plant-based foods, food processing to increase bioavailability of nutrients. Investigating bioactive compounds in underutilized fruits in Mozambique and fostering further collaborations with low- and middle-income countries in multidisciplinary development projects. She has published various peer-reviewed scientific papers, book chapters and technical reports.

Abbreviations

1-MCP	1-Methylcyclopropene
3,4,5-triCQA	3,4,5-Tri-O-Caffeoylquinic Acid
5-LOX	5-Lipoxygenase
ACE	Angiotensin Converting Enzyme
ADF	Acid Detergent Fiber
AG	α-Glucosidase
AGRs	Agricultural Genetic Resources
AMMI	Additive Main Effect and Multiplicative Interaction
AP	Active Packaging
APGRs	Agricultural Plant Genetic Resources
APH	Amaranth Protein Hydrolysates
API	Amaranth Protein Isolate
BAP	Biodiversity Action Plan
BARI	Bangladesh Agricultural Research Institute
BAU	Bangladesh Agricultural University
BBS	Bangladesh Bureau of Statistics
Bcl-2	B-Cell Leukemia/Lymphoma 2
BFN	Biodiversity for Food and Nutrition
CA	Controlled Atmospheric
CBD	Convention on Biological Diversity
CBM	Community based Biodiversity Management
CCL_4	Carbon Tetrachloride
CGRFA	Commission on Genetic Resources for Food and Agriculture
CIRAD	French agricultural research and international cooperation organization working for the sustainable development of tropical and Mediterranean regions
CMC	Carboxymethyl Cellulose
CSA	Climate Smart Agriculture
CTX	Cyclo-phosphamide
CVD	Cardiovascular Diseases
DFSs	Diversity Field Schools

DFTQC	Department of Food Technology and Quality Control
DNA	Deoxyribonucleic Acid
DU145	Human prostate cancer cell line
EAC	Ehrlich's Ascites Carcinoma
FAO	Food and Agricultural Organization
FCT	Food Composition Tables
FDA	Food and Drug Administration
FPGR	Food Plant Genetic Resources
FPOs	Farmer Producer Organizations
FSFs	Future Smart Foods
Ft	Feet
GAP	Good Agricultural Practices
GCC	Gulf Cooperation Council
GDP	Gross Domestic Product
GEF	Global Environment Facility
GFR	Growth Factor Receptor Protein
GHI	Global Hunger Index
GID	Gastrointestinal Digestion
GLP-1	Glucagon Like Peptide-1
GLUT4	Glucose Transporter Type 4
GMOs	Genetically Modified Organisms
GPA	Global Plan of Action
GPCSR	G Protein Coupled Receptors
GSH	Glutathione: a tripeptide, γ-L-glutamyl-L-cysteinylglycine
HepG2	Hepatoma G2
HFE	Hydrolyzed Fenugreek Extract
IAPS	Invasive Alien Plant Species
IBPGR	International Board for Plant Genetic Resources
IBS	Irritable Bowel Syndrome
ICAR	Indian Council of Agricultural Research
IDA	Iron Deficiency Anemia
IFAD	International Fund for Agricultural Development
Ig	Immunoglobulin
IGF-1	Insulin-like Growth Factor 1
IIMR	Indian Institute of Millet Research
IIMR	Institute of Millets Research
IL	Interleukins
IMISAP	Implementation Strategy and Action Plan
INGO	International Non-Governmental Organization
INOS	Inducible Nitric Oxide Synthase
IPCC	Intergovernmental Panel on Climate Change
IPGRI	International Plant Genetic Resources Institute
IPM	Integrated Pest Management
IRRI	*International Rice Research Institute*

ITPGRFA	International Treaty on Plant Genetic Resources for Food and Agriculture
IU	International Undertaking
JBF	Jackfruit Bulb Flour
JSF	Jackfruit Seed Flour
KHG	Kandyan Home Gardens
KNMS	Kenyan National Micronutrient Survey
LDL	Low Density Lipids
LI-BIRD	Local Initiatives for Biodiversity, Research and Development
LNCaP	Lymph Node Carcinoma of the Prostate
LPS	Lipopolysaccharide
MA	Modified Atmospheric
MAD	Minimum Acceptable Diet
MAP	Modified Atmosphere Packaging
MCF-7	Michigan Cancer Foundation-7
MDA-MB-231	M.D. Anderson – Metastatic Breast 231
MDD	Minimum Dietary Diversity
MDD-W	Minimum Dietary Diversity Women
MLS	Multilateral System
MMF	Minimum Meal Frequency
MoALD	Ministry of Agriculture and Livestock Development
MOFA	Monounsaturated Fatty Acids
MPI	*Mantegazzianus* Isolated Protein
MPUAT	Maharana Pratap University of Agriculture & Technology
MSG	Monosodium Glutamate
MT	Metric Tons
MT	Million Tons
n-3	Omega 3
NAPA	National Adaptation Programs of Action
NARC	Nepal Agricultural Research Council
NARS	National Agricultural Research Systems
NASA	National Aeronautics and Space Administration
NCDs	Non Communicable Diseases
NCLs	Neglected Crop Landraces
NCS	National Conservation Strategy
NDF	Neutral Detergent Fiber
NDMA	National Disaster Management Authority NDMA
NF-κB	Nuclear Factor Kappa – light-chain-enhancer of activated B cells
NFP	Neglected Food Plants
NFPs	Neglected Food Plant Species
NGO	Non-Governmental Organization
NPFs	Neglected Plant Foods
NSAIDs	Non-Steroidal Anti-inflammatory Drugs
NUCS	Neglected and Underutilized Crops Species
NUE	Nitrogen Use Efficiency

NUFPs	Neglected and Underutilized Food Plants
NUS	Neglected and Underutilized Species
NWFP	Northwest Frontier Province
PAR	Platform on Agrobiodiversity Research
PDS	Public Distribution System
PEM	Protein Energy Malnutrition
PGE2	Prostaglandin E2
PSI	Pound per Square Inch
PUFA	Polyunsaturated Fatty Acids
QF	Quinoa Flakes
RBC	Red Blood Cell
RDA	Recommended Dietary Allowance
RH	Relative Humidity
RQC	Royal Quinoa Catalogue
RTE	Ready to Eat
RTS	Ready to Serve
SAARC	South Asian Association for Regional Cooperation
SDGs	Sustainable Development Goals
SFA	Saturated Fatty Acids
SHR	Spontaneously Hypertensive Rats
SPECIES	Spatial Evaluator of Climate Impacts on the Envelope of Species
SPG	Sweet Potato Glycoprotein
SSF	Solid State Fermentation
T2DM	Type 2 Diabetes Mellitus
TBARS	Thiobarbituric Acid Reactive Substance
TNF	Tumor Necrosis Factor
TNF-α	Tumor Necrosis Factor alpha
UHPLC	Ultra-High-Performance Liquid Chromatography
UN	United Nations
UNDP	United Nations Development Programme
UNEP	United Nations Environment Programme
UNICEF	United Nations International Children's Emergency Fund
USDA	United State Department of Agriculture
VLDL	Low-Density Lipoproteins
VP	Vacuum Packaging
WF	Wheat Flour
WFP	World Food Program
WHO	World Health Organization
WRI	*World Resources Institute*
WUE	Water Use Efficiency
ZHC	Zero Hunger Challenge

Chapter 1
Agricultural Biodiversity and Food Security: Opportunities and Challenges

Tariq Ismail, Muhammad Qamar, Maria Khan, Sadaf Rafique, and Attiya Arooj

1.1 Introduction to Agrobiodiversity and Food Security

Agrobiodiversity or agriculture biodiversity is the key characteristic of the farming system and comprises of genetic resources, edible plants and the food crops, livestock, soil microflora, biological insect pest control system and the wild resources that can ensure ecosystem function and the services. Agrobiodiversity also includes the conventional practices opted by the farmers to produce crops and manage farm resources. Traditional practices such as use of particular plant species for pest control and integrating woody shrubs and the trees into the farming system are also considered integral components of agrobiodiversity (Thrupp 2000). Considering its significance to the quality of life of farming communities, agrobiodiversity is termed crucial for delivering agroecosystem services (Sibhatu et al. 2015). Biodiversity, agrobiodiversity in particular, has been realized as fundamental to sustainable agro food production system and the food security. Changes in agriculture production system as evident from existing agriculture growth patterns, have drastically affected biodiversity in plant genetic resources, farm animals, insects and soil microflora. Climate change and limited availability of external inputs have generated need for developing a sustainable agriculture system by practicing and promoting agrobiodiversity. Asia in hand with Africa and Latin America are ranked as the centre of diversity since majority of the staple food crops had originated in one of these continents. Traditional farming systems like poly cultural system or the home gardening which provide basis for high biodiversity are also in existence in South East Asia, Central America, and Sub-Saharan Africa.

A major loss to agro-biodiversity is attributed to human-led activities. According to the estimates from Food and Agriculture Organization (FAO), 30–70% losses

T. Ismail (✉) · M. Qamar · M. Khan · S. Rafique · A. Arooj
Department of Food Science & Technology, Faculty of Food Science & Nutrition, Bahauddin Zakariya University, Multan, Pakistan

T. Ismail et al. (eds.), *Neglected Plant Foods of South Asia*,
https://doi.org/10.1007/978-3-031-37077-9_1

have been made to the genetic diversity of livestock and the field crops. Many of the livestock breeds are at the verge of extinction (FAO 2015). Such drastic losses to biodiversity may be referred as footprints of the human activities such as opting high yielding crop varieties and the animal breeds at the cost of a diverse range of low yielding but resilient native crops and livestock breeds (Zimmerer 2013). Varietal replacement being a narrow approach to increase productivity is the key cause of loss in genetic biodiversity. Faulty planning to enhance farm productivity by varietal replacement or genetic erosion is happening at a higher pace in the developing countries than observed in developed economies (Sahai 2010).

1.2 Food Security Situation of South Asia

South Asia provides ~5% of the global agricultural land and ~20% of the world food requirements. The region is one amongst the densely populated part of the world and is considered center to hunger, food insecurity, climate change, and conflicts. Nearly one third of the world's poor live in South Asia, 70% of whom belong to the rural areas and their sustenance is met from agriculture (Rasul 2021). Global food security statistics recorded prior to COVID – 19 pandemic indicate nearly 649 million peoples in South Asia as moderately food insecured and 271 million heads as severely food insecured. The situation was expected to get worsen in post pandemic period (Sumner et al. 2020). Despite the hopes that world will emerge from the crises of pandemic and that food security situation will recover to normal, COVID-19 tighten its grip over the GDP growth of many countries, and did not translate to any gain in food security. After remaining unchanged since 2015, prevalence of undernourishment increased from 8.0 to 9.8 between 2019 and 2021. Due to unequal recovery of the economy of many countries, hunger affected the poor nations from Asia, Africa and Latin America at a larger extent. Nearly 425 million peoples were affected by hunger in Asia in the year 2021. The region although witnessed a slight decrease in moderate to severe food insecurity from 2020 to 2021, the situation of sever food insecurity was bit higher than recorded in 2020 (FAO 2022).

According to the Food and Agriculture Organization (FAO) estimates for the year 2019, nearly 14.9% of the peoples in the South Asian Association for Regional Cooperation (SAARC) countries are suffering from hunger. There is also a continued upsurge in the number of malnourished peoples and the rate of prevalence of various forms of malnutrition in many of the South Asian countries (FAO 2019).

Agriculture is amongst the most vulnerable sector to the changing climate while the rate of this change is too rapid. This is likely to affect agro food production, food process, economic accessibility of peoples to diversified range of foods and the situation of food security. Though the recent upsurge in food prices are mainly linked to increased food demands by increasing number of consumers and better income levels, food prices are expected to further rise with climate changes associated reduction in food production (Bandara and Cai 2014). Cereals including Wheat, Rice, Maize and Barley are the main staple food crops of Asia. During the last

3 years, a significant decline has been noticed in production of wheat crop in wheat producing South Asian countries including India, Pakistan, Afghanistan and Nepal. Rice produced in South Asia accounts approximately 90% of the global rice supply, and is cultivated in the areas with high rate of precipitation. Higher yield (million metric tons) was projected for rice in India and Bangladesh while expected rate of rice production in the year 2023 was estimated lower than one recorded in 2020/21 in Pakistan, Nepal and Sri Lanka (USDA 2023).

Achieving the challenge of zero hunger in South Asia by the year 2030 requires collaborative support to resolve present era issues including climate change, loss in agrobiodiversity, water scarcity, poverty, social inequality, and the issues like food safety and nutrition (IRRI 2023). Current agriculture and food production system, also known as intense agriculture though has contributed a significant improvement in food production to meet current population requirements however; it has also anticipated negative impact on environment. Excessive use of water, agrochemicals and tillage linked losses to soil structure have contributed challenges including environmental pollution, loss in biodiversity, scarcity of useable water and energy resources, and has intensified risks to human health (Pingali 2012; Lam et al. 2017). Such a situation calls for immediate actions to neutralize the negative effects of climate change and making food production system more resilient and friendly to the ecosystem and the human health.

1.3 Agrobiodiversity and Sustainable Food System

Sustainability of agriculture and agro-food production system without significant loss to the ecosystem can be ensured by promoting agrobiodiversity with better use of highly valuable but neglected, underutilized food crops or species. Promoting agrobiodiversity can create an array of development opportunities like smart farming system, environment sustainability, better health and nutrition and prosperity of the farming communities.

According to the United Nations, twentieth century has witnessed upto 75% loss in diversity of world food crops as high yielding genetically uniform food crops were opted by the farmers at the cost of local cultivars (FAO 2009). Farmers in Indian subcontinent were reported to cultivate nearly 30,000 wild food crops by the year 1950 however; merely 50 were projected to thrive by the year 2015 (Gonzalez 2011). This also indicates many of the countries in South Asia as center to genetic biodiversity. A diverse range of edible plant species under domestication by the small farmers are vital to global food security. Diverse plant species have abilities to prevent conventional crops failure and serve as the raw germplasm for developing plant varieties that may better thrive to emerging environmental challenges. During centuries of their cultivation, traditional crops have developed a range of valuable characteristics to better stand against different types of stress. It's the reason that breeders often choose wild relatives of commonly cultivated crops for breeding, and to develop traits enabling plant resilient and sustainable to changing ecosystem.

Historically, wild species of different cultivated crops are used to breed wheat, potatoes, maize, sugarcane, tomato, strawberries, black pepper, cacao and peanuts (Gonzalez 2011). Nevertheless, extinction of the wild or traditional plant species due to extensive degradation of natural habitats is a threat to biodiversity and ecosystem stability.

Approach to produce more food to feed the current and future needs of the world is no more a sole workable idea to address multipronged issues of food security. The renewed interest in food systems focuses on increasing food production by simultaneously addressing the issues like diet quality, environmental impact of food production and distribution, and socioeconomic impact on food growers and producers (Béné et al. 2019). With the new concepts of the food systems, current system of food production, consumption and the governance is facing a systematic failure, and demands a paradigm shift (Hajer et al. 2016; Scott 2017). This narrative is likely built on the evidences that our food system is not able to feed the future demands, do not address the issues of nutrition and healthy diets production, unable to deliver equal and equitable benefits to all stakeholders, anticipate more loss to the environment. Consequently, it may threaten situation of the food and nutrition security, preservation of natural resources and agrobiodiversity, and social justice (Béné et al. 2019). Over the last 50 years, food production system changed from diversified to ecologically simple cereals based system that resulted in substantial loss to agro-biodiversity at species to allelic levels while over exploitation of the natural resources drastically affected cultivable land environment. Under given circumstances, key agroecological principle to work upon is sustainable use of natural resources and creating compatibility between agricultural practices and the agroecological system (Fanzo et al. 2012; Scott 2017). Cultivation and rearing of a diverse range of animals and plant species buffers agriculture against the key threats such as extreme weathers, plant diseases, pest outbreaks, conflicts or hazards mediated interruption in seeds supply. Preserving and promoting natural diversity in a specific region thus supports transforming food system to a more adoptive and resilient to unsolicited shocks and changes (Lin 2011) (Fig. 1.1).

Despite of the evidence based reports on benefits of the agrobiodiversity for the mankind and the ecosystem, nearly 75% of the global biodiversity was lost in twentieth century. In China, nearly 10,000 varieties of wheat were under cultivation in 1940s which were abandoned in next 30 years. Habitat destruction coupled with loss of life diversity in many of the regions of South Asia and peoples are witnessing mass extinction of different species of plants and animals at historically highest pace. Globally there are 34 biodiversity hotspots among those 8 are in Asia. To be considered as a biodiversity hotspot, the area must have atleast 1500 vascular plant species as endemic and have witnessed at least 70% loss of its habitat (Conservation International 2012).

Fig. 1.1 Agrobiodiversity explained in one frame – a selection from poster competition on role of agrobiodiversity in rural food security. (Courtesy: *Sana Noor*)

1.4 Biodiversity and Agrobiodiversity Situation of South Asia

1.4.1 India

Despite the fact that a large pool of biodiversity is underreported from India, the region is still house to nearly 8% of the global biodiversity and is listed among 17 megadiverse countries which maintain the criteria of having more than 5000 endemic species. The region maintains nearly 190,000 species of plants, animals and fungi out of 800,000. India is one amongst the mega diverse countries with nearly 7% (45,000) of the world plant species 33% of whom are endemic. There are nearly 15,000 flowering plants and 91,000 animal species that account for 6–6.5% of the world fauna and flowering plant species. Out of 36 regional biodiversity hotspots, four hotspots namely Himalaya, Western Ghats, Sundaland and Indo-Burma are located in India. Still a sizeable biodiversity of the country is undocumented while mounting pressure of population density, deforestation, and the climate changes like challenges has generated unprecedented threats to status of the India's biodiversity (Sengupta and Dayanandan 2022). Reference to the "State of India's Environment in Figures 2021", the Indo-Burma hotspot witnessed a worst loss of vegetation area from 23.7 to 1.2 lacs sq. km. Nearly 25 native plant species have became extinct in the listed four hotspots of the country. Alongside many other factors, swift changes in climatic conditions have badly affected situation of biodiversity in India. In 2016, an average 0.7 °C temperature was increased that led to 5.41 lacs forest fire in 16 states of the country anticipating a significant loss to the forests volume and their biodiversity (Dhawan 2021).

1.4.2 Pakistan

Pakistan has variety of landscape. The country in its nine agro-ecological zones has a rich diversity of indigenous crops with approximately 3000 taxa and their 500 wild relatives. A majority of the wild crops belong to the northern areas of the country (Anwar et al. 2005). Principle crops of the region include wheat, rice, barley and maize, sugarcane, oilseeds, pulses, vegetables and the fruits of temperate and sub-tropical regions. Plant Genetic Resource Institute of the country maintains nearly 16, 899 accessions of 40 crops (GOP 1999; Anwar et al. 2005).

According to the UNDP (1999) estimates, about 240 species of the plant which are endemic to Pakistan belong to mountainous areas of the country. Gilgit Baltistan – a high mountainous area of the country is considered diverse in a range of ecosystem. The diverse flora of the region include but not limited to a wide range of fruiting plants and herbs such as wild cumin, thyme, apricot, plant nuts, hawthorne, buckthorn and the major food crops such as rice, wheat, barley, maize and potato (Khan 2014). Thar is the largest desert of Pakistan and peoples of this region meet their livelihood needs with agriculture and animals rearing. Agrobiodiversity of Thar region furnishes food, pastures for milking animals, herbs for traditional culinary and medicinal uses and animals' wools for preparing warm clothing. Irrespective of the fact that agrobiodiversity has remained a source of sustenance for the peoples of Thar region, the issues of water scarcity is resulting in a rapid loss to agrobiodiversity with every passing day. Excessive grazing is another challenge anticipating damage to native vegetations and loss to the soil fertility (Usman et al. 2022). Failure to pass on traditional knowledge on health promotive uses of many indigenous medicinal plants and herbs to the generations also anticipated a significant damage to the biodiversity in this region (Husain et al. 2008). Conservation of traditional knowledge thus can be considered a vital component of biodiversity conservation strategy.

Government of Pakistan developed National Conservation Strategy (NCS) in the year 1992 and conservation of the biodiversity was an essential part of the NCS. The country is also a part of Convention on Biological Diversity (CBD) and to comply with the CBD, a Biodiversity Action Plan (BAP) was developed and approved by the government in 1999. BAP deals with conservation and sustainable exploitation of biodiversity in all agroecological zones of the country (Baig and Ahmed 2007). Plant genetic resources for food and agriculture provide the basis for improved rural food and nutritional security and support peoples' livelihood. Political will of the government and related institutions to develop and implement policies on promoting agrobiodiversity, incentivizing farmers and growers to prioritize cultivation of native plant species, establishing a value chain for marketing indigenous food plants and their products, conservation of the genetic resources and the traditional knowledge too are amongst the few important considerations for conserving and promoting agro-biodiversity in Pakistan.

1.4.3 Bangladesh

Bangladesh is situated on Ganges – Brahmaputra delta that is world's largest delta plain. Geographically, country falls in one of the largest global biodiversity hotspot i.e., Indo – Burma region with an endemic plant species count of more than 7000 (Mittermeier et al. 1998). Bangladesh is also rich in biodiversity however; like other countries of the region and the world, biodiversity and the ecosystem of Bangladesh has also a critical period of decline. Three major vegetations are covered by the forest of Bangladesh which include hill forest, plain land forest and mangrove forests. Although these regions remained very rich in biodiversity however; since last few decades, forests of the country and their ecosystems have been heavily degraded (Mukul et al. 2008). Around 2260 plant species have been reported from the hill tracts of Chittagong. There are 5700 species of angiosperm, 68 woody legumes, 500 plants of medicinal significance, 29 orchids, 130 fiber producing plants, 3 angiosperms and 1700 pteridophytes (Firoz et al. 2004). Country has more than 8000 varieties of rice alongside 3000 varieties of other crops. Bangladesh Agriculture Research Council has recorded 3000 pulses varieties, 3516 of vegetables, 781 of oilseed crops, 89 varieties of fruits, 156 of the spices and 475 varieties of tea (Chowdhury 2012). A large pool of this agrobiodiversity is under threat due to mass scale cultivation of rice and other high yielding conventional crops, and injudicious application of agrochemicals for insect pests control and yield improvement (Mukul et al. 2018).

1.4.4 Sri Lanka

Heterogenity of topography and the climate makes Sri Lanka rich in ecosystem diversity. Due to high specie endemism, Western Ghats of Sri Lanka and the India too were listed among first 18 global biodiversity hotpots. The list has now been increased to 34 indicating a severe threat to the biodiversity. A severe forest lost from an area of 182, 500 km^2 to merely 12,450 km^2 has been reported from the Western Ghats of Sri Lanka only (Gunawardene et al. 2007). Nearly 1414 Sri Lankan plan species have ethnic uses as traditional medicine, 50 plant species among these are heavily used while 79 are experiencing extinction threat. Country has also considerable biodiversity of major food crops under cultivation, their wild relatives and traditional varieties. Man made ecosystem in the form of home gardening is of particular interest in Sri Lanka. Kandyan Home Gardens (KHG) is the home garden system in central Sri Lanka and are unique example of home gardens (Gunawardena and Fernando 2022). KHG are distributed over the Kandy, Matale, Kegalle, Kurunegala and Rathnapura districts of Sri Lanka with an average size of 0.4 acres and have a total count of 315 species including those of food crops, spices, fruits, medicinal plants, timber and ornamental plants. KHGs are good example of self sustainability and resilience of the small farmers since every home garden has a sizeable number of plant species like clove, pepper and nutmeg which ensure livelihood needs of the household (Warnasooriya et al. 2023). Biodiversity of Sri Lanka, particularly the ecosystems and the habitats quality is considered at high risk of deterioration due to invasive alien plant species (IAPS) demanding placement of a mechanism for a better control over IAPS cultivation in the country (Gamage 2022).

1.4.5 Nepal

Nepal ranks 49th in the world for biodiversity however, the native genetic resources have never been prioritized in agriculture research and development that resulted in severe genetic erosion. Around 28% of the 24,300 species of the country belongs to agriculture genetic resources which may also be referred as agrobiodiversity. The country has 12 agroecosystems which support 1026 crops species, 510 forage species, 35 livestock, 250 aquatic animal species, 17 aquatic plants species and 3500 insects' species. Over the last few decades, Nepal has experienced an average 40% loss of its agro-biodiversity while some farmers of the country have reported nearly 100% extinction of some endemic species in certain regions. Under agrobiodiversity conservation plan, 18,765 accessions of agriculture genetic resources have been conserved so far. Likewise a sum of 24,683 accessions of crops of Nepal, forages and the microbs have been conserved in different international genetic resource conservation banks (Joshi et al. 2020). Among the leading challenges to Nepal agrobiodiversity, rapid erosion of agriculture genetic resources and 95–100% dependency on foreign germplasm are the leading one. Reliance on native genetic resources and their products, enabling ecosystem to support cultivation of diversity rich plant species, identify and develop markets for consumption of native agriculture resources and commodities are amongst the possible opportunities to promote and strengthen agrobiodiversity situation in Nepal (Joshi 2017).

1.4.6 Bhutan, Hindu Kush Himalaya

Bhutan – a Himalayan kingdom has forest coverage of 72.5% of its total land area while the area under cultivation is merely 7.7%. Agriculture is considered as the backbone of Bhutan economy and more than 80% Bhutanese rely on agriculture farming to meet their livelihood and food requirements. Among 36 biodiversity hotspots of the world, four are hosted by Hindu Kush Himalaya that makes this region a storehouse of agrobiodiversity. Kangchenjunga Landscape is a transboundary with 25,086 km^2 area and is home to around 7.2 million peoples. Kangchenjunga Landscape is rich in plants diversity which include wild edible species, crops and products of non-timber forests. Local residents of Kangchenjunga Landscape use nearly 750 cultivated or wild species of plants such as cereals and pseudo-cereals, oil seeds, pulses, beans, vegetables, fruits, tuber crops, condiments, wild edible plant species and agro-forest trees for food, health, trade and other purposes (Aryal et al. 2021). Agroecosystem in Hamalayan region is rich in diversity and is constituted with strongly interconnected forest, crop farming and animal husbandry system thus providing services and goods to the human and other life in agro-ecosystem.

With an increasing challenge of population growth, Bhutan is undergoing transition in agriculture development deploying high yielding cultivars of agricultural crops, extensive use of chemicals to control pests and weeds, reduced crop rotation and development of infrastructure to meet recent era requirements (Wangda 2008). Such changes in agriculture food production system are expected to result in reduction in agrobiodiversity.

1.5 Climate Change and Agrobiodiversity, Governments' Commitments to Agrobiodiversity Conservation

Depending on future growth scenarios, the Intergovernmental Panel on Climate Change (IPCC) has projected an estimated temperature rise from 0.88 to 3.16 °C by the year 2050 and the recent report indicate a century of burning the fossil fuels and land use has led to global warming of 1.1 °C above the pre-industrial level (IPCC 2007, 2023). This long-term change in temperature and precipitation patterns is anticipated to influence cropping seasons, cultivation appropriateness, crop yields, output, and food markets. In Pakistan, spring maize growing season has moved 4.6 days earlier per decade from 1980 to 2014. Fall maize planting has been delayed to 3 days per decade in the same period resulting in a substantial loss to crop productivity (Abbas et al. 2017). The major concern is aforementioned changes affecting millions of farmers around the globe having no resources to withstand climate change.

Wheat, maize, and rice are three major crops of South Asia and effect of climate change on said crops have been studied for decades. The region's temperature and precipitation patterns or variations in rainfall are strongly affecting productivity of the staple cereals. In the given context, Parry et al. (2004) in their study has predicted 5–30% decline in crop yield by 2050s as compared to production rates observed in 1990s. Agricultural productivity of the summer crops is expected to drop by the year 2050 due to high rate of climate variability and pest infestation. Winter crops may be more harmed by average 2 °C temperature rise. If the temperature rises by 3 °C, South Asia's net cereal production will drop by 4–10% by the end of the century. Knox et al. (2012) reported that climate change will reduce average yield of eight important South Asian crops by 8% by the mid of twenty-first century. Aggressive changes in climate of South Asia are also predicted to reduce yield of maize and sorghum by 11–16%. In India, regional warming reduced wheat, rice, and maize productivity by 5%, 6–8%, and 10–30%, respectively leading to high rate of uncertainties such as failure in loan repayment attributed suicides of the poor farmers (Carleton 2017).

Mitigation and adaptation may lower climate change risks, increase socio-economic system resilience, and reduce vulnerabilities (Adger et al. 2014). Farm-scale adaptation efforts influence ecological, social, and economic systems (Sardaro et al. 2015; Ignaciuk 2015). These actions involve crop components (implementation of drought or heat-resistant crops and varieties) and cultivation systems (use of innovative management practices, cropping calendar change, multi-crop system, irrigation equipment investment) (Jarvis et al. 2011). Varietal shifting from modern and more productive ecotypes to landraces like local varieties and breeds of domesticated species that can better adapt to the natural and cultural local environment. Farm level adaptation action can improve sustainability in a wide range of climatic conditions (Pascual et al. 2013). Indigenous species are important due to their genetic variety and adaption to ecological niches (Padulosi et al. 1999). Neglected and underutilised crop species are typically well-adapted to local growing circumstances (Padulosi et al. 1999), making them viable food producers (Idowu 2009). Many Amaranthus species (Laker 2007), wild mustard (*Brassica spp.*) and other wild edible leafy vegetables (Modi et al. 2006), sweet potatoes (*Ipomoea batata*), wild melon (*Curcubita spp.*), taro (*Colocasia esculenta*), and bambara groundnut (*Vigna subterranea*) are neglected crop species. These crops supported indigenous diets. Neglected and underutilized plant species farming has become uncompetitive and unappealing compared to "major" crops, which are encouraged even in unsuitable locations due to differences in productivity. A vast majority of the neglected plant species has high adaptability to a variety of ecological niches, low-input agriculture, and abiotic and biotic stresses, and can better fight poverty and hunger (Zeven 1998). Commitments to conservation of agrobiodiversity have been witnessed across the world and are better reflected by the initiatives such as:

- *The Establishment of the International Board for Plant Genetic Resources (IBPGR)*
- *The International Undertaking (IU)*
- *The Convention on Biological Diversity (CBD)*
- *Global Plan of Action (GPA)*
- *International Treaty on Plant Genetic Resources for Food and Agriculture (ITPGRFA)*
- *International Network of Ex Situ Collections*
- *The Institutional and Capacity Building Framework*
- *Global Crop Diversity Trust*

The International Plant Genetic Resources Institute (IPGRI) protects and uses plant genetic resources for future generations. IPGRI collaborates with other organizations on research, training, and scientific and technical assistance. Since 1974, IPGRI has led national and international efforts to conservation and utilization of neglected species (Padulosi 1998). IPGRI's goal for promoting underused and neglected species aims at using agriculture crops genetic diversity to achieve more balanced and sustainable development. The IPGRI has four major targets to its approachtowardconservationandbetterutilizationofneglectedplantspeciesandincludes:-

- Enhance the conservation
- Strengthen the work of other actors
- Strengthen research on the choice of species based on strategic factors for conservation, development, and food security
- Identify criteria for research, development, and conservation actions on neglected and underutilized species

1.6 Conservation of Plant Agrobiodiversity

With the increase in size of world population, there will be an exigency of more food to satisfy the food requirements of the population. Worldwide, agriculture sector is facing a myriad of issues to ensure global food security under the drastic challenges of changing climate, high water demands in agriculture production, reduced soil fertility, loss of agriculture lands due to urbanization, and consumers' safety concerns to technological advancements in food production and processing. Agricultural diversity and conservation under the given global food security challenges is highly demanded to achieve sustainable food supply and human development (Kahane et al. 2013).

According to the statistics in 2019 more than 650 million heads residing on planet earth were undernourished and these figures are expected to rise with every passing year (FAO et al. 2021). South Asia comes under the umbrella of poorest regions of the world. Before the pandemic of COVID-19, about 271–649 million South Asians were facing moderate to severe food insecurity. Likewise, 36% South Asian children were reported stunted and 16% as malnourished. Despite implementation of global sustainable development initiatives, situation has become even worse during COVID-19 pandemic era. These figures indicate that the nutritional status of developing world is alarmingly poor. Inadequate food consumption coupled with lack of dietary diversity and low meal qualities were reasoned as the root causes of malnutrition in South Asia indicating nutritional status of the peoples of this region as an issue of global concern (Rasul 2021).

Regular consumption of plants based foods is a sign of healthier dietary pattern, and is considered important to ensure health and wellness (Ford et al. 2009). Plant foods have been proven as rich carriers of dietary fibers, essential nutrients and non-nutrient health promoting factors like antioxidants (Steinbrecher et al. 2009). Plant derived foods are plentiful reserves of essential vitamins like vitamin A, E, C and folates, essential minerals, phenolics (flavonoids, phenolic acids, anthocyanins) and glucosinolates. Natural antioxidants of plants origin inhibit production of free radicals in food matrix and inside the human body, and provide protection against oxidative stress and related diseases (Lopez et al. 2007).

Despite the fact that there are more than 30,000 plant species worldwide which can be utilized as food, about 90% of human energy needs are fulfilled from only 82 different plant species. Irrespective of the fact that more than 7000 food plant species are cultivated globally, reliance on a limited number of plants to feed hunger seems unjustified and unwise (Garn and Leonard 1989; Prescott-Allen and Prescott-Allen 1990; Khoshbakht and Hammer 2008). Currently, less than 150 species of crops are being cultivated commercially. Only 103 food crops species are fulfilling ~90% caloric needs of humans while 60% of human energy requirements are met with wheat, maize, potato and rice (Bailey 2016). Thus thousands of plants species are neglected and underutilized which can be considered useful in fulfilling global food requirements of the day, and the future (Chivenge et al. 2015).

Unlike domesticated or staple food crops, underutilized crops are referred to as minor, neglected, orphan or crops of little-use (Li et al. 2020). Many of the neglected food plant species in South Asia are also called as future smart foods (FSFs) for their significant role in eliminating dependency of mainstream agriculture or staple crops (Li and Siddique 2020). Neglected food plants species if incorporated into our conventional farming system, can generate a climate resilient and sustainable food system against the issues of food insecurity, hunger and rural poverty. Many of the underutilized and neglected plant species possess a high nutrients profile since they are a good source of protein, fiber, micronutrients and dietary energy. Unlike conventional food crops, neglected or underutilized food crops being native and resilient to specific ecosystem are very easy to cultivate and produce (Li et al. 2020).

1.7 In Situ Conservation of Agrobiodiversity

This method of conservation involves conservation of natural habitats and ecosystems and maintenance of different specie populations in their natural ecosystem. On the other hand cultivated and domesticated plant species are given an environment that better help them evolve their unique properties. The above mentioned definition thus consists of two techniques which are (a) conservation of genetic reserves, (b) conservation on farm. Both of these concepts involve maintaining the genetic diversity in places where it was evolved. Conservation of genetic resources includes monitoring and management of genetic diversity (natural wild species) within specified locations for the purpose of active and long term conservation. Conservation on farm includes managing genetic diversity of crop varieties (landraces) that are developed locally and management of weed species in traditional agricultural system (Maxted et al. 2011).

1.8 Ex Situ Conservation of Agrobiodiversity

Method of ex situ conservation involves storing the material of interest outside its natural habitat. The techniques used for the preservation of germ plasm include tissue and pollen storage, seed banks, vegetative cloning and plant maintenance (Volis and Blecher 2010). Other ex situ methods of conservation are botanical gardens, conservation gene, seeds, pollens, tissue culture, seedling and DNA banks (Jaisankar et al. 2018).

Seeds banks are used for the purpose of ex situ conservation. Seed banks are the places where seeds are stored under controlled conditions of temperature and humidity (Volis and Blecher 2010). Botanical gardens are the facilities in which endangered and threatened species are conserved. Such botanical gardens are managed at different levels by governments and educational institutes with international assistance. More than 1500 arboretas and botanical gardens are present worldwide having about 80,000 plant species. These botanical gardens have different ex situ conservation facilities like tissue culture and seed banks. Ex situ conservation of plant species and their wild relatives will provide genetic engineers and breeders with required genetic material for creating new phenotypes and genotypes with desired characteristics (Jaisankar et al. 2018).

1.9 Challenges to Agrobiodiversity

New crops other than normally utilized are not easily accepted by the consumers. There is a famous saying that 'United we stand divided we fall' but in case of new and underutilized crops this saying will become 'United crops stand! Divided new species fail'. This elaborates that when neglected and underutilized species of plants are introduced in the market unfortunately they face rejection by the consumers. In recent era, consumer acceptability for underutilized or neglected plant foods is a great hurdle in utilization of new species. Availability of limited information regarding significance of these species, their traditional uses, methods of cultivation and propagation, and comparatively low output are reasoned for poor consumers' acceptability. Moreover, there is very low funding for research for the improvement of neglected, underutilized species, and their utilization. Likewise, global market demand for a limited number of food crops is another challenge to underutilized plant species production (Padulosi et al. 2002).

1.10 Conservation of Agrobiodiversity in South Asian Countries

Sri Lanka is the country comprising of rich agrobiodiversity that is a source of diversified commodities for livelihoods of farmers and other population. A lot of underutilized crop species in Sri Lanka are being neglected because of so many reasons like urbanization, changed food habits and lack or no awareness. But, these underutilized and neglected crops have so many benefits like they can be cultivated with minimal inputs and can be a source of higher incomes for farmers. Information about underutilized crops of Sri Lanka is very limited. A study was conducted to gather information about underutilized crops of different regions of Sri Lanka. Agricultural farms were used for identification of underutilized crops and impact of these crops on food security and household income was observed. Three types of agricultural farms (home gardens, chena and off season paddy lands) were observed for this study. About 37 crops were identified as underutilized among these finger millet was most common followed by cowpea, cassava, sweet melon and cashew nut. Farmers preferred four crops which were cashew (*Anacardium occidentale* L.), finger millet (*Eleusine coracana* L.), mung bean/green gram (*Vigna radiata* L. R. Wilczek), and cowpea (*Vigna unguiculata* L. Walp.) as underutilized crops with best potential. Different strategies can be considered to conserve, sustain and promote underutilized and neglected plant species present in different regions of Sri Lanka. This will benefit the farmers, households, community, region and the country at large (Bandula and Nath 2020).

In Bangladesh, peoples of the hilly areas use shifting cultivation, locally known as jhum, for agriculture farming. This practice is done by burning and clearing the natural vegetation. This system involves the cultivation of different crops like legumes, cereals, tubers, oil seeds, herbs, vegetables, bulbs, spices and rhizomes. Crops of jhum include *Benincasa hispida, Hibiscus sabdariffa, Lagenaria siceraria, Seseli daucifolium, Cucumis melo var. utilissimus, Ocimum basilicum var. purpurence, Elsholtzia incia, Gossypium herbaceum, Pachyrrhizus erosus, Solanum melongena, Maranta arundinacea*, some varieties of *Capsicum annum, Spilanthes olereca and Plectranthus mollis.*

Shifting cultivation can be used for conservation of land races and different varieties. It can be said that this is a traditionally performed in-situ conservation method. Traditionally conservation of agrobiodiversity is done by the following methods

- Farmers gather seeds of best fruit crops and conserve them for future as result diversity is maintained for generations.
- Farmers save seeds for their own use a little more than their requirement and exchange the extra amount of seeds with other farmers of same village and neighbor villages. This eventually eliminates the risk of extinction.
- Women play a very vital role in preservation, selection and storage of seeds. In Nepal about 60% decisions on selection of seeds for plantation are made by women (Alam 2020).

Nepal is one of the most important hotspot of world's biodiversity. Nepal is a country with high rates of climatic and eco-geographic variations therefore, it consist of about 1506 species of forage genetic resources and agricultural species. Out of these 1506 species, 93 are introduced species, 670 wild plants which are edible, 224 wild relatives of crops, 35 semi domesticated species and 484 native cultivated species. Out of 484 natively cultivated species, 275 forage species, 64 agronomic and 145 are horticultural plants.

In Nepal there are four agrobiodiversity conservation strategies which include (a) ex-situ, (b) on – farm, (c) in – situ, and (d) breeding. Within Nepal, about 120 accessions of 2 species (Sweet potato, large cardamom and potato) are conserved in two tissue banks. Fifteen gene banks are present in the country which have about 650 accessions of thirty crops maintained in them. DNA bank is maintaining about 550 accessions of ten edible plant species (rice, sugarcane, cardamom, garlic, maize, mango, chayote, broad leaf, mustard, pea and wheat). Thirty accessions of nine forage plant species are also saved in seed bank (Joshi et al. 2020).

Following methods have been proposed by the researchers from India- for conservation of wild and underutilized plant species:

- Gene banks and research centers can be used for the conservation of land races and traditional crops. Awareness and motivation can be given to farmers for using in situ conservation method.
- Local people should be made aware of the importance of use of such plant species. Because when the use of underutilized wild plant species will increase that will automatically help in their conservation (Pande et al. 2016).

1.11 Role of Local Agrobiodiversity in Promoting Food and Nutritional Security

Sustainability of a viable food and agriculture production system guarantees food and nutritional security, and is determined by safely managed biological resources in an ecosystem. Agrobiodiversity or agricultural biodiversity is also known as biodiversity of animals, plants and the microorganism which are one way or the other used to satisfy food or agriculture needs. In a broader sense, agrobiodiversity includes activities such as cultivating crops and rearing animals of different breeds, the wild relatives, and the species that interact and support every single life in this typical ecosystem. All these species work simultaneously in ecosystem to keep equilibrium, to increase food production and food security (FAO 1999a). Contrary to its role in conservation of natural ecosystem and sustainability of the life on planet earth, extensive loss in agrobiodiversity manifested by disappearance of a large number of plants and animal species is a matter of critical concern.

1.12 Alarming Scenario Over Years of Agricultural Change

Agrobiodiversity expeditious loss is occurring worldwide. Rapidly growing population rate of the developing world, high per capita consumption of natural resources in industrialized nations and non-sustainable agriculture leading to extensive losses in soil structure are the key factors affecting food production globally. Under the recent era innovations and mechanization in agro-food production system, farmer's conventional food production skills, traditions and local knowledge for farm species are under threat, and are yielding a significant decline in agricultural biodiversity (WRI 2005). According to estimates, genetic diversity of some native plants has been lost to about 75%, approximately 30% of the livestock breeds are at extinction risk. Nearly 250,000 to 300,000 edible plant species are known, out of which only 4% are consumed by humans. Wheat, rice and maize are the main contributor plants for humans as they nearly satisfy 60% of calories and proteins needs of the consumers. Nearly 30% of the human's food and agriculture requirements are fulfilled by animals while 12% of the entire world population relies on ruminant's products (FAO 1999b). According to the Food and Agriculture Organization estimates, 2022 world's cereal production has increased from 9 million tons to 2774 million tons while that of milled rice production has shifted from 4.9 million tons to 517 million tons. Alarmingly, many traditional and local food crops having high nutritive value are gradually disappearing. At an instant, high mineral content local grain crops i.e., *Panicum miliaceum* (proso millet) or *Digitaria exilis* (fonio) are not adequately marketed and have been progressively replaced with modern era's high yield crops like wheat and maize (Adoukonou-Sagbadja et al. 2006). This change is leading towards steady loss of agricultural biodiversity associated with loss of traditional knowledge. Human's food dependency on different food resources to achieve food and nutritional security of the populations can be ensured by promoting agrobiodiversity. Within a food system, agricultural biodiversity is the basis of dietary diversity. While being sourced locally, promoting agrobiodiversity can help mitigating health and nutrition linked challenges by providing a variety of nutrient dense foods and the dietary components having potential health benefits.

1.13 Agrobiodiversity Importance with Respect to Food and Nutritional Security

Agricultural biodiversity generates basis for human food production systems. Since it is linked to community's food culture, practices and traditions, agrobiodiversity provides cultural, religious, spiritual and aesthetic values to human societies (Brush 2004). It offers diverse range of nutritious foods, preserve ecosystem health, increase the resilience of the farmers, improve soil fertility and adoptability of the indigenous plants and animals to changing climate. Agrobiodiversity increases food productivity which in turn improves situation of food security, nutrition and health of the peoples, and contributes to their economic growth (Thrupp 1997). Intercropping is proposed as one amongst the many strategies to increase agriculture biodiversity in an ecosystem. Intercropping helps balancing the ecosystem, better utilization of the farm resources, increase crops yield by preventing damage by weeds, pests and plant diseases. Mixed cropping, intercropping and crop rotations with minor crops like legumes have been extensively discussed as promising techniques to boost soil fertility and break disease cycles. Increased consumption trends have been cited for the local or the traditional food crops which contribute to economic stability, dietary diversity and improved health outcomes of the farming communities. A range of food crops have been identified with prospects of improving food security situation at local and regional levels, and include cereals such as fonio (*Digitaria exilis*) and tef (*Eragrostis tef*); non cereal grains like quinoa (*Chenopodium quinoa*), minor millets (*Setaria italica, Eleusine coracana, Panicum miliaceum, Paspalum scrobiculatum, Echinocha utilis, Panicum sumatrense*) and amaranth (*Amaranthus caudatus*); pulses such as lentils (*Lens culinaris*), different species of Vigna such as adzuki bean (*Vigna angularis*), ricebean (*Vigna umbellata*), and mungbean (*Vigna radiate*); oilseeds like noug (*Guizotia abyssinica*); roots and tubers such as yams (*Dioscorea spp.*), cassava, ulluco (*Ullucus tuberosus*) or yacon (*Smallanthus sonchifolius*); fruits such as plantain or cooking bananas, breadfruit (*Artocarpus altilis*), jujube (*Ziziphus mauritiana*) or baobab (*Adansonia digitata*); various edible seeds such as Malabar chestnut (*Pachira aquatica*) or Bambara groundnut (*Vigna subterranea*); and vegetables such as leaf amaranth (*Amaranthus spp.*), African eggplant (*Solanum aethiopicum*), the greens from Brassica rapa varieties or various seeds sprouts (mung bean, wild mustard) (Chadha et al. 2007). Amaranth is a well-known non-cereal food grain crop carrying promising levels of important nutrients such as protein (13.56 g/100 g), calcium (159 mg/100 g) and iron (7.16 mg/100 g) (USDA 2010). Mung bean is also good carrier of plant protein (14.6 to 33 g per 100 g) and Fe (7.6 mg per 100 g) (Dahiya et al. 2015). Yam is ranked amongst most nutritious tropical root crops and is a good source of carbohydrates, vitamin A, vitamin C, and is low in antinutrients such as phytates and trypsin inhibitor (O'Sullivan 2010). Cassava is the drought tolerant crop and may be considered as a future food crop for the countries experiencing higher risks of water scarcity and food insecurity. Amaranthus leaves contain protein in a range between 17.5–38.3 g/100 g, Ca (44.15 mg/100 g), and high concentration of biologically available Fe (13.58 mg/100 g) (Kumar et al. 2010). Considering their attractive food potential, indigenous or the local, underutilized or neglected food crops if considered as main stream food crops in a diverse agro-ecological system, can help alleviate the burden of food insecurity and various forms of malnutrition.

1.14 Role of Wild Plant Species of South Asia in Food and Nutritional Security

Historically, wild plants and animals had been major source of human sustenance. People use to satisfy their food needs by fishing, hunting and gathering plants such as roots, stems, leaves, flowers, fruits, seeds and buds which were considered safer for consumption. Domestication of wild food animals and plants forms the basis of our traditional food system, and constitute a large pool of foods produced in recent era. Non domesticated wild foods are still contributing to the food requirements of human especially during the period of food scarcity or shortage. Likewise, diverse range of wild food plants is the best suited survival strategy for the peoples having insufficient household budget for the food (Ojelel et al. 2019). Wild food plants refer to the plant species which are neither cultivated nor domesticated, and are sourced from the natural habitat while their application as food or food ingredient is validated by local knowledge. Though the caloric contribution of wild food plants is less compared to the staple foods however; their nutrient density makes them superior choice to neutralize the burden of malnutrition and improve food and nutritional security. Accessions of a variety of wild plants like *Cassia fistula, Cassia auriculata, Abutilon indicum, Pavetta indica* and *Oroxylum indicum* are used as vegetables. Tubers of certain species are cooked and are eaten as curries. Tender root of *Decalepis hamiltonii* is cut into small pieces and is used to make pickles; these species are edible mainly in Asian countries especially in India. In studies from Benin, Tanzania and the Philippines, wild foods of plant and animal origin were used to compliment 1–19% of the dietary iron requirements, 5–45% of the calcium and ~31% of the vitamin A needs of the children and women (Powell et al. 2015). Despite wild food plants have a significant role in overcoming various forms of malnutrition, modernization, globalization, habitat loss and promotion of the modern crops is anticipating a significant decline in interests of the growers and the consumers to promote wild foods production and consumption, respectively. If prioritized, wild food plant species being culturally acceptable and nourishing could serve an integral constituent of the regional or national diet of the peoples at higher risks of food and nutritional insecurity.

1.15 Agrobiodiversity and Malnutrition

World population is predicted to reach over nine billion by the year 2050. Around 16% of the world's population that counts approximately 900 million heads is malnourished. Persistent growth in number of malnourished people on planet earth is itself an ultimatum for the governments and the stakeholders to improve adequate and balanced supply of nutritious foods to achieve food and nutritional security. With tremendous population pressure, agriculture of this recent era is facing challenges in satisfying food requirements of the current and future populations. Today's food production system is emphasizing more on minimizing the risks of food shortage and in ensuring sustainable supply of adequate, safe and nutritious food to satisfy food security situation of the world population.

Despite of improvement in global food production statistics, still hundreds millions of people in world are experiencing caloric insufficiency and micronutrient deficiencies especially among children who are at greater risk of stunting and wasting (FAO 2022). Hence, hunger accompanied with various forms of malnutrition is still the greatest challenge of the world. Global nutrition surveillance data report 462 million adults as underweight while nearly 149 million and 45 million children below the age of 5 as stunted and wasted, respectively. Nearly 45% of the deaths in children under 5 are linked to undernutrition. Undernourished children are always at higher risk of getting short heighted and weaker socioeconomic status due to low IQ and poor academic performance. Though mass scale food fortification programs across many nations have larger coverage area still their acceptability and cost of implementation are the challenges to nations-wide adoptability of the fortification programs. Parallel to the conventional food crops, interests to produce alternate plant sources carrying substantial concentrations of micronutrients can however help alleviating the risks of malnutrition. Some neglected crops have been mentioned in Table 1.1 which can be a preventive approach in reducing malnutrition.

Dietary diversity is major concern in many of the developing countries where starchy staple constitute more than 70% of peoples diet. Studies among Kenyan

Table 1.1 Examples of neglected plant foods of South Asia, their nutritional value and role in malnutrition prevention

NUS	Scientific Name	Nutrient and non-nutrient components	Role in malnutrition prevention and health promotion	References
Soybean	*Glycine max*	Protein, Isoflavones, Phytochemicals, Ca, Fe, Fibera	Management of PEM, iron deficiency anemia, and supports bone health	Martin et al. (2010)
Mungbean	*Vigna Radiata*	Protein, Vit C, Iron, Ca, Mg, Zn, Cu, fiber	Management of PEM, iron deficiency anemia, osteoporosis, improve gut health, anti-toxic effect	Anwar et al. (2007)
Quinoa	*Chenopodium quinoa Willd*	Protein, Mg, Zn, Iron, omega 3 fatty acid, anti –oxidants, fiber, gluten free	Ensure quality protein adequacy, address gluten intolerance and micronutrient deficiencies, anti- inflammatory	Pathan and Siddiqui (2022)
Pearl Millet	*Pennisetum glaucum*	Protein, Ca, Fe, Zn, Phosphoros, Riboflavin, Folic acid, Fiber, antioxidants	Helps alleviate anemia, constipation, celiac disease, diarrhea, and has anti-cancer activity	Patni and Agrawal (2017)
Sorghum	*Sorghum bicolor*	Carbohydrates, Protein, PUFA, MUFA, Vit E, Fe, Zn, fiber, flavonoids	Reduce risk of inflammation, Zinc and iron deficiency	Mohamed et al. (2022)
Yam	*Dioscore spp.*	Vit C, Potassium, Vit B6, fiber, Manganese	Improves digestion, blood sugar level, gut health, boost immunity	Lawal et al. (2012)
Kale	*Brassica oleracea var. sabellica*	Fiber, potassium, calcium	Anticipate better bone health, cancer protection, and improves gut health	Satheesh and Workneh Fanta (2020)
Moringa	*Moringa Oleifera*	Vit C, Vit A, Ca. K, Fe, Protein, antioxidant	Fights inflammation, lowers cholesterol, prevent micronutrient deficiencies, promote bone density	Gopalakrishnan et al. (2016)

(continued)

Table 1.1 (continued)

NUS	Scientific Name	Nutrient and non-nutrient components	Role in malnutrition prevention and health promotion	References
Black Mulberry	*Morus nigra L.*	Na, Ca, P, K, Antioxidants, Flavonoids, Anthocyanins	Fights inflammation, lowers cholesterol, prevent micronutrient deficiencies	Koyuncu et al. (2014)
Jammun fruit	*Syzygium cumini*	Ca, Mg, Iron, Vit C, Flavonoids, Antioxidants	Diarrhea, improves GIT health, decrease oxidative stress, chemo-preventive effect	Ghosh et al. (2017)
Breadfruit	*Artocarpus altilis*	Essential amino acids, Ca, K, Fe, Mg, B1, B3, Vit C	Prevents bacterial infection, aid in digestion, improves muscle tone	Turi et al. (2015)
Velvet apple	*Diospyros blancoi*	Vit C, Vit E, Mg, Ca, Fiber	Management of micronutrient deficiencies, diarrhea, eczema and blood sugar level	Yadav et al. (2018)
Pomelo	*Citrus maxima*	Protein, fiber, Vit C, potassium, antioxidant, antibacterial	Support digestive health, regulate blood pressure, fulfills micronutrients needs, scavenge free radicals	Lan-Phi and Vy (2015)
Jujube	*Ziziphus jujuba Mill.*	Vit C, thiamine, riboflavin, Ca, Mg, P, K, folic acid, amino acids, phenolics, flavonoids, fiber	Improve bone health, prevent osteoporosis, anticancer, antiaging improves gut health, blood cells generation	Gao et al. (2013)
Common Fig Moraceae	*Ficus Carica*	Fe, Ca, K, B1, B2, fiber, antioxidants	Prevents constipation, cures reproductive problems, strong joints increase bone density, control blood pressure, improves digestion	Solomon et al. (2006)
Lychee	*Litchi chinensis*	Vit C, Fe, Mg, Cu, K, fiber	Neuroprotection, anticancer, antioxidant, anti-obesity, hepato-protective effect	Cabral et al. (2014)
Phalsa/ Fruit berry	*Grewia asiatica*	Carbs, protein, Ca, Fe, Vit C, K, fiber, Phytochemicals	Regulates electrolyte imbalance, treats iron deficiency anemia, controls blood sugar levels, treats urination issues	Khan et al. (2019)

children had shown children's diet to be limited in variety and nutrient composition, and could be a reason of high prevalence rate of stunting. In Western Kenya, a week food recall reported ~45% of the children consumed very low diversified diets. The results report nearly 22% children didn't take pulses and nuts (Ekesa et al. 2008). Malnutrition is widespread in developing countries, increasing fruits and vegetables consumption can improve dietary diversification and net intake of micronutrients intake in peoples of such weaker economies. Agriculture industry should also introduce strategies to increase production and consumption of local, traditional and staple plant foods. Value addition of the available plant food resources may also help meet future food requirements. With expected surge in global population size in the next two decades and increasing risks of famines, global food production is suggested to be diversified with a variety of underutilized, neglected, non-domesticated wild plant foods alongside staple crops. This will certainly promote agrobiodiversity and improve global situation of nutritional security.

1.16 Agrobiodiversity Management for Food and Nutritional Security

Globally, agrobiodiversity management is of foremost importance. There is a growing realization that agrobiodiversity is fundamental to food security. Management of agrobiodiversity for improving food security, nutrition and health includes:

- Livestock management, crop breeding, genetic manipulation, genetic resources conservation, soil management, agronomy, crop diversification especially non staple minor crops and crop protection.
- Genetic improvement of indigenous low yield food crops bearing better natural pest control system. This will certainly minimize dependency on alien and or hybrids crops cultivation to gain higher yields.
- Efficient use of external inputs; fertilizer and water. Adaptation of green revolution strategies for sustainable agrobiodiversity which should be ecologically appropriate, socially desirable and economically justified.
- Preservation of indigenous knowledge of food production and transfer of the food crops / food production technologies to the stakeholders.
- Collaboration among farmers, food manufacturers, food scientists and agriculture stakeholders is needed to improve technology utilization and for diverse accessibility and affordability of locally grown foods.
- Home gardens promotion either urban or rural, enables the provision of variety of foods to people and is directly linked to prosperity of the population and optimal nutrition.
- Policies and laws needed to influence behavior and practices in the cultivation of high level of diversity in an agriculture system can be the best approach for agrobiodiversity conservation of local, traditional and neglected plants of every region as they can promote food and nutritional security.

1.17 Conclusion

Conventional food production system supports development of high yielding varieties to maximize production and profitability. Expecting this system to support sustainability of environmental health and consumers' food and nutritional requirements and safety seems out of place. Ensuring food security of the poor and vulnerable populations of the developing countries spontaneously conserving natural resources of their habitat is a challenge of the recent era. Though the issue has been globally taken-up by a multitude of stakeholders, yet there is no one size fits all approach for the issues of food security and conservation of agroecosystems. Food security and agricultural biodiversity are strongly interconnected to each other and have always been advocated by the researchers, farmer organizations, development experts and the civil society as a sustainable alternate to resource-extractive agriculture of the era. Agrobiodiversity has been proven essential to enhanced global food productivity, food security, better agroecosystem functionality and adaptability and resilience of the small farmers who contribute a larger share of global food demands. Subject to the commitment of the governments and other key stakeholders, promoting agrobiodiversity without sacrificing the life in agroecosystems can bridge the gaps between future demands and supply of wholesome foods.

Conflicts of Interest All authors read and approved the final manuscript and declare that they have no conflict of interest.

References

Abbas G, Ahmad S, Ahmad A et al (2017) Quantification the impacts of climate change and crop management on phenology of maize-based cropping system in Punjab, Pakistan. Agric For Meteorol 247:42–55. https://doi.org/10.1016/j.agrformet.2017.07.012

Adger WN, Pulhin JM, Dabelko BJ (2014) Human security. In: Field CB, Barros VR, Dokken DJ, Mach KJ, Mastrandrea MD, Bilir TE, Chatterjee M, Ebi KL, Estrada YO, Genova RC (eds) Climate change: impacts, adaptation, and vulnerability. Part A: global and sectoral aspects. Cambridge University Press, Cambridge/New York, pp 755–791

Adoukonou-Sagbadja H, Dansi A, Vodouhe R et al (2006) Indigenous knowledge and traditional conservation of fonio millet (Digitaria exilis, Digitaria iburua) in Togo. Biodivers Conserv 15(8):2379–2395

Alam MK (2020) Traditional food plants and agro-biodiversity in Chittagong Hill Tracts, Bangladesh: food security, nutrition and conservation strategy. J Bangladesh Agric Univ 10(1):7–21

Anwar M, Jasra AW, Sultani MI et al (2005) Conservation and sustainable use of biodiversity in Pakistan – a review. Pakistan Agric 1(1):56–65

Anwar F, Latif S, Przybylski R et al (2007) Chemical composition and antioxidant activity of seeds of different cultivars of mungbean. J Food Sci 72(7):S503–S510. https://doi.org/10.1111/j.1750-3841.2007.00462.x

Aryal K, Partap U, Chaudhary RP et al (2021) Agrobiodiversity in the Kangchenjunga landscape: status, threats, and opportunities. http://www.tourismklindia.com/publication/books_technical_manual/3.pdf

Baig MB, Ahmed M (2007) Biodiversity in Pakistan: status challenges and strategies for its conservation. Int J Biol Biotechecnol 4(4):283–292

Bailey A (2016) Mainstreaming agrobiodiversity in sustainable food systems: scientific foundations for an agrobiodiversity index-summary. Bioversity International. https://hdl.handle.net/10568/80360

Bandara JS, Cai Y (2014) The impact of climate change on food crop productivity, food prices and food security in South Asia. Econ Anal Policy 44(4):451–465. https://doi.org/10.1016/j.eap.2014.09.005

Bandula A, Nath TK (2020) Underutilized crops in the agricultural farms of southeastern Sri Lanka: farmers' knowledge, preference, and contribution to household economy. Econ Bot 74:126–139

Béné C, Oosterveer P, Lamotte L et al (2019) When food systems meet sustainability–current narratives and implications for actions. World Dev 113:116–130. https://doi.org/10.1016/j.worlddev.2018.08.011

Brush SB (2004) Farmers bounty: locating crop diversity in the contemporary world. Yale University Press, New Haven

Cabral TA, de Morais Cardoso L, Pinheiro-Sant'Ana HM et al (2014) Chemical composition, vitamins and minerals of a new cultivar of lychee (Litchi chinensis cv. Tailandes) grown in Brazil. Fruits 69(6):425–434. https://doi.org/10.1051/fruits/2014031

Carleton TA (2017) Crop-damaging temperatures increase suicide rates in India. Proc Natl Acad Sci 114:8746–8751. https://doi.org/10.1073/pnas.1701354114

Chadha ML, Guo G, Gowda CLL et al (2007) Proceedings of the first international conference on indigenous vegetables and legumes—prospects for fighting poverty, hunger and malnutrition. Acta Hortic (ISHS):752–623

Chivenge P, Mabhaudhi T, Modi AT et al (2015) The potential role of neglected and underutilised crop species as future crops under water scarce conditions in Sub-Saharan Africa. Int J Environ Res Public Health 12(6):5685–5711. https://doi.org/10.3390/ijerph120605685

Chowdhury MKA (2012) Conservation and sustainable use of plant genetic resources in Bangladesh, vol 175. Bangladesh Agricultural Research Council (BARC), Dhaka, p 4

Conservation International (2012) Biodiversity hotspots. http://www.conservation.org/where/priority_areas/hotspots/Pages/hotspots_main.aspx

Dahiya PK, Linnemann AR, Van Boekel MAJS et al (2015) Mung bean: technological and nutritional potential. Crit Rev Food Sci Nutr 55(5):670–688. https://doi.org/10.1080/10408398.2012.671202

Dhawan B (2021) Biodiversity loss: India has lost 90% of area under four biodiversity hotspots, 25 species extinct, finds study. https://www.financialexpress.com/lifestyle/science/biodiversity-loss-india-has-lost-90-of-area-under-four-biodiversity-hotspots-25-species-extinct-finds-study/2267618/. Accessed 27 Apr 2023

Ekesa BN, Walingo MK, Abukutsa-Onyango MO et al (2008) Influence of agricultural biodiversity on dietary diversity of preschool children in Matungu Division, Western Kenya. Afrrican J Food Agric Nutr Dev 8(4):390–404. https://doi.org/10.4314/ajfand.v8i4.19200

Fanzo J, Cogill B, Mattei F et al (2012) Metrics of sustainable diets and food systems. https://www.bioversityinternational.org/e-library/publications/detail/metrics-of-sustainable-diets-and-food-systems/

FAO (1999a) Agricultural biodiversity, multifunctional character of agriculture and land conference, background paper 1, Maastricht Netherlands. https://www.fao.org/3/X2775E/X2775E02.htm

FAO (1999b) Women: users, preservers and managers of agrobiodiversity. www.fao.org/FOCUS/E/Women/Biodev-e.htm

FAO (2009) First fruits of plant gene pact. http://www.fao.org/news/story/0/item/20162/icode/en/. Accessed 21 June 2009

FAO (2015) Meeting the 2015 international hunger targets: taking stock of uneven progress. https://www.fao.org/agrifood-economics/publications/detail/en/c/288368/

FAO (2019) Strengthening food security and nutrition in South Asia. https://www.fao.org/policy-support/tools-and-publications/resourcesdetails/en/c/1195525/. Accessed 24 Apr 2023
FAO (2022) The state of food security and nutrition in the world 2022. https://www.fao.org/3/cc0639en/online/sofi-2022/food-security-nutritionindicators.html. Accessed 24 Apr 2023
FAO, IFAD, UNICEF, WFP, WHO (2021) Transforming food systems for food security, improved nutrition and affordable healthy diets for all. https://www.fao.org/3/cb4474en/cb4474en.pdf
Firoz R, Mobasher S M, Waliuzzaman M et al (2004) Proceedings of the regional workshops on national biodiversity strategy and action plan. https://portals.iucn.org/library/node/8489
Ford ES, Bergmann MM, Kroger J et al (2009) Healthy living is the best revenge: findings from the European prospective investigation into cancer and nutrition-potsdam study. Arch Int Med 169(15):1355–1362. https://doi.org/10.1001/archinternmed.2009.237
Gamage T (2022) Seen but not addressed: inefficacy of the present legal framework on invasive alien plant species in the face of looming biodiversity extinction in Sri Lanka. KDU Law J 2:45
Gao QH, Wu CS, Wang M et al (2013) The jujube (Ziziphus jujuba Mill.) fruit: a review of current knowledge of fruit composition and health benefits. J Agric Food Chem 61(14):3351–3363. https://doi.org/10.1021/jf4007032
Garn SM, Leonard WR (1989) What did our ancestors eat? Nutr Rev 47(11):337–345. https://doi.org/10.1111/j.1753-4887.1989.tb02765.x
Ghosh P, Pradhan R, Mishra S et al (2017) Physicochemical and nutritional characterization of jamun (Syzygium cuminii). Curr Res Nutr Food Sci 5(1):25–35. https://doi.org/10.12944/CRNFSJ.5.1.04
Gonzalez CG (2011) Climate change, food security, and agrobiodiversity: toward a just, resilient, and sustainable food system. Fordham Environ Law Rev 22(3):493–522
GOP (1999) Biodiversity action plan of Pakistan. https://portals.iucn.org/library/efiles/documents/2000-081.pdf
Gopalakrishnan L, Doriya K, Kumar DS et al (2016) Moringa oleifera: a review on nutritive importance and its medicinal application. Food Sci Human Wellness 5(2):49–56. https://doi.org/10.1016/j.fshw.2016.04.001
Gunawardena AR, Fernando TT (2022) The Kandyan Home Garden–sustainable agricultural ecosystem in Sri Lanka. In: Remote sensing of agriculture and land cover/land use changes in South and Southeast Asian countries. Springer, Cham, pp 471–485
Gunawardene NR, Daniels AE, Gunatilleke IAU et al (2007) A brief overview of the Western Ghats--Sri Lanka biodiversity hotspot. Curr Sci 93(11):1567–1572
Hajer MA, Westhoek H, Ingram J et al (2016) Food systems and natural resources. https://www.fao.org/policy-support/tools-and-publications/resources-details/en/c/469455/
http://courseresources.mit.usf.edu/sgs/ph6934/webpages/CC/module_5/read/Linking_agricultural_biodiversity_Linking_agricultural_biodiversity_thrupp.pdf
https://alliancebioversityciat.org/publications-data/mainstreaming-agrobiodiversity-sustainable-food-systems-scientific-foundations-0
Husain SZ, Malik RN, Javaid M et al (2008) Ethonobotanical properties and uses of medicinal plants of Morgah biodiversity park, Rawalpindi. Pak J Bot 40(5):1897–1911
Idowu O (2009) Contribution of neglected and underutilized crops to household food security and health among rural dwellers in Oyo State, Nigeria. ISHS (International Society for Horticultural Science), Leuven
Ignaciuk A (2015) Adapting agriculture to climate change: a role for public policies. OECD Food, Agriculture and Fisheries OECD, Paris
IPCC (2007) Synthesis report-intergovernmental panel on climate change. IPCC Secretariat, World Meteorological Organization, Geneva
IPCC (2023) AR6 Synthesis report: climate change. https://www.ipcc.ch/report/ar6/syr/downloads/press/IPCC_AR6_SYR_PressRelease_en.pdf. Accessed 29 Apr 2023
IRRI (2023) International Rice Congress. https://www.irri.org/IRC2023-teaser.html

Jaisankar I, Velmurugan A, Sivaperuman C et al (2018) Biodiversity conservation: issues and strategies for the tropical islands. In: Biodiversity and climate change adaptation in tropical islands. Elesvier, pp 525–552, United States.
Jarvis A, Lau C, Cook S et al (2011) An integrated adaptation and mitigation framework for developing agricultural research: synergies and trade-offs. Exp Agric 47(2):185–203. https://doi.org/10.1017/S0014479711000123
Joshi BK (2017) Conservation and utilization of agro-biodiversity advanced from 1937 to 2017 in Nepal. In: Devkota F (ed) Krishi Sanchar Smarika. Agricultural Information and Communication Center (AICC), MoAD, Khumaltar, pp 181–208
Joshi BK, Gorkhali NA, Pradhan N et al (2020) Agrobiodiversity and its conservation in Nepal. J Nepal Agric Res Counc 6:14–33. https://doi.org/10.3126/jnarc.v6i0.28111
Kahane R, Hodgkin T, Jaenicke H et al (2013) Agrobiodiversity for food security, health and income. Agron Sustain Dev 33:671–693
Khan MZ (2014) Agrobiodiversity in Gilgit-Baltistan on the verge of extinction. https://www.researchgate.net/profile/Muhammad_Zafar_Khan/publication/238722793_Agrobiodiversity_in_Gilgit-Baltistan_on_the_Verge_of_Extinction/links/545b4cd00cf2c46f66440276.pdf?origin=publication_detail
Khan RS, Asghar W, Khalid N et al (2019) Phalsa (Grewia asiatica L) fruit berry a promising functional food ingredient: a comprehensive review. J Berry Res 9(2):179–193. https://doi.org/10.3233/JBR-180332
Khoshbakht K, Hammer K (2008) How many plant species are cultivated? Genet Resour Crop Evol 55:925–928
Knox J, Hess T, Daccache A et al (2012) Climate change impacts on crop productivity in Africa and South Asia. Environ Res Lett 7:034032
Koyuncu F, Çetinbas M, Erdal İ et al (2014) Nutritional constituents of wild-grown black mulberry (Morus nigra L). J Appl Bot Food Qual 87:93–96
Kumar BA, Lakshman K, Jayaveera KN et al (2010) Evaluation of the antioxidant activity of Amaranthus spinosus Linn by non-enzymatic haemoglycosylation. Sains Malays 39(3):413–415
Laker M (2007) Introduction to the special edition of Water SA on indigenous crops, water and human nutrition. Water SA 33(3):311–316. https://doi.org/10.4314/wsa.v33i3.180588
Lam S, Pham G, Nguyen-Viet H et al (2017) Emerging health risks from agricultural intensification in Southeast Asia: a systematic review. Int J Occup Environ Health 23(3):250–260. https://doi.org/10.1080/10773525.2018.1450923
Lan-Phi NT, Vy TT (2015) Chemical composition, antioxidant and antibacterial activities of peels' essential oils of different pomelo varieties in the south of Vietnam. Int Food Res J 22(6):2426
Lawal OO, Agiang MA, Eteng MU et al (2012) Proximate and anti-nutrient composition of white Guinea yam (Dioscorea rotundata) diets consumed in Ibarapa, South West region of Nigeria. J Nat Prod Plant Res 2(2):256–260
Li X, Siddique KH (2020) Future smart food: harnessing the potential of neglected and underutilized species for zero hunger. Matern Child Nutr 16:13008. https://doi.org/10.1111/mcn.13008
Li X, Yadav R, Siddique KH (2020) Neglected and underutilized crop species: the key to improving dietary diversity and fighting hunger and malnutrition in Asia and the Pacific. Front Nutr 7:593711. https://doi.org/10.3389/fnut.2020.593711
Lin BB (2011) Resilience in agriculture through crop diversification: adaptive management for environmental change. Bioscience 61(3):183–193. https://doi.org/10.1525/bio.2011.61.3.4
Lopez V, Akerreta S, Casanova E et al (2007) In vitro antioxidant and anti-rhizopus activities of *Lamiaceae* herbal extracts. Plant Foods Hum Nutr 62:151–155
Martin H, Laswai H, Kulwa K (2010) Nutrient content and acceptability of soybean based complementary food. Afr J Food Agric Nutr Dev 10(1). https://doi.org/10.4314/ajfand.v10i1.51482
Maxted N, Dulloo ME, Ford-Lloyd BV, Frese L, Iriondo J, de Carvalho MAP (2011) Agrobiodiversity conservation securing the diversity of crop wild relatives and landraces. CABI, Wallingford

Mittermeier RA, Myers N, Thomsen JB et al (1998) Biodiversity hotspots and major tropical wilderness areas: approaches to setting conservation priorities. Conserv Biol 12:516–520
Modi M, Modi A, Hendriks S et al (2006) Potential role for wild vegetables in household food security: a preliminary case study in KwaZulu-Natal, South. Africa. Afr J Food Agric Nutr Dev 6:1–13. https://doi.org/10.4314/ajfand.v6i1.19167
Mohamed HI, Fawzi EM, Basit A et al (2022) Sorghum: nutritional factors, bioactive compounds, pharmaceutical and application in food systems: a review. Phyton 91(7):1303–1325
Mukul SA, Uddin MB, Uddin MS et al (2008) Protected areas of Bangladesh: current status and efficacy for biodiversity conservation. Proc Pak Acad Sci 45(2):59–68
Mukul SA, Biswas SR, Rashid AM et al (2018) Biodiversity in Bangladesh. In: Global biodiversity, vol 1. Apple Academic Press, Oakville, pp 93–103
O'Sullivan JN (2010) Yam nutrition: nutrient disorders and soil fertility management. Australian Centre for International Agricultural Research (ACIAR), Canberra
Ojelel S, Mucunguzi P, Katuura E et al (2019) Wild edible plants used by communities in and around selected forest reserves of Teso-Karamoja region, Uganda. J Ethnobiol Ethnomed 15:1–14
Padulosi S (1998) The underutilized Mediterranean species project (UMS): an example of IPGRI's involvement in the area of underutilized and neglected species third regional workshop of MEDUSA, Coimbra, Portugal, 27–28 April 1998
Padulosi S, Eyzaquirre P, Hodgkin T et al (1999) Challenges and strategies in promoting conservation and use of neglected and underutilized crop species. In: Janick J (ed) Perspectives on new crops and new uses. ASHS Press, Alexandria, pp 140–145
Padulosi S, Hodgkin T, Williams J T et al (2002) Underutilized crops: trends, challenges and opportunities in the 21st century. In: Managing plant genetic diversity. Proceedings of an international conference, Kuala Lumpur, Malaysia, 12–16 June 2000
Pande PC, Awasthi P, Bargali K et al (2016) Agro-biodiversity of Kumaun Himalaya, India: a review. Curr Agric Res J 4(1):16. https://doi.org/10.12944/CARJ.4.1.02
Parry ML, Rosenzweig C, Iglesias A et al (2004) Effects of climate change on global food production under SRES emissions and socio-economic scenarios. Glob Environ Chang 14:53–67
Pascual U, Jackson LE, Drucker AG et al (2013) Economics of agrobiodiversity. In: Encyclopedia of biodiversity. Academic Press, Amsterdam, pp 31–44
Pathan S, Siddiqui RA (2022) Nutritional composition and bioactive components in quinoa (Chenopodium quinoa Willd) greens: a review. Nutrients 14(3):558. https://doi.org/10.3390/nu14030558
Patni D, Agrawal M (2017) Wonder millet–pearl millet, nutrient composition and potential health benefits-a review. Int J Innov Res Rev 5(1):6–14
Pingali PL (2012) Green revolution: impacts, limits, and the path ahead. Proc Natl Acad Sci 109(31):12302–12308. https://doi.org/10.1073/pnas.0912953109
Powell B, Thilsted SH, Ickowitz A et al (2015) Improving diets with wild and cultivated biodiversity from across the landscape. Food Secur 7:535–554
Prescott-Allen R, Prescott-Allen C (1990) How many plants feed the world? Conserv Biol 4:365–374. https://doi.org/10.1111/j.1523-1739.1990.tb00310.x
Rasul G (2021) Twin challenges of COVID-19 pandemic and climate change for agriculture and food security in South Asia. Environ Chall 2:100027. https://doi.org/10.1016/j.envc.2021.100027
Sahai S (2010) The role of genetic diversity in ensuring food security in South Asia. South Asian Surv 17(1):111–129. https://doi.org/10.1177/097152311001700109
Sardaro R, Fucilli V, Acciani C et al (2015) Measuring the value of rural landscape in support of preservation policies. Ital J Reg Sci 14(2):125–138
Satheesh N, Workneh Fanta S (2020) Kale: review on nutritional composition, bio-active compounds, anti-nutritional factors, health beneficial properties and value-added products. Cogent Food Agric 6(1):1811048. https://doi.org/10.1080/23311932.2020.1811048
Scott P (2017) Global panel on agriculture and food systems for nutrition: food systems and diets: facing the challenges of the 21st century. Springer, London, p 132
Sengupta A, Dayanandan S (2022) Biodiversity of India: evolution, biogeography, and conservation. Biotropica 54(6):1306–1309. https://doi.org/10.1111/btp.13168

Sibhatu KT, Krishna VV, Qaim M et al (2015) Production diversity and dietary diversity in smallholder farm households. Proc Natl Acad Sci 112(34):10657–10662. https://doi.org/10.1073/pnas.1510982112

Solomon A, Golubowicz S, Yablowicz Z et al (2006) Antioxidant activities and anthocyanin content of fresh fruits of common fig (Ficus carica L.). J Agric Food Chem 54(20):7717–7723

Steinbrecher A, Nimptsch K, Husing A et al (2009) Dietary glucosinolate intake and risk of prostate cancer in the EPIC-Heidelberg cohort study. Int J Cancer 125:2179–2186. https://doi.org/10.1002/ijc.24555

Sumner A, Hoy C, Ortiz-Juarez E et al (2020) Estimates of the impact of COVID-19 on global poverty. https://www.wider.unu.edu/publication/estimates-impact-covid-19-global-poverty

Thrupp LA (1997) Linking biodiversity and agriculture: challenges and opportunities for sustainable food security. World Resources Institute, USA

Thrupp LA (2000) Linking agricultural biodiversity and food security: the valuable role of agrobiodiversity for sustainable agriculture. Int Aff 76(2):265–281. https://doi.org/10.1111/1468-2346.00133

Turi CE, Liu Y, Ragone D et al (2015) Breadfruit (Artocarpus altilis and hybrids): a traditional crop with the potential to prevent hunger and mitigate diabetes in Oceania. Trends Food Sci Technol 45(2):264–272. https://doi.org/10.1016/j.tifs.2015.07.014

UNDP (1999). Project document of the Pakistan's mountain areas conservancy project. pp. 7,8

United States Department of Agriculture Nutrient Data Laboratory (USDA) (2010) http://www.nal.usda.gov/fnic/foodcomp/cgi-bin/measure.pl. Accessed 26 Dec 2011

USDA (2023) World agricultural production. https://apps.fas.usda.gov/psdonline/circulars/production.pdfz. Accessed 24 Apr 2023

Usman F, Hussain F, Leghari SK et al (2022) Possible threats to agrobiodiversity of Thar Desert in Pakistan. GU J Phytosciences 2(1):60–67

Volis S, Blecher M (2010) Quasi in situ: a bridge between ex situ and in situ conservation of plants. Biodivers Conserv 19:2441–2454

Wangda P (2008) Preservation of agro-biodiversity landscape in a typical rural Bhutan. Preservation of biocultural diversity–a global issue. BOKU, Vienna

Warnasooriya PGAS, Weerakkody WAP, Bandaranayake PRSD et al (2023) Assessment of Agro-biodiversity in Kandyan Homegardens; a case study. https://www.bfnsrilanka.org/images/symposium/posters/session3/Assessment_of_Agro_biodiversity.pdf. Accessed 28 Apr 2023

WRI (World Resources Institute) (2005) The wealth of the poor: managing ecosystems to fight poverty. http://pdf.wri.org/wrr05fullhires.pdf. Accessed Jan 2007

Yadav M, Srilekha K, Maheswari KU et al (2018) Potential health benefit of underutilized fruits: a review. J Pharmacogn Phytochem 7(5):1417–1420

Zeven AC (1998) Landraces: a review of definitions and classifications. Euphytica 104:127–139

Zimmerer KS (2013) The compatibility of agricultural intensification in a global hotspot of smallholder agrobiodiversity (Bolivia). Proc Natl Acad Sci 110(8):2769–2774. https://doi.org/10.1073/pnas.1216294110

Chapter 2
Role of Neglected Plant Foods in Achieving Dietary Diversity, Zero Hunger and Good Health

Mahbubjon Rahmatov and Claudia E. Lazarte

2.1 Introduction: The State of Food Security in the World

The definition of Food Security was coined in the World Food Summit, 1996, it states that "Food security exists when all people, at all times, have physical and economic access to sufficient, safe and nutritious food that meets their dietary needs and food preferences for an active and healthy life" (FAO 1996) . Since then, there have been many attempts and suggestions to modify this definition, those however were not fully endorsed thus, and up to date the 1996's definition remains valid. This definition implies that food security covers the following four dimensions: physical availability, economic and physical access to food, food utilization and stability of the other three dimensions over time. Food utilization refers to the way the body makes the most nutrients content in food. Sufficient energy, protein and micronutrient intake are the result of adequate diversity in the diet, feeding practices, food preparation (FAO 1996). Thus, it becomes clear that dietary diversification is a key factor to achieve healthy diets and better health.

To achieve food security in the world remains being a challenge, the current statistics shows that food security has worsened in recent years, intensified by the COVID-19 pandemic as it is shown in the Report on the State of Food Security and Nutrition in the World, 2022 (FAO 2022). In 2020 the COVID-19 pandemic spread across the world and created major setbacks towards achieving the Sustainable Development Goals (SDGs) 2030. The current situation indicates that malnutrition in all its forms continues, and child and maternal malnutrition has been exacerbated. It is globally reported that the number of undernourished people in 2005 was 805.5

M. Rahmatov
Department of Plant Breeding, Swedish University of Agricultural Sciences, Alnarp, Sweden

C. E. Lazarte (✉)
Department of Food Technology, Engineering and Nutrition, Lund University, Lund, Sweden
e-mail: Claudia.lazarte@food.lth.se

T. Ismail et al. (eds.), *Neglected Plant Foods of South Asia*,
https://doi.org/10.1007/978-3-031-37077-9_2

million, in 2019 was 618.4 million which indicated that the progress towards reducing malnutrition was on track (FAO 2022). However, in 2021 the projected number of undernourished people increased to 767.9 million due to the adverse effects of COVID-19. The ongoing Russia-Ukraine conflict has led to an increase in the price of food, fuel, and fertilizer on a global scale, which may lead to food shortages by 2023 and beyond (Behnassi and El Haiba 2022).

There is a global call for Food Systems 2030 to concert efforts to bring up agricultural and nutritional interventions to fight against malnutrition and to improve progress in achieving all 17 SDGs by 2023. Long-term food security for the increasing global population requires dietary diversity, farming diversification and intensification along with food governance and improvement of traditional diets (Siddique et al. 2021; Baldermann et al. 2016).

2.2 All Forms of Malnutrition

Many countries, at all levels of development, present multiple forms of malnutrition and some of them face what it is called a triple burden of malnutrition which refers to the coexistence of undernutrition (i.e., underweight, child wasting and stunting), micronutrient deficiencies and overweight, obesity, and diet-related non-communicable diseases (Popkin et al. 2020). The challenges are exacerbated in low- and middle- income countries where the highest percentages of people affected by diverse forms of malnutrition are reported. Data from 2021 indicated that 767.9 million of people in the world were undernourished, of which 55% are in Asia (424.5 million), 36% in Africa (278.0 million) and 7% (56.5 million) in Latin America and the Caribbean (FAO 2022). Data from 2020 indicated that globally an estimated 22% children (149 million) were stunted, 6.7% (45 million) were wasted while 5.7% (39 million) were overweight (FAO 2022), progress was made in stunting and wasting whereas childhood obesity has worsened.

Micronutrient deficiencies affects around one-third of the global population, this is about 2.5 billion people suffering from one or more micronutrient deficiencies, being the main iron, vitamin A, iodine and zinc (WHO 2022b; Ritchie and Roser 2017). Children and women of reproductive age are the most vulnerable groups to micronutrient deficiencies. Data from 2019 showed that nearly one in three women aged 15–49 years (29.9%, 570.8 million women worldwide) were affected by iron deficiency anemia. The progress on fighting micronutrient deficiencies is very slow, between 2012 and 2019, there was only 1.4% improvement (FAO 2022). In addition, zinc and vitamin A deficiencies persist at high levels in South-Asia and Sub-Saharan Africa (Popkin et al. 2020). The direct health consequences of these deficiencies are night blindness, impaired physical and cognitive development and impaired immune system. In turn, this will be reflected in high social and public costs, reduced work capacity, and a tragic loss of human potential (Darnton-Hill et al. 2005).

In low-and middle-income countries, where overweight and obesity rates are rapidly increasing, food insecurity among families with medium to high incomes manifests in low-quality diets, diets of low diversity, usually dominated by high consumption of energy-rich foods, snacks, and beverages high in energy, sugar, fat, and salt (Drewnowski 2012). This type of diets also known as empty calories have excessive contents of energy and low levels of protein and micronutrients, which is conducive to diet related Non-Communicable Diseases (NCDs) and deficiencies of iron, zinc, folate, vitamin A, D, C, B6 and B12 (Ritchie and Roser 2017).

Overweight and obesity in children has immediate and long-term detrimental health effects, higher risk for NCDs later in life. The prevalence of overweight among children under five years has increased in the last years, data from 2020 showed that 38.9 million (5.7%) of children around the world suffers from overweight, the statistics for 2012 were 33.3 million children (5.6%), such rising trends have been seem in around half of the counties worldwide (FAO 2022).

Concerning adult obesity, prevalence has also been rising in all regions. Data from 2012 showed that 343.1 million (11.8%) people worldwide suffered from obesity, latest data from 2020 indicated that adult obesity has almost doubled affecting 675.7 million people (13.1%) (FAO 2022). The rising prevalence in overweight and obesity are considered a global health concern as evidence showed that overweigh and obesity are very much linked to higher risk of NCDs such as type 2 diabetes and hypertension which are leading contributors to adult morbidity and mortality (Popkin and Ng 2022; Harb et al. 2022).

By the above information, it can be clearly seen that to combat all forms of malnutrition, integrated approaches are necessary, strategies that require a comprehensive perspective at individual and community level alongside the adoption of measures to engage decision makers and governments.

2.3 Neglected Plant Foods and Dietary Diversity

There is a crucial link between food security and nutritional outcomes, according to WHO (2022a), a healthy diet protects against all forms of malnutrition, from undernourished to overweight and obesity, a healthy diet lowers the risks of NCDs such as stroke, diabetes, heart disease and even cancer. By definition a healthy diet contains a balanced, diverse and appropriate selection of foods eaten over a period of time and according to physiological state and level of physical activity. A healthy diet warrants that the individual dietary requirements for macronutrients (i.e., proteins, fats and carbohydrates including fibers) and essential micronutrients (i.e., vitamins and minerals) are met (WHO 2022a; Cena and Calder 2020).

Healthy diets must include less than 30% of total energy intake from fat intake, with less than 10% of saturated fat, the recommendation encourages to shift saturated fat to unsaturated fats and the reduction to minimum industrial trans fats (WHO 2022a). Unsaturated fats have been associated with lower risk of cardiovascular and mortality, whereas trans fats and saturated fats are associated with unfavorable impacts on health and higher mortality risk (Ricci et al. 2018). Dietary recommendations includes reducing the intake of added sugars to less than 5% as it is associated with higher risk of overweight, obesity and their derived NCDs (Rippe and Angelopoulos 2016). Additionally, increasing the consumption of fruits and vegetables is encouraged in a healthy diet to at least 400 g of fruits and vegetables per day (WHO 2022a). High intake of fruits and vegetables was shown to be inversely correlated with the risk of metabolic syndrome (Tian et al. 2018) and other NCDs (Li et al. 2016; Gan et al. 2015). Moreover, fruits and vegetables are rich source of bioactive compounds such as polyphenols, carotenoids and phytosterols which are associated with various positive health effects such reducing risk of NCDs, and cancers (Probst et al. 2017).

Dietary recommendations may vary between countries, as they consider cultural context, dietary habits, and locally available foods. Although there are differences in recommendations, the basic principles of what comprises a healthy diet remain the same. There is common agreement that dietary diversity is a very important dimension to ensure healthy diets for better nutrition and health.

Dietary diversity also known as food variety, has historically been recognized as an important component of healthy eating as it helps to ensure adequate amounts of nutrients needed for normal human physiology (Elizabeth Randall et al. 1985).

The indicator Minimum Dietary Diversity (MDD) is used as a measure of micronutrient adequacy in children and women of reproductive age (FAO 2021; WHO et al. 2008). For children aged 6–23 months old, it is considered that they met a MDD if they have consumed foods from five out of eight food groups in the previous 24 hours. Table 2.1 presents the food groups included in the MDD for children. For women of reproductive age (between 15 and 49 years old) FAO has updated the indicator and named it Minimum Dietary Diversity Women (MDD-W) which is defined as a dichotomous indicator of whether women have consumed foods from at least five out of ten food groups in the previous 24-hours (FAO 2021), Table 2.1 presents the 10 groups included in the MDD-W.

Table 2.1 Food groups included in the indicators of dietary diversity for young children (WHO et al. 2008) and women of reproductive age (FAO 2021)

Dietary Diversity – Food Groups	
MDD – children (6–23 months old)	MDD-W (15–49 years old)
1. Grains, roots, and tubers	1. Grains, roots and tubers and plantains
2. Legumes and nuts	2. Pulses (beans, peas, and lentils)
3. Dairy products	3. Nuts and seeds
4. Flesh foods, including meat, poultry, and fish	4. Milk and milk products
5. Eggs	5. Meat, poultry, and fish
6. Vitamin A-rich fruits and vegetables	6. Eggs
7. Other fruits and vegetables	7. Dark green leafy vegetables
8. Breastmilk	8. Other vitamin A-rich fruits and vegetables
	9. Other vegetables
	10. Other fruits

In a cross-sectional study conducted in India, it was found that dietary diversity of children (6–59 months) was a significant predictor of child underweight, wasting and stunting ($p < 0.001$) (Borkotoky et al. 2018). Likewise, in a study conducted in Cambodia it was reported that consumption of a diverse diet was associated with reduction of stunting (AOR, 0.95; CI 95%, 0.91–0.99; p = 0.01) it was however not significantly associated with underweight and wasting (Darapheak et al. 2013). A multi-country study in the Asia Pacific region provided information on low dietary diversity and dietary quality among young children and infants (6–23 months old), where only 50% of children met the dietary diversity in 15 out of the 20 countries where the survey was conducted. The countries with the lower percentages were India 20% and Myanmar 21% these low dietary diversity scores were associated to the high prevalence of wasting and stunting among this child population (FAO 2019). Another multi-country study conducted in Bangladesh, Nepal, India and Pakistan reported that there was a significant association ($p < 0.001$) between stunting in children (6–23 months) and dietary diversity, a higher risk of stunting was recorded in children with the lowest dietary diversity (OR, 1.47; 95% CI. 1.28–1.69) (Krishna et al. 2018).

Dietary diversity can be, however, more than an indicator of nutritional adequacy, as it can have a considerable impact on health (Cooper et al. 2012) for example populations may not only consume inadequate quantities of different foods but they may also consume a low dietary diversity diet with only foods of certain food groups. The consequences can be seen in low protein intake and in inadequate intake of micronutrients (FAO 2021; Roy et al. 2022). Studies indicated that adequate dietary diversity, and the inclusion of foods of certain food groups (i.e. fruits and vegetables) can reduce the risk of overweight, obesity and certain NCDs by improving the intestinal microbiota (Sonnenburg et al. 2016; Heiman and Greenway 2016). It has been reported that low dietary diversity affect gut microbiota, a diet that excludes animal-based foods or plant-based foods of diverse groups could potentially eliminate crucial microbiota by avoiding essential nutrients needed to supply enough energy for microbiota to survive their habitat (Sonnenburg et al. 2016).

Thus, improving dietary diversity has the potential to reduce all forms of malnutrition from undernourished to micronutrient deficiencies and obesity. Production and consumption of nutrient-rich neglected foods should be encouraged to increase dietary diversity and thus nutrition. Nonetheless, there is a current trend of higher availability of empty-foods in the market, these are ultra-processed foods rich in fats and added sugar but with low levels of protein and essential micronutrients. Due to the increased availability of ultra-processed foods dietary diversity has been greatly reduced during the past 50 years (Heiman and Greenway 2016). Furthermore, agrobiodiversity defined as the 'variety and variability of plants, animals and microorganisms use directly or indirectly for food and agriculture' (FAO 2004) is under treat and reducing at a rapid rate. According to FAO (2004), 75% of plant genetic diversity has been lost. Farmers worldwide were pushed to leave their multiple local varieties for high-yielding varieties instead and thus reducing the production of already neglected foods, all these will have negative impact on the dietary diversity (Leite et al. 2022; Heiman and Greenway 2016).

It has been reported that the increase availability and consumption of ultra-processed foods reduced the dietary diversity in Mexico, where 30% of the population's energy intake comes from ultra-processed foods, foods high in added sugars and saturated fats. The MDD score indicated that only 46.3% of the population was above the minimum dietary diversity for good nutrition and health. Moreover, low intakes of micronutrients (i.e., niacin, folate, vitamin B12, vitamin C, zinc, calcium, etc.) were associated to the low MDD (Marrón-Ponce et al. 2023).

Increasing dietary diversity and micronutrient intake is essential. It should be pursued by increasing production of nutrient-rich neglected foods and discouraging ultra-processed foods consumption. The section below presents and discusses the potentiality of neglected and underutilized food to achieve better dietary diversity and therefore better nutrition and health.

2.4 Nutritional Potential of Neglected Plant Foods (NPFs)

There is a clear link between the relation of neglected plant foods, dietary diversity and nutrition. According to the FAO (2004) dependence on limited number of crops negatively affects ecosystems, food diversity, and health. As described in the previous section, low dietary diversity is associated with micronutrient deficiencies and other forms of malnutrition. Neglected foods often have superior nutrient content than those crops that are currently dominating the food systems (Singh et al. 2022; Hunter et al. 2019), thus neglected foods hold great potential to improve dietary diversity and nutrition as presented in Fig. 2.1.

Worldwide there are about 30,000 edible plants identified, from which around 7000 are cultivated for food, however only 150 crops are cultivated for commercialization (Khoshbakht and Hammer 2008). From all edible species, it is known that only about one hundred (i.e., varieties of rice, wheat, potato and maize) are staple and provide up to 90% of energy intake (Bailey 2016). In consequence there are left thousands of plant species being neglected and underutilized and they could potentially increase dietary diversity of populations. NPFs are mostly used at local level, they could be wild or semi-domesticated species growing in local environments and climate conditions (Li and Siddique 2020). With the growing agricultural technologies, NPFs cultivars could be expanded to other regions and potentially contribute to dietary diversity and nutrition. However, with the increasing production and consumption of ultra-processed foods, NPFs are becoming more and more neglected with the time. NPFs stand up against multidimensional challenges such as sociocultural acceptance, policy, economy and sustainability, challenge that need to be faced for NPFs to be part of healthy diets. NPFs has huge potential to fight poverty, hunger, and all forms of malnutrition, their incorporation into food systems could lead to greater availability of nutrient rich, climate-resilient, and sustainable agriculture (Li et al. 2020). Many NPFs are nutrient rich containing higher levels of protein,

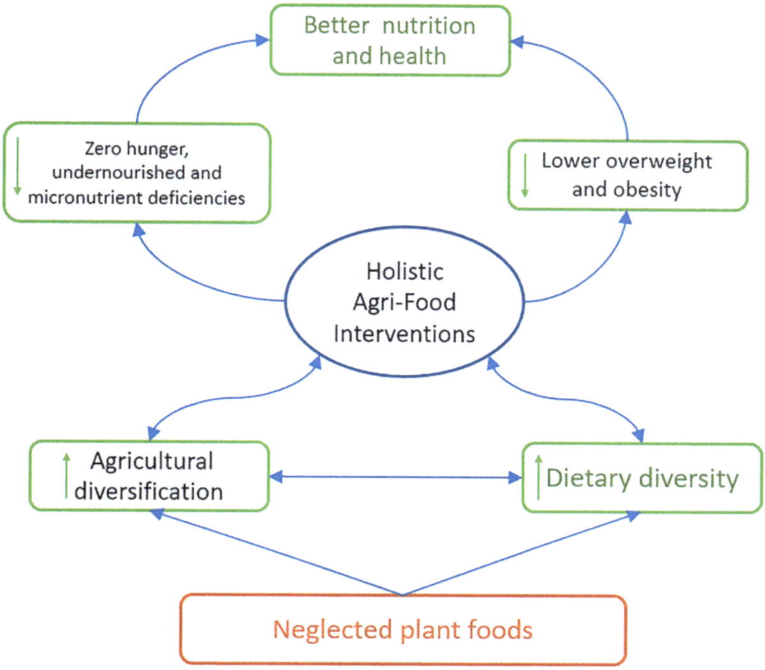

Fig. 2.1 Potential of neglected plant foods to achieve dietary diversity for better nutrition and health

fiber, and essential micronutrients than conventional staple food. Pseudocereals are a big group of NPFs that has proven to have better nutritional quality than conventional cereals, for example, quinoa, amaranth and canihua have a complete profile of essential amino acids and are considered complete proteins (Malik and Singh 2022). There are various varieties of millets that have shown to have superior nutritional qualities than other cereals, they also content vitamin A and unsaturated fats that can provide health benefits (Kumar et al. 2022; Tiwari et al. 2023). Another positive trait of NPFs, such pseudocereals and millets is that they are gluten free, which is very important for people suffering from celiac disease and gluten intolerances (Woomer and Adedeji 2021). Pseudocereals and millets are low-glycemic index foods, several studies reported that consuming low-glycemic index foods can reduce the glycemic response and decrease the risk of type 2 diabetes (Baggio et al. 2023). A human study in an overweight/obese group showed that replacing high glycemic index foods by alternatives with glycemic index reduced hepatic fat (Basset-Sagarminaga et al. 2023). In addition various NPFs (i.e. sorghum, quinoa, yam bean, etc.) have shown to be rich in bioactive compounds such as polyphenols, anthocyanins, carotenoids, bioactive peptides etc. that can confer multiple health benefits (Rashmi and Negi 2020; Shahbaz et al. 2022; Fernández-Tomé et al. 2023). Table 2.2 presents a list of nutrient rich NPFs, with a description of their nutritional qualities and major areas or production. In addition, various NPFs are climate

Table 2.2 List of nutrient-dense underutilized crops with potential for agricultural, dietary diversification and improved nutrition

Food Group	Common and (scientific name)	Nutritional characteristics	Main areas of cultivation	References
Millets	Pearl millet (*Pennisetumglaucum*), Finger millet (*Eleusinecoracana*), Proso millet (*Panicummiliaceum*), Foxtail millet (*Setaria italica*), Barnyard millet (*Echinochloa colona*), Kodo millet (*Paspalumscrobiculatum*), Guinea millet (*Urochloa deflexa*), Raishan millet (*Digitaria cruciata*), Little millet (*Panicum sumatrense*), Browntop millet (*Brachiaria ramosa*), Job's tears (*Coix lacryma-jobi*)	Rich in minerals including Fe, Ca, K, Mg, Zn; vitamins; antioxidants; dietary fibers and resistance starch	Semi-arid and warm regions of Asia, Africa, and America	Singh et al. (2022)
	Tef (*Eragrostis tef*)	Rich in protein, fiber, magnesium, zinc, calcium, and vitamin B6	Ethiopia and Eritrea	Singh et al. (2022)
	Fonio (*Digitaria exilis*)	Rich source of dietary fiber, B vitamins, and minerals (Fe and Zn)	Cultivated in West Africa (Senegal, Guinea, Mali, Burkina Faso, and Nigeria)	Deriu et al. (2022)
	Sorghum (*Sorghum bicolor L.*)	Protein, lipids, non-starch polysaccharides and phytochemicals	Africa, Asia, and Central America	Stefoska-Needham et al. (2015)
Pseudo-cereals	Buckwheat (*Fagopyrum esculentum*)	High-calorie seeds, rich source of soluble and insoluble dietary fiber, vitamins (B and E), Fe, and Zn	Hilly regions of Southeast Asia	Singh et al. (2022)
	Chia (*Salvia hispanica*)	Rich source of fiber, minerals, omega fatty acids, and minerals	Central and Southern Mexico and Southwest America	Singh et al. (2022)

(continued)

Table 2.2 (continued)

Food Group	Common and (scientific name)	Nutritional characteristics	Main areas of cultivation	References
	Quinoa (*Chenopodium quinoa*)	Grains rich in high quality protein, essential aminoacids, dietary fiber, B vitamins, Mg, K, and Ca	South America, Bolivia and Peru are the main producers. The cultivars are expanding to Europe, and Asia	Singh et al. (2022)
	Canihua (*Chenopodium pallidicaule*)	Grains rich in high quality protein, essential aminoacids, dietary fiber, minerals such as Ca, Fe, Mg and Zn	South America	Farooq et al. (2023)
	Amaranth (*Amaranthus caudatus*)	Grains rich in high quality protein, essential aminoacids, dietary fiber, minerals such as Ca, Fe, Mg and Zn	South America	Farooq et al. (2023)
Beans and legumes	Pearl Lupin (*Lupinus mutabilis*)	Rich in protein, minerals and vitamins	Asian and African regions	Farooq et al. (2023)
	Horse gram (*Macrotyloma uniforum*)	Rich source of protein, carbohydrate, vitamins, iron, and molybdenum	Southeast Asia	Singh et al. (2022)
	Petai (*Parkia speciosa*)	Rich source of protein, fibers, minerals, vitamins, and flavonoids	Southeast Asia	Singh et al. (2022)
	Grass pea (*Lathyrus sativus*)	Rich source of protein, essential amino acids, and vitamins	Asia, East Africa, and South America	Singh et al. (2022)
	Moth bean (*Vignia aconitifolia*)	Rich source of protein, essential amino acids, and vitamins	South Asia	Farooq et al. (2023)

(continued)

Table 2.2 (continued)

Food Group	Common and (scientific name)	Nutritional characteristics	Main areas of cultivation	References
	Mung bean (*Vigna radiata*)	Source of plant-based protein, dietary fiber, vitamins (B6 and C), antioxidants, and minerals.	Cultivated in most countries worldwide	Farooq et al. (2023)
	Cowpea (*Vigna unguiculata*)	Source of plant-based protein, dietary fiber, vitamins (A, B, and C), antioxidants, and minerals.	Cultivated in Sub-Saharan Africa, South Asia, Southeast Asia, the Americas, and Southern Europe	Abebe and Alemayehu (2022)
	Adzuki bean (*Vignia angularis*)	Rich in protein, Fe, Mg and Zn	East Asia	Farooq et al. (2023)
	Groundnut (*Arachis hypogaea*)	Rich source of protein, vitamin E, folate, Mg, and Cu	Tropical and subtropical countries in the world	Singh et al. (2022)
	Bambara groundnut (*Vigna subterranea*)	Rich source of protein, carbohydrate, and fiber	West Africa	Singh et al. (2022)
	Yam bean (*Pachyrhizus erosus*)	Rich in carbohydrate, proteins, minerals, and vitamin B, and carotene	Mexico and South America	Singh et al. (2022)
Roots and tubers	Mashua tuber (*Tropaeolum tuberosum*)	Rich in carbohydrate, proteins, vitamin C, and ß-carotene	Peru, Bolivia, Colombia, and Ecuador	Singh et al. (2022)
	Yam (*Dioscorea alata*)	Rich in carbohydrate, proteins, vitamin B6, Cu, and Mn	West Africa, Southeast Asia, and New Guinea	Singh et al. (2022)
	Taro or dasheen (*Colocasia esculenta*), Swamp Taro (*Cyrtosperma merkusii*)	Rich in carbohydrate, proteins, vitamins, Ca and Mn	Bay of Bengal region of Southeast Asia.	Xuan and Siddique (2018)

(continued)

Table 2.2 (continued)

Food Group	Common and (scientific name)	Nutritional characteristics	Main areas of cultivation	References
	Purple yam (*Dioscorea alata*)	Rich in starch, proteins, lipids, vitamins, and minerals, anthocyanins	Southeast Asia, New Guinea	Padhan and Panda (2020)
	Yam (*Dioscorea spp.*)	Rich source of essential nutritional components such as starch, proteins, lipids, vitamins, and minerals	Southeast Asia and Indo-Malayan region	Padhan and Panda (2020)
Fruits and vegetables	Malabar spinach (*Basella alba*)	Rich in vitamins, Fe, Ca, antioxidants including ß-carotene, and lutein	Southeast Asia and Australia	Singh et al. (2022)
	Sparrow grass (*Asparagus ofcinalis*)	Rich in vitamins, Mn, Mg, P, and K	Asia and Europe	Singh et al. (2022)
	Chinese leek (*Allium tuberosum*)	Rich in essential flavonoids, amino acids, and has therapeutic value	Southeast Asia	Singh et al. (2022)
	Wild watermelon (*Citrullus lanatus*)	Rich in Ca, Fe, P, Mg, vitamins, and carotenoid including lycopene	South America, Africa, Middle East, and South Asia	Singh et al. (2022)
	Bayberry (*Myrica esculenta*)	Rich in vitamin C and polyphenolic compounds, has therapeutic values	Southeast Asia	Singh et al. (2022)
	Sugar apple (*Annona squamosa*)	Rich in essential amino acids, fibers, vitamins, and Mg	South America, Australia, and Asia	Singh et al. (2022)
	Indian jujube (*Ziziphus mauritiana*)	Rich in carotene, vitamins (A and C), and fatty oils	Southeast Asia and Eastern Africa	Singh et al. (2022)
	Baobab (*Adansonia digitata*)	Rich in fiber, Calcium, Potassium, Vitamin C, etc.	Africa	Rahul et al. (2015)
	Drumstick (*Moringa olifera Lam*)	Rich in protein, lipids, minerals and dietary fiber	Tropics and subtropics of Asia, Africa, South America	Ebert (2014)

resilient and grow in marginal land (Bailey 2016), this could be an asset to incorporate NPFs into production lands and combined crops to achieve sustainable and resilient food systems must needed for the future generations.

There are various studies that showed the nutritional and health effects of inclusion of supplementation of diets with NPFs. For example, bambara groundnut (*Vigna subterranean*) is considered a nutrient rich and healthy food with outstanding values of dietary fiber, protein high in content and quality, carbohydrate and unsaturated fat, minerals, vitamins and antioxidants (Ramatsetse et al. 2023). The effect of antioxidant and dietary fiber in bambara groundnut has shown to decrease blood pressure and serum cholesterol as well as to reduce the risk of cardiovascular diseases (Tan et al. 2021). Lupin (*Lupinus mutabilis*) besides of having high content of protein and dietary fiber has shown to have several health-promoting properties such as anti-inflammatory, antioxidant, hypoglycemic and hypolipidemic (Okagu et al. 2021). Likewise pseudocerals such as quinoa and amaranth are crops that are rich in high quality protein, they provide a full profile of essential amino acids, thus the biological value of these pseudocereals is quite similar to that of beef, and they are rich in minerals in particular iron (Liu et al. 2022). In addition, these are crops that are climate-resistant which is a very valuable feature towards achieving resilient and sustainable food systems (Farooq et al. 2023). All these agricultural, nutritional and health characteristics of NPFs make them very valuable for the fight against all forms of malnutrition and to achieve the goal of zero hunger.

2.5 Asian Forgotten Food Crops to Fight Hunger and Improving Nutritional Security

Crop farming practices in Asia often rely on a limited number of crops, resulting in monocultures and decreased resilience to biotic and abiotic stresses (Campbell et al. 2017). This situation is aggravated in countries with large populations and high levels of food insecurity in Asia, where the lack of agrobiodiversity poses a significant problem (Fatima et al. 2020). The global food system heavily depends on three crops, wheat, maize, and rice, which provide over 60% of all plant-based foods consumed daily (Palacios-Rojas et al. 2020). As plant-based products are expected to provide over 80% of calories and nutrition by 2050, the current food production will have to be increased by 50% to meet future demands (FAO 2017). Thus, diversifying crop production and promoting alternative crops with high nutritional value and environmental resilience can enhance a sustainable food production system. There is a wealth of diverse crops in Asian countries that have served as food sources since ancient times, but today they have been replaced mainly by a few mainstream staple crops (Li and Siddique 2020).

The neglected and underutilized Asian crops include cereals, roots, tubers, pulses, fruits, vegetables, and nuts, among others, which are often overlooked despite their potential nutritional and economic benefits (Li and Siddique 2020).

Cultivating and consuming neglected and underutilized plant foods can contribute to diversifying and sustaining the food system while preserving traditional knowledge and local biodiversity. Importantly, these crops often have deep cultural and traditional significance, reflecting centuries of human knowledge and environmental interaction. The cultivation and consumption of underutilized crops can help Asian countries improve their food security and preserve their cultural heritage. As a result, more sustainable and resilient agricultural systems can be developed that are better able to meet a changing climate and rising demand for food and nutrition.

SDGs 2, 3, and 12 are interrelated and aim to eliminate hunger, promote healthy life and well-being for all age groups, and ensure sustainable consumption and production patterns. Achieving these SDGs requires diversifying agricultural productivity, ensuring sustainable food systems with nutritional characteristics and health benefits, and maintaining genetic diversity for sustainable consumption and production. Promoting NPFs may present a viable solution to achieving sustainable solutions for all while minimizing environmental impact and enabling producers to earn a decent living. By promoting NPFs, Asian countries can enhance the resilience and diversity of their food systems, reducing their dependence on a few staple crops and improving nutrition and health outcomes. These crops have unique nutritional and ecological properties and can thrive in challenging environments, making them a valuable resource for promoting food security and sustainable agriculture, as it can be seen in Table 2.2, many of the nutritious NPFs were originated in Asian countries. However, promoting NPFs requires a multi-sectoral approach that includes research and development, policy support, market development, and capacity building. Governments, civil society organizations, and the private sector must collaborate to create an enabling environment for NPFs promotion and ensure that small-scale farmers and local communities benefit from the increased demand for these crops.

The United Nations has declared several International Years (International Years | United Nations) to raise awareness and promote the cultivation and consumption of NPFs with nutritional and health benefits. As an example, the International Year of Quinoa in 2013 was devoted to promoting the nutritional value and economic potential of quinoa as a staple crop, while the International Year of Pulses in 2016 was dedicated to raising awareness about the benefits of legumes for food security, nutrition, and environmental sustainability. These efforts are crucial to diversifying the global food system, promoting sustainable agriculture, and improving food security and nutrition. Millets are nutritious neglected foods that have a lot of potential to improve, agriculture and nutrition in Asia, they contain high levels of minerals, vitamins, antioxidants, fiber, and protein, which can help prevent a wide range of non-communicable diseases, such as cardiovascular disease, cancer, and type 2 diabetes. Figure 2.2 shows the nutritional and health benefits that millets can provide, these benefits make them a novel "well-being" food. Therefore, the UN General Assembly has declared 2023 as the "International Year of Millets" to raise public awareness of their multipurpose properties as a food-feed-bioenergy crop, including their health benefits and use in crop rotation (Home | International Year of Millets 2023 | Food and Agriculture Organization of the United Nations (fao.org)).

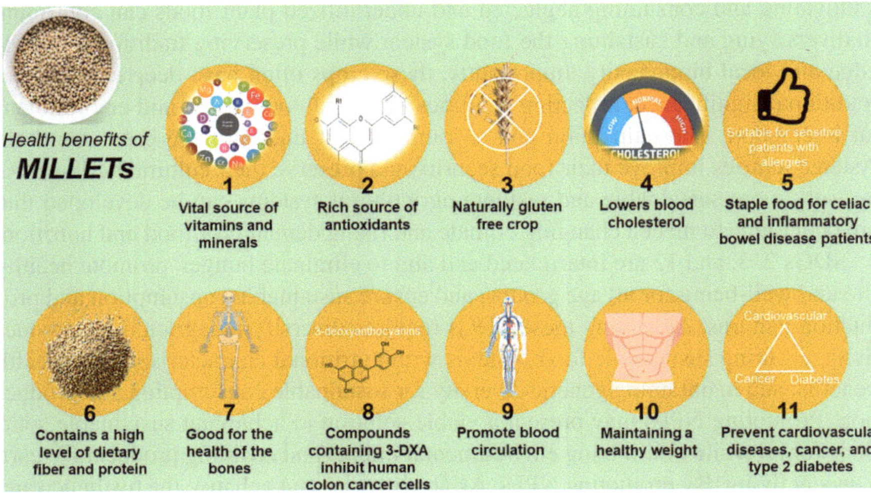

Fig. 2.2 Nutrition and health benefits of millets

Diversifying crop production in Asia by cultivating and consuming NPFs can benefit human well-being and the environment. Ultimately, integrating NPFs into agricultural schemes can result in a number of benefits, including:

1. NPFs are often high in nutrients and vitamins which are lacking in mainstream staple crops, thereby alleviating the serious nutritional deficiencies that affect millions of people;
2. NPFs cultivation contributes to the preservation of traditional knowledge and local biodiversity, which is often threatened by monocultures and industrial farming practices;
3. Increasing food security and nutrition by diversifying the crop range available and reducing reliance on a limited number of crops vulnerable to biotic and abiotic stresses;
4. Improved land use through NPFs cultivation, since marginal lands inaccessible to mainstream staple crops can be brought into cultivation;
5. Reducing the use of agricultural inputs, such as pesticides and fertilizers, since NPFs require less intensive cultivation practices and has a higher resistance to pests and diseases.

2.6 Neglected Plant Foods and Management of Malnutrition

As it was presented above, malnutrition is a significant global public health issue that affects millions of people, especially in developing countries (Malnutrition (who.int)). Primary causes of malnutrition are inadequate intake of vitamins, minerals, and proteins, which often lack in the daily diet due to a lack of access to a variety of crops (Deshmukh et al. 2023). NPFs are often overlooked despite their potential to improve food security and combat malnutrition. Table 2.2 demonstrates a wide range of essential nutrients, proteins, fiber, vitamins, and minerals in NPFs, often lacking in commonly consumed staple crops. Thus, these NPFs can significantly improve the nutritional status of malnourished individuals, particularly children who are vulnerable to the adverse effects of malnutrition. Table 2.3 presents the role of NPFs in malnutrition management.

NPFs can also contribute to food security and economic growth by promoting sustainable agriculture practices, increasing crop yields, and generating income for smallholder farmers (Mustafa et al. 2019). A well-balanced diet incorporating NPFs can effectively combat malnutrition and increase biodiversity.

Quinoa and pulses have gained popularity as nutritious and healthy crops worldwide in recognition of the International Years of Quinoa and Pulses to improve food security, malnutrition, and poverty. Millet crops (e.g., sorghum, pearl millet, finger millet, etc.) are vital for ensuring food and nutrition security in Africa and Asia, where they serve as staple crops and primary sources of nutrition for millions (Patil 2016). Climate-resilient millet crops thrive in harsh environments, such as drought and low soil fertility, and provide a reliable food source in areas with limited resources.

Table 2.3 Role of NPFs in malnutrition management

NUC	Nutrients	Management of Malnutrition
Millets	Protein, fiber, iron, calcium, zinc, etc.	The use of millets as a staple food can provide essential nutrients to malnourished individuals. Millets are also high in fiber, aiding digestion and preventing malnutrition-induced diarrhea.
Pseudo-cereals	Protein, iron, calcium, dietary fiber, polyphenols, etc.	Protein and other essential nutrients are abundant in pseudo-cereals, making them suitable for managing malnutrition. As an alternative to rice or wheat, they may be incorporated into meal preparations or added to salads and soups as a nutrient-dense addition to the diet.
Beans and Legumes	Protein, fiber, iron, zinc, etc.	Beans and legumes are excellent sources of protein and other essential nutrients, making them ideal for addressing malnutrition. They can be incorporated into meals such as soups, stews, and curries or used as a meat substitute in dishes such as tacos and burgers.
Roots and Tubers	Vitamin A, fiber, potassium, etc.	Roots and tubers are rich sources of Vitamin A, fiber, and potassium, making them an essential part of a balanced diet for malnourished individuals. They can be prepared in various ways, including boiling, roasting, and baking, and used in dishes such as stews and casseroles.
Fruits and Vegetables	Vitamins, fiber, minerals, antioxidants etc.	Fruits and vegetables are an essential part of a balanced diet, providing a range of vitamins, minerals, and fiber. Incorporating a variety of fruits and vegetables into meals can help manage malnutrition by providing essential nutrients. They can be consumed fresh, cooked, or as juices and smoothies.

Plant-based protein, fiber, vitamins, and minerals are found in beans and legumes, making them important components of a healthy diet and managing malnutrition (Table 2.3). Incorporating these NPFs into local diets can effectively address malnutrition, promote food biodiversity, and contribute to the sustainability of the food supply (Li and Siddique 2020). This, in turn, can promote climate-smart agriculture practices and enhance farming systems' resilience to environmental challenges. NPFs' cultivation and consumption may also provide socioeconomic benefits, such as improving soil fertility, increasing crop yields, and generating income for smallholder farmers. Throughout history, NPFs have been integral to indigenous cultures' diets and culinary traditions, providing various tastes, textures, and nutritional profiles that enrich and diversify meals and cuisines (Akinola et al. 2020). Unfortunately, these crops are often under-appreciated and underutilized in modern cuisine, despite their potential to offer a more sustainable and flavorful alternative to more common ingredients. A combination of NPFs and modern cuisine can provide unique and exciting taste profiles and promote sustainability and biodiversity in the food system. Therefore, recognizing the value of NPFs in culinary traditions and incorporating them into modern cuisine can enhance our food supply diversity and sustainability. The presence of biodiversity in NPFs provides significant opportunities for improving food security, managing malnutrition, and preventing undernutrition. Adequate nutrition is crucial for individuals' and nations' health, survival, growth, development, learning, and productivity. This means that nutrition plays an important role in achieving all 17 SDGs.

Currently, NPFs are cultivated on a small scale and have poor yields. To improve their yields, production conditions and consumption, multi efforts are demanded to establish national policies linked with value chains which could lead to sustainable food systems for zero hunger and improved nutrition.

2.7 Policies to Incentivize Nutrient-Rich Neglected Plant Foods Production

There are increasing calls to better integrate SDGs such zero hunger and, good health and wellbeing into food and agriculture programs and projects. Policies to incentivize diversification and promote the production, utilization and consumption of nutrient-rich neglected and underutilized foods provide the opportunity to increase production and intake of NPFs for better nutrition.

India, Mali, Sri Lanka, Turkey, Kenya, Brazil and Guatemale are taking the lead for setting up policies to explore and promote locally available, climate-resilient and affordable underutilized crops and wild species towards achieving sustainable local food systems (Borelli et al. 2020).

The Brazil's Ordinance No 284/2018 on 'Brazilian Socio diversity Native Food Species of nutritional value' was endorsed in 2018 with the aim to set the market price for many underutilized and neglected foods in Brazil, it also contributed to dissemination of knowledge and understanding. Moreover, the policy intents to enhance the promotion of these foods and achieve a sustainable use and consumption and to be included in different activities as for example school feeding (Hunter et al. 2019). Kenya recently endorsed the policy 'Biodiversity Conservation Policy for the County of Busia', which recognize the importance of neglected and underutilized foods for improved nutrition and food security. As per policy, more resources were allocated to conserve regional food biodiversity, inclusion of underutilized foods in school meals and connecting farmers to institutional markets (Beltrame et al. 2019; Hunter et al. 2019).

The global recognition of the potential -nutritional, resilience, sustainability- of NPFs is slowly increasing, for example, school feeding programs that link local farmer groups to include nutrient-rich NPFs are being advocated in various countries with the aim to incentivize the local production of NPFs. Programs that link NPFs to sustainable gastronomy and culinary tourism are also encouraged to increase awareness, as part of the international efforts to promote NPFs (Hunter et al. 2019; Borelli et al. 2020).

A holistic approach to promoting NPFs from production to consumption should include the following actions discussed by various authors (Hunter et al. 2019; Beltrame et al. 2019; Borelli et al. 2020):

- Increase research investments to explore NPFs; their nutritional characteristics and climate resilience.
- Increasing the information and database of the nutritional value and health benefits of NPFs.
- Connecting research to decision makers, including policy makers.
- Raising awareness among farmers and consumers, information is very powerful, and it should be disseminated in a responsible way to all stakeholders including consumers.
- Providing greater support, through policies and actions, for the NPFs mainstreaming in food security policies and programs.
- Increase policy support for production, marketing and commercialization of NPFs.

2.8 Conclusion

NFPs have great potential to improve dietary diversity of populations, which in turn would be reflected in better nutrition and health. Many NPFs such as millets and pseudocereals are nutrient-rich foods, they have better content and quality of protein as compared with conventional staple foods. In addition, they are also good sources of dietary fiber, essential vitamins, minerals and bioactive compounds that are good for health. Policies that promote and advocate the production, usage and consumption of NPFs are very much needed to improve the current food systems and achieve the SDGs 2030 towards zero hunger and good health and wellbeing for all.

Conflicts of Interest Authors read and approved the final manuscript and declare that they have no conflict of interest.

References

Abebe BK, Alemayehu MT (2022) A review of the nutritional use of cowpea (Vigna unguiculata L. Walp) for human and animal diets. J Agric Food Resh 10:1–14
Akinola R, Pereira LM, Mabhaudhi T et al (2020) A review of indigenous food crops in Africa and the implications for more sustainable and healthy food systems. Sustainability 12:1–30
Baggio A, Federici E, Gebtilucci V et al (2023) Brown rice and pulses for the development of shelf-stable and low glycemic index ready-to-eat meals. J Funct Foods 100:1–9
Bailey A (2016) Mainstreaming agrobiodiversity in sustainable food systems: scientific foundations for an agrobiodiversity index-summary. Bioversity International, Rome, p 32
Baldermann S, Blagojevic L, Frede K et al (2016) Are neglected plants the food for the future? CRC Crit Rev Plant Sci 35:1–14
Basset-Sagarminaga J, Roumans KH, Havekes B et al (2023) Replacing foods with a high-glycemic index and high in saturated fat by alternatives with a low glycemic index and low saturated fat reduces hepatic fat, even in Isocaloric and macronutrient matched conditions. Nutrients 15:735
Behnassi M, El Haiba M (2022) Implications of the Russia–Ukraine war for global food security. Nat Hum Behav 6:754–755
Beltrame D, Eliot G, Gunner B et al (2019) Mainstreaming biodiversity for food and nutrition into policies and practices: methodologies and lessons learned from four countries. ANADOLU Ege Tarımsal Araştırma Enstitüsü Dergisi 29:25–38
Borelli T, Hunter D, Padulosi S et al (2020) Local solutions for sustainable food systems: the contribution of orphan crops and wild edible species. Agro 10:231
Borkotoky K, Unisa S, Gupta AK (2018) State-level dietary diversity as a contextual determinant of nutritional status of children in India: a multilevel approach. J Biosoc Sci 50:26–52
Campbell BM, Beare DJ, Bennett EM et al (2017) Agriculture production as a major driver of the Earth system exceeding planetary boundaries. Ecol Soc 22:1–12
Cema H, Calder PC (2020) Defining a healthy diet: evidence for the role of contemporary dietary patterns in health and disease. Nutrients 12:334
Cooper AJ, Sharp SJ, Lentjes MA et al (2012) A prospective study of the association between quantity and variety of fruit and vegetable intake and incident type 2 diabetes. Diabetes Care 35:1293–1300
Darapheak C, Takano T, Kizuki M et al (2013) Consumption of animal source foods and dietary diversity reduce stunting in children in Cambodia. Int Arch Med 6:1–11
Darnton-Hill I, Webb P, Harvey PW, Hunt JM et al (2005) Micronutrient deficiencies and gender: social and economic costs. AJCN 81:1198S–1205S
Deriu AG, Vela AJ, Ronda F (2022) Techno-functional and gelling properties of Acha (Fonio) (Digitaria exilis stapf) flour: a study of its potential as a new Gluten-free starch source in industrial applications. Foods 11:183
Deshmukh R, Nadaf A, Ansari WA et al (2023) Correction to: biofortification in cereals. Biofortification in cereals: Progress and prospects. Springer Nature, Singapore
Drewnowski A (2012) The economics of food choice behavior: why poverty and obesity are linked. Obesity treatment and prevention: new directions. Karger AG; Basel Switzerland
Ebert AW (2014) Potential of underutilized traditional vegetables and legume crops to contribute to food and nutritional security, income and more sustainable production systems. Sustainability 6:319–335
Elizabeth et al (1985) Diet diversity and nutrient intake. J Am Diet Assoc 85:830–836
FAO (1996) World food summit, declaration on World Food Security. Rome, Italy

FAO (2004) What is agrobiodiversity? https://www.fao.org/3/y5609e/y5609e01.htm Accessed 5 Oct 2022

FAO (2017) The future of food and agriculture Trends and challenges Rome https://www.fao.org/3/i6583e/i6583e.pdf Accessed 27 April 2023

FAO (2021) Minimum dietary diversity for women, Rome. https://doi.org/10.4060/cb3434en

FAO I, UNICEF, WFP, WHO (2022) The state of food security and nutrition in the world. Repurposing food and agricultural policies to make healthy diets more affordable. FAO, Rome. https://doi.org/10.4060/cc0639en

FAO U, WFP, WHO (2019) Placing nutrition at the Centre of Social Protection. Asia and the Pacific regional overview of food security and nutrition. FAO, Bangkok

Farooq M, Rehman A, Li X, Siddique KH (2023) Neglected and underutilized crops and global food security. In: Farooq M, Siddique KH (eds) Neglected and underutilized crops, Academic Press, USA, pp 3–19

Fatima A, Farid M, Safdar K et al (2020) Loss of agro-biodiversity and productivity due to climate change in continent Asia: a review. In: Plant ecophysiology and adaptation under climate change: mechanisms and perspectives I: general consequences and plant responses, pp 51–71

Fernandez-Tome S, Ashaolu TJ, Hernandez-Ledesma B (2023) Exploration of the nutritional and functional properties of underutilized grains as an alternative source for the research of food-derived bioactive peptides. Nutrients 15:351

Gan Y, Tong X, Li L, Cao S et al (2015) Consumption of fruit and vegetable and risk of coronary heart disease: a meta-analysis of prospective cohort studies. Int J Cardiol 183:129–137

Harb AA, Shechter A, Koch PA et al (2022) Ultra-processed foods and the development of obesity in adults. Eur J Clin Nutr 77:619–627

Heiman ML, Greenway FL (2016) A healthy gastrointestinal microbiome is dependent on dietary diversity. Mol Metab 5:317–320

Hunter D, Borelli T, Beltrame DM, Oliveira CN (2019) The potential of neglected and underutilized species for improving diets and nutrition. Planta 250:709–729

Khoshbakht K, Hammer K (2008) How many plant species are cultivated? Genet Resour Crop Evol 55:925–928

Krishna A, Mejia-Guevara I, Mcgovern M (2018) Trends in inequalities in child stunting in South Asia. Matern Child Nutr 14:1–12

Kumar A, Kumari P, Kumar M (2022) Role of millets in disease prevention and health promotion. In: Functional foods and nutraceuticals in metabolic and non-communicable diseases. Academic Press, USA, pp 341–357

Leite FHM, Khandpur N, Andrade GC (2022) Ultra-processed foods should be central to global food systems dialogue and action on biodiversity. BMJ Glob Health 7:1–4

Li B, Li F, Wang L (2016) Fruit and vegetables consumption and risk of hypertension: a meta-analysis. J Clin Hypertens 18:468–476

Li X, Siddique KH (2020) Future smart food: harnessing the potential of neglected and underutilized species for zero hunger. Matern Child Nutr 16:1–22

Li X, Yadav R, Siddique KHM (2020) Neglected and underutilized crop species: the key to improving dietary diversity and fighting hunger and malnutrition in Asia and the Pacific. Front Nutr 7:1–12

Liu C, Ma R, Tian Y (2022) An overview of the nutritional profile, processing technologies, and health benefits of quinoa with an emphasis on impacts of processing. Crit Rev Food Sci Nutr:1–18. https://doi.org/10.1080/10408398.2022.2155796

Malik AM, Singh A (2022) Pseudocereals proteins-a comprehensive review on its isolation, composition and quality evaluation techniques. Food Chem 1:1–14

Marron-Ponce JA, Sanchez-Pimienta TG, Rodriguez-Ramirez S (2023) Ultra-processed foods consumption reduces dietary diversity and micronutrient intake in the Mexican population. J Hum Nutr Diet 36:241–251

Mustafa M, Mayes S, Massawe F (2019) Crop diversification through a wider use of underutilised crops: a strategy to ensure food and nutrition security in the face of climate change. In:

Sustainable solutions for food security: combating climate change by adaptation. Springer, Cham, pp 125–149
Okagu IU, Ndefo JC, Aham EC et al (2021) Lupin-derived bioactive peptides: intestinal transport, bioavailability and health benefits. Nutrients 13:1–18
Padhan B, Panda D (2020) Potential of neglected and underutilized yams (Dioscorea spp.) for improving nutritional security and health benefits. Front Pharmacol 11:496
Palacios-Rojas N, Mcculley L, Kaeppler M et al (2020) Mining maize diversity and improving its nutritional aspects within agro-food systems. Compr Rev Food Sci Food Saf 19:1809–1834
Patil JV (2016) Millets and Sorghum: biology and genetic improvement. John Wiley & Sons, Inc., UK, p 463
Popkin BM, Corvalan C, Grummer-Strawn LM (2020) Dynamics of the double burden of malnutrition and the changing nutrition reality. Lancet 395(10217):65–74.
Popkin BM, Ng SW (2022) The nutrition transition to a stage of high obesity and noncommunicable disease prevalence dominated by ultra-processed foods is not inevitable. Obes Rev 23:1–18
Probst YC, Guan VX, Kent K (2017) Dietary phytochemical intake from foods and health outcomes: a systematic review protocol and preliminary scoping. BMJ Open 7:1–9
Rahul J, Jain MK, Singh SP (2015) Adansonia digitata L.(baobab): a review of traditional information and taxonomic description. Asian Pac J Trop Biomed 5:79–84
Ramatsetse KE, Ramashia SE, Mashau ME (2023) A review on health benefits, antimicrobial and antioxidant properties of Bambara groundnut (Vigna subterranean). Int J Food Prop 26:91–107
Rashmi HB, Negi PS (2020) Health benefits of bioactive compounds from vegetables. In: Plant-derived bioactives: production, properties and therapeutic applications. Springer, Singapore, pp 115–166
Ricci C, Baumgartner J, Zec M et al (2018) Type of dietary fat intakes in relation to all-cause and cause-specific mortality in US adults: an iso-energetic substitution analysis from the American National Health and Nutrition Examination Survey linked to the US mortality registry. Br J Nutr 119:456–463
Rippe J, Angelopoulos T (2016) Added sugars and risk factors for obesity, diabetes and heart disease. Int J Obes 40:S22–S27
Ritchie H, Roser M (2017) Micronutrient deficiency. Our World in data, https://ourworldindata.org/micronutrient-deficiency
Roy D, Zulfiqar F, Tsusaka TW et al (2022) Household food insecurity and dietary diversity of women of reproductive age among smallholder farming households in northwest Bangladesh. Ecol Food Nutr 61:460–483
Shahbaz M, Raza N, Islam M et al (2022) The nutraceutical properties and health benefits of pseudocereals: a comprehensive treatise. Crit Rev Food Sci Nutr:1–13. https://doi.org/10.1080/10408398.2022.2071205
Siddique KH, Li X, Gruber K (2021) Rediscovering Asia's forgotten crops to fight chronic and hidden hunger. Nat Plants 7:116–122
Singh RK, Sreenivasulu N, Prasad M (2022) Potential of underutilized crops to introduce the nutritional diversity and achieve zero hunger. Funct Integr Genom 22:1459–1465
Sonnenburg ED, Smits SA, Tikhonov M et al (2016) Diet-induced extinctions in the gut microbiota compound over generations. Nature 529:212–215
Stefoska-Needham A, Beck EJ, Johnson SK et al (2015) Sorghum: an underutilized cereal whole grain with the potential to assist in the prevention of chronic disease. Food Rev Int 31:401–437
Tan WC, Tan CH, Nyam KL et al (2021) Nutritive bambara groundnut powdered drink mix: characterization and in-vivo assessment of the cholesterol-lowering effect. J Food Sci Technol 58:2992–3000
Tian Y, Su L, Wang J et al (2018) Fruit and vegetable consumption and risk of the metabolic syndrome: a meta-analysis. Public Health Nutr 21:756–765
Tiwari H, Naresh R, Bhatt R et al (2023) Underutilized nutrient rich millets: challenges and solutions for India's food and nutritional security: a review. Int J Plant Soil Sci 35:45–56

WHO (2022a) Healthy diet. https://www.who.int/en/news-room/fact-sheets/detail/healthy-diet. Accessed 20 Jan 2023
WHO (2022b) Vitamin and Mineral Nutrition Information System (VMNIS). https://www.who.int/teams/nutrition-and-food-safety/databases/vitamin-and-mineral-nutrition-information-system. Accessed Jan 2023
WHO U, USAID A, AED U (2008) Indicators for assessing infant and young child feeding practices. World Health Organization, Geneva
Woomer JS, Adedeji AA (2021) Current applications of gluten-free grains–a review. Crit Rev Food Sci Nutr 61:14–24
Xuan L, Siddique K (2018) Future smart food: rediscovering hidden treasures of neglected and underutilized species for zero hunger in Asia. FAO, Bangkok, pp 1–241

Chapter 3
Exploring Neglected and Underutilized Plant Foods to Fight Malnutrition and Hunger in South Asia

Iqra Akram, Anam Layla, and Tariq Ismail

3.1 Introduction

During the Green Revolution and following decades, the goal of agricultural research was to boost crop yields to ensure adequate supply of calories to the peoples. Providing a sufficient amount of food was prioritized over providing food that was nourishing, therefore nutritional quality received less emphasis. As a result, a large segment of populations from different continents including South Asia, still consume diets low in crucial micronutrients and protein. In 2019, South Asia was reported as region with highest estimated number of undernourished individuals (257.3 million) in the world (FAO et al. 2021). According to WHO statistics, anemia affects 40% of pregnant women and 42% of children below 5 years worldwide while in South Asia prevalence rate of anemia in women of reproductive age (15–49 years) and children (aged 6–59 months) was 48.2% and 50.7%, respectively (WHO 2023). Neglected and underutilized plant species (NUS) of food potential are mostly wild or semi-domesticated species that have adapted to their surroundings. These traditional foods were used for centuries but became increasingly

I. Akram
Department of Human Nutrition, Faculty of Food Science & Nutrition,
Bahauddin Zakariya University, Multan, Pakistan

A. Layla
National Institute of Food Science & Technology, University of Agriculture,
Faisalabad, Pakistan

T. Ismail (✉)
Department of Food Science and Technology, Faculty of Food Science & Nutrition,
Bahauddin Zakariya University, Multan, Pakistan

T. Ismail et al. (eds.), *Neglected Plant Foods of South Asia*,
https://doi.org/10.1007/978-3-031-37077-9_3

obsolete as more productive (or easier to process, or prestigious, or profitable) crops took their place in farming systems. Future Smart Foods, another name of neglected or underutilized foods are nutritionally dense, locally available or adaptable, economically viable and climate resilient (FAO 2018). NUS have the potential to contribute to food security, health, dietary and culinary diversification, and income generation. Cereals and pseudo cereals, roots and tubers, vegetables, legumes, oilseeds, nuts, aromatic and medicinal plants, and many other plant species of different origin were among the crops that were overlooked and underutilized. NUS are less well-known plant species in terms of marketing and research, but they are well adapted to marginal abiotic and biotic stress conditions. NUS have traditionally been grown and inextricably linked to food security and cultural identity of local rural communities, and were considered as valuable resource to earn livelihood. They have higher nutritional value dominated with adequate supply of essential micronutrients and protein.

3.2 State of Food Insecurity and Nutrition in South Asia

South Asia is the most densely populated regions of the world. Approximately 20% of the global population is fed from 5% world agricultural land of South Asia. According to the estimates, one-third of the world's poor population live in South Asia while 70% of this population belongs to the rural South Asia making rural communities as the poorest households of the world (FAO et al. 2020). Climate change has become a critical concern in South Asia annihilating agriculture and threatening food insecurity. Climate change is affecting agriculture through change in weather, temperature, rainfall and by escalating risks and episodes of droughts, floods and heat stress. Alike climate change, pandemics also pose a global risk. Recent COVID-19 pandemic interrupted agriculture production system and food supply chain, and amplified the challenges of sustainable livelihoods, food and nutrition security (Rasul 2021).

Regardless of hopes that the world would rise from the pandemic in 2021 and the situation of food security will start to improve, a sharp increase has been witnessed in global hunger index. Prevalence of undernourishment being relatively unchanged in 2015, rose from 8.0% in 2019, 9.3% in 2020 and 9.8% in the year 2021. Around 702–828 million people were affected by hunger in 2021 while 297.8 million people in South Asia were undernourished. In 2021, approximately 2.3 billion people i.e., 30% of the global population were reported moderately or severely food insecure. The estimate claims, 764.3 million food insecure population belong to the South Asia (Table 3.1) (FAO et al. 2022).

Table 3.1 Prevalence of undernourishment, moderate or severe food insecurity

Regions/ countries	Number of undernourished people 2004–2006 (%)	2019–2021 (%)	Food insecurity in total population[a] 2014–2016 (%)	2019–2021 (%)	Moderate or severe food insecurity in total population[a] 2014–2016 (%)	2019–2021 (%)	Population 2010 (1000 persons)	2019 (1000 persons)
World	12.2	9.0	7.7	10.7	21.8	28.1	6,869,582	7,668,812
South Asia	19.9	15.3	13.2	18.8	27.7	39.4	1,711,871	1,918,211
Afghanistan	36.1	29.8	14.8	22.5[b, c]	45.1	70.0[b, c]	29,186	38,928.35
Bangladesh	14.2	11.4	13.3	10.7	32.2	31.7	147,575	164,689.4
India	21.6	16.3	–	–	–	–	1,234,281	1,380,004
Iran	5.2	4.1	9.5	7.7	48.0	42.4	73,763	83,992.95
Maldives	N/A	N/A	N/A	2.2	N/A	13.4	366	540.54
Nepal	16.8	5.5	10.4	13.6	29.5	37.8	27,013	29,136.81
Pakistan	17.6	16.9	0.9[b, c]	8.7[c, d]	14.0[b, c]	32.6[c, d]	179,425	220,892.3
Sri Lanka	14.0	3.4	0.7[b, c]	1.1[b, c]	5.9[b, c]	10.0[b, c]	20,262	21,413.25

Source: FAO et al. (2022) and FAOSTAT (2022)

[a]Estimates of the percentage of people in the total population living in households where at least one adult has been found to be food insecure

[b]Based on official national data

[c]Values are projected using FAO data or estimates

[d]Data provide food insecurity estimates for year 2020 collected from a 3 months reference period COVID-19 national impact assessment survey

3.3 Food Production Statistics and Nutritional Gaps, Burden of Food Insecurity on Conventional Food Crops

Looking into the perspective of food demand, emerging urbanization, population growth and rise in size of middle income class have brought changes to consumer attitudes and food preferences, particularly linked to food safety and quality. This trend is likely expected to affect the future dietary patterns of peoples in South Asia. Moreover, from food supply perspective, low rate of growth in yield of staple food crops is an indicator of insufficiency of the food resources for future needs of populations. Increased rate of food crops production demands expansion of cropping area that is already scarce in Asia. Decline in agricultural biodiversity, climate change, degradation of natural resources and water scarcity are the major causes of food insecurity. Relative to the normal climate conditions, climate change had been predicted to anticipate reduction in rice production by 14%, wheat by 44–49% and 9–19% in maize production in South Asia (Nelson et al. 2009).

In last one decade, the world population approximately increased from 6.8 billion to 7.6 billion while rate of increase in population was from 1.7 billion to 1.9 billion in Southern Asia. This may be equivalent to 24.8% of the total world population (Worldmeter 2022; FAOSTAT 2022). Food production staistics for the South Asian countries are given in Table 3.2. Subsequent to rise in regional and global

Table 3.2 Food production statistics

	Year	World	South Asia	Afghanistan	Bangladesh	India	Iran	Maldives	Nepal	Pakistan	Sri Lanka
Food supply	2010	2807.58	2420.00	2294.03	2397.74	2378.80	2960.35	2581.18	2544.01	2205.28	2458.67
(kcal/capita/day)	2019	2920.05	2592.00	2289.40	2586.81	2549.85	2878.29	2328.11	2860.65	2408.18	2678.10
Cereals	2010	27,108	147	0	0	106	–	–	8	33	0
(1000 tonnes)	2019	29,818	113	–	0	30	–	–	11	68	–
Pulses	2010	36,412	13,094	53	159	11,672	270	0	209	718	12
(1000 tonnes)	2019	45,573	16,903	44	310	15,421	308	0	356	455	8
Fruits	2010	182,012	44,013	298	2513	31,646	2932	4	925	5539	157
(1000 tonnes)	2019	234,380	64,599	401	3288	48,857	5404	6	866	5588	181
Vegetables	2010	806,989	95,288	1267	2627	73,102	11,527	2	3084	3141	538
(1000 tonnes)	2019	972,585	972,585	972,585	972,585	972,585	972,585	972,585	972,585	972,585	972,585
Nuts & products	2010	14,248	14,248	14,248	14,248	14,248	14,248	14,248	14,248	14,248	14,248
(1000 tonnes)	2019	17,610	17,610	17,610	17,610	17,610	17,610	17,610	17,610	17,610	17,610
Roots & tubers	2010	15,568	15,568	15,568	15,568	15,568	15,568	15,568	15,568	15,568	15,568
(1000 tonnes)	2019	20,473	20,473	20,473	20,473	20,473	20,473	20,473	20,473	20,473	20,473
Oil crops	2010	10,734	10,734	10,734	10,734	10,734	10,734	10,734	10,734	10,734	10,734
(1000 tonnes)	2019	11,456	11,456	11,456	11,456	11,456	11,456	11,456	11,456	11,456	11,456

Source: FAOSTAT (2022)

population, demand for adequate, safe and nutritious foods is tremendously increasing, and may cause a burden on conventional agriculture and food production systems. Climate change is also a factor effecting the food production. Considerable decline i.e., from 147 to 113 thousand tons in the production of cereal crops has been observed in South Asia. Cereals production status for the year 2010 to 2019 indicate highest rate of decline in production of cereals from India where cereals yield was decreased from 106 to 30 thousand tons.

By the year 2050, two major gaps in food production are expected to be countered by existing food and agriculture system. Global food production will need to be increased by 50% to feed 9.7 billion people by the year 2050 (FAO et al. 2017). It is unlikely that the production of conventional staple crops will meet the growing food demands since majority of the staple food crops like rice, wheat and maize have already attained ~80% of the yield potential. Hence, reliance on conventional staple food crops will not be sufficient to decrease the gap between the food demand and supply. Conventional staple crops although provide adequate calories to prevent hunger, but are not enough to supply every essential nutrient for a balanced and healthy diet. Unbalanced diets with inadequate nutrients diversity are among the leading causes of micronutrients malnutrition making it quite difficult to assure adequate nutrients delivery to the additional 2 billion peoples by the year 2050 (Li and Siddique 2018). Conclusively, closing food production and nutrition gaps require a paradigm shift from conventional towards a sustainable food production system emphasizing more on diversification in food resources to address existing and the future global food challenges.

Holistic food production system can help mediating the burden of malnutrition under existing and forecasted threats to the climate and the environment. NUS are nutritious, economical, climate resilient and adaptive specially in marginal areas. Previously, NUS have never been included in agricultural extension curricula and had been discriminated for becoming a part of organized food value chains. Likewise, research on NUS has also been badly neglected at the cost of staple food crops. Given a considerable interest, NUS can contribute to raise nutrition security, food availability and affordability.

3.4 Role of Dietary Diversity in High Burden of Malnutrition

Though global hunger index score for Asia has reduced by 34% compared to the score recorded in 1990, still the region bears a highest count of hungry people i.e., 512 million (12% of the total population of Asia). Highest number of hungry peoples and the malnourished children i.e., 70% of the total malnourished children is reported from Asia. Within the region highest count of nutritionally compromised populations is reported from population dense South Asian countries like Bangladesh, India and Pakistan (WHES 2023). Persistent high magnitude of malnutrition in South Asia may be associated with lesser dietary diversity including lower agro food production diversity.

Sustainable development goals (SDGs) endorsed by the United Nations cover a broad range of global issues. SDG 2 – Zero Hunger mainly focuses on ending hunger by assuring access to safe and nutritious food for all, and promoting sustainable agriculture by 2030. Even though substantial progress has been made toward achieving SDGs, ending malnutrition and hunger continue to be a major concern in South Asia. Magnitude of the problem is better manifested from the highest number of undernourished peoples i.e., 257.3 million (FAO et al. 2021).

Dietary diversification is a balanced, diverse, and healthier approach to ensures adequate or optimal nutrition. Positive correlation had been reported in dietary diversity and the mean micronutrients intake by children to ensure proper growth and development (FANTA 2006). Dietary quality of children diets is better measured by assessment of minimum meal frequency (MMF), minimum dietary diversity (MDD) and minimum acceptable diet (MAD). A child consuming five out of eight food groups is expected to meet the minimum dietary diversity (FAO et al. 2020). Lower dietary diversity is mostly reflected from the diets high in conventional cereals and low in fruits, vegetables and pulses.

Poor food consumption behaviors coupled with faulted agriculture and food production system limit access to diversified food resources thus increasing the magnitude of malnutrition. Diversification in existing food production system by promoting neglected and underutilized plant foods may prove to be formidable tools in achieving Zero Hunger and alleviating various forms of malnutrition (Fanzo et al. 2013). Rice and wheat are cultivated as major staple food crops in South Asia and thus constitute a large share of nutrient energy for the South Asian populations. Only a few staple crops are cultivated in the region of Asia mainly rice and wheat, contributing as a major component of people's diets (Fairulnizal et al. 2015). Neglected and underutilized plant foods are nutrient-dense, locally available, easily adaptable, climate-resilient, profitable and may play a fundamental role in improving dietary and production diversity (Li et al. 2020). Complimenting conventional staple food crops with neglected, underutilized food plants may therefore not merely help achieving the Zero hunger index but also improve the situation of micronutrient malnutrition.

3.5 Concept of Future Smart Foods and Their Role to Fight Malnutrition

Future Smart Food Initiative is led by FAO's Regional Initiative on Zero Hunger which aims to utilize the enormous benefits of neglected and underutilized plant species to fight against malnutrition and hunger. Future smart foods (FSFs) are neglected and underutilized species (NUS) which are nutrient-dense, locally available, easily adaptable, and climate-resilient (Li and Siddique 2018). In sustainable agriculture, NUS are the important elements of the agrobiodiversity. Globally, out of approximately five hundred thousand plant species, only thirty thousand have been identified as edible species. Though the number of edible plants species is enormous yet ~7000 crop species have been domesticated from the wild and adopted

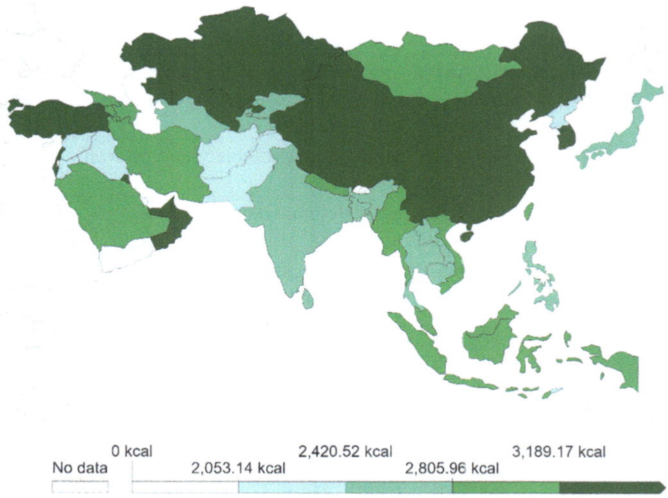

Fig. 3.1 Per capita kilocalorie supply from foods per day in Asia, 2019 (OWID 2022). This measures the quantity that is available for consumption at the end of the supply chain. It does not account for consumer waste, so the quantity that is actually consumed may be lower than this value

for cultivation as food crop (Garn and Leonard 1989). Among the domesticated food crops, barely 150 crops are commercially cultivated; of these 103 provide up to 90% of the calories in human diet. Maize, rice and wheat are the three main crops which satisfy ~60% of the global dietary energy requirements. Per capita kilocalorie supply from foods per day in Asia is given in Fig. 3.1. This indicates an enormously large number of edible plants which are still 'underutilized' even though they could have a significant role in decreasing too much reliance on mainstream staple food crops. NUS are largely superior in nutritional value, can grow on marginal soils and easily fit into integrated crop production system in small holdings. NUS being more tolerable to environmental stresses, make food production systems more diverse, climate resilient and sustainable. FAO has set crops as an entry point for NUS to address malnutrition and hunger (Li and Siddique 2018).

3.6 Nutritional Attributes of Underutilized Future Smart Plant Foods

Asians have grown and enjoyed a diverse range of healthful foods for ages. Unfortunately, subsequent generations' diets have slowly but steadily shifted away from many traditional foods. Majority of such traditional neglected and underutilized crops are referred as Future Smart Food (FSF) crops. Previously, FSF crops were characterized as poor people's crops, marginal crops or third order crops and received virtually little attention in agricultural research and development. Because of their multiple applications and relevance, many formerly neglected crops have

Table 3.3 Nutritional traits of some Future Smart Foods

FSF	Nutritional traits/medicinal use	Source
Bael	Good carrier of carbohydrates and provides minerals and vitamins including calcium, iron, phosphorous, thiamine, riboflavin, carotene and ascorbic acid	Lakht-e-Zehra et al. (2015)
Buckwheat	Good source of essential fatty acids, soluble vitamins, iron and zinc	Kim et al. (2004)
Moringa (drumstick)	Rich in protein, potassium, calcium, vitamin A & C Well-known for medicinal value	Liang et al. (2019)
Fenugreek	Relief from anemia, diabetes, stomach problems, fever, insomnia, respiratory disorders, and appetite stimulant	Singletary (2017)
Fiddlehead fern	Caloric rich and good source of protein	Seal (2012)
Foxtail millet	Good carrier of fiber and carbohydrates Helps control blood-sugar levels and reduces risk of heart attacks	Amadou et al. (2013)
Goosefoot	Green matter is rich in protein Contains high degree of potassium, calcium, zinc, iron, magnesium & fiber	Poonia and Upadhayay (2015)
Lentils	Second-highest ratio of protein per calorie of any legume, after soybeans.	Li and Siddique (2018)
Mung bean	High in protein, dietary fiber, resistant starch and iron	Dahiya et al. (2015)
Quinoa	Rich source of fiber, antioxidants, protein, iron, and zinc Good for controlling blood sugar	Singh et al. (2014)
Snake gourd	Contain vitamin A, vitamin B, vitamin C, calcium, iodine, iron, manganese, magnesium, and potassium. Good source of dietary fiber and protein and lower in calories	Idowu et al. (2019)
Taro	Rich in carbohydrates and high levels of calcium and vitamin A	Temesgen and Retta (2015)

now become globally significant and received space as Future Smart Food crops due to increased awareness on importance of smart foods. Compared to mainstream staple crops, NUS as future smart foods usually contain higher contents of different nutrients. They are richer in essential vitamins, minerals, phytochemicals like flavonoids and have good composition for macronutrients including protein, carbohydrates, and dietary lipids. Nutritional traits of some future smart foods are highlighted in Table 3.3. With regards to their health benefits, NUS like pulses are considered as superior foods. When comparing with white rice, chickpeas, mung beans and lupins are 44–78 times richer in folate, four times higher in Fe, nearly 4–8 times higher in dietary fibers, and 3–5 times dense in proteins (Sathe 1996). Lentils have the potential to provide adequate amount of some essential minerals such as zinc, iron and selenium (Migliozzi et al. 2015).

Earlier to the rice crop, Millets (*Panicum*) were one of the promising smart staple foods which played a key role in ensuring dietary diversity and household food security in many Asian countries. Millets contain protein contents similar to wheat and are good source of B vitamins especially folate, pyridoxine and niacin, and among the minerals iron, zinc, calcium, magnesium and potassium (Chang et al. 2012). *Moringa oleiferia* or drumstick is also underutilized plant species with exceptional nutritional value and health benefits. Seed pods and leaves of *M. olifera* are rich in protein. Fresh leaves contain ascorbic acid while the dried leaf powder is a rich source of nutrients especially vitamin A. They also contain good concentrations of B vitamins, iron, magnesium, calcium, and potassium. A study by Asghari et al. (2015) suggests daily intake of moringa seeds and leaves to provide recommended daily allowance (RDAs) of vitamin A & C, calcium, and potassium. Table 3.3 shows some examples of future smart foods along with their nutritional significance. Thus, neglected and underutilized plants species offer exceptional nutritional value for improving micronutrient and protein contents in daily diet, also prevents non-communicable diseases in millions of people. NUS can provide food when other crops fail, as they are climate resilient. That is why new global trend is observed in affluent societies with various NUS like chia, quinoa, açaí and moringa, now coming back to the dining tables.

3.7 Underutilized and Wild Food Plants, Ethnic Knowledge on Food Uses, a Critical Comparison of Neglected Plant Foods as Substituent to Conventional Food Crops

Food and Agriculture Organization (FAO) of United Nations has been working with eight countries of Asia-Pacific to mitigate the challenges of achieving Zero Hunger and climate change by encouraging transformation of mono-cropping to diversified agriculture production system, and increased use of future smart foods. Diversified food production system guarantees dietary diversification and strongly influential in achieving the Zero Hunger goal. Under the aforementioned program of FAO, 39 neglected and underutilized species have been identified by national and international partners and are ranked as future smart food based on their ability to address nutritional needs of the masses and adoptability to climate change while improving social conditions and farmer income (Table 3.4).

Table 3.4 Some neglected and underutilized plants in South Asia

Cereals	Roots & tubers	Pulses	Fruits	Vegetables, shrubs, herbs	Nuts, seeds & spices
Amaranth Buckwheat Finger millet Foxtail millet Proso millet Quinoa Sorghum Specialty rice	Air yam Elephant's foot yam Indian Kudzu Lady tulip Purple yam Swamp taro Sweet potato Taro	Black gram Cow pea Faba bean Grass pea Horse gram Mung bean Rice bean Soybean	Chayote Date palm Indian gooseberry Jack fruit Java plum Jujube Mulberry Roselle Toothbrush tree Wood apple	Camel's foot Drumstick Fenugreek Fiddlehead fern Goosefoot Pumpkin Snake gourd White mugwort	Linseed Nepali butter tree Nepali pepper Perilla Prickly ash Walnut

NUS are often associated with the cultural heritage of their place of origin. They are well adapted to harsh environments and marginal areas, grow well with traditional production systems with least to no external inputs. Nevertheless, underutilized plant species have gained limited attention by the extension service organizations, researchers, technology providers and consumers, donors, policy and decision markers. Limited knowledge about cultivation methods of NUS, their nutritional value, mode of consumption and the dietary habits of consumers serve as the baseline information necessitating change in conventional food production systems. To conserve the NUS and agricultural ecosystems, indigenous knowledge is considered vital (Chivenge et al. 2015). Farmers of recent agriculture age are not well aware of the role of underutilized crops in agro-diversity and their cultivation methods due to lack of indigenous knowledge. Consumers are not aware of the nutritional value of NUS which may be strong contributor to nutritional security (Johns and Sthapit 2004; Fassil et al. 2000) and being rich in micronutrients are helpful in eradicating hidden hunger. Earlier, nutrition programs and policies were not much focused on creating consumer awareness about the dietary diversity and inclusion of neglected and underutilized foods in their food basket. Within the households, nutritional knowledge of mothers is of immense significance in keeping a balanced dietary pattern of the family, children and herself (Osei et al. 2010). Thus, generating a data base on food and nutritional significance of NUS and transfer of the knowledge to the peoples like mothers can create a real impact in ensuring household nutritional security.

Many of the underutilized plant species of food significance also hold medicinal value. For example, the people in remote areas of Hindu-Kush Himalaya and the plains of India and Pakistan have been using the fruit and dried seed of Java plum indigenously known as *Jamun* to treat diabetes. Likewise, local inhabitants of Gilgit-Baltistan province of Pakistan have realized the nutritional and medicinal significance of sea-buckthorn and expanded its cultivation. To establish its value chain, some NGOs have helped the local communities particularly women to prepare a range of sea-buckthorn derived products (Rasul et al. 2014). Similarly, people in Pamir Mountains of Afghanistan and Tajikistan use local neglected food crops as conventional remedy to different ailments and sicknesses. Nearly 58 cultivated and non-cultivated plants have been identified by the farming communities as conventional preventive and therapeutic plant based ethnopharmacological resources in 310 districts of Afghanistan and Tajikistan (Kassam et al. 2010).

During last decade, several neglected food plants species have gained status of essential commercial crops. For instance, lentils and quinoa have gained global interest. Quinoa was once a subsistence crop but its production is now increased due to rising consumer appreciation. Cultivation of quinoa has expanded to more than 70 countries globally, including India and other countries of South Asia. Production of quinoa has almost doubled in lower income countries from 2010 to 2020 (FAOSTAT 2022).

3.8 An Example of Success in South Asia

In South Asia, Nepal is the major success story in implementing NUS at various levels. Underutilized food crops constitute nearly 85% of the 570 species of cultivated crops of Nepal (Delêtre et al. 2013). Future Smart Food crops were reintroduced lately through coordinated efforts of national and international organizations. Over the last two decades, Nepal has promoted agricultural biodiversity through a variety of approaches, including community-based biodiversity management (CBM), which encourages community members to share and exchange seeds and other planting materials. CBM also includes components such as establishment of community seed banks, CBM trust fund, running campaigns for plant breeding and seed exchange, holding awareness activities at community and school levels, arranging crop and food diversity fairs, and the buy-back programs. The Nepalese government has also established offices dedicated to strengthening underutilized FSF crops such as naked barley, buckwheat, amaranth, and various millet species (Siddique et al. 2021).

Currently, Nepal is targeting nearly 300 varieties of underutilized crops, with 60 varieties identified as promising and 500 accessions of local crops supplied for long-term ex situ conservation (Gauchan et al. 2020). Some eminent achievements include:

- One species of amaranth and three local varieties registered at National Seed Board (Joshi et al. 2017),
- From 35 crop species, 232 landraces are targeted for conservation (Gauchan et al. 2018).
- With the support of national and local governments, farmers are managing four community seed banks and a CBM trust fund.
- Diversity field schools (DFSs) have instructed 120 classes showing evident increased in participation of local women (Pudasaini et al. 2020).
- Women are representing more than half of the management committees for community seed banks and instigating the way in seed management.

3.9 Role of Neglected Foods in Addressing Various Forms of Malnutrition

Despite the fact that many people are hungry, nutritional health of peoples in developing countries has been less of a concern. Worldwide, impact of diets low in essential nutrients on consumers' health has become evident with recent changes in living standards of people. Neglected and underutilized food plant species have the potential to reverse the hidden hunger trend of the recent era. In order to achieve higher food production levels and lower the food prices, countries of the continent Asia have made significant transformation to their agricultural systems during the Green

Revolution period (Wieringa et al. 2019). As a result, most of the South and Southeast Asian countries have adopted an agricultural model based on two or three high yielding staple crops, such as wheat, rice or maize. Researchers and other stakeholders have recently realized that over reliance on these staple crops decreased dietary diversity and nutrients availability subsequently contributing toward higher prevalence levels of micronutrient malnutrition (Pingali 2012).

Hidden hunger is characterized by inadequate dietary exposure to essential vitamins and minerals. Hidden hunger or micronutrients malnutrition is widely distributed in developing countries while Westernized diets are reasoned as one amongst the many factors attributing significant loss to the nutrients pool of peoples from the developed countries (Gödecke et al. 2018). NUS when compared to conventional plant foods have higher levels of nutrients including energy, carbohydrates, proteins, dietary fibers, essential fats, minerals, vitamins and non-nutrient phytochemicals such as flavonoids (Sathe 1996). NUS are also rich carriers of dietary fibers as prebiotics likewise a majority of NUS are allergen free i.e., gluten free.

Some of the neglected food plants are rich source of carbohydrates and can serve as substitute to mainstream food crops to meet dietary energy gaps. Taro (*Colocasia esculenta*) belongs to the group of roots & tubers, and has been widely cultivated and consumed in the various areas of Bangladesh, Nepal and India. Raw taro is processed to taro powder for preparing energy dense and gluten free noodles and cookies. Raw taro provides 112 kcal of energy and 26.4 g of carbohydrates per 100 g. Taro tubers contain small granules of starch which are gluten free, hypoallergenic and easily digestible (Rashmi et al. 2018).

3.9.1 Protein Energy Malnutrition (PEM)

NUS such as non-conventional legumes, beans, quinoa and buckwheat are promising sources of plant proteins which may help prevent or treat protein energy malnutrition. Mainstream cereals which have largely been adapted to Asian agriculture system are low in essential amino acids such as lysine and tryptophan, many of the B vitamins and minerals. Over-reliance on staple cereals in South Asia and South East Asia are expected to risk consumers' nutritional and overall health. In fact, increased daily cereal consumption is frequently associated with severe protein malnutrition and may cause kwashiorkor, a chronic protein and energy imbalance that increases susceptibility to life-threatening diseases such as gastroenteritis and tuberculosis (Rolfes et al. 2009; Rizwana et al. 2015). Kwashiorkor is also known as "weaning disease" due to the symptoms which largely appear in young children during the transition from breast milk to complimentary feeding (Lee et al. 2008). Marasmus, on the other hand, is derived from the Greek word marasmus, which means wasting or withering, and is characterized by an insufficient intake of calories and protein, resulting in emaciation. Childhood malnutrition has also been linked to adult-onset chronic degenerative diseases like cancer (Sawaya et al. 2003).

The group of leguminous food crops is one of the best plant protein sources and has longstanding dietary significance on account of significant nutritional value, long storage times, and low cost in comparison to animal products (Sanchez-Chino et al. 2015). Leguminous foods are easier and cheaper way to increase the level of dietary protein in foods of vulnerable groups (Broughton et al. 2003). Sorghum is another important and neglected cereal in Asia. Several studies on nutritional significance of sorghum-based foods have suggested supplementing legumes in sorghum to improve nutritional quality of the consumer goods. Kayitesi et al. (2012) demonstrated that combining sorghum with marama bean (an underutilized legume) improved overall protein content, type and concentration of essential amino acids in composite flour when compared to the porridges made with sorghum alone. The energy values of the sorghum marama bean porridges improved significantly as compared to sorghum porridge. Asma et al. (2006) had also developed low-cost, protein rich food supplements for weaning infants using sorghum, cowpea/pigeon pea, and groundnut/sesame seed blends.

Proteins from many of the neglected plant foods have higher biological value than cereals. Protein quality of quinoa is comparable to casein – a high quality protein of milk origin. Biological value of meat protein is 74% while that of quinoa, corn, wheat and white rice are 73%, 36%, 49% and 56%, respectively (Gordillo-Bastidas et al. 2016). Quinoa protein is considered complete since it contains all essential amino acids however; amount of some essential amino acids are not comparable to proteins of animal origin. Many of the staple grains lacks in adequate amount of some essential amino acids like lysine and isoleucine. Contrary to the staple cereals, the pseudo-cereal quinoa has far higher amounts of lysine and isoleucine (Singh and Singh 2016).

Chenopodium album locally known as "*bathu*" in Pakistan and India is reported to possess medicinal properties and a higher nutritional value. *Bathu* is a good source of protein. The total concentration in vegetative matter is around 203 g/kg. Lysine concentrations are high due to a combination of its production and accumulation in soluble and protein forms. Green matter protein contains more phenylalanine, isoleucine, leucine, threonine, and valine than seed protein (Poonia and Upadhayay 2015). Buckwheat is another source of plant-based protein containing 12 different amino acids. Buckwheat has more protein than rice, wheat, millet, or corn. The grains contain approximately 11–14 g of protein per 100 g, which is less than quinoa or beans and legumes but more than most many whole grains. Buckwheat is an excellent vegetarian or vegan food with promising levels of essential amino acids to include on a regular basis because it contains two types of essential amino acids (lysine and arginine), which our body cannot produce and must obtain from the foods we consume (Mahata 2018). Complimenting conventional cereals with buckwheat thus ensure adequate supply of essential nutrients to the consumers. Likewise, finger millet or the foxtail millet have also been presented promising in improving various nutritional indices of school children fed on minor millets from school feeding program (Chang et al. 2012).

3.9.2 Anemia

Plant based foods are considered promising in alleviating the burden of micronutrients deficiencies. Among different micronutrients deficiency disorders, anemia is ranked as highly prevalent nutritional disorder in many countries of continent Asia. Lentils as plentiful carrier of iron, selenium and zinc can help achieving the DRIs of essential micronutrients (Migliozzi et al. 2015). Findings of randomized controlled studies suggest short term intake of lentils to enhance serum iron levels and the level and severity of anemia. In a study by Wijesuriya et al. (2013), 33 mildly anemic children were were fed with 50 g lentils daily for a period of 60 days. The results suggest a significant change in serum Fe contents with a visible improvement in mean Hb levels (Table 3.5).

Likewise, substituting rice with foxtail or finger millet in meals of 60 undernourished and anemic adolescent school children were found to significantly improve Hb and the weight. Children consuming millet-based school meals had improved their Hb levels by 32.0–37.6% in a 3 months intervention period. More than 85% of the children enrolled in this study reported millet-based meals as more delicious. Millets have thus demonstrated the potential to improve school children's nutritional status (Fanzo et al. 2013). Seeds of garden cress seeds are conventionally used in preparation of local sweets *"laddu"* to treat anemia (Singh et al. 2015). In a study by Sarkar et al. (2014), 7 administration of 25 g powdered garden cress seeds in coconut, molasses based *laddu* to young girls significantly improved their mean Hb levels. Likewise, another report by Jain and Grover (2017) suggests administrating roasted seeds of garden cress to school children in biscuit formulations, to improve Hb from 10.6 to 11.1 g/dL. Combine therapy of iron folates supplemented with garden cress seeds to anemic pregnant women has also been suggested to increase Hb levels by 1.3 g/dL (Nair et al. 2014).

3.9.3 Diarrhea and Malnutrition

Worldwide, diarrhea and malnutrition, alone or in combination, are major causes of morbidity and mortality among children below 5 years. Diarrheal illnesses impair normal process of weight and height gains while the recurrent illnesses severely hamper catch-up growth period. Malnutrition in children measured as impaired

Table 3.5 Improvement in mildly anemic children after a 60-day red lentil feeding trial (n = 33)

Indicator	0 Day	60 Day	% Improvement
Hemoglobin (g/dL)	11.1	11.8	6.3
Serum Fe (μg/dL)	51.5	89.8	74.4
Total Fe binding capacity (μg/dL)	405.3	377.6	−6.8
Trans ferritin saturation (%)	12.8	24.3	89.8
Serum ferritin (ng/dL)	29.5	41.2	39.7

Source: Wijesuriya et al. (2013)

weight or height for age, increases the frequency and duration of diarrheal illnesses by 37% and 73%, respectively (Guerrant et al. 1992).

Anti-diarrheal properties have been documented for many underutilized or neglected plant foods. The beal fruit tree (*Aegle marmelos*) is the most valuable medicinal plant in the Siddha system of medicine, because of its numerous medicinal properties. The unripe fruit is astringent, digestive, stomachic, and slightly constipating. Bael is an important medicinal herb that is widely used in Ayurveda, Siddha, Unani, and traditional medicine. Bael fruit has anti-diarrheal properties (Rakulini and Sounthararajan 2019). According to Brijesh et al. (2009) a decoction of *A. marmelos* un-ripened fruit pulp anticipate reduction in bacterial colonization, adherence to the gut epithelium, and reduction in production and action of certain enterotoxins. Diverse role of bael against infectious diarrhoea is validated from the ancient Indian texts and has a traditional application as a complimentary medicine to treat diarrheal diseases in various cultures. Among some other neglected food plants, anti-diarrheal activity has also been documented for garden cress (Azene et al. 2022), *Rhus chinensis* (nutgall tree), *Rubus elliptiens* Sm. (yellow raspberry) and *Rubus moluccanus* (Ceylon blackberry) (Singh et al. 2014).

3.9.4 Bone Health

Calcium and phosphorus account for approximately 80–90% of the bone mineral content. Normal bone metabolism or the bone formation process also require protein, iron, magnesium, manganese, copper, zinc, fluoride and essential vitamins including vitamin A, C, D & K from the regular dietary sources (Ilich and Kerstetter 2000). NUS are plentiful carriers of essential nutrients that may aid in bone health in one or the other way.

Manganese in buckwheat ensures development of healthy bone structure by producing essential enzymes required for bone formation, and also acts as a co-enzyme to aid various metabolic processed. Manganese also supports formation of connective tissues and improves calcium absorption, and fat and sugar metabolism. Magnesium is another important micronutrient for bone health and has plentiful availability in NUS including buckwheat, jackfruit, quinoa etc. Magnesium from buckwheat may also aids in calcium absorption and prevention onset of osteoporosis and other bone disorders (Mahata 2018). Magnesium is also abundant in the seed (54 mg/100 g) and young fruit of the jackfruit (27 mg/100 g) (Swami et al. 2012). Quinoa is also a good carrier of magnesium (249.6 mg/100 g) and calcium (148 mg/100 g) on dry weight basis (Singh and Singh 2016).

Tripathi et al. (2014) attempted to determine whether finger millet flour and hibiscus flower powder can increase the nutritional value of instant health beverage powder without compromising their sensory features, and if it can be used as a calcium supplement in women with sedentary lifestyle. When compared to other health drinks on the market, the results demonstrated that beverage powder had a very high calcium (96.5%) and protein (12.25%) contents, as well as extremely advantageous nutraceutical qualities.

3.9.5 Celiac Disease

Celiac disease is an autoimmune disorder which is triggered with ingestion of the trace amounts of "gluten," a term used to describe the storage proteins found in wheat, rye, and barley. Although removing gluten or replacing gluten carrying foods with non-gluten carrier ingredients temporarily heals the intestine, the intolerance still exists and the damage revives if gluten is reintroduced. Persistent gluten immunological response or gluten allergy may consequently result in mal-digestion and malabsorption of many essential nutrients thus increasing the burden of malnutrition (Barton et al. 2007).

In botanical perspective, amaranth, buckwheat and quinoa are not considered as true cereals as they are dicotyledonous plants contrary to most cereals (e.g., barley, rice and wheat), which are monocotyledonous. They are known as pseudo cereals because their seeds operate and are composed similarly to those of true cereals. Pseudo-cereals are not only gluten free but are rich in micronutrients especially thiamin, riboflavin, pyridoxine, calcium, magnesium and iron which may help replenish their levels. Multiple studies have reported to investigate the utilization of pseudo cereals in preparation of nutrient dense gluten-free products such as pasta, bread and confectionary products (Alvarez-Jubete et al. 2010). Using these as a gluten-free ingredient substitute for refined rice flour, corn, potato and tapioca can significantly increase the antioxidant and nutritional value of the diet (Jubete et al. 2010; Lee et al. 2009).

3.9.6 Prebiotics and Probiotics

A prebiotic is a "selectively fermented ingredient that allows specific changes in the composition and/or activity of the gastrointestinal microflora that confers benefits upon host well-being and health" (Roberfroid 2007). Such changes in specific microbial colonization in the human gut can have a variety of beneficial effects, including prevention of diarrhea, constipation, and increased satiety function by increasing production of short-chain fatty acids. Prebiotics also regulate intestinal motility and ensure reduction in pathogen colonization within the short bowl. Furthermore, prebiotics may also improve mineral absorption and stimulate the immune system (Johnson et al. 2013).

Millets as wonderful carrier of dietary fibers are similar to prebiotics since they are easily digestible, soothing, and promote the health of gut microflora (Banerjee et al. 2017). Many researchers have investigated the pro- and pre-biotic properties of millets and concluded that using millets as a prebiotic is both beneficial and acceptable (Kunchala et al. 2016). Many of the neglected lentils are high in prebiotic carbohydrates, which are beneficial to human health. Beneficial gut microflora ferment lentil prebiotic carbohydrates and soothes consumers with multiple gut health related benefits (Johnson et al. 2020).

3.10 Food Safety and Toxicological Aspects of Underutilized Plant Foods

Prior issuing recommendations for their food uses or as ingredient of choice in food, nutrients, nutraceutical formulations, neglected or underutilized food plants may be evaluated for safety and toxicological aspects. With the increased use of traditional foods and medicine as a complementary or alternative approach to satisfy dietary or health needs in a broader perspective, the safety and efficacy of traditional plants has become a public health concern. It is quite difficult to make reliable estimates if NUS as traditional foods or health promoting products do not anticipate any risk of illness to the consumers.

Many of the NUS also carry significant amount of anti-nutrients that may hinder biological availability of the essential nutrients. At an instance, presence of a significant amount of anti-nutrients compromises the nutritional quality of legumes. Consumption of improperly processed legumes may disrupt normal metabolism and have a negative impact on human health. Effective processing is required to lessen these anti-nutritional factors prior to consumption (Sharma 2021). Garden cress seeds contain goitrogens that interfere with iodine absorption and can lead to hypothyroidism. Seeds of garden cress are therefore not recommended to the hypothyroid patients in particular. Furthermore, it may also act as an abortifacient (contains components that induce abortion) if taken in higher doses. Consumption of a single large dose of the seeds of garden cress by pregnant women may mimic uterine contractions and induce spontaneous abortion (Singh and Paswan 2017). Anti-nutrients found in taro root may ensure better growth of the root without necessarily high doses of fungicides and the pesticides, however these anti-nutrients may compromise nutritional attributes of this neglected food crop. Taro contains oxalic acid, lectins, mucilage, tannins, proteases (trypsin & chymotrypsin) and α-amylase inhibitor, which are all listed as anti-nutrients (Ramanatha et al. 2010).

3.11 Conclusion

Despite notable progress, high prevalence rates of food insecurity, hunger and malnutrition in changing climate are matter of critical concern for countries in South Asia. Over-reliance on staple crops presented with lack of dietary diversity is one amongst the major causes of chronic malnutrition. Though they are heavily underutilized, NUS can play a critical role in the fight against malnutrition and hunger. NUS as future smart foods with their four dimensions i.e., good nutrition, climate resilient, economic stable and locally adoptable can potentially contribute to Zero Hunger. Various NUS with their generic features to withstand a variety of stresses make food production systems not merely diverse, but also more sustainable and climate resilient. Because of their resistance to the climate change, NUS can be used as food when other crops fail.

Promoting NUS cultivation and consumption can help improve dietary diversity and alleviating various forms of malnutrition including micronutrient deficiencies. Well documented medicinal value of the many neglected plant foods and their derived components may improve gut health & digestion, and severity of ulcers, diabetes, hyperlipidemia, atherosclerosis, cancer, aging etc. Further research is mandated to broaden the list of plant foods with indigenous and wild species of neglected food plants comprehensively explored for their nutritional, safety and toxicological properties. Policies may be developed to strengthen food systems with NUS, and improve their availability and access to peoples. Ethnic knowledge of local communities for cultivation and traditional use of neglected food plants and their foods may be preserved and disseminated among the stakeholders to promote agrobiodiversity and food security.

Conflicts of Interest All authors read and approved the final manuscript and declare that they have no conflict of interest.

References

Alvarez-Jubete L, Arendt EK, Gallagher E (2010) Nutritive value of pseudocereals and their increasing use as functional gluten-free ingredients. Trends Food Sci Technol 21(2):106–113

Amadou I, Gounga ME, Le GW (2013) Millets: nutritional composition, some health benefits and processing-a review. Emir J Food Agric 25(7):501–508

Asghari G, Palizban A, Bakhshaei B (2015) Quantitative analysis of the nutritional components in leaves and seeds of the Persian *Moringa peregrina* (Forssk). Pharm Res 7(3):242–248

Asma MA, El-Fadil EB, El-Tinay AH (2006) Development of weaning food from sorghum supplemented with legumes and oil seeds. Food Nutr Bull 27(1):26–34

Azene M, Habte K, Tkuwab H (2022) Nutritional, health benefits and toxicity of underutilized garden cress seeds and its functional food products: a review. Food Prod Process Nutr 4(1):1–13

Banerjee DP, Chowdhury R, Bhattacharya P (2017) Sustainability of the probiotic *lactobacillus casei* in fortified indian milk cakes under different preservation conditions-effects of co-immobilization of *L. casei* and commercial prebiotic inulin (chicory based) and millet inulin. Int J Pharm Pharm Sci 9(1):152–157

Barton SH, Kelly DG, Murray JA (2007) Nutritional deficiencies in celiac disease. Gastroenterol Clin N Am 36(1):93–108

Brijesh S, Daswani P, Tetali P, Antia N, Birdi T (2009) Studies on the antidiarrhoeal activity of Aegle marmelos unripe fruit: validating its traditional usage. BMC Complement Altern Med 9:1–12

Broughton WJ, Hernández G, Blair M, Beebe S et al (2003) Beans (Phaseolus spp.)–model food legumes. Plant Soil 252:55–128

Chang WC, Wahlqvist ML, Chang HY et al (2012) A bean-free diet increases the risk of all-cause mortality among Taiwanese women: the role of the metabolic syndrome. Public Health Nutr 15(4):663–672

Chivenge P, Mabhaudhi T, Modi AT, Mafongoya P (2015) The potential role of neglected and underutilized crop species as future crops under water scarce conditions in Sub-Saharan Africa. Int J Environ Res Public Health 12:5685–5711

Dahiya PK, Linnemann AR, Van Boekel MA, Khetarpaul N, Grewal RB, Nout MJ (2015) Mung bean: technological and nutritional potential. Crit Rev Food Sci Nutr 55(5):670–688

Delêtre M, Gaisberger H, Arnaud E (2013) Agrobiodiversity in perspective: a review of questions, tools, concepts and methodologies in preparation of SEP2D. Bioversity International

Fairulnizal MM, Norhayati MK, Zaiton AH et al (2015) Nutrient content in selected commercial rice in Malaysia: an update of Malaysian food consumption database. Int Food Res J 22:768–776

FANTA (2006) Developing and validating simple indicator of dietary quality and energy intake of infants and young children in developing countries. FANTA, Washington, DC

Fanzo J, Hunter D, Borelli T, Mattei F (2013) Diversifying food and diets: using agriculture biodiversity to improve nutrition and health. Routledge, New York

FAO (2018) Future Smart Food: rediscovering hidden treasure of neglected and underutilized species for Zero Hunger in Asia. FAO, Bangkok

FAO, IFAD, UNICE et al (2022) The state of food security and nutrition in the world 2022. Repurposing food and agricultural policies to make healthy diets more affordable. FAO, Rome

FAO, IFAD, UNICEF et al (2017) The state of food security and nutrition in the world: building resilience for peace and food security. FAO, Rome

FAO, IFAD, UNICEF et al (2020) The state of food security and nutrition in the world 2020. Transforming food systems for affordable healthy diets. FAO, Rome. https://doi.org/10.4060/ca9692en

FAO, UNICEF, WFP, WHO (2021) Asia and the Pacific regional overview of food security and nutrition 2020: maternal and child diets at the heart of improving nutrition. FAO, Bangkok. https://doi.org/10.4060/cb2895en

FAOSTAT Statistical Database (2022) Food and Agriculture Organization of the United Nations. https://www.fao.org/faostat/en/#data/QCL. Accessed 26 Dec 2022

Fassil H, Guarino L, Sharrock S et al (2000) Diversity for food security: improving human nutrition through better evaluation, management, and use of plant genetic resources. Food Nutr Bull 21:497–502

Garn SM, Leonard WR (1989) What did our ancestors eat? Nutr Rev 47(11):337–345. https://doi.org/10.1111/j.1753-4887.1989.tb02765.x

Gauchan D, Joshi BK, Bhandari B (2018) Farmers' rights and access and benefit sharing mechanisms in community seed banks in Nepal

Gauchan D, Joshi BK, Bhandari B et al (2020) Traditional crop biodiversity for mountain food and nutrition security in Nepal. Tools and research results of the UNEP GEF Local Crop Project, Nepal

Gödecke T, Stein AJ, Qaim M (2018) The global burden of chronic and hidden hunger: trends and determinants. Glob Food Sec 17:21–29

Gordillo-Bastidas E, Diaz-Rizzolo DA, Roura E et al (2016) Quinoa (*Chenopodium quinoa* Willd), from nutritional value to potential health benefits: an integrated review. J Nutr Food Sci 6(3):1000497

Guerrant RL, Schorling JB, McAuliffe JF, De Souza MA (1992) Diarrhea as a cause and an effect of malnutrition: diarrhea prevents catch-up growth and malnutrition increases diarrhea frequency and duration. Am J Trop Med Hyg 47:28–35

Idowu DO, Fashina AB, Kolapo OE, Awolusi OM (2019) Snake gourd (Trichosanthes cucumerina L.): an underutilized crop with great potential. Int J Curr Microbiol Appl Sci 8(9):1711–1717

Ilich JZ, Kerstetter JE (2000) Nutrition in bone health revisited: a story beyond calcium. J Am Coll Nutr 19(6):715–737

Jain T, Grover K (2017) Nutritional evaluation of garden cress chikki. Agric Res Technol 4(2):555–631

Johns T, Sthapit BR (2004) Biocultural diversity in the sustainability of developing-country food systems. Food Nutr Bull 25:143–155

Johnson CR, Thavarajah D, Combs GF, Thavarajah P (2013) Lentil (*Lens culinaris L.*): a prebiotic-rich whole food legume. Food Res Int 51(1):107–113

Johnson N, Johnson CR, Thavarajah P, Kumar S, Thavarajah D (2020) The roles and potential of lentil prebiotic carbohydrates in human and plant health. Plants People Planet 2(4):310–319

Joshi BK, Bhatta MR, Ghimire KH et al (2017) Released and promising crop varieties of mountain agriculture in Nepal (1959–2016)
Jubete LA, Wijngaard H, Arendt EK, Gallagher E (2010) Polyphenol composition and in vitro antioxidant activity of amaranth, quinoa, buckwheat and wheat as affected by sprouting and baking. Food Chem 119(2):770–778
Kassam KA, Karamkhudoeva M, Ruelle M, Baumflek M (2010) Medicinal plant use and health sovereignty: findings from the Tajik and Afghan Pamirs. Hum Ecol 38:817–829
Kayitesi E, de Kock HL, Minnaar A, Duodu KG (2012) Nutritional quality and antioxidant activity of marama–sorghum composite flours and porridges. Food Chem 131(3):837–842
Kim SL, Kim SK, Park CH (2004) Introduction and nutritional evaluation of buckwheat sprouts as a new vegetable. Food Res Int 37(4):319–327
Kunchala R, Banerjee R, Mazumdar SD et al (2016) Characterization of potential probiotic bacteria isolated from sorghum and pearl millet of the semi-arid tropics. Afr J Biotechnol 15(16):613–621
Lakht-e-Zehra A, Dar NG, Saleem N, Soomro UA, Afzal W, Naqvi B, Jamil K (2015) Nutritional exploration of leaves, seed and fruit of bael (Aegle marmelos L.) grown in Karachi region. Pak J Biochem Mol Biol 48(3):61–65
Lee HJ, Km KH, Park HJ et al (2008) A case of lethal kwashiorkor caused by feeding only with cereal grain. Korean J Pediatr 51(3):29–34
Lee AR, Ng DL, Dave E, Ciaccio EJ, Green PH (2009) The effect of substituting alternative grains in the diet on the nutritional profile of the gluten- free diet. J Hum Nutr Diet 22(4):359–363
Li X, Siddique KHM (2018) Future Smart Food – rediscovering hidden treasures of neglected and underutilized species for Zero Hunger in Asia. FAO, Bangkok
Li X, Yadav R, Siddique KH (2020) Neglected and underutilized crop species: the key to improving dietary diversity and fighting hunger and malnutrition in Asia and the Pacific. Front Nutr 7:593711
Liang L, Wang C, Li S, Chu X, Sun K (2019) Nutritional compositions of Indian Moringa oleifera seed and antioxidant activity of its polypeptides. Food Sci Nutr 7(5):1754–1760
Mahata D (2018) Importance of buckwheat (*Fagopyrum esculentum* Moench). Int J Chem Stud 6(5):2121–2125
Migliozzi M, Thavarajah D, Thavarajah P, Smith P (2015) Lentil and kale: complementary nutrient-rich whole food sources to combat micronutrient and calorie malnutrition. Nutrients 7(11):9285–9298
Nair S, Meghani S, Tripathi V (2014) Some solutions to intra uterine growth deficiencies and iron deficiency anemia using garden cress seeds. Int J Food Nutr Sci 3(3):1–7
Nelson GC, Rosegrant MW, Koo J et al (2009) Climate change: impact on agriculture and costs of adaptation. Food policy report. International Food Policy Research Institute (IFPRI), Washington, DC
Osei A, Pandey P, Spiro D et al (2010) Household food insecurity and nutritional status of children aged 6 to 23 months in Kailali district of Nepal. Food Nutr Bull 31:483–494
Our World in Data (2022) Agricultural production. https://ourworldindata.org/agricultural-production#explore-data-on-agricultural-production. Accessed 26 Dec 2022
Pingali PL (2012) Green Revolution: impacts, limits, and the path ahead. Proc Natl Acad Sci U S A 109(31):12302–12308
Poonia A, Upadhayay A (2015) Chenopodium album Linn: review of nutritive value and biological properties. J Food Sci Technol 52:3977–3985
Pudasaini N, Bhandari B, Gurung R (2020) 15. Diversity Field School (DFS) for managing agrobiodiversity
Rakulini R, Sounthararajan K (2019) A review of anti–diarrheal activity of Aegle marmelos. Orig Res Artic 7:1–10
Ramanatha RV, Matthews PJ, Eyzaguirre PB, Hunter D (2010) The global diversity of Taro: ethnobotany and conservation. Bioversity International, Rome
Rashmi DR, Raghu N, Gopenath TS, Palanisamy P et al (2018) Taro (*Colocasia esculenta*): an overview. J Med Plants Stud 6(4):156–161

Rasul G (2021) Twin challenges of COVID-19 pandemic and climate change for agriculture and food security in South Asia. Environ Chall 2:100027

Rasul G, Hussain A, Khan MA, Ahmad F, Jasra AW (2014) Towards a framework for achieving food security in the mountains of Pakistan, ICIMOD Working Paper 2014/5. ICIMOD, Kathmandu

Rizwana B, Sadiq Butt M, Tauseef Sultan M, Farhan S, Rabia N (2015) Protein-energy malnutrition: a risk factor for various ailments. Crit Rev Food Sci Nutr 55(2):242–253

Roberfroid M (2007) Prebiotics: the concept revisited. J Nutr 137(3):830–837

Rolfes SR, Pinna K, Whitney E (2009) Protein: amino acids. In: Understanding normal and clinical nutrition. Wadsworth, Belmont

Sanchez-Chino X, Jimenez-Martinez C, Davila-Ortiz G, Alvarez-Gonzalez I, Madrigal-Bujaidar E (2015) Nutrient and nonnutrient components of legumes, and its chemopreventive activity: a review. Nutr Cancer 67(3):401–410

Sarkar S, Datta S, Ghosh I (2014) Experimental studies on nutritional and medicinal-role of garden cress seed on animal and human being-a review. Int J Pharm Chem Anal 4(1):41–45

Sathe SK (1996) The nutritional value of selected Asiatic pulses: chickpea, black gram, mung bean and pigeon pea, Boston

Sawaya AL, Martins P, Hoffman D, Roberts SB (2003) The link between childhood undernutrition and risk of chronic diseases in adulthood: a case study of Brazil. Nutr Rev 61(5):168–175

Seal T (2012) Antioxidant activity of some wild edible plants of Meghalaya state of India: a comparison using two solvent extraction systems. Int J Nutr Metabol 4(3):51–56

Sharma A (2021) A review on traditional technology and safety challenges with regard to antinutrients in legume foods. J Food Sci Technol 58(8):2863–2883

Siddique KH, Li X, Gruber K (2021) Rediscovering Asia's forgotten crops to fight chronic and hidden hunger. Nat Plants 7(2):116–122

Singh CS, Paswan VK (2017) The potential of garden cress (*Lepidium sativum L.*) seeds for development of functional foods. In: Jimenez-Lopez JC (ed) Advances in seed biology. Intech, Coarotia, pp 279–294

Singh KV, Singh R (2016) Quinoa (*Chenopodium quinoa* Willd), functional superfood for today's world: a review. World Sci News 58:84–96

Singh SR, Phurailatpam AK, Wangchu L, Ngangbam P, Chanu TM (2014) Traditional medicinal knowledge of underutilized minor fruits as medicine in Manipur. Int J Agric Sci 4(8):241–247

Singh CS, Paswan VK, Naik B (2015) Exploring potential of fortification by garden cress (Lepidium sativum L.) seeds for development of functional foods—a review. Indian J Nat Prod Resour 6(3):167–175

Singletary KW (2017) Fenugreek: overview of potential health benefits. Nutr Today 52(2):93–111

Swami SB, Thakor NJ, Haldankar PM, Kalse SB (2012) Jackfruit and its many functional components as related to human health: a review. Compr Rev Food Sci Food Saf 11(6):565–576

Temesgen M, Retta N (2015) Nutritional potential, health and food security benefits of taro Colocasia esculenta (L.): a review. Food Sci Qual Manag 36:23–30

Tripathi J, Gupta A, Prasad R, Puranik V (2014) Enhancing micronutrient content of beverage powder by incorporating malted finger millet. Indian J Commun Health 26(2):339–342

WHES (2023) Asia hunger facts. https://www.worldhunger.org/asia-hunger-facts/. Accessed on 6 Apr 2023

WHO (2023) The Global Health Observatory Indicators. https://www.who.int/data/gho/data/indicators. Accessed on 12 Mar 2023

Wieringa FT, Dijkhuizen MA, Berger J (2019) Micronutrient deficiencies and their public health implications for South-East Asia. Curr Opin Clin Nutr Metab Care 22(6):479–482

Wijesuriya AP, Pallemulla S, Thavarajah D et al (2013) Nutritional impact of iron rich lentil diet on iron deficient anemic children in Sri Lanka. Lady Ridgeway Hospital, Colombo

Worldmeter (2022) Population: world. https://www.worldometers.info/population/world/. Accessed on 29 Dec 2022

Chapter 4
Inclusion Criteria for Underutilized Food Plants in Nutrition-Sensitive Programming

Teresa Borelli, Jai Rana, Devendra Gauchan, Sharon Mendonce, and Danny Hunter

4.1 Introduction

Eating varied and balanced amounts of fruits, vegetables, grains, protein foods, and dairy as part of healthy diets can provide our bodies with all the necessary macro- and micro-nutrients for optimal health as well as lowering the incidence of diet-related chronic diseases such as obesity, heart disease and diabetes (Afshin et al. 2019; Stubbendorff et al. 2022; Xu et al. 2022). However, it doesn't all boil down to personal behavior and unhealthy choices can result from systemic issues that are rooted in national food systems and structural barriers to an optimal diet, such as over-reliance on a narrow food base and a small number of major crops, lack of information on what constitutes a healthy balanced diet, and access to affordable and healthy food. The volatility of food production and food prices have also increased in recent years, with conflict (e.g., the Russia-Ukraine war), climate shocks, disasters, resource conflicts, pandemics as well as disease and pest outbreaks exacerbating existing problems.

South Asian Countries
1. Afghanistan
2. Bangladesh
3. Bhutan
4. India
5. Maldives
6. Nepal
7. Pakistan
8. Sri Lanka

T. Borelli (✉) · J. Rana · D. Gauchan · S. Mendonce · D. Hunter
Bioversity International, Rome, Italy
e-mail: T.Borelli@cgiar.org; J.Rana@cgiar.org; D.Gauchan@cgiar.org; S.Mendonce@cgiar.org; D.Hunter@cgiar.org

T. Ismail et al. (eds.), *Neglected Plant Foods of South Asia*,
https://doi.org/10.1007/978-3-031-37077-9_4

A perfect example is the South Asian enigma, illustrating how strong economic growth and rising disposable income levels do not necessarily translate into healthy dietary choices and reduced malnutrition. Despite increased rates of per capita gross domestic product (GDP) over the last 60 years (Macrotrends.net) and the region being the center of diversity of many important species of crops (rice, soybean, banana, coconut) and varieties therein, animals and livestock breeds, as well as forest biodiversity (FAO 2019), South Asians appear to consume extremely low quantities of fruit and vegetables, lower than the World Health Organization (WHO) recommendations (Kalmpourtzidou et al. 2020), resulting in limited progress towards both regional and global hunger and nutrition targets (Bishwajit 2015; Development Initiatives 2021).

According to the 2022 Global Hunger Index (GHI) report, which tracks world progress in tackling hunger, malnutrition is still a serious problem in South Asia, putting many countries at risk during future shocks and crises (Von Grebmer et al. 2022). Despite a reduction in overweight among children below 5 years of age (2.5%) in 7 countries in the region, and reduced stunting among children below 5 years of age in 3 countries, South Asia has the highest child wasting rate of any world region, spearheaded by India (19.3%) (Von Grebmer et al. 2022). In all South-Asian countries more than 1 in 10 (14%) children aged 0–5 years are wasted (Table 4.1), while anemia affects an estimated 48.4% of women of reproductive age (Table 4.2), and men and women aged 18 and over struggle with diabetes (10–11%) and obesity (4.8–8.4%) (Development Initiatives 2021; UNICEF 2021). In 2021, diabetes caused the death of over one million South Asians, with India and Pakistan leading this grim record with over 647,000 and 396,000 deaths respectively, and many people remaining undiagnosed (Boyko et al. 2021). This may be a gross underestimation, however, as the Global Nutrition Report recounts that five countries in the subregion have insufficient data to adequately monitor progress towards the global nutrition targets (Development Initiatives 2021). The good news is that the average prevalence of overweight is 2.5% – the lowest compared to other subregions for which data is available.

As a region, South Asia has the first highest GHI score (27.4), higher than that of Sub-Saharan Africa. The global average is 18.2 based on an assessment of 122 countries in the world. The highest GHI in South Asia is 29.6 for Afghanistan, 29.1 for India, 26.1 for Pakistan and the more modest GHI for Bangladesh of 19.6, 19.1 for Nepal and 13.6 for Sri Lanka. Alarmingly, the malnutrition situation is stagnating. Since 2000, no country in South Asia has been able to halve its score by 50% or more, except for Nepal which has reduced it by 48.4% (Von Grebmer et al. 2022).

Research into the region's food habits shows that South Asians increasingly consume cheap, highly processed, energy dense foods (high in energy per unit weight) that are poor in micronutrients while consuming extremely low quantities of fruit and vegetables (Kalmpourtzidou et al. 2020).

Of the South Asian countries, Bhutanese adults consume the highest amounts of fruits with 2.4 servings/day while lowest fruit consumption was reported in Sri Lanka and Bangladesh (0.43 servings/day) (Jayawardena et al. 2020). Indians had

Table 4.1 Levels of stunting, wasting and overweight expressed as a percentage of the population of children under 5 years of age in Bangladesh, India, Pakistan, and Sri Lanka, as well as country progress towards the 2030 global nutrition targets

	Malnutrition among children under 5 years of age									
	Modelled estimates			Survey estimates			Modelled estimates			Survey estimates
	Stunted (%)* (Moderate and severe)			Wasted (%)*			Overweight (%)* (Moderate and severe)			Free from stunting, wasting, overweight (%)
	Baseline	Latest		Moderate and severe (Latest)		Severe (Latest)	Baseline	Latest		
Countries	2012	2020	Progress to 2030 target**	2014–2020*	Progress to 2030 target**	2014–2020*	2012	2020	Progress to 2030 target**	2014–2020*
Bangladesh	38	30	Some progress	10	Off track	2	2	2	On track	64
India	42	31	On track	17	Off track	5	2	2	On track	47
Pakistan	43	37	Some progress	7	On track	2	5	3	On track	57
Sri Lanka	17	16	Some progress	15	Off track	3	1	1	On track	69

Source: UNICEF (2021)

* The collection of household survey data on child height and weight was limited in 2020 due to the physical distancing measures required to prevent the spread of COVID-19. Only four national surveys included in the database were carried out (at least partially) in 2020. The Joint Child Malnutrition Estimates are therefore based almost entirely on data collected before 2020 and do not take into account the impact of the COVID-19 pandemic.

** Progress was assessed using the 2030 targets proposed as an extension of the 2025 global nutrition targets. For further information, see: World Health Organization and United Nations Children's Fund. 'WHO/ UNICEF discussion paper: The extension of the 2025 maternal, infant and young child nutrition targets to 2030', WHO and UNICEF, Geneva and New York, 2017.

Table 4.2 Levels of malnutrition among school-aged children, adolescents, and women in Bangladesh, India, Pakistan, and Sri Lanka

	Malnutrition among school-aged children (5–19 years)						Malnutrition among women					Households
	Low birthweight (%)			Unweighed at birth (%)	Thinness (%)	Overweight (%)	Underweight 18+ years (%)	Overweight 18+ years (%)	Anemia 15–49 years (%)			Households consuming salt with iodine (%)
	Baseline	Latest	Progress to 2030 targets**						Baseline	Latest	Progress to 2030 targets**	
Countries	2012	2015		2014–2020*	2016	2016	2016	2016	2012	2019		2014–2020*
Bangladesh	29	28	Some progress	49	18	9	23	21	36	37	No progress	76
India	-z	-z		22	27	7	24	21	53	53	No progress	92y
Pakistan	-z	-z		84	19	10	15	29	43	41	No progress	69x
Sri Lanka	17	16	No progress or worsening	3	15	13	13	29	34	35	No progress	92

Source: UNICEF (2021)

Data refer to years or periods other than those specified in the column heading. Such data are not included in the calculation of regional and global averages. Estimates from data years prior to 2000 are not displayed; **y**- Data differ from the standard definition; if they fall within the noted reference period, such data are included in the calculation of regional and global averages; **z**- Country modelled estimates not presented but has been used for regional and global aggregates

* The collection of household survey data on child height and weight was limited in 2020 due to the physical distancing measures required to prevent the spread of COVID-19. Only four national surveys included in the database were carried out (at least partially) in 2020. The Joint Child Malnutrition Estimates are therefore based almost entirely on data collected before 2020 and do not take into account the impact of the COVID-19 pandemic.

** Progress was assessed using the 2030 targets proposed as an extension of the 2025 global nutrition targets. For further information, see: World Health Organization and United Nations Children's Fund. 'WHO/ UNICEF discussion paper: The extension of the 2025 maternal, infant and young child nutrition targets to 2030', WHO and UNICEF, Geneva and New York, 2017.

Table 4.3 Average per capita vegetable and fruit consumption for South Asian countries in 2017

Country/ region	Average per capita vegetable consumption (kg/year)	Average per capita fruit consumption (kg/year)	Average per capita vegetable intake (g/ day)	Average per capita fruit intake (g/day)	GDP per capita measured in constant international $
World	90.52	78.49	248	215.04	16,211
South Asia	63.04	45.17	172.71	123.75	5774
Afghanistan	48.34	38.23	132.45	104.75	2058
Bangladesh	35.09	27.62	96.15	75.68	4161
Bhutan	No data	No data	No data	No data	–
India	79.86	57.69	218.82	158.07	6183
Maldives	79.33	62.47	217.36	171.17	18,058
Nepal	128.04	65.48	**350.83**	179.42	3565
Pakistan	20.83	22.64	57.07	62.03	4571
Sri Lanka	49.83	42.08	136.53	115.3	12,584

Source: Our World in data (Accessed 16 November 2022). Datasets are sourced from the UN Food and Agriculture Organization (FAO) FAOSTAT and combine data from its Food Balance Sheets into a complete series from 1961 to 2017

the lowest reported intake of vegetables (0.9 servings/day), while the highest value, 3.8 servings/day, was reported in Bhutan (Jayawardena et al. 2020).

In 2017, the per capita consumption of vegetables in the world (172 countries) was 90.52 Kg per year while for the South Asia sub region, the average was 67.705 kg/year (Table 4.3). No data was available for Bhutan for that year. The per capita consumption for India was 61.12 kg/year, while Bangladesh and Sri Lanka consumption levels are much lower at just 35.09 kg/year and 49.83 kg/year, respectively. Nepalese score the highest record, with people consuming 128.04 kg/year. Nepal was also the only country in South Asia to meet the WHO dietary guidelines for vegetable intake (above 250 g per day). All other South Asian countries failed to meet the WHO dietary recommendations for that year. Similarly, in 2017, no South Asian country met the WHO dietary guidelines for fruit intake (above 200 g per day), regardless of per capita GDP. Sri Lanka, with one of the highest per capita GDP for 2017, scored average in terms of per capita fruit consumption (kg/year).

According to world dietary guidelines, eating five servings (80 g each) of fruits and vegetables per day as part of a varied and balanced diet is associated with lower risks of many health conditions, particularly of coronary heart diseases (Afshin et al. 2019; Branca et al. 2019; Lock et al. 2005). Combined with a reduction in physical activity and increased sedentary lifestyles, current South Asian diets fail to deliver sufficient nutrients at critical stages of human development.

Agricultural biodiversity, and the thousands of crop species and wild food plants that are part of this diversity yet are largely underutilized, represents a strategic resource in ensuring food and nutrition security for humankind (Bioversity International 2017; Frison et al. 2011; Thrupp 2000). Neglected and underutilized food plants (NUFPs) often contain copious amounts of important vitamins and

Fig. 4.1 Sri Lankan rice varieties (*Oryza sativa*) on display at the 2014 traditional food fair held in Colombo, Sri Lanka. (Credit: Bioversity/S. Landersz)

micronutrients (Padulosi et al. 2021) that can help curb diet-related non-communicable diseases (diabetes, heart attacks, overweight, obesity and cancer) – the leading cause of death at the global level (Branca et al. 2019, 2020). For example, Sri Lankan local rice varieties are generally higher in vitamins and minerals, and could substantially improve the intake of nutrients, such as iron, if consumed as an alternative to common white rice (Hunter et al. 2019) (Fig. 4.1). Finding efficient ways of reintroducing NUFPs into production systems and diets could reduce humanity's current over-reliance on a handful of major crops. Further, being locally available and generally adaptable to growing in difficult climates and resource-poor settings, NUFPs are accessible and affordable for local consumers providing them with a greater choice of ingredients for adopting nutritious and healthy diets (Li et al. 2020; Padulosi et al. 2021).

Like most countries worldwide, the diversity of NUFPs and the numerous traditional varieties existing in many South Asian countries are declining rapidly due to climate change, market trends and changes in food culture brought about by globalization, migration, and modernization. Particularly in South Asia, the Green Revolution first and then the release of improved varieties of cereals, roots and tubers and pulses in the early 2000s caused the reduction of crop varietal diversity and the spatial displacement of traditional landraces by a handful of modern cultivars known as mega-varieties (Gatto et al. 2021). It is estimated, for example, that approximately 50% of traditional crop varieties grown in Nepal have been lost in the last few decades of agricultural development and many more are disappearing (Gauchan et al. 2019; Joshi et al. 2017). Key driving factors behind the declining production trend include limited awareness of the nutrient content and climate resilience value of these crops, accompanied by low crop productivity of NUFPs compared to improved varieties, low value addition, difficult post-harvest operations, high labour processing requirements and lack of markets in rural areas (Adhikari et al. 2017; Gauchan et al. 2020a, 2020b; Joshi et al. 2020).

Internationally, there is growing interest in new foods and plant-based products that can contribute to human health, nutrition and sustainable food systems (Hunter et al. 2020). This attention can be exploited to develop markets for non-staple local, traditional crop varieties and minor crops from which poor communities can benefit, providing incentives for farmers to produce and conserve these species as well as to reintegrate them in their diets. Further, the potential of NUFPs to generate income, address food security and provide options for climate change adaptation links with a number of Sustainable Development Goals (SDGs) addressing social, economic and environmental issues (Aryal et al. 2020).

4.2 Key Nutritional Aspects of Select Neglected and Underutilized Food Plants in India, Nepal, Sri Lanka

Domesticated since time immemorial but now mostly forgotten in farming and food systems, South Asian NUFPs include pseudocereals and millets, grain legumes, root and tubers, leafy vegetables and fruits (Siddique et al. 2021; Singh et al. 2022). Some of these are listed in Table 4.4.

4.2.1 India

In India included among neglected pseudocereals are grain amaranth, buckwheat, and quinoa (Table 4.5a). Among minor millets are finger millet, foxtail millet, pearl millet, proso millet, kodo millet, little millet, and barnyard millet. These are a group of highly nutritious, small-grained cereal food crops, whose use in India dates to before the Indian Bronze Age (4500 BP), when they are referred to in some of the oldest Yajurveda texts indicating their widespread consumption at the time. Until the mid-1960s millets were a major staple in India, integral to local and indigenous food cultures. Like many forgotten crops worldwide, millets gradually came to be disdained by urban consumers and looked upon as "unmodern" and "coarse" – to be replaced by more modern and "refined" foods (ICRISAT 2017). Minor pulses include horse gram, rice bean, adzuki bean, moth bean, faba bean and cowpea.

Minor cereals like Job's tears (*Coix lacryma-jobi*) are considered nutritionally relevant due to their higher content of protein, oil, dietary fiber, and minerals compared to common cereals and pulses (Tables 4.5a and 4.5b). However, their use has remained limited. Pseudo-cereals, minor cereals and millets have lower starch content, high amounts of dietary fiber and low or no gluten. Although these crops are nutritionally superior when evaluated for nutrients of concern for human health,

Table 4.4 Examples of South Asian NUFPs with the potential to offset current nutrient deficiencies when included as part of a varied and balanced diet

Cereals/pseudocereals	Root and Tubers	Pulses	Fruits	Vegetables	Nuts, seeds, and spices
Foxtail millet (*Setaria italica*), Finger millet (*Eleusine coracana*), Little millet (*Panicum miliare*), Pearl millet (*Pennisetum typhoideum*), Sorghum (*Sorghum vulgare*), Traditional rice varieties (e.g., red, brown) (*Oryza* spp.), Water lily seeds (*Nymphaea pubescens*), Naked barley (*Hordeum vulgare* var. *nudum*)	Imperial taro (*Colocasia antiquorum*), Lasia (*Lasia spinosa*), Sacred lotus (*Nelumbium nelumbo*), Cassava (*Manihot esculenta*), Elephant yam (*Amorphophallus campanulatus*), Wild yam (*Dioscorea alata*)	Bengal gram (*Cicer arietinum*), Black gram (*Phaseolus mungo*), Green gram (*Vigna radiata*), Horse gram (*Macrotyloma uniflorum*)	Mango (*Mangifera indica*), Jackfruit (*Artocarpus heterophyllus*), Tamarind (*Tamarindus indica*), Ceylon olive (*Elaeocarpus serratus*), Bengal quince (*Aegle marmelos*), Indian gooseberry (*Phyllantus emblica*), Bengal currant (*Carissa carandas*), Rambutan (*Nephelium nappaceum*), Noni (*Morinda citrifolia*)	Amaranth (*Amaranthus* spp.), Ambarella (*Spondias dulcis*), Ash gourd (*Benincasa hispida*), Bitter gourd (*Momordica charantia*), Bottle gourd (*Lagenaria vulgaris*), Winged bean (*Psophocarpus tetragonolobus*), Snake gourd (*Trichosanthes cucumerina* var. *anguina*), Moringa leaves and stalk (*Moringa oleifera*), Okra (*Abelmoschus esculentus*), Long red bean (*Vigna unguiculata sesquipedalis*), Golden leather fern (*Acrostichum aureum*)	Jackfruit seeds (*Artocarpus heterophyllus*), Sesame (*Sesamum indicum*), Indian almond (*Terminalia catappa*)

Table 4.5a Proximate, minerals and vitamin content of minor and major Indian food crops

	Crops	Moisture (g/100 g)	Protein (g/100 g)	Ash (g/100 g)	Total fat (g/100 g)	Dietary fiber (g/100 g)	Carbohydrates (g/100 g)	Energy (KJ)	Ca (mg/100 g)	Fe (mg/100 g)	Zn (mg/100 g)	Thiamin (mg/100 g)	Total folates (μg/100 g)
Millets	Pearl millet (*Pennisetum typhoideum*)	8.9	10.9	1.37	5.43	11.5	61.8	1456	27.3	6.42	2.76	0.25	36.1
	Finger millet (*Eleusine coracana*)	10.9	07.2	2.04	1.92	11.2	66.8	1342	364	4.62	2.53	0.37	34.7
	Little millet (*Panicum miliare*)	11.4	10.1	1.34	3.89	7.72	65.5	1449	16.1	1.26	1.82	0.26	36.2
	Kodo millet (*Paspalum scrobiculatum*)	14.2	8.9	1.72	2.55	6.39	66.2	1388	15.3	2.34	1.65	0.29	34.5
	Foxtail millet (Setaria italica)[1]	10.9	7.68	1.37	3.86				12.9				
	Proso millet (*Panicum miliaceum* L.)[1]	10.5	10.6	1.16	3.08				10.8				
	Barnyard millet (*Echinocloa esculenta*)[2]	8.7	10.5	2.0	3.6	12.6	62.6	398					
	Sorghum (*Sorghum vulgare*)	9.01	9.97	1.39	1.73	10.2	67.7	1398	27.6	3.95	1.96	0.35	39.4

(continued)

Table 4.5a (continued)

	Crops	Moisture (g/100 g)	Protein (g/100 g)	Ash (g/100 g)	Total fat (g/100 g)	Dietary fiber (g/100 g)	Carbohydrates (g/100 g)	Energy (KJ)	Ca (mg/100 g)	Fe (mg/100 g)	Zn (mg/100 g)	Thiamin (mg/100 g)	Total folates (µg/100 g)
Pseudo-cereals	Quinoa (*Chenopodium quinoa*)	10.4	13.1	2.65	5.50	14.7	53.6	1374	198	7.51	3.31	0.83	173
	Amaranth (grain type) (*Amaranthus* spp.)	9.20	13.3	3.05	5.56	7.47	61.5	1489	162	8.02	2.50	0.04	24.6
	Buckwheat (*Fagopyrum esculentum*)[3]	13.5	12	1.8	3.1	4.3	69.6	1510	17	2.8	2.4	0.46	51
Minor cereals	Barley (*Hordeum vulgare*)	9.77	10.9	1.06	1.30	15.6	61.3	1321	28.6	1.56	1.50	0.36	31.6
	Job's tears (*Coix lacryma-jobi*)	11.3	15.9	3.34	4.66	5.53	63.5						
Common cereals	Maize (*Zea mays*)	9.26	8.80	1.17	3.77	12.2	64.8	1398	8.91	2.49	2.27	0.33	25.8
	Wheat (*Triticum* spp.)	10.6	10.6	1.42	1.47	11.2	64.7	1347	39.4	3.97	2.85	0.46	30.1
	Rice (*Oryza sativa*)	9.93	7.94	0.56	0.52	2.81	78.2	1491	7.49	0.65	1.21	0.05	9.32

Minor pulses	Horsegram (*Macrotyloma uniflorum*)	9.28	21.7	3.24	0.62	7.88	57.2	1379	269	8.76	2.71	0.32	163
	Rice bean (*Vigna umbellatum*)	11.1	20.0	3.54	0.74	13.4	51.3	1265	200	4.76	2.29	0.46	122
	Moth bean (*Vigna aconitifolia*)	8.14	19.7	3.14	1.76	15.1	52.1	1291	154	7.90	1.92	0.45	349
	Cowpea (*Vigna unguiculata*)	9.42	20.4	2.90	1.15	11.5	54.6	1340	84.1	5.04	3.41	0.33	231
	Faba bean (*Vicia faba*)[4]	12.4	26.2	2.47	1.22	10.3	50.0						
Major pulses	Bengal gram (*Cicer arietinum*)	8.56	18.8	2.78	5.11	25.2	39.6	1201	150	8.78	3.37	0.37	233
	Red gram (*Cajanus cajan*)	9.30	20.5	3.53	1.38	22.8	42.5	1146	139	5.37	2.99	0.74	229

(continued)

Table 4.5a (continued)

	Crops	Moisture (g/100 g)	Protein (g/100 g)	Ash (g/100 g)	Total fat (g/100 g)	Dietary fiber (g/100 g)	Carbohydrates (g/100 g)	Energy (KJ)	Ca (mg/100 g)	Fe (mg/100 g)	Zn (mg/100 g)	Thiamin (mg/100 g)	Total folates (µg/100 g)
Minor oils seeds	Linseed (*Linum usitatissimum*)	5.48	18.6	3.15	35.7	26.2	11.0	1857	257	5.44	4.86	0.28	86.5
	Sesame (*Sesamum indicum*)	4.51	19.2	5.78	43.1	17.2	10.3	2124	1664	13.9	8.59	0.34	127
	Perilla (*Perilla frutescens*)[5]	7.4	17.4	3.6	51.7				209	9.0	4.7		
	Safflower (*Carthamus tinctorius*)	5.24	17.7	2.57	30.9	13.5	30.18	1981	211	4.06	3.90	0.85	82.4
Major oil seeds	Mustard (*Brassica* spp.)	5.67	19.5	3.73	40.2	14.1	16.8	2132	402	13.5	4.03	0.55	94.9
	Soybean (*Glycine max*)	5.51	35.6	4.74	19.8	21.6	12.8	1596	239	8.29	4.01	0.59	297

Sources: Longvah et al. (2017), [1]Choi et al. (2019), [2]Ugare et al. (2014), [3]MEXT (2015), [4]Singh et al. (2014), [5]Longvah and Deosthale (1991)

Table 4.5b Fatty acids' profile of minor and major Indian oil seeds expressed ad mg/100 g edible portion

	Crops	Myrisitic (C14:0)	Palmitic (C16:0)	Stearic (C18:0)	Arachidic (C20:0)	Behenic (C22:0)	Lignoceric (C24:0)	Palmitoleic (C16:1)	Oleic (C18:1n9)	Eicosenoic (C20:1n9)	Erucic (C22:1n9)	Nervonic (C24:1n9)	Linoleic (C18:2n6)	Eicosadienoic (C20:2)	α-Linolenic (C18:3n3)
Minor oils seeds	Linseed (*Linum usitatissimum*)		1503	1323	59.7	51.2	31.9	25.6	5049	36.6			3191		12,956
	Sesame (*Sesamum indicum*)		3735	2230	263	51.3	38.4	62.2	15,475	67.7			18,966		112
	Perilla (*Perilla frutescens*)		14,030	4460					1885				8805		28,380
	Safflower (*Carthamus tinctorius*)	28.1	1789	546	90.7	58.6	36.1	17.1	3487	49.3		36.4	18,760		21.9
Major oil seeds	Mustard (*Brassica* spp.)		856	389	289	348	230	69.7	4012	1838	14,468	644	4932	245	3341
	Soybean (*Glycine max*)		2338	523	43.3	65.5	31	25.4	4723	68.1			9754		1318

they also contain anti-nutrients like phytates, oxalates, tannins, protease inhibitors that often limit their bioavailability. Suitable processing methods like malting, parboiling, partial fermentation, and enzymatic hydrolysis techniques can be employed to improve nutrient bioavailability as well as processing characteristics for their enhanced use. Minor oilseeds (like sesame, linseed, safflower and perilla) are another group of NUFPs that are known to contain high amounts of polyunsaturated fatty acids, particularly omega-3 fatty acids. Perilla (*Perilla frutescens*), for example, can contain more than 50% of total fatty acids. However, this characteristic makes the plant highly susceptible to oxidative and hydrolytic rancidity, which results in the low shelf life of oils extracted from this crop. Oil from this crop can be used for blending with common oils from palm, soybean, mustard, groundnut, and sunflower to improve the fatty acids' profile of blended oil.

Among the most interesting neglected native fruits of India is the jackfruit (*Artocarpus heterophyllus* Lam.). It is a multi-purpose crop used for food, traditional medicine, timber, fuel, and fodder. Considerable variation is observed in the fruit's shape and size, in the density of its tubercles, in rind color, bulb size and color, flake thickness, fiber content, fruit quality and maturity. Phytochemical analysis revealed high carotenoids (4.43 mg/100 g) and lycopene content (1.12 mg/100 g). Total flavonoids and phenols, antioxidant activity was also higher compared to yellow and white colored flakes indicating its potential as health promoting and functional food. Another interesting fruit is tamarind (*Tamarindus indica*), cultivated for its fruit pulp, which is used to prepare soft drinks, confectionary, to flavor curries and sauces, and is accepted as traditional medicine in many parts of the world (Gitanjali and Shashi 2020). Recently, research undertaken into the nutrient profile of the fruit's kernel flour showed that 100 g of the flour are rich in calcium (109.25 mg), magnesium (247.51 mg), iron (11.37 mg) and potassium (1245.11 mg) and can play a key role in dietary diversity and enrichment (Gitanjali and Shashi 2020). The fruit tree is also widely adaptable to extreme climates and grows well on marginal lands due to its large tap root system that protects it from strong winds and cyclones.

4.2.2 Sri Lanka

A wide range of fruits, vegetables, tubers, pulses, and grains, both cultivated and wild, exist in Sri Lanka. One of the world's 34 biodiversity hotspots, Sri Lanka's edible diversity includes approximately 4500 plant and animal species (20% of which are endemic) that are adapted to growing in the country's three distinct climatic zones: wet, dry and intermediate. Using agrobiodiversity-friendly traditional agricultural practices – such as the *chena*, *owita* and Kandyan home garden systems – Sri Lankan farmers have maintained this rich agricultural diversity for thousands of years. Of the plants, many traditional species are health promoting and feature in Sri Lanka's tradition of Ayurvedic medicine. The leafy green *Centella asiatica*, for instance, is used to improve cognitive function and resolve skin and digestive disorders. When it comes to nutrient content, food composition analysis carried out as part of the GEF-supported *Biodiversity for Food and Nutrition* Project (2012–2019) highlighted, for example, high levels of inter- and intra-species diversity in Sri Lankan native rice varieties compared to more widely consumed white rice. Traditional fragrant rice (Suwandel), dark rice (Kalu heenati), red rice varieties *Kurulutuda*, *Madathavalu*, *Pachchaperuma'l*, *Pokkali* and *Sudu heena*ti rice were shown to contain significantly higher levels of iron and fiber (Fig. 4.1) to potentially offset common nutrient deficiencies (Hunter et al. 2019).

Similarly for finger millet (*Eleusine coracana*), two local varieties *Bala kurakkan* and *Wadimal kurakkan* differ substantially in their calcium content, which is more than twice as high in the former (13.98 mg/100 g of raw seed) compared to the latter (6.39 mg/100 g). Notwithstanding, both varieties are rich sources of iron (144.28 mg and 124.99 mg/100 g respectively) and vitamin C (454.86 mg/100 g and 438.06 mg/100 g) (Borelli et al. 2020). Where fruits are concerned, wood apple (*Limonia acidissima*) contains large amounts of protein (3.66%) and minerals including calcium (8.17 mg) and magnesium (267.40 mg). While among vegetables, centella (*Centella asiatica*) that grows in swampy areas in Sri Lanka and is consumed in salads, juices and traditional dishes to accompany rice and curry, or used in traditional medicine (stems and roots), contains large amounts of vitamins B1 and B2 (88.10 μg & 79.36 μg respectively) (Biodiversity for Food and Nutrition (BFN) 2018).

The recognition that the inherent nutritional qualities of different varieties within the same crop, and lesser known wild and cultivated fruit and vegetable species may contribute to improving diet quality has been acknowledged in the recently published Sri Lanka food composition tables (FCT) (Jayatissa et al. 2021). At its second edition, the FCT celebrates Sri Lanka's incredible food biodiversity and complex culinary heritage by including species such as the wild yam (*Dioscorea alata* L.) and the indigenous Ceylon olive (*Elaeocarpus serratus*). The fruit, which, is being explored for its nutritional properties and health food potential (Sarananda et al. 2017; Gee et al. 2020), represents a good source of flavonoids, condensed tannins, carotenoids and Vitamin C (de Lima et al. 2019).

4.2.3 Nepal

Situated at the foot of the Himalayas and endowed with a rich diversity of climates and agroecological zones, Nepal harbors 3.2% of the world's flora, including 790 food plant species. Of the 587 cultivated plants, 494 are indigenous (74 crops, 145 horticultural crops and 275 forages) while 93 are introduced (MOAD 2017). About 85% of the total cultivated crop species are neglected and underutilized in Nepal

(Joshi et al. 2020). NUFPs in Nepal include nutrient dense and climate resilient crops such as amaranths, buckwheat, barley, common beans, minor millets (finger millet, proso millet, foxtail millet, sorghum), grain legumes (horse gram, black grams, rice beans etc.) as well as indigenous fruits and vegetables. NUFPs have been cultivated in Nepal over millennia by farmers, particularly in remote rainfed locations such as in mid- and high mountain areas and some parts of the interior *terai* – a lowland region in southern Nepal characterized by tall grasslands, scrub savannah, sal forests and clay rich swamps. Adapted to growing in poor soils and under extreme climates, these crops help marginalized communities meet their food and nutrition requirements and cope with the effects of climate change on agricultural production. Many are nutrient dense, rich in micronutrients, dietary fiber, rare amino acids and vitamins, often containing higher protein, calcium and iron compared to major food staples such as rice, wheat and maize (Table 4.6) as well as being gluten free (DFTQC 2017). For instance, flour from buckwheat (*Fagopyrum esculentum*) and black naked barley (*Hordeum vulgare* var. *nudum*) – a traditional, culturally important, climate-resilient winter cereal crop of the high mountain region of Nepal– contain double and three times the amount or iron compared to maize flour. The difference with wheat flour is lower (1.5 times) but still significant for black naked barley (DFTQC 2017). Marked diversity in morphology and agronomic traits have been reported between and within Nepalese landraces (Karkee et al. 2020; Yadav et al. 2018), but only recently has there been interest in determining their nutritional traits (Lyu et al. 2022; Mut et al. 2023). The protein content of highland barley was found to be higher than that of most grains, while the crop contains 18 kinds of amino acids and a large amount of dietary fiber, especially soluble *ß*-glucans, which act as cholesterol inhibitors and thus have the potential to reduce the incidence of chronic diseases such as heart disease and diabetes (Lyu et al. 2022).

Table 4.6 Nutrient content of selected traditional cereals (per 100 g of edible portion) compared to major food staples such as rice, wheat and maize

Crops	Protein (g)	Fat (g)	Carbohydrate (g)	Fiber (g)	Energy (Kcal)	Calcium (mg)	Phosphorous (mg)	Iron (mg)
Proso millet	11.0	4.2	72.9	1.0	378	8	285	3.0
Amaranth grain	9.4	7.2	68.1	2.2	375	37	529	5.2
Black naked barley (Uwa)	12.6	1.7	70.9	2.5	340	20.3	–	7.5
Naked barley flour	9.6	2.6	76.7	2.0	369	–	–	1.4
Buckwheat flour	6.1	1.3	69.2	7.8	313	–	–	5.6
Rice (milled)	6.8	0.5	78.2	0.2	345	10	160	0.7
Maize flour	9.2	3.9	72.1	1.6	360	20	256	2.4
Wheat flour	12.1	1.7	69.4	1.9	341	48	355	4.9

Source: DFTQC (2017)

In the mountain regions of Nepal, naked barley grains are roasted, milled, and consumed as a flour known as *satu*, which is used to prepare flat breads, dumplings, and noodles to make *tukpa* (Tibetan noodle soup). *Satu* can also be mixed with wheat, buckwheat, and finger millet flour (Yadav et al. 2018). Lately, naked barley and other traditional crops are facing a revival and being sought after by urban consumers who are willing to pay premium prices for organically produced foods that grow in mountain agroecosystem and are perceived as being healthier and nutritious (Gauchan et al. 2020a).

4.3 Determinants and Barriers/Constraints Preventing Better Use of These Foods in Food Security and Nutrition Strategies, Policies, and Interventions

Despite the above opportunities, traditional, nutrient rich underutilized crops remain undervalued. They continue to be a major source of food and nutrition security for smallholder farmers living in remote, inaccessible areas across South Asia, but are generally neglected by policy, research and development (Gauchan et al. 2019). For instance, in Nepal, the use of NUFP landraces in research and breeding is negligible (less than 5%). For this reason, many of these crops, particularly numerous traditional farmer varieties, are disappearing rapidly and some of them are critically endangered (Gauchan et al. 2019; Joshi et al. 2020).

In Nepal, surveys carried out in 2014–2015 among farming households in high mountain areas as part as part of a GEF UNEP project[1] identified several market constraints for improved use of local NUFPs. Major production constraints included lack of quality seeds and drudgery reducing processing technologies. Limited awareness of the nutritional value of NUFPs was also a major constraint for market promotion, as was the lack of proper market infrastructure, market linkages and market information (Gauchan et al. 2020a). Further along the value chain, in terms of value addition, major limitations included the lack of food processing technologies available to communities as well as capacity to comply with food quality and safety standards and regulations. Poor research investments and weak seed regulatory frameworks were key policy constraints (Gauchan et al. 2019). Lastly, when positive outcomes do occur, in terms of nutritional or monetary gains particularly for smallholder growers, there is limited capacity to document these impacts.

[1] UNEP GEF project "Integrating Traditional Crop Genetic Diversity into Technology: Using a Biodiversity Portfolio Approach to buffer against Unpredictable Environmental Change in Nepal Himalayas".

4.4 Mapping, Assessing and Prioritizing NUFPs for Inclusion in Food Security and Nutrition Planning: Experiences from Relevant Projects and Initiatives. Methods, Tools, Criteria for Prioritization of NUFPs and Species Selection

Given the large number of high potential NUFPs that exist in biodiversity-rich South Asian countries, it will be important to map existing information, fill information gaps by identifying additional species that have untapped potential but had not been previously considered and finally prioritize this long list of species to create a manageable shortlist that can be used to achieve national food security and nutrition objectives. Using this stepwise approach, countries will avoid potential overlaps and maximize time and resources.

4.4.1 Mapping Existing Information

Much information on high potential NUFPs is probably already available but exists scattered across different national resources and databases. The mapping will include identifying national experts and data holders that can share this information. Useful starting points for mapping information include national agricultural and biodiversity surveys that are generally used in country reporting to global conventions, such as the national reports that have been used to date to monitor progress on the implementation of the Aichi Biodiversity Targets of the Convention on Biological Diversity (CBD)[2] or country reporting to the Commission on Genetic Resources for Food and Agriculture (CGRFA) in preparation for its global assessments on the status of biodiversity.[3] Other important sources of information to consider are ethnobotanical studies and national food composition tables (FCTs) and databases. The latter, however, are often either outdated or tend to disregard NUFPs as they may be only locally important. Of the South Asian countries considered in this chapter, only India, Nepal and Sri Lanka have updated their food composition tables in the last 5 years (Jayatissa et al. 2021; Longvah et al. 2017), and are the only countries that have included nutrient values for several select NUFPs. These FCTs include nutrient values for underexploited cereals, such as the seeds of the water lily *Nymphaea* spp., for Lasia stalks (*Lasia spinosa*) among roots and tubers as well as for different types of uncommon vegetables and fruits, such as golden leather fern (*Achrosticum aureum*), and Ceylon olive (*Elaeocarpus serratus*), a fruit endemic to Sri Lanka and which thrives in the wild as well as in homegardens across the Indian subcontinent (Raji and Siril 2021). These FCTs go one step further to include

[2] https://www.cbd.int/reports/search/

[3] www.fao.org/cgrfa/assessments/global-assessments/sow/en/

nutrient values at the varietal level for several NUFPs, raising awareness of the fact that nutrient content may differ among varieties of the same species, for example between the short and different elongated varieties of bitter gourd (*Momordica charantia*) in Sri Lanka, or between white and black naked barley (*Hordeum vulgare* var. *nudum*) in Nepal.

4.4.2 Filling Information Gaps

When desk reviews fail to produce comprehensive information, agrobiodiversity surveys at the farm, landscape and market level can help inventory NUFP diversity. Working with national and local experts, as well as with local communities, focus group discussions are a quick and cost-efficient appraisal method. But if time and funding are not an issue, key informant interviews, direct observations, and participatory methods may also be used to gain additional insights into available biodiversity. The latter are particularly important when the aim is to capture traditional knowledge from communities on NUFPs, such as local names, uses and characteristics, dietary and cultural significance, seasonal availability, and collecting sites in the case of wild species. The *How to do note #1 on Crop selection for diet quality and resilience* contains a complete list of rapid and alternative assessment methodologies (IFAD and Bioversity International 2021) as does the *Agrobiodiversity Assessment Compendium* compiled by the Platform on Agrobiodiversity Research (PAR 2018).

4.4.3 Prioritizing Species

Criteria are normally established to reduce the long list of species uncovered from the desk review and agrobiodiversity assessments to a manageable size. Factors such as nutritional importance, economic and market value, production or collection constraints, conservation status, use and cultural acceptance would normally be evaluated by a broad range of stakeholders and experts (Gee et al. 2020). However, when the intent is to identify high potential NUFPs for household food security, to increase the availability of food groups and/or specific nutrients in local diets as well as on farm resilience, considerations might include:

Nutrition

- Does the NUFP contribute to diet quality year-round, or in specific times (e.g., during the lean season)?
- Can the NUFP fill important nutrient gaps? Is it from an under-consumed food group? Can it provide limiting micronutrients?

Production

- Are there any production constraints (e.g., seed/planting material availability?)
- What is the cost of production?
- Is there capacity to grow/collect the NUFP?
- What is the NUFP's growth rate? Vegetation period? Annual growth rate?

Climate change resilience

- Is the NUFP adapted to growing in resource-poor settings under variable climatic conditions?
- Does the NUFP possess important traits such as drought tolerance, early maturation, resistance to pests and diseases?

Income opportunities

- Do market opportunities exist for the NUFP?
- Is the NUFP attractive to consumers, and suitable for processing and storage?
- Does the NUFP offer specific market opportunities for women and youth empowerment?

CASE STUDY: Scoping, Prioritizing and Mapping of NUS in Nepal and India

A scoping study carried out among the indigenous communities of Nepal as part of the Future Smart Food Initiative in Asia helped identify a long list of neglected and underutilized species (NUS) with the potential to contribute to food security and nutrition, dietary and culinary diversification, health, income generation, as well climate change adaptation (Joshi et al. 2020). As part of the study, 800 plant species with medicinal use, 440 wild food plants, 100 fodder species, 71 fiber plants, and 30 spices were listed by indigenous communities who are deeply connected with nature and hold important traditional knowledge of their uses (Dangol 2008). Using a multi-stakeholder, multi-sectoral approach to prioritize the species based on nutritional, agronomic and climate resilience characteristics, the long list of species was reduced to six, one from each of the main food groups. The selection included Tartary buckwheat (*Fagopyrum tataricum*) for cereals, grass pea (*Vicia sativa*) for pulses, taro (*Colocasia esculenta*) among roots and tubers, moringa (*Moringa oleifera*) for horticultural crops, jackfruit (*Artocarpus heterophyllus*) for fruits and lastly Nepal butter tree (*Bassia latifolia*) among oilseeds.

Climate adaptation and local food and nutrition security, on the other hand, were the prime concerns guiding crop selection as part of the UNEP GEF *Integrating Traditional Crop Diversity for Mountain Food Security* project, implemented in the Himalayan region between 2014 and 2019 by Bioversity International and partners. With national level consultations and background field studies, the project selected eight NUS that were important to local food and nutrition security, local diversity and adaptation to changing climate in the mountains of Nepal. These include buckwheat (*Fagopyrum esculentum* and *F. tataricum*), a cold tolerant rice variety (*Oryza sativa*), common bean (*Phaseolus vulgaris*), finger millet (*Eleusine coracana*), foxtail millet (*Setaria italica*), grain amaranth (*Amaranthus caudatus* and *A. leucocarpus*), naked barley (*Hordeum vulgare* var. *nudum*), and proso millet (*Panicum miliaceum*). The species, harboring globally important cold-tolerant genes, were chosen to help mountain farmers cope with their marginal and heterogeneous environments that are likely to be affected by climate change in the future (Gauchan et al. 2020a, 2020b; GEF UNEP 2013). The intra-specific diversity of these crops is very high as most of them have evolved or are located at the center of diversity in the Nepal Himalayas. These traditional underutilized crops are used intensively and cultivated under default organic conditions by local communities in many of the remote hills and mountainous regions, and contribute considerably to their site-specific food security, nutrition and adaptation (Gauchan et al. 2020a). They are nutrient dense and climate resilient crops, and provide food, fodder, nutrition, livelihood, and ecological security to smallholder farmers with potential for value chain development and income generation. Considering their great value for nutrition, climate resilience and local adaptation and risk diversification, they are considered Himalayan Superfoods (www.himalayancrops.org).

The *Tropical Fruit Tree* project, carried out in India, Indonesia, Malaysia, and Thailand between 2009 and 2015 to promote fruit crops for health and nutrition, used a prioritization process to identify the most favorable traits for widespread promotion. Working with 36 communities across the four countries, mixed methods helped identify farmers' preferred varieties or genotypes with unique traits (e.g., colour, taste, shape, quality, aroma, local adaptability and disease resistance). Across the four countries, a total of 95 varieties of mango (*Mangifera indica*), 32 citrus (*Citrus* spp.), 5 mangosteen (*Garcinia mangostana*) and 2 rambutan (*Nephelium lappaceum*) were identified. The seeds of best-performing trees were collected and multiplied in 126 fruit tree nurseries and subsequently made available to over 77,000 farming households.

Source: IFAD and Bioversity International (2021).

4.5 Identification of Entry Points to Better Harness Use of NUFPs for Nutrition-Sensitive Approaches for Improved Household Food Security and Nutrition

4.5.1 Strengthening Policies and Enabling Environments

Information is key to gain policy support for the widespread adoption of NUFPs and their greater integration into sustainable production and national food systems. In Chap. 4 of this publication, Sri Lanka's national government research organizations (such as the Plant Genetic Resources Centre) and policy documents (such as the National Biodiversity Strategic Action Plan 2016–2022) that support and promote the mainstreaming of NUFPs into national food systems were highlighted. Similar national efforts to mainstream NUFPs are observed in Nepal and India.

The Indian Institute of Millets Research (IIMR), established in 1958 under the umbrella of the Indian Council of Agricultural Research (ICAR), undertakes research, development, extension and value chain development of millets. Efforts have led to the millet revolution in India and given these neglected crops a new lease of life. The United Nations declared 2023 the "International Year of Millets" and policy support was provided under the India Millet Initiative. Support is being provided for post-harvest value addition, enhancing domestic consumption, and for branding millet products nationally and internationally. Efforts are being made to increase efficiency along the entire millet supply-chain for better millet production, processing and distribution. NUFPs are also being promoted under the Paramparagat Krishi Vikas Yojana Krishi Vikas Yojana (Traditional Farming programs) as part of the Government of India's Mission on Sustainable Agriculture. Finger millet, proso millet, amaranth, buckwheat, adzuki bean, kulthi and rice bean are also being considered for inclusion in the Public Distribution System (PDS), which should hopefully help counteract the bias towards wheat and rice in both domestic production and consumption. A stable and reliable market such as PDS would be a major incentive for farmers to continue growing NUFPs.

Conversely to India, Nepal has no government-led process in place to ensure that food is provided to the most food insecure. However, sustained effort and lobbying by farmer organizations, the NGO LI-BIRD, the NARC-Gene Bank, local government agencies and Bioversity International, has led the Nepalese Government, via the Department of Agriculture, to give priority to the promotion of indigenous NUFPs. This has led to implementation of the *Indigenous Crop Promotion Program* – focusing on minor millets, buckwheat, barley and amaranths – and the *Organic Mission Program* promoting NUFPs (Shrestha 2022). Similarly, an easing in the Seed Regulation Act (2013) and a simplification in the procedures for registering farmers' varieties accompanied by technical facilitation and training from the Seed Quality Control Centre of the Government of Nepal has led to the registration in 2021 of six unique indigenous landraces including of proso millet (*Panicum miliaceum*), foxtail millet (*Setaria italica* L. *Beauv*), amaranth (*Amaranthus hypochondriacus*), finger millet (*Eleusine coracana*), and common bean (*Phaseolus vulgaris* L.). This has opened up opportunities for the promotion and marketing of registered varieties of NUFPs (Pudasaini 2021).

4.5.2 Nutrition-Sensitive Markets, Value Chains

NUFPs can play a significant role in nutrition-sensitive markets and value chains. This involves due consideration to the entire NUFP value chain, from production to processing, distribution, marketing, and consumption. At present, value chains for traditional NUFPs are weak, fragmented, and delinked, in India and Nepal as elsewhere (Gauchan 2019), with limited returns for those producers, mostly small-scale farmers, who continue to grow this diversity, mostly for home consumption. Investments into biodiversity-based value chain upgrading could improve the performance, efficiency, and interlinkages between the different value chain components and elevate NUFPs from their condition of abandonment as well as provide an important source of income for producers (Padulosi et al. 2015).

A successful example of value chain upgrading backed by government support is ICAR-IIMR's Nutrihub[4] – an innovation center that undertakes value addition research and development on both improved and traditional millet varieties. Funded by India's Department of Science & Technology, Nutrihub commercializes state-of-the-art processing and value addition technologies developed by public-funded institutes through start-ups managed by young agri-preneurs while providing them with mentoring, training and financial facilitation. To date, more than 175 start-ups were supported, and more than 190 millet technologies were transferred to industries and entrepreneurs. ICAR-IIMR also supports 31 farmer producer organizations (FPOs) that help millet farmers in growing, processing, and marketing their produce. These efforts are helping to popularize millet products particularly among health-conscious rural and urban consumers. These efforts are also adding to the livelihoods of small farmers by increasing their disposable income and providing them with better nutrition. Unfortunately, most of these positive outcomes are undocumented and would require investments into impact evaluation for further promotion of nutritious NUFPs.

Also in India, value chain development for custard apple (*Annona squamosa*) is providing increased income for communities living in the tribal belts of Southern Rajasthan, where the species grows wild (Shailza Sharma et al. 2020; Tripathi 2016). The fruits, which are collected by local communities for sale suffer heavy harvest and post-harvest losses, with resulting missed revenues (Kaushik 2013). Training on harvesting and value addition provided by the Maharana Pratap University of Agriculture & Technology (MPUAT) as well as the establishment of a village collection center, a processing and storage unit have reduced post-harvest losses by 20%. At the same time, raw custard apple transformation into pulp and powder, has reduced waste and increased the marketability of the product (Shailza Sharma et al. 2020). The estimated value following the transformation of custard apple production for the 2017–2018 season is more than 65,000 USD (Shailza Sharma et al. 2020).

4.6 Conclusions and Recommendations

This chapter, and others in this book, make abundantly clear that the countries comprising South Asia have a wealth of underutilized, nutrient-rich food plants which not only contribute to positive outcomes for food security and nutrition but also confer climate resilience and other benefits. In more instances than not, these local and seasonal food products from sustainable, predominantly

[4] https://www.nutrihubiimr.com/about-nutrihub

agroecological production, are significantly superior in nutrients compared to more commonly found exotic food plants which tend to dominate value chains and market outlets. This natural wealth and abundance of underutilized food plants can often seem overwhelming when it comes to focusing on which crops or species to target for promotion, for markets, for improving nutrition or livelihoods or other nutrition-sensitive programming. Where does a researcher or development practitioner or another actor start? What this chapters makes abundantly clear is that there are tested approaches to assist with this challenge. We know how to map and assess available food biodiversity, and we know how to determine its nutritional value and other positive attributes. All criteria that have been successfully used in many contexts to help prioritize which underutilized food plants present the greatest potential to promote especially for enhancing food security and nutrition. Furthermore, we know clearly what the entry points and opportunities are to mainstream or better harness these underutilized food plants, be it for improving nutrition or livelihoods through market creation. Yet, when we closely examine our current food system it becomes quickly apparent that few of these food plants are part of pathways that contribute to improving food security and nutrition and livelihoods. It is not that these underutilized food plants are not fit for purpose. Rather they do not fit into a food system that is geared towards specialization, uniformity and concentration, which continues to supply large volumes of foods to global markets, but which generates negative outcomes on multiple fronts. Policies that are no longer fit for purpose mean that the region continues to be locked-in to unhealthy and unsustainable food systems that deliver cheap, highly processed, energy dense foods that are poor in micronutrients. Addressing these lock-ins, and the wider political economy and policy space, is central to reversing this scenario. As a starting point, there is urgent need to reorient policies and subsidies which continue to support a narrow range of cereals, to focus more on harnessing nutrient-rich, underutilized food plants. For example, creating policy incentives for diversification and agrobiodiversity could create a space for more agroecological production linked with the power of public food procurement, sustainable and healthy sourcing of food and short supply chains to integrate underutilized food plants. These are but a few examples where real difference could be made. The lack of action and political will to challenge these lock-ins and the broader political economy of food systems that maintains the *status quo* means we have another example of the 'knowing and doing gap' or the disconnect between knowledge that something is inherently good and the necessary action to achieve it. Until this incoherence is addressed, the food system transformation that is needed to create a level playing field for much of the world's NUFPs will remain elusive.

Conflicts of Interest All authors read and approved the final manuscript and declare that they have no conflict of interest.

References

Adhikari L, Hussain A, Rasul G et al (2017) Tapping the potential of neglected and underutilized food crops for sustainable nutrition security in the mountains of Pakistan and Nepal. Sustainability 9(2):291. https://doi.org/10.3390/su9020291

Afshin A, Sur PJ, Fay KA et al (2019) Health effects of dietary risks in 195 countries, 1990–2017: a systematic analysis for the Global Burden of Disease Study 2017. Lancet 393(10184):1958–1972. https://doi.org/10.1016/S0140-6736(19)30041-8

Aryal JP, SapkotaT B, Khurana R et al (2020) Climate change and agriculture in South Asia: adaptation options in smallholder production systems. Environ Dev Sustain 22(6):5045–5075. https://doi.org/10.1007/s10668-019-00414-4

Béné C, Bakker D, Chavarro MJ et al (2021) Global assessment of the impacts of COVID-19 on food security. Glob Food Sec 31:100575. https://doi.org/10.1016/j.gfs.2021.100575

Biodiversity for Food and Nutrition (BFN) (2018) Harvesting the benefits of local foods: from healthy diets to sustainable growth Bioversity International. http://www.b4fn.org/fileadmin/templates/b4fn.org/upload/documents/Country_profiles/SRI-LANKA.pdf

Bioversity International (2017) Mainstreaming agrobiodiversity in sustainable food systems: scientific foundations for an Agrobiodiversity Index, 1st edn. Bioversity International. https://www.bioversityinternational.org/fileadmin/user_upload/online_library/Mainstreaming_Agrobiodiversity/Mainstreaming_Agrobiodiversity_Sustainable_Food_Systems_WEB.pdf

Bishwajit G (2015) Nutrition transition in South Asia: the emergence of non-communicable chronic diseases. F1000Research 4:8. https://doi.org/10.12688/f1000research.5732.2

Borelli T, Hunter D, Padulosi S et al (2020) Local solutions for sustainable food systems: the contribution of orphan crops and wild edible species. Agronomy 10(2):231. https://doi.org/10.3390/agronomy10020231

Boyko EJ, Magliano DJ, Karuranga S et al (2021) IDF diabetes atlas, 10th edn. International Diabetes Federation. https://diabetesatlas.org/idfawp/resource-files/2021/07/IDF_Atlas_10th_Edition_2021.pdf

Branca F, Lartey A, Oenema S et al (2019) Transforming the food system to fight non-communicable diseases. BMJ 1296:364. https://doi.org/10.1136/bmj.l296

Branca F, Demaio A, Udomkesmalee E et al (2020) A new nutrition manifesto for a new nutrition reality. Lancet 395(10217):8–10. https://doi.org/10.1016/S0140-6736(19)32690-X

Choi JM, Kwak DY, Choe ME et al (2019) Nutritional compositions of varieties of Foxtail Millet and Proso Millet Developed in Korea. Korean J Crop Sci 64(2):137–143. https://doi.org/10.7740/kjcs.2019.64.2.137

Dangol DR (2008) Traditional uses of plants of common land habitats in Western Chitwan, Nepal. J Inst Agric Anim Sci 29:71–78

de Lima FF, Caroline AB, Claudia ALC et al (2019) Evaluation of nutritional composition, bioactive compounds and antimicrobial activity of Elaeocarpus serratus fruit extract. Afr J Food Sci 13(1):30–37. https://doi.org/10.5897/AJFS2018.1760

Department of Food Technology and Quality Control (DFTQC) (2017) Nepalese food composition table (3rd edn). http://www.dftqc.gov.np/downloadfile/FoodCompositionTable2017_1572781821_1590383994.pdf

Development Initiatives (2021) Global nutrition report: the state of global nutrition (1st edn). Development Initiatives. file:///C:/Users/tborelli/Downloads/2021_Global_Nutrition_Report.pdf

FAO (2019) The state of the world's biodiversity for food and agriculture. https://doi.org/10.4060/ca3129en

FAO, IFAD, UNICEF, WFP, WHO (2022). The state of food security and nutrition in the world 2022. https://doi.org/10.4060/cc0639en

Frison EA, Cherfas J, Hodgkin T et al (2011) Agricultural biodiversity is essential for a sustainable improvement in food and nutrition security. Sustainability 3(12):238–253. https://doi.org/10.3390/su3010238

Gatto M, De Haan S, Laborte A et al (2021) Trends in varietal diversity of main staple crops in Asia and Africa and implications for sustainable food systems. Front Sustain Food Syst 5:12. https://doi.org/10.3389/fsufs.2021.626714

Gauchan D (2019) Exploiting biodiversity of traditional crops for mainstreaming nutrition sensitive agriculture in Nepal. https://cgspace.cgiar.org/bitstream/handle/10568/106496/policy-research-brief-5-2019.pdf?sequence=1&isAllowed=y

Gauchan D, Joshi BK, Bhandari B et al (2019) Value chain development of underutilized food crops for nutrition sensitive agriculture in the mountains of Nepal. In: Sharma R, Gauchan D, Ghimire R, Bhandari T, Khanal M, Thapa S, Shrestha SG (eds) 5th International agricultural marketing conference. Nepal Agricultural Economics Society (NAES), Kathmandu, pp 130–144

Gauchan D, Joshi BK, Bhandari B et al (2020a) Traditional crop biodiversity for mountain food and nutrition security in Nepal, 1st edn. Alliance of Bioversity International and CIAT, NAGRC, LI-BIRD

Gauchan D, Joshi BK, Sthapit S et al (2020b) Traditional crops for household food security and factors associated with on-farm diversity in the Mountains of Nepal. J Agric Environ 21:31–43. https://cgspace.cgiar.org/bitstream/handle/10568/109041/Traditional_Gauchan_2020.pdf?sequence=1&isAllowed=y

Gee E, Borell T, Beltrame DMO et al (2020) The ABC of mainstreaming biodiversity for food and nutrition: concepts, theory and practice. In: Hunter D, Borelli T, Gee E (eds) Biodiversity, food and nutrition, a new agenda for sustainable food systems, 1st edn. Routledge, UK

GEF UNEP (2013) Integrating traditional crop genetic diversity into technology: using a biodiversity portfolio approach to buffer against unpredictable environmental change in Nepal Himalayas. https://publicpartnershipdata.azureedge.net/gef/PMISGEFDocuments/Biodiversity/Nepal-%284464%29-IntegratingTraditional CropGeneticDiversityint/4–20-11-Revised PIFdoc.pdf

Gitanjali VS, Shashi J (2020) Nutritional properties of tamarind (Tamarindus Indica) kernel flour. Int J Curr Microbiol App Sci 9(5):1359–1364. https://doi.org/10.20546/ijcmas.2020.905.153

HLPE (2020) Food security and nutrition: building a global narrative towards 2030. http://www.fao.org/fileadmin/user_upload/hlpe/2020._Global_Narrative/HLPE_2020._FSN__building_a_Global_Narrative_towards_2030.pdf

Hunter D, Borelli T, Beltrame DMO et al (2019) The potential of neglected and underutilized species for improving diets and nutrition. Planta 250(3):709–729. https://doi.org/10.1007/s00425-019-03169-4

Hunter D, Borelli T, Gee E et al (2020) Biodiversity, food and nutrition a new agenda for sustainable food systems. Routledge

ICRISAT (2017) Bringing back millets – the super crop of our ancestors. https://www.icrisat.org/a-short-history-of-millets-and-how-we-are-recognising-their-importance-in-the-modern-context/

IFAD, Bioversity International (2021) How to do Note #1 Crop selection for diet quality and resilience International Fund for Agricultural Development. https://www.ifad.org/documents/38714170/43559125/HTDN_NUS_1.pdf/6fa73ff4-ae5d-8e7f-c94f-6f81e8450607?t=1629384629095

IPCC (2022) Summary for policymakers. In: Pörtner H-O, Roberts DC, Poloczanska ES, Mintenbeck K, Tignor M, Alegría A, Craig MM, Langsdorf S, Löschke S, Möller V, Okem A (eds) Climate change 2022: impacts, adaptation and vulnerability. Contribution of working group II to the sixth assessment report of the intergovernmental panel on climate change. Cambridge University Press. https://doi.org/10.1017/9781009325844.001

Jayatissa R, Perera AG, Silva BG et al (2021) Sri Lanka food composition tables (2nd edn). Department of Nutrition, Medical Research Institute. Ministry of Health. Colombo, Sri Lanka. ISBN: 978-624-97488-0-4

Jayawardena R, Jeyakumar DT, Gamage M et al (2020) Fruit and vegetable consumption among South Asians: A systematic review and meta-analysis. Diabetes Metab Syndr Clin Res Rev 14(6):1791–1800. https://doi.org/10.1016/j.dsx.2020.09.004

Joshi BK, Acharya AK, Gauchan D et al (2017) Agrobiodiversity status and conservation options and methods. In: Joshi BK, Bahadur HK, Acharya AK (eds) Conservation and utilization of agricultural plant genetic resources in Nepal. Proceedings of 2nd national workshop 22–23 May 2017, Dhulikhel, pp 21–38

Joshi BK, Shrestha R, Gauchan D et al (2020) Neglected, underutilized, and future smart crop species in Nepal. J Crop Improv 34(3):291–313. https://doi.org/10.1080/15427528.2019.1703230

Kalmpourtzidou A, Eilander A, Talsma EF et al (2020) Global vegetable intake and supply compared to recommendations: a systematic review. Nutrients 12(6):1558. https://doi.org/10.3390/nu12061558

Karkee A, Ghimire KH, Joshi BK et al (2020) Evaluation of Naked barley landraces for agro-morphological traits. J Nepal Agric Res Counc 6:34–43. https://doi.org/10.3126/jnarc.v6i0.28112

Kaushik R (2013) Technology on custard apple processing and value chain: a step towards doubling the income of tribals. Manual RCA Alumni Assoc Udaipur 1(3):72–75

Li X, Yadav R, Siddique KHM (2020) Neglected and underutilized crop species: the key to improving dietary diversity and fighting hunger and malnutrition in Asia and the Pacific. Front Nutr 7:593711. https://doi.org/10.3389/fnut.2020.593711

Lock K, Pomerleau J, Causer L et al (2005) The global burden of disease attributable to low consumption of fruit and vegetables: implications for the global strategy on diet. Bull World Health Organ 83(2):100–108. https://www.ncbi.nlm.nih.gov/pmc/articles/PMC2623811/pdf/15744402.pdf

Longvah T, Deosthale YG (1991) Chemical and nutritional studies on hanshi (Perilla frutescens), a traditional oilseed from northeast India. J Am Oil Chem Soc 68:781–784. https://link.springer.com/article/10.1007/BF02662172

Longvah T, Ananthan R, Bhaskarachary K et al (2017) Indian food composition tables (1st edn). https://www.researchgate.net/profile/Thingnganing-Longvah/publication/313226719_Indian_food_Composition_Tables/links/589328db92851c5457486b19/Indian-food-Composition-Tables.pdf

Lyu Y, Ma S, Liu J et al (2022) A systematic review of highland barley: ingredients, health functions and applications. Grain Oil Sci Technol 5(1):35–43. https://doi.org/10.1016/j.gaost.2021.12.002

MEXT (2015) Standard tables of food composition in Japan (7th edn). https://www.mext.go.jp/en/policy/science_technology/policy/title01/detail01/1374030.htm

MOAD (2017) In: Joshi BK, Acharya AK, Gauchan D, Chaudhary P (eds) The state of Nepal's biodiversity for food and agriculture, 1st edn. Ministry of Agricultural Development

Mut Z, Erbaş Köse ÖD, Kardes YM et al (2023) Determination of Grain Yield and Nutrition Traits in Hull-less Barley Genotypes. Gesunde Pflanzen 75, 1369–1377. https://doi.org/10.1007/s10343-022-00767-3

Padulosi S, Mal B, King O et al (2015) Minor millets as a central element for sustainably enhanced incomes, empowerment, and nutrition in rural India. Sustainability 7(7):8904–8933. https://doi.org/10.3390/su7078904

Padulosi S, Meldrum G, Oliver King IED et al (2021) NUS what they are and why we need them. In: Padulosi S, Oliver King IED, Hunter D, Swaminathan MS (eds) Orphan crops for sustainable food and nutrition security: promoting neglected and underutilized species, 1st edn. Routledge, UK, pp 3–18

PAR (2018) Assessing agrobiodiversity: a compendium of methods (Mijatović D, Hodgkin T (eds)). Platform for Agrobiodiversity Research. https://static1.squarespace.com/static/5cd00f2c94d71a802807347f/t/5e431f667e88114bfbbee386/1581457302378/Assessing-Agrobiodiversity-A-Compendium-of-Methods-lowres.pdf

Pudasaini N (2021) Six farmers' varieties of neglected and underutilized crop species officially registered in Nepal. Himalayan Crops. https://himalayancrops.org/2021/06/01/six-farmers-varieties-of-neglected-and-underutilized-crop-species-officially-registered-in-nepal/

Raji R, Siril EA (2021) Genetic diversity analysis of promising Ceylon olive (Elaeocarpus serratus L.) genotypes using morphological traits and ISSR markers. Curr Plant Biol 26:100201. https://doi.org/10.1016/j.cpb.2021.100201

Sarananda KH, ThillakawardaneT U, Alexander B et al (2017) Production of health-friendly, ready-to-serve fruit drinks from under-utilized local fruits from Sri Lanka. Sri Lanka J Food Agric 3(2):37–48. https://doi.org/10.4038/sljfa.v3i2.50

Shailza Sharma L, Burark SS, Kaushik RA et al (2020) Prospects of custard apple value chain development in Rajasthan. Econ Aff 65(2):207–212. https://doi.org/10.46852/0424-2513.2.2020.11

Shrestha RK (2022) Mainstreaming farmers' variety. NUS promotion endeavours of the government of Nepal. In: 9th ITPGRFA governing body meeting

Siddique KHM, Li X, Gruber K et al (2021) Rediscovering Asia's forgotten crops to fight chronic and hidden hunger. Nat Plants 7(2):116–122. https://doi.org/10.1038/s41477-021-00850-z

Singh AK, Bharadwaj R, Singh IS et al (2014) Assessment of nutritional quality of developed faba bean (Vicia faba L.) lines. J AgriSearch 1(2):96–101. https://jsure.org.in/journal/index.php/jas/article/view/26/16

Singh RK, Sreenivasulu N, Prasad M et al (2022) Potential of underutilized crops to introduce the nutritional diversity and achieve zero hunger. Funct Integr Genomics 22:1459–1465. https://doi.org/10.1007/s10142-022-00898-w

Stubbendorff A, Sonestedt E, Ramne S et al (2022) Development of an EAT-Lancet index and its relation to mortality in a Swedish population. Am J Clin Nutr 115(3):705–716. https://doi.org/10.1093/ajcn/nqab369

Thrupp LA (2000) Linking agricultural biodiversity and food security: the valuable role of agrobiodiversity for sustainable agriculture. Int Aff 76(2):265–281. https://doi.org/10.1111/1468-2346.00133

Tripathi S (2016) Livelihood dependency of Garasia tribes utilizing non timber forest products in Abu Road area of Sirohi district in Rajasthan, India. Int J Adv Res 4(7):498–504. https://doi.org/10.21474/IJAR01/1144

Ugare R, Chimmad B, Naik R et al (2014) Glycemic index and significance of barnyard millet (Echinochloa frumentacae) in type II diabetics. J Food Sci Technol 51(2):392–395. https://doi.org/10.1007/s13197-011-0516-8

UNICEF (2021) Fed to fail. The crisis of children's diets in early life. https://data.unicef.org/wp-content/uploads/2021/09/Fed-to-Fail-FULL-REPORT-Child-Nutrition-Report-2021-FINAL.pdf

Von Grebmer K, Bernstein J, Wiemers M et al (2022) Global Hunger Index. Food systems transformation and local governance (1st edn). https://www.globalhungerindex.org/pdf/en/2022.pdf

Xu C, Cao Z, Yang H et al (2022) Association between the EAT-Lancet diet pattern and risk of type 2 diabetes: a prospective cohort study. Front Nutr 8:1153. https://doi.org/10.3389/fnut.2021.784018

Yadav RK, Gautam S, Palikhey E et al (2018) Agro-morphological diversity of Nepalese naked barley landraces. Agric Food Sec 7(1):86. https://doi.org/10.1186/s40066-018-0238-5

Chapter 5
Neglected and Underutilized Food Plants of India

Debabrata Panda, Prafulla K. Behera, Suraj K. Padhi, Aloukika Panda, and Jayanta K. Nayak

5.1 Introduction

One of the major difficulties that humanity is currently facing is to ensure food and nutrition security in a sustainable way (Caron et al. 2018). With the world's population growing and natural resources fast depleting, it became vital to investigate the possibility of utilizing indigenous/traditional plant resources. Agriculture is considered to be the world's most essential renewable resource in today's scenario (Thakur 2014). There are many plant genetic resources still categorized under unexplored and unexploited. Thus, researchers are underway focusing on using alternative genetic resources or underutilized plant varieties for numerous uses (Thakur 2014). Food crops or plants can be categorised into two major types such as (i) Staple crop and (ii) Underutilised or neglected crop, largely wild or semi-domestic in nature and are cultivated by the local people (Li et al. 2020). Underutilized crops are defined as "those species with under exploited potential for contributing to food security, health, income generation and environmental services" (Jaenicke and Hoeschle 2006). In general, "Underutilised plants" are those plant species that exist in extreme environmental conditions and threatened habitats as life support species, possess genetic potential to exist in difficult environments, and have nutritional value and industrial application for a number of purposes (Li et al. 2020).

There are around 30,000 plant species thought to be edible; but only 7000 species are farmed for human use (Garn and Leonard 1989). Only 20 plant species out of 7000 are capable of meeting 90% of the global population's dietary needs (Collins and Hawtin 1999). Food crops like rice, maize etc. consisting of 60% of the diet, are

D. Panda (✉) · P. K. Behera · S. K. Padhi · A. Panda
Department of Biodiversity and Conservation of Natural Resources, Central University of Odisha, Koraput, India

J. K. Nayak
Department of Anthropology, Central University of Odisha, Koraput, India

T. Ismail et al. (eds.), *Neglected Plant Foods of South Asia*,
https://doi.org/10.1007/978-3-031-37077-9_5

referred to as "staple crops" (Chivenge et al. 2015). The rest of the crop varieties are "underutilized", despite their ability to meet the global population's food and nutrition demands (Muthamilarasan et al. 2019). Underutilized crops may be unknown outside of a specific region, and there is a perception these underutilized crops mostly cultivated or farmed in rural areas. Some people find it socially unacceptable to consume these plants in some locations because they are considered poor people's food (Jaenicke and Hoeschle 2006). Although underutilized crops are important in their local communities, they often lack national importance, thus it requires national and global recognition, attention and appreciation (Mayes et al. 2011). Food, fiber, fodder, oil, and medicinal characteristics have all been traditionally employed for these crops. These species, on the other hand, have untapped potential for ensuring food security, nutrition, health, revenue generation, and environmental protection (Thakur 2014). Except these qualities they have the super ability of life sustainability in extremely difficult environmental circumstances and threatened ecosystem, as well as the genetic tolerance to survive in adverse environmental situations (Muthamilarasan et al. 2019).

There are different names available to describe the underutilized plants. They are often called as 'orphan', 'neglected', 'local', 'underused', 'lost', 'traditional', 'abandoned', 'minor', 'forgotten' and 'underdeveloped' (Thakur 2014). For many years, these plants were used to make traditional cuisines, but when more productive crops started to appear in farming systems, they started to get less attention. Neglected and underutilised species (NUS) remain underutilised as a result of the multifaceted issues they face, including agro-technical, economical, policy and institutional issues (Li et al. 2020). The marginalisation of NUS, which have a minimal function in the present farming and food systems, is the result of modernisation of agriculture, extensive monoculture, and the promotion of high-yielding plants (Li et al. 2020). Cultural stigmatisation of NUS due to the belief that, it is "food for the poor" or "poverty food" deters its production and consumption (Li et al. 2020). Further, NUS could not be incorporated into current diets since; there is dearth of an environment that made it possible for NUS to be produced, marketed, distributed, and consumed (Li et al. 2020). In the context of global climate change, it is the need-of-the-hour in the development of climate-resilience crop to sustain rapid population explosion and climate change. However, keeping in view the mandate to feed the ever-growing population, it is also critical to recognize and explore the potential of underutilized and neglected plant and varieties available around the India. Therefore the present chapter reviews some of the important underutilized and neglected plants of India, their potential role in livelihood, processed products, conservation strategies and development of valuable resources to meet ever increasing requirements of the global population during this recent era of climate change.

5.2 Concept of 'Neglected and Underutilized Plants'

Agro-biodiversity is necessary to sustainable agriculture of which neglected and underutilised plant species are vital elements (Li et al. 2020). The world is consisting of an average 300,000–500,000 different species of plants, of which 30,000 are thought to be edible, but only about 7000 edible plant species are found to be cultivated for human use (Muthamilarsan et al. 2019; Li et al. 2020). However, only 20 plant species meet the 90 percent of the world's population's dietary needs (Collins and Hawtin 1999; Muthamilarsan et al. 2019). The 60% of the world dietary requirements are fulfilled by major staple crops like rice, wheat and maize considered and more than 50% of the global population depend on these food crops. The remaining plant species are continued to be "underutilised" despite their capacity to supply food and nourishment to the world's population (Muthamilarsan et al. 2019). Hence, tens of thousands of different edible plant species are comparatively "underutilised and neglected" and might be employed to increase the global food need (Li et al. 2020). The underutilised or neglected foods may be referred as "the foods which are less available, less used or rarely utilized or region specific" (Thakur 2014). Indigenous ancient plant species that are still used to some extent in local, national, or even international populations are frequently underutilised or neglected species that have the potential to add more to the diversity of food sources than they now do (Thakur 2014). The underutilised crops of India has been described by Thakur (2014) and has illustrated basic characteristics and features of a plant should be consists of to be categorised as "Underutilised and neglected plant or crop" such as: the crop or plant must have (i) a scientific or ethno-botanical evidence of food value; (ii) week seed supply system; (iii) least cultivated as compared to other indigenous crops; (iv) traditional uses at local level; (v) high nutritional, therapeutics and medicinal value etc. (Fig. 5.1) (Thakur 2014). In this regard, there are so many neglected food plants available around the world such as millets, sweet potato, taro, bambara groundnut, cowpea, common bean etc. (Muthamilarsan et al. 2019).

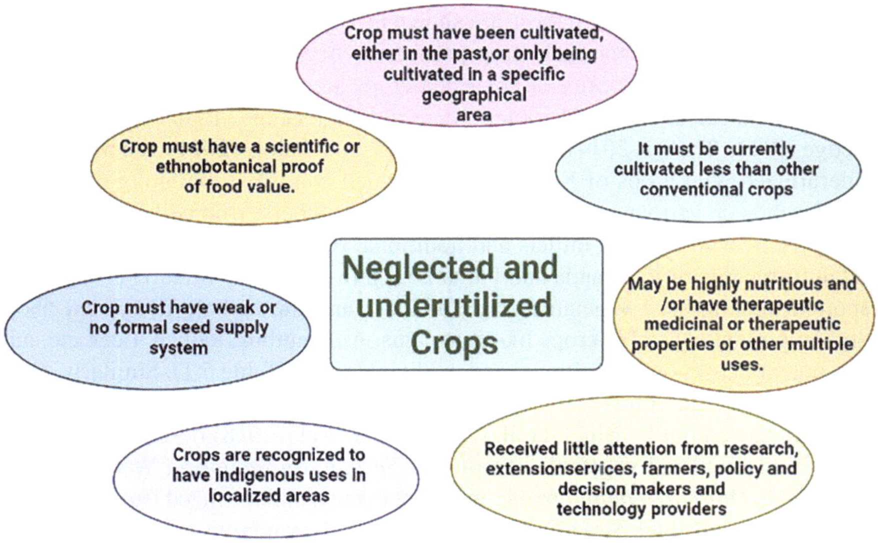

Fig. 5.1 Characteristic features of underutilised and neglected plants

5.3 Need to Explore Underutilized and Neglected Plants

Due to rising population pressure, India as a developing country faces difficulties such as food security, unemployment, and environmental deterioration. About 65 percent of the Indian population living in rural areas from which 85 percent of them depend on agriculture and agro-based pursuits for their livelihood (Williams and Haq 2002). Essential crops are expected to face considerable challenges in the near future and agriculture food diversification will be critical to ensure ever increasing wholesome and nutritious food demands. About 40% of the world's food consumption is accounted by just three crops, 'rice, maize and wheat' (Muthamilarasan et al. 2019). Underutilized plants have always played an important part in the subsistence and economy of poor people in developing countries. Despite their promise for nutritional diversity and micronutrient supplementation, like vitamins and minerals, they continue to receive little research and development attention. As a result, emerging countries such as India are urged to diversify their food exports by developing new resources (Thakur 2014).

5.4 Neglected and Underutilized Food Plants Available in India

In India, there are numerous crops that are underutilized. The majority of the species are unknown or poorly described and documented. The ethno-botanical characterization of underutilized and neglected species and its related data must be examined and scrutinized. The under-utilized plant species such as fruits, legumes, millets, vegetables, oil crops found in India are shown in Table 5.1. Marginal and wasteland in India are unsuitable for basic crop production because lack of soil quality and water resources. The majority of neglected crops are resistant to difficult agro-climatic circumstances and highly potent for establishment on marginal and wasteland (Hedge 2002; Thakur 2014). Recently, Panda and Palita (2021) reported about underutilized wild crops of Koraput region which is a tribal inhibiting dominated area depends on various wild crops like wild edible tubers, wild edible leaves, wild edible fruits, wild flowers, millets and traditional rice for their livelihood and food and nutritional security (Panda and Palita 2021). In a study by Madhvi et al. (2020) reported underutilised vegetable crops of India and the authors illustrated about some important vegetable crops like asparagus, Amaranthus, lettuce, Leek etc. and their potential uses and, nutritional and medicinal value (Table 5.1). Similarly, characterisation of underutilized vegetable crops and its diversity in North-East India has been carried out by Singh et al. (2012). Jena et al. (2018) described food features of underutilised vegetables crops like *Sesbania grandiflora*, Water spinach, Clove bean, *Dioscorea bulbifera* etc. In another study, underutilized fruits of Punjab subtropics such as Bael, *Syzygium cumini*, Jackfruit, Caronda etc. have been revealed by Bal (2003) in their exploration. Many underutilized fruit crops, such as ber,

Table 5.1 Neglected and underutilised plants of India and its uses and health benefits

Type of crop	Common name/ Scientific name/Family	Uses and health benefits	References
Fruits	Java plum *Syzygium cumini* Myrtaceae	All parts of the jamun tree are used in traditional medicine and in general, have astringent, antidiarrheal, antimicrobial, and anti-inflammatory properties. The bark is acrid, sweet, digestive, bowel astringent, anthelmintic, and used to treat sore throat, asthma, and ulcers. The fruit eliminates bad breath, biliousness, stomachic, astringent, diuretic, and antidiabetic properties.	Jena et al. (2018)
	Anola *Emblica officianalis* Euphorbiaceae	In Ayurvedic medicine, it is used to make Triphala and Chyavanprash. Aonla powder, which is superior to synthetic vitamin C in treating deficiencies, is also made from fruit. It has antimicrobial, antioxidant, anti-inflammatory, analgesic, antipyretic, adaptogenic, hepatoprotective, antitumor, and antiulcerogenic properties, as well as immunomodulatory properties.	Singh et al. (2019)
	Bael *Aegle marmelos* Rutaceae	All plant parts are valuable and have different medicinal properties. It has antifungal, hypoglycemic, wound healing, and insecticidal properties etc. Bael is also used to treat chronic diarrhoea, and to recover from respiratory ailments.	Singh et al. (2019)
	Jackfruit *Artocarpus heterophyllus* Moraceae	Ripe Jackfruit is used to make jams, jellies, chips, and other products. Fruit aids in the reduction of high blood pressure and the promotion of a healthy cardiovascular system. Furniture is made from trees. The root and its extract are used to cure asthma, skin diseases, diarrhea, and fever.	Tamang et al. (2021)
	Karonda *Carissa carandus* Apocynaceae	The pickles, jelly, chutney and sauces are made from fruits of karonda, among other things, dry fruits could be used in place of raisins. The fruits have traditionally been used to treat malaria, epilepsy, nerve disorders, pain and headache relief etc.	Singh et al. (2019)
	Carambola *Averrhoa carambola* Oxalidaceae	It has a variety of nutritional and medicinal applications such as antioxidant, anti-inflammatory, anti-tumor etc. The carambola can be eaten raw or juices, salads, or pickels made from it.	Lakmal et al. (2021)
	Fig *Ficus carica*; Moraceae	Fruits are eaten fresh or dried, and they are also used to make jam. Some people have been reported to consume mature fig inflorescences. Fig is valued for its laxative properties and is frequently used to treat skin infections.	Tamang et al. (2021)

(continued)

Table 5.1 (continued)

Type of crop	Common name/ Scientific name/Family	Uses and health benefits	References
	Custard apple *Annona squamosa* Annonaceae	The fruits are effective against dysentery, and the bark is vermifuge and astringent. Unani, a Southeast Asian medicine system, uses the leaves as an emmenagogue and aphrodisiac. Fruits high in potassium assist the body in regulating electrolytic balance, enhancing muscle growth, and improving waste processing ability.	Tamang et al. (2021)
	Loquat *Eriobotrya japonica* Rosaceae	It is transformed into a variety of products such as jam, jellies, juice, wine, canned fruit, dried slices, and so on. The loquat plant has a high medicinal value. Loquat extracts have traditionally been used to treat cough, chronic bronchitis (CB), inflammation, diabetes, and cancer.	Liu et al. (2016), Dhiman et al. (2020)
	Tamarind *Tamarindus indica* Leguminosae	It has many pharmacological properties such as hypolipidemic, analgesic, weight loss, antioxidant etc. Tamarind fruit Pulp is used to make a refreshing acid drink and syrup and other food products.	Singh et al. (2019)
	Assyrian plum *Cordia myxa* Boraginaceae	Unripe green organic products are commonly used to make vegetables and pickles. The leaves have been used in traditional medicine. It has antibacterial, calming, anti-hypersensitive, antiviral, and anti-neoplastic properties.	Kamboj et al. (2021)
	Phalsa *Grewia subinaequalis* Tiliaceae	Used to make pastries or invigorating soda pops such as squash, Sherbet, and so on. Phalsa unripe products are used to treat inflammation and blood problems, as well as to reduce fever. It has cooling tonic and sexual enhancer properties that combat thirst and sensation.	Kamboj et al. (2021)
	Wood apple *Feronia elephantum* Rutaceae	Value added products can be made from fruit pulp such as preserves, candy, sherbet, juice, chutneys, jam, jelly, and squash. It has curative properties for a variety of bone and joint diseases, bilious diseases, capillary bleeding prevention, cold, influenza, piles, dysentery, habitual constipation, and scurvy. The fruit was useful for diarrhea, dysentery, as well as liver and cardiac tonics.	Singh et al. (2019)
	Emetic nut *Randia duementorum* Rufiaceae	The different plant parts have been applied in Ayurvedic medicine, and fruits have been reported to be used in both medicine and food. The plant is used to treat food poisoning, induce vomiting, and to treat allergies and inflammation. Pharmaceutical activities include piscicidal, molluscicidal, antioxidant, anti-inflammatory, diabetic, and anti-hyperlipidemic effects.	Timalsina and Devkota (2021)

Vegetables	Amarnathus *Amarnathus spp.* Amaranthaceae	Important source of Vitamins, leaves are consists of carotene, minerals, phytochemicals are responsible for antioxidant properties, and aid in the suppression and prevention of cancer etc.	Madhvi et al. (2020)
	Lettuce *Lactuca sativa* Asteraceae	It contains high vitamins A, calcium and iron. It has antimicrobial characteristics, and aids in preventing cancer, heart disease, and controlling anxiety.	Madhvi et al. (2020)
	Celery *Apium graveolens* Apiaceae	It is commonly utilized in preparation of traditional medicine. It can help to prevent cardiovascular and kidney diseases, jaundice, gout etc. Celery leaf extracts in ethanol boost spermatogenesis and fertility. Celery has antifungal and anti-inflammatory properties. Bronchitis, asthenopia, asthma, chronic skin disorders, including psoriasis, vomiting, and fever can all be treated with the seeds of celery.	Kooti and Daraei (2017)
	Elephant Foot Yam *Amorphophallus campanulatus* Araceae	Used as a vegetable and also used to make pickles, chips, and other foods. It is also raised for its therapeutic qualities. Additionally, it has blood purifying qualities and is used to treat piles and rheumatism etc.	Ramalingam et al. (2010)
	Pointed gourd *Trichosanthes dioica* Cucurbitaceae	Leaves can be used as a vegetable and are particularly nutrient-dense. It is also good source of vitamin A and C, and minerals. Its juice is utilized in subacute cases of liver enlargement, baldness, and as a tonic.	Nadkarni (1996)
	Parsnip *Pastinaca sativa* Apiaceae	Fresh leaves and buds are also used as vegetables for food and soups in some countries. Parsnip is also used as an appetizer, digestive, and diuretic in some countries. It is also used to treat pulmonary diseases like coughing and liver disorders like obstructive liver disease and ascites.	Kenari et al. (2021)
	Rhubarb *Rheum rhabarbarrum* Polygonaceae	Rhubarb stems are eaten in pies and other dishes. Rhubarb is also used as a flavoring agent and is sometimes used to treat cold sores on the skin. Used as an antibacterial, anti-inflammatory, anti-fibrotic, and anticancer medicine, antitumor, gastrointestinal flora regulation, intestinal mucosal barrier protection, and fibrosis inhibition.	Xiang et al. (2020)

(continued)

Table 5.1 (continued)

Type of crop	Common name/ Scientific name/Family	Uses and health benefits	References
Cereals and millets	Proso millet *Panicum miliaceum* Poaceae	It is the most affordable manganese source. It helps in cholesterol reduction, best for bone health, and the prevention of breast cancer and cardiovascular disorders. Proso millets are more effective at inhibiting the proliferation of human liver cancer cells.	Rao et al. (2018)
	Kodo milet Paspalum scrobiculatum Poaceae	In comparison with rest of the millets, it contains the largest concentration of dietary fibers, making it an excellent food for diabetics and postmenopausal women who have high blood pressure, dyslipidemia, or other lipid-related issues may find it beneficial, beri-beri patients can use the stem as a poultice.	Rao et al. (2018)
	Buckwheat *Fagopyrum esculentum* Polygonaceae	Buckwheat is used to make muffins, crackers, cookies, pasta, noodles, cakes, bread, and pancakes. It prevents the genesis of disorders like excessive cholesterol, hypertension, atherosclerosis, and diabetes, and is utilized medically in many countries to stop capillary deformations.	Yilmaz et al. (2020)
	Pearl millet *Pennisetum glaucum* Poaceae	Flour is used in a variety of bakeries, aids in the treatment of migraines, eases asthmatic patients' respiratory issues, aids in the prevention of breast cancer and cardiovascular disease, and also has hypoglycemic effects.	Ambati and Sucharitha (2019), Sukar et al. (2020)
	Sorghum *Sorghum vulgare* Poaceae	More nutritious than rice, it lowers esophageal cancer risk, has anti-carcinogenic properties, and aids in the prevention of arthritis, cardiovascular diseases, obesity, malnutrition etc.	Dykes and Rooney (2006)
	Finger millet *Eleusine coracana* Poaceae	Good source of dietary fibers, amino acids, and vitamins. Finger millet extract or health beverage powder supplement can be used to treat diseases caused by a calcium deficiency, it is anti-microbial and antitumorigenic in nature, helps control intestinal cancer, regulates blood sugar stability.	Michaelraj and Shanmugam (2013), Tripathi et al. (2014)
	Foxtail milet *Setaria italica* Poaceae	Increases disease resistance ability, steadily releases carbohydrates into the blood without impairing body metabolism, fights diabetes and cardiovascular disease, anti-microbial, anti-tumorigenic, and maintaining dyslipidemia also aid in the body's cleansing.	Rao et al. (2018), Gupta et al. (2013)
	Barnyard millet *Echinochloa spp.* Poaceae	Barnyard millet is highly helpful in lowering blood sugar levels and because it is gluten-free; prevents against celiac illnesses.	Thakur and Tiwari (2019)
	Little millet *Panicum miliar;* Poaceae	Used in processed foods, baby foods, snacks, and other items, prevents cancer, diabetes, celiac disease, heart disease, and high cholesterol levels.	Rao et al. (2018)

Legumes	Grass pea *Lathyrus sativus* Fabaceae	Good source of protein, it is extremely suitable for human consumption and offers both aesthetic and therapeutic advantages, including the treatment of hypoxia and cardiovascular disease.	Grela et al. (2010), Rehman et al. (2022)
	Winged bean *Psophocarpus tetragonolobus* Fabaceae	Pods can be eaten unripe, cooked or raw vegetable, all of the plant parts of this bean are edible, can be consumed as bean curd and nut in snacks. Winged bean has also been reported as an effective remedy for smallpox.	Reddy (2019)
	Jack bean *Canvalia ensiformis* Fabaceae	In different regions of India, seeds are ingested. Jack bean seeds are eaten by different tribal's after cooking. Jack bean is a rich source of protein and carbohydrates.	Jena et al. (2018), Patel et al. (2016)
	Bambara groundnut *Vigna subterranean* Fabaceae	More than virtually any other bean, seed proteins are a good source of the crucial amino acid like methionine. It is believed that seeds can lower the risk of heart disease, helps to prevent colon cancer, treat some malignancies, and alleviate inflammatory conditions. Seeds are also used to treat diarrhoea and other gastrointestinal problems.	Kone et al. (2011)
	Lima bean *Phaseolus lunatus* Fabaceae	In order to treat fever, seeds are used in the diet. It exhibits anti-diabetic, anti-fungal, anti-proliferative, cysteine proteinase inhibition, Hypocholesterolemia, and trypsin and chymotrypsin inhibitory properties.	Chopra et al. (1986)
	Horsegram *Macrotyloma uniflorum* Fabaceae	Seeds are used as pulse and are useful for curing bronchial asthma, piles, abdominal lump, hiccup in causing and regulating perspiration, used as medicine in treatment of asthma, urinary diseases and kidney stones etc.	Bharatiya et al. (2015)
	Moth bean *Vigna aconitifolia* Fabaceae	Seeds can be cooked or fried; pods can be eaten after cooked, seed helps in treating fever, used as forage for animals, can be used in the preparation of *Papad*, *Bhujia*, *Vada*, etc.	Brink and Jansen (2006)
	Velvet bean *Mucuna pruriens* Fabaceae	Roasted seeds used to make a coffee substitute, cooked as a vegetable, used in various medicinal purposes and externally applied to ulcers.	Lampariello et al. (2011)
	Adzuki beans *Vigna angularis* Fabaceae	Dried seeds used as food Beans may be popped like corn, can be used as sugar candy substitute. Flour is used for making shampoos and face creams. Used in the treatment of kidney trouble, constipation, boils, leaves helps in lowering fever.	Phogat and Sharma (2000)
Oil crops	Niger *Guizotia abyssinica* Asteraceae	Mainly used for oil extraction and its oil in contraception and syphilis treatment. Cough is treated with a mixture of Niger sprouts and garlic. Niger seeds are also used as poultices, helps in relieving pain, itching, boils and curing rheumatoid arthritis etc.	Ramadan (2009)
	Safflower *Carthamus tinctorius* Asteraceae	It is a multipurpose crop that can be used as a vegetable, cut flower, fodder, medicinal plant, textile dye crop, and safflower oil is used in the manufacture of fine paints. Safflower leaves can also be used to treat fractures, contusions, and strains as well as to improve blood circulation and reduce mucus.	Emongor et al. (2015)

gooseberry, jamun, tamarind and others are in high demand, but are not popular among various farmers (Thakur 2014). Now-a-days, millet is considered as underutilized crops such as finger millet, kodo millet, proso millet, foxtail millet (*Setaria italica*), and little millet etc. (Gupta et al. 2013; Rao et al. 2018; Thakur and Tiwari 2019; Muthamilarasan et al. 2019; Panda et al. 2022) (Table 5.1). The importance of neglected legumes and Pulses such as Grass pea (Grela et al. 2010; Rehman et al. 2022), Winged bean (Reddy 2019), Bambara groundnut (Kone et al. 2011), Moth bean (Brink and Jansen 2006) etc. and their nutritional value and health benefits have been presented in Table 5.1. Further, oil seed crops like Niger (*Guizotia abyssinica*) are used for oil. The Niger oil is used for contraception and syphilis treatment (Ramadan 2009).

5.5 Importance of Neglected and Underutilized Plants in Food and Nutritional Security

The primary force underlying the selection, characterisation and improvement of staple crop was domestication, which is thought to have started about 8–10 million years before. The minor underused plant species which were grown only in a few places around the world, had nor undergone substantial domestication (Harlan 1992; Muthamilarsan et al. 2019). However, these minor underutilised plants found having superior extraordinary properties and many advantages over staple food crops. There are many neglected underutilized crops available in India and most of them are not documented properly and not well known (Thakur 2014). Millets including the minor millet are one amongst the underutilised crops having higher nutritional value and health benefits. The small millets viz., finger millet, little millet etc. contain a variety of essential amino acids, and are rich in minerals like calcium, zinc, iron (Gupta et al. 2017; Ceasar et al. 2018; Sood et al. 2019). Chandel et al. (2014) reported that "Rau-8" cultivar of foxtail millet consists of 13% of seed protein which is about three to five fold more than other crops such as rice, maize etc. The resistant starch contents of yellow and white foxtail millet are about 13.35% and 14.56%, respectively which suggest millets to be used as efficient dietary ingredient to help manage postprandial hyperglycemia in diabetics and also lower their plasma cholesterol and triglyceride levels (Bangoura et al. 2012; Muthamilarsan et al. 2019).

5.6 Value Addition and Some Processed Product from Underutilized Plants

Value addition is the process of altering a product from its actual state to a more profitable state. Value addition to agricultural products is the technique of improving the financial value and consumer interest in agricultural output (Srivastava et al. 2017). Different value addition technology like processing, protection techniques, dehydrating, drying, freezing, packing, and labelling can be assigned to the agricultural procedure to improve its value and identity (Srivastava et al. 2017). The total value of any agricultural strategy is enhanced by applying some of the post-harvesting processing operations instead of selling as such post-harvest. Value addition to such crops would not only facilitate the economic status of people but also improves the health and nutrition status of people consuming such crops (Thakur 2014). Value addition through processing and protection has to be considered as a promising approach to reduce post-harvest losses of nourishing and healthful underutilized food crops and for serving in the off season. There is a huge prospective for processing and value addition to the minor fruit crops. Minor fruits not only have acceptable sensory profiles but are also treasure troves of many health-promoting compounds. In addition to economic gains, adding value to these fruits would also benefit consumers' health and nutritional status (Srivastava et al. 2017). The underutilized plants can be processed to products with better sensory appeal for the consumers. The basic processed products produced from underutilized crops being used by people in India were jam, RTS-fruit drinks, juices, syrups, candies, pickles, etc. which can improve economical sustainability and ensure better livelihood standards (Singh et al. 2008; Thakur 2014). Bael pulp can be used to make a variety of processed products like nectar, squash, toffee powder, and so on. Ber can be made into murrabha, candies, jam, and a ready-to-drink beverage (Srivastav et al. 2017). Jamun fruits are used in making of high quality jelly; jam etc. The seeds can be ground into a powder that can be used to treat diabetes (Thakur 2014). Figure 5.2 explains few processed products that can be obtained from underutilized plants.

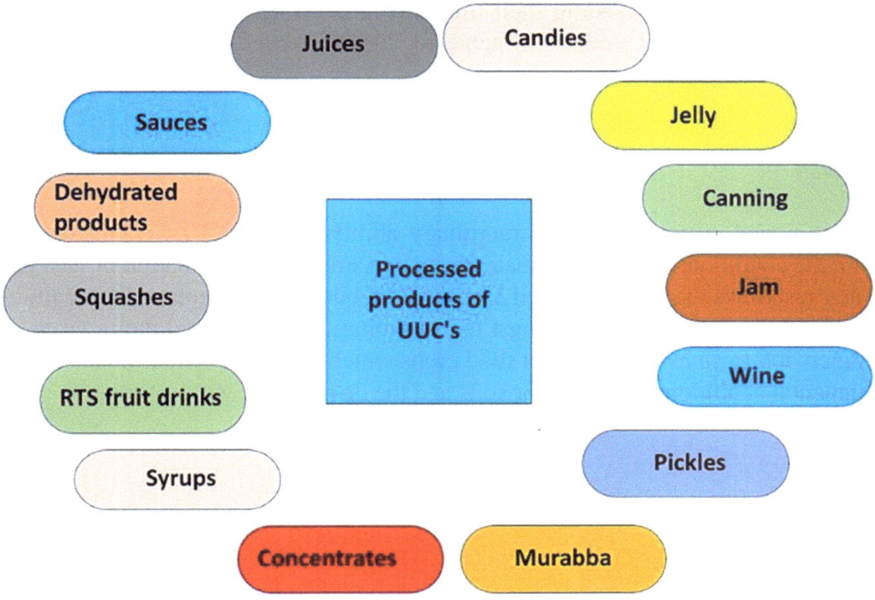

Fig. 5.2 Processed products from underutilised plants

5.7 Potential Role of Underutilized Plants

5.7.1 Food Security and Nutritional Value

Neglected crops are considerably rich in nutrients and are highly modified to low intake agriculture. The sudden effects by the decline of the species can affect the food security and wellbeing of the needy people. Promoting production of neglected or minor food crops thus can bring promising changes in food and nutritional security of the vulnerable populations. Paying more attention to this sector guarantee dietary diversification, adoptability and sustainability of neglected food crops in existing food system, and improved nutrition and health indices of the peoples at risk of food insecurity. Neglected plant foods are also considered helpful in fighting micronutrient deficiencies or the 'hidden hunger' as well as other nutritional deficits in rural people of developing nation (Mayes et al. 2011).

5.7.2 Increased Income for Rural People

In existing world scenario of climate change and food security, the pressure for food diversity and originality from the consumers is evolving new market and destinations for the neglected crops. These market opportunities produce additional income to the rural households' especially poor farming communities thriving on resilient food crops in existing climatic stress. These crops are considered more advantageous over the major staple crops. The modern technology helps to modify the quality of the crops and extending shelf life, thus farmers can be benefited by market these species and their products (Singh et al. 2008; Thakur 2014).

5.7.3 Ecosystem Stability

Underutilized crops have the extraordinary ability to stand against unfavourable climatic conditions. Hence if climate change occurs or the degradation of land and water resources then underutilized crops will be boon for conventional agriculture. It will further built interest amongst the agricultural scientists and the progressive farmers too to generate and adopt food crops which can thrive better in harsh environmental conditions such as poor soils and the droughts.

5.7.4 *Cultural Biodiversity*

Since ancient times, many neglected and underutilized crop species are playing crucial role in religious rituals and tradition. Neglected food plants are intimate part of local culture and traditions. They are also used in protecting cultural integrity associated with food habits, health practices and public interaction.

5.8 Constraints in Utilization and Market Value of Underutilized Plants

Overall, a variety of limitations, which are outlined below, contribute to the slow growth and low popularity in the beneficial development and use of underutilised crops (Thakur 2014):

- Lack of awareness about the nutritional and health benefits, and medicinal significance of underutilized crops among the farmers.
- Absence or little research on underutilized crops.
- There is a shortage of suitable seeds and planting material.
- Lack of producer interest due to low yield.
- A shortage of innovative technology in enhancing productivity of NUS.
- Limitations in the post-harvest management practices.
- Lack of marketing supports and limitation in infrastructure facilities.
- Bad recognition of NUS in horticulture publicity programs.
- Improper institutional agreements and the uncommon role of financial institutions played in the establishment of horticulture- and agro-based industrial units.
- Lack of interest by researchers, and agriculturalist.

5.9 Strategies for the Development of Underutilized Food Plants

To restrict and control over-exploitation of natural resources, domestication of wild species should be stimulated. Supports are needed for popularisation and promotion of the neglected food crops and establishing marketing networks in local and international market. NUS are high in nutrition and hold ability to stand against unfavourable climatic conditions. National strategies are desired to be developed aimed at concentrating or supporting preservation and utilization of existing species of underutilized crops. Indigenous knowledge on key characteristics of the neglected food crops may be brought in line with the botanical studies to generate the species level database of NUS. Likewise policies should be developed at regional and national levels to improve production of NUS with more resilient and productive varieties, satisfying local requirements, stimulating household markets, and support income generation. The immediate development of infrastructure facilities for transport, communications needs to be carried out. The profit and integrity of these crops must be improved for commercial exploitation (Sharma 2003; Thakur 2014). The agencies can organize awareness camps, exhibition at local, regional, sub-national and national level. Similarly, better exploitation of electronic media like TV, radio and printed media like newspapers can play a significant role in establishing awareness on NUS among farmers and the consumer market.

5.10 Improving Resources for Conservation of Neglected and Underutilized Plants

Several native crops are underutilized or neglected in India, and are reported in various researches. Some crops like bael, koranda, phalsa are native to India but they are hidden due to some reasons. With their outstanding flavours and impressive sensory appeal, many of the NUS are also considered highly therapeutic and nutritious. For this reason, they might earn a great repute in food and pharma industry, and the future food market. Since last decade, tremendous response has been generated from the consumers for availability and supply of healthful and nutritious foods. Untapped underutilised plant resources might therefore be considered crucial in supplying consumers with wholesome, delightfully flavoured, and aesthetically appealing natural foods with significant medicinal benefits (Thakur 2014). Now-a-days many species are under threat of becoming extinct; therefore they should be conserved before wiped out from the face of earth. The resources of some crops are well collected while other crops somewhat have yet to be executed (Arora 1994). The main component of the conservation of underutilized crop resources are identification and collection of live samples of neglected crop varieties, research on multiplication of the seeds for the usefulness of upcoming generations, promoting awareness among the stakeholders on nutritional and health promoting properties of the neglected plants, performing field exploration and investigations aimed at identifying novel techniques for improved cultivation neglected and underutilised plant (Thakur 2014).

5.11 Conclusion and Future Prospects

Underutilized and neglected plants are highly nutritious foods and hold the ability to withstand against unfavourable climatic situations. This can be a blessing and in interest of all stakeholders such as those of growers, consumers, and environmentalists, if utilized in a proper way. Historically, many underutilized crops once cultivated widely but now-a-days, they are declining due to many factors such as agronomic, genetic, economic, and cultural modifications. NUS could attract least interests of the farmers and the consumers because they are not competitive with conventional food plant species in same farming system. Such a depletion of neglected plant species resources may deteriorate the genetic base and prevent the useful traits in crop improvement. In spite of their identified significance, least utilisation of neglected plants is linked to limited knowledge on nutritious and medicinal value, lack of availability of planting material, and scarcity of information on production techniques. In these circumstances, there arise the needs to develop strategy on genetic resources assessment, supervision, application, and enhancement of neglected food plants production to better secure food and nutrition of the future generations, more specifically in the developing nations. Furthermore, a handsome share of NUS in agriculture can also play vital role in the national economy of India. The Government of India should take necessary steps towards highlighting and promoting production of underutilized food plants.

Conflicts of Interest All authors read and approved the final manuscript and declare that they have no conflict of interest.

References

Ambati K, Sucharitha KV (2019) Millets- review on nutritional profiles and health benefits. Int J Recent Sci Res 10(7):33943–33948

Arora RK (1994) Promoting conservation and use of tropical fruit species in Asia. Proceedings of expert consultation on tropical fruit species of Asia, Kuala Lumpur, Malaysia, pp 19–30

Bal JS (2003) Genetic resources of underutilized fruits in Punjab subtropics. Acta Hortic 623:325–331

Bangoura ML, Nsor-Atindana J, Ming ZH, Wei P, Mothibe KJ, Xue ZK (2012) Starch functional properties and resistant starch from foxtail millet [*Setaria italica* (L.) P. Beauv] species. Pak J Nutr 11:821–830

Bhartiya A, Aditya JP, Kant L (2015) Nutritional and remedial potential of an underutilized food legume horsegram (*Macrotyloma uniflorum*): a review. J Anim Plant Sci 25:908–920

Brink M, Jansen PCM (2006) *Vigna aconitifolia* (Jacq.) Maréchal. In: Brink M, Belay G (eds) PROTA (Plant Resources of Tropical Africa). Wageningen

Caron P, Ferrero y de Loma-Osorio G, Nabarro D, Hainzelin E, Guillou M, Andersen I, Arnold T, Astralaga M, Beukeboom M, Bickersteth S, Bwalya M, Caballero P, Campbell BM, Divine N, Fan S, Frick M, Friis A, Gallagher M, Halkin J-P, Hanson C, Lasbennes F, Ribera T, Rockstrom J, Schuepbach M, Steer A, Tutwiler A, Verburg G, (2018) Food systems for sustainable development: proposals for a profound four-part transformation Agronomy for Sustainable Development 38(4). https://doi.org/10.1007/s13593-018-0519-1

Ceasar SA, Maharajan T, Krishna TPA, Ramakrishnan M, Roch GV, Satish L, Ignacimuthu S (2018) Finger millet [*Eleusine coracana* (L.) Gaertn.] Improvement: current status and future interventions of whole genome sequence. Front. Plant Sci 9:1054

Chandel G, Meena RK, Dubey M, Kumar M (2014) Nutritional properties of minor millets: neglected cereals with potentials to combat malnutrition. Curr Sci 107:1109–1111

Chivenge P, Mabhaudhi T, Modi AT, Mafongoya P (2015) The potential role of neglected and underutilized crop species as future crops under water scarce conditions in Sub-Saharan Africa. Int J Environ Res Public Health 12:5685–5711

Chopra R, Nayar SL, Chopra I (1986) Glossary of Indian medicinal plants. Council of Scientific and Industrial Research, New Delhi

Collins WW, Hawtin GC (1999) Conserving and using crop plant biodiversity in agroecosystems. Biodiversity in agroecosystems. CRC Press, London, pp 215–236
Dhiman A, Suhag R, Thakur D, Gupta V, Prabhakar P (2020) Current status of Loquat (*Eriobotrya Japonica* L): bioactive functions, preservation approaches, and processed products. Food Rev Int. https://doi.org/10.1080/87559129.2020.1866007
Dykes L, Rooney LW (2006) Sorghum and millet phenols and antioxidants. J Cereal Sci 44(3):236–251
Emongor V, Oagile O, Kedikanetswe B (2015) Effects of plant population and season on growth and development of safflower (*Carthamus tinctorius* L.) as an ornamental plant. Acta Hortic 1077:35–45
Garn SM, Leonard WR (1989) What did our ancestors eat? Nutr Rev 47:337–345
Grela ER, Rybinski W, Klebaniuk R, Matras J (2010) Morphological characteristics of some accessions of grass pea (*Lathyrus sativus* L.) grown in Europe and nutritional traits of their seeds. Genet Resour Crop Evol 57:693–701
Gupta A, Sood S, Agrawal PK, Bhatt JC (2013) Under-utilized food crops of Himalayan region: utilization and prospective. In: Newer approaches to biotechnology, pp 101–120
Gupta SM, Arora S, Mirza N, Pande A, Lata C, Puranik S (2017) Finger millet: a "certain" crop for an "uncertain" future and a solution to food insecurity and hidden hunger under stressful environments. Front Plant Sci 8:643
Harlan JR (1992) Crops and man. American Society of Agronomy, Madison
Hedge NG (2002) Promotion of underutilised fruit crops. In: Haq N, Hughes A (eds) Fruits for the future in Asia. Proceedings of the regional consultation meeting, Bangkok, Thailand. International Centre for Underutilised Crops, University of Southampton, Southampton, UK, pp 45–53
Jaenicke H, Hoeschle-Zelodon I (2006) Strategic frame work for global research and development of underutilized plant species: a contribution to the enhancement of indigenous vegetables and legumes. International Conference on Indigenous Vegetables and Legumes 752:103–110
Jena AK, Deuri R, Sharma P, Singh SP (2018) Underutilized vegetable crops and their importance. J Pharmacogn Phytochem 7(5):402–407
Kamboj V, Tripathi S, Joshi U, Tewari V (2021) Underutilized fruits crops and technologies for meeting their market needs. J Postharvest Technol 8:64–76
Kenari HM, Kordafshari G, Moghimi M, Eghbalian F, TaherKhani D (2021) Review of pharmacological properties and chemical constituents of *Pastinaca sativa*. J Pharmacopunct 24(1):14–23
Kone M, Paice A, Yaya T (2011) Bambara groundnut [*Vigna subterranea* (L.) Verdc: (Fabaceae)] usage in human health. In: Nuts and seeds in health and disease prevention. Academic, pp 189–196
Kooti W, Daraei N (2017) A review of the antioxidant activity of Celery (*Apium graveolens* L). J Evid Based Complement Alternat Med 22(4):1029–1034
Lakmal K, Yasawardene P, Jayarajah U, Seneviratne SL (2021) Nutritional and medicinal properties of Star fruit (*Averrhoa carambola*): a review. Food Sci Nutr 9(3):1810–1823
Lampariello LR, Cortelazzo A, Guerranti R, Sticozzi C, Valacch G (2011) The magic velvet bean of Mucuna pruriens. J Trad Complement Med 2:331–339
Li X, Yadav R, Siddique KHM (2020) Neglected and underutilized crop species: the key to improving dietary diversity and fighting hunger and malnutrition in Asia and the Pacific. Front Nutr 7:593711
Liu Y, Zhang W, Xu CJ, Li X (2016) Biological activities of extracts from Loquat (*Eriobotrya japonica* L): a review. Int J Mol Sci 17(12):1983
Madhvi, Sharma D, Chauhan A, Banyal SK, Divya (2020) Underutilized vegetable crops: a lost treasure. Int J Curr Microbiol Appl Sci 9(04):2172–2178
Mayes S, Massawe F, Alderson PG, Roberts JA (2011) The potential of underutilized crops to improve security of food production. J Exp Bot 63(3):1075–1079
Michaelraj PSJ, Shanmugam A (2013) A study on millets based cultivation and consumption in India. Int J Mark Financial Serv Manag Res 2(4):49–58

Muthamilarasan M, Singh NK, Prasad M (2019) Multi-omics approaches for strategic improvement of stress tolerance in underutilized crop species: a climate change prospective. Adv Genet 103:0065–2260

Nadkarni AK (1996) Indian materiamedica, vol 3, 3rd edn. Popular Prakashan, Mumbai, p 1236

Panda D, Behera PK, Panda A, Nayak JK (2022) Advancement in Omics technologies for enhancing abiotic stress tolerance in finger millet. In: Roychoudhury A et al (eds) Omics approach to manage abiotic stress in cereals. Springer Nature, Singapur Pte, pp 569–574

Panda D, Palita SK (2021) Potential of underutilized wild crops in Koraput, Odisha, India for improving nutritional security and promoting climate resilience. Curr Sci 120(6):989–996

Patel R, Rkr S, Tyagi V, Raju P (2016) Nutritional evaluation of *Canavalia ensiformis* (Jack bean) cultivated in North East region of India. Int J Botan Stud 1(6):18–21

Phogat BS, Sharma GD (2000) Underutilized food crops: their uses, adaptation and production technology. National Bureau of Plant Genetic Resources, New Delhi

Ramadan MF (2009) Niger seed oil. In: Gourmet and health-promoting specialty oils. AOCS Press, pp 283–289

Ramalingam R, Hima BK, Bindu MB, Ravinder NA, Banji D (2010) Phytochemical and anthelmintic evaluation of corm of Amorphophallus campanulatus. Int J Pharma Bio Sci 4:1–9

Rao BD, Ananthan R, Hariprasanna K, Bhatt V, Rajeswari K, Sharma S, Tonapi VA (2018) Nutritional and health benefits of nutri cereals. Nutri hub TBI, ICAR-Indian Institute of Millets research (IIMR), Rajendranagar, Hyderabad

Reddy KJ (2019) Nutritional uses of three underutilized legume vegetables. Agric Food e-Newslett 1:16–19

Rehman U, Bahar FA, Saad AA, Bhat MA, Singh L, Mahdi SS, Mir MS, Bashir M, Bhat TA, Palmo T (2022) Trends, challenges and prospects of underutilized food crops under temperate ecologies: a review. Pharma Innov J 11(2S):503–512

Sharma DV (2003) Transfer of technology for increasing the scope of underexploited horticultural crop. Department of Horticulture, College of Agriculture, MPUAT, Rajasthan, pp 313–320

Singh A, Singh S, Saroj PL, Mishra DS, Singh P, Singh R (2019) Aonla (*Emblica officinalis*) in India: a review of its improvement, production and diversified uses. Indian J Agric Sci 89(11):1773–1781

Singh D, Wangshu L, Prahalad VC (2008) Processing and marketing feasibility of underutilized fruit species of Rajasthan. IAMO Forum, pp 1–12

Singh SJ, Batra VK, Singh SK, Singh TJ (2012) Diversity of underutilized vegetable crops species in NorthEast India with special reference to Manipur: a review. NeBIO 3(2):87–95

Sood S, Joshi DC, Chandra AK, Kumar A (2019) Phenomics and genomics of finger millet: current status and future prospects. Planta 250(3):731–751

Srivastava A, Bishnoi SK, Sarkar PK (2017) Value addition in miinor fruits of Eastern India: an opportunity to generate rural employment. In: Dutta AK, Mondal B (eds) Fruits for livelihood: production technology and management practices. Agrobios (India), Jodhpur, pp 395–417

Sukar KAO, Abdalla RI, Humeda HS, Alameen AO, Mubarak EI (2020) Effect of pearl millet on glycemic control and lipid profile in streptozocin induced diabetic wistar rat model. Asian J Med Health Sci 18(3):40–51

Tamang A, Subba SK, Kumar S, Chhetri S (2021) Wild edible and minor fruits of Odisha. J Pharm Innov 10:609–613

Thakur M (2014) Underutilized food crops: treasure for the future India. Food Sci Res J 5(2):174–183

Thakur M, Tiwari P (2019) Millets: the untapped and underutilized nutritious functional foods. Plant Arch 19(1):875–883

Timalsina D, Devkota H (2021) *Catunaregam spinosa* (Thunb.) Tirveng: a review of traditional uses, phytochemistry, pharmacological activities, and toxicological aspects. Evid Based Complement Alternat Med 2021:Article ID 3257732

Tripathi J, Gupta A, Prasad R, Puranik V (2014) Enhancing micronutrient content of beverage powder by incorporating malted finger millet. Indian J Community Health 26(2):339–342
Williams J, Haq N (2002) Global research on underutilized crops: an assessment of current activities and proposal for enhanced cooperation. International Centre for Underutilized Crops, Southampton. ISBN 92-9043-545-3
Xiang H, Zuo J, Guo F, Dong D (2020) What we already know about rhubarb: a comprehensive review. Chin Med J 15:88
Yılmaz H, Nurcan Y, Meric C (2020) Buckwheat: a useful food and its effects on human health. Curr Nutr Food Sci 16:29–34

Chapter 6
Underutilized Plant Foods of Pakistan

Majid Hussain, Marium Nazir, Muhammad Azam, and Muhammad Waseem

6.1 Introduction

Since ages, mankind has over-relied on the plant-based staple food crops for thousands of years to satiate hunger and meet the economic needs. Earlier literature reveals, of the naturally existing 300,000 plant species only 5000 have been used by humans as food or feed (Teye et al. 2020). Globally, there is an over-dependence on to the use of staple food crops consisting of wheat, maize, rice and potato. The available statistical data on dietary intake of these staple foods unveils the provision of about 60% energy of the total energy requirements. Likewise, despite of a wide dietary diversity of the plant food species majority of the global population is being fed by the only 20–30 plant food species. These crops are the basis of the food systems and human existence, contributing a significant amount of energy (Grote et al. 2021). However, given the current state of the world, it is pertinent to consider both the increase in global population and un-avoidable impacts of climate change on the production of staple food crops. Recent scientific literature has ranked Pakistan at 12th position in the list of countries affected by climate change, drought and land infertility and less agricultural production (Syed et al. 2022) which could be linked with the geographical position of Pakistan. The rising temperature due to greenhouse gases is an alarming threat for majority of the staple food crops' productivity.

M. Hussain (✉) · M. Nazir
Faculty of Food Science & Nutrition, Bahauddin Zakariya University, Multan, Pakistan
e-mail: mhussain@bzu.edu.pk

M. Azam
National Institute of Food Science & Technology, University of Agriculture, Faisalabad, Pakistan

M. Waseem
Department of Food Science and Technology, Faculty of Agriculture and Environment, The Islamia University, Bahawalpur, Pakistan

T. Ismail et al. (eds.), *Neglected Plant Foods of South Asia*,
https://doi.org/10.1007/978-3-031-37077-9_6

These un-avoidable issues enlightens the significance of the neglected plant foods with better cropping patterns, nutrition and health promoting capacities (Syed et al. 2022). The strategic approaches to uplift the use of the most neglected plant foods will not only improve the nutritional status of vulnerable population groups, will also create appreciable opportunities for sustainable agriculture.

The underutilized plant species are the plant food commodities which has been historically used as food, feed, fodder, and/or medicine. Earlier studies have portrayed the neglected plant food crops as pseudo-cereals and cereals, grains, legumes, roots, tubers, greens, and fruit species consisting of peanuts, yams, lablab beans, barley, oats, quinoa, and yam. These plant food species may contribute towards improving the food and health related challenges in several ways; including (i) reducing the over -dependency of consuming the staple crops only, (ii) enhancing food quality, (iii) improving the dietary diversification and (iv) improving livelihood of the low-income generating population groups (Grote et al. 2021). However, less preference is given to these plant food species owing to the presence of some anti-nutrients bland flavor, and/or unpleasant taste. However, these challenges could be mitigated using the natural processing techniques such as fermentation, blanching and soaking (Samtiya et al. 2020). Therefore, there is dire need to focusing on the research for enhancing the productivity of underutilized plant food crops, and their food applications to develop value added novel food products for overcoming the several health challenges such as malnutrition, food insecurity and health maladies (Teye et al. 2020).

6.2 Neglected Food Plants of Pakistan

6.2.1 Cereals and Pseudo Cereals

(a) Barley

Barley (*Hordeum vulgare* L.) is a vital edible cereal crop which belongs to the family Poaceae. The plant is grown in the temperate regions of the globe. The barely crop belongs to Kharif Crop cultivated between May and Oct under different conditions of soils, including sodic, light and saline soil (Abebaw 2021). However, barely crop is also favored by moderately fertile, neutral to saline soils ranging from thick loam to sandy soils. Barley is nutritionally rich grain crop cultivated around the globe and in Pakistan primarily due to its remarkable nutritional potential and health promoting features (Naheed et al. 2015). The nutritional composition of barley exhibits the presence of health featuring dietary fibers i.e., 17.0%, fats: 2.3%, carbohydrates: 73.0%, protein 12%, sodium: 12 mg, potassium: 452 mg and energy 354 kcal/100 g (Abebaw 2021). Worldwide, about 30% of the total global production of barley is used in preparation of the fermentable sugars as ingredients for beer and distilled beverages, while the remaining 70% is used as feed for animals. Barley based bread, stews, and soups are also developed in different cultures. Earlier literature reports the immense potential of barely against several health malignancies such as gallstones and gallbladder related issues (Matin et al. 2018; Abebaw 2021). Barely, also embodies antinutrients such as phytates, tannins, protease inhibitors which could be linked with certain health issues of food digestion and nutrients absorption (Matin et al. 2018). Also, the retrospective studies have reported the consumption of barely containing the anti-nutrients may result in development of stomach gas, bloating, and/or feeling of fullness. The load of anti-nutrients in barely could be reduced by adopting some viable techniques such as soaking and sprouting (Samtiya et al. 2020).

(b) Sorghum

Sorghum (*Sorghum bicolor*), a cereal grain with starchy seeds often called as gigantic millet, Indian Millet, Milo, Durra, Orshallu. From the time of flowering to physiological maturity, it takes around 40–45 days to grow. The growth period of sorghum in Pakistan range from March to August and June and July for feed and grain, respectively. Rainfed crops are seeded at the beginning of monsoon season. The ideal soils for cultivating sorghum contain between 10–30% clay. Warm-weather crops like sorghum require high temperatures for proper germination and growth. The nutritional composition of sorghum reveals the presence of proteins, carbohydrates, and starches between 8–15%, 5–15%, and 32–57%, respectively. However, the elemental composition of sorghum elucidates the presence of the iron (41–127 mg/kg), zinc (14–35 mg/kg), phosphorus (1498–3797 mg/kg), calcium (207–447 mg/kg), potassium (1150–2569 mg/kg), manganese (10–24 mg/kg), sodium (12–54 mg/kg), and magnesium (750–1506 mg/kg), respectively (Abah et al. 2020).

Sorghum grain powder have been utilized in development of numerous supplemented food products such as leavened bread, couscous, dumplings, and fermented and non-fermented porridges as appetising recipes. It is the grain of choice for brewing traditional African beers (Abah et al. 2020). Sorghum is used as a common ingredient of traditional recipes to remediate anaemia in the developing countries (Adhikari et al. 2017). Sorghum contains antinutrient factors as tannins, cyanogenic glucoside, phytic acid, trypsin inhibitors, and oxalates. Anti-nutrients levels in the sorghum grains are decreased by adopting different food processing techniques such as boiling, soaking, grinding, roasting, and fermentation (Gunawan et al. 2022).

(c) Millets

Millet (*Panicum miliaceum)*, is a cereal grain which belong to family Poaceae. Millet are cultivated all around the globe and is widely consumed in the developing countries including Africa and Asia (Amadou et al. 2013). In Pakistan, the millet crop is cultivated in kharif/monsoon season. It is cultivated in most areas south of latitude that receive more than 800 mm of precipitation. Reghional distribution for the growth of the millet reveals Gujrat, Gujranwala, Chakwal, Mianwali, and Bahawalnagar in Punjab; Hyderabad, Khairpur, and Dadu in Sindh; and Lorali, Khuzdar, and Sibbi in Baluchistan, respectively (Ullah et al. 2017). Nutritional composition of millet consist of 65–75% carbohydrates, 5.6–12% proteins, 2–5% fats, and 15–20% dietary fibres (Amadou et al. 2013). Traditional fermented foods and drinks produced from millet or millet based composite flours consist of koko (millet porridge), fura, mangishi, jandh, uji, burukutu, kunu-zaki, ogi, and bushera (corn and sorghum). Similarly, several non-fermented products of sorghum include Roti, Masvusvu, and Dambu (Adebiyi et al. 2018). Anti-nutrients found in millets frequently include polyphenols, phytic acid, and tannins. These anti-nutrients interfere with macro-molecules and other food constituents, whether they are synthetic or organic, affecting the ingestion, absorption, and digestion. Millet seeds can be soaked in water to reduce the amount by 28% of phytates (Amadou et al. 2013).

(d) Oat

Oat (*Avena sativa* L.), is a cereal crop grown in mountainous and temperate regions of the globe (Mao et al. 2022). Oats is used in the development of forage which however is grown as a primary source of winter and spring pasture from the lowlands, both irrigated and rainfed, to high altitudes (1000–2300 m). In Pakistan, oats occupy more than 35% of the land used for forage. Oat is favored for its growth at loamy soils with plenty of organic matter and excellent drainage. Oat grow at an optimum pH range of 5–6.5 (Shoaib et al. 2018). The nutritional profile of oats exhibits the occurrence of 51% carbohydrates, 13% proteins, 5% dietary fats, and 8% fibers, and 303 kcal/100 g energy, respectively. Statistical data has revealed the oats as the most nutrient dense foods among other cereals. Oats have been in use for the development of porridge or bircher-style muesli. Earlier practices have shown the reduction of anti-nutrients in oat on soaking in clean drinking water overnight and to develop the porridge (Angelov et al. 2018). Pakistani cuisine the oats are also known as traditional/old fashioned oats as these are soft and less cooking time (Adhikari et al. 2017). Such other grains and pseudo grains, oats include anti nutrients like phytic acid, lectins, and avenin (Decker et al. 2014). Lectins can be inactivated by cooking, especially by wet heat techniques such as boiling, stewing and/or water soaking for several hours (Samtiya et al. 2020).

(e) Buckwheat

Buckwheat (*Fagopyrum esculentum*) is a pseudo cereal which is a member of Polygonaceae family, order Polygonales, and genus Fagopyrum. The psedo-cereal is devoid of gluten protein which makes it a pseudo-cereal (Babu et al. 2018). Earlier literature has revealed the occurrence of about 15 species in the genus Fagopyrum, 9 species of which have agronomic and nutritional importance (Rodríguez et al. 2020). Common (*Fagopyrum esculentum Moench*) and Tartary buckwheat are the only two species that are grown the most frequently (*Fagopyrum Tataricum* L.). Due to the brief growth season, buckwheat is grown in Pakistan as a summer crop. There is also significant heterogeneity among genotypes in terms of yield and yield components, which showed Pakistan would be able to breed and choose high-yielding buckwheat cultivars (Rodríguez et al. 2020). Buckwheat grain is considered as a nutrient dense food source. Buckwheat grains are used as natural food ingredients which play their role in reducing the plasma cholesterol levels, and work in neuroprotection. The studies have also shown its role against cancer, inflammation and diabetes. Buckwheat has also reportedly been found to exhibit prebiotic activities and also contain unique set of antioxidants (Babu et al. 2018).

Data on the nutritional composition of buckwheat reveals the presence of about 5.7% proteins, 1.1% fats, 33.5% carbohydrates, 4.5% dietary fibers. Likewise, the mineral profile of the buckwheat portrays about 148 mg of potassium, 118 mg of phosphorus, 86 mg of magnesium, 12 mg of calcium, and 1.34 of mg iron in it. Buckwheat also contains vitamins such as thiamin, riboflavin, niacin, folate, vitamin K, and vitamin B-6 (Lu et al. 2013). Traditionally the hulled grains of buckwheat are primarily used to make breakfast cereals for humans or as flour to make a variety of baked goods (i.e., bread, cakes, and snacks), noodles enriched with buckwheat flour (0.3–60%), tea, honey, and tarhana and pillows, quilts, mattresses, collars, eye masks, and children's toys (Babu et al. 2018). Buckwheat sprouts are being used in bakery items in addition to flour and groats. Buckwheat also contains some anti-nutrient elements consisting of protease inhibitors, tannins, and phytates and interferers the protein digestibility in humans and animals. Anti-nutritional activity can be significantly decreased by the use of several processing approaches such as soaking and germination techniques could be viable techniques to minimize these anti nutrients (Samtiya et al. 2020).

(f) Quinoa

Quinoa is a pseudo cereal because the morphology of grass and grain similarity. Based on geographic adaptation, it has five ecotypes. It is a vital source of commodities that are gluten-free and of the essential amino acids (Anaya et al. 2022). The quinoa crop is ideal for the marginal area because of its resistance to salt stress and drought. It is susceptible to high temperatures even though it can endure temperatures between –8 and 35 °C (Rodríguez et al. 2020). Pakistan has good potential for the cultivation of quinoa crop. Quinoa is known to contain appreciable magnitudes of proteins, carbohydrates, and fats with the balanced amino acid profile. On comparison with the other plant-based diets, quinoa has higher concentrations of lysine i.e., essential amino acid (Anaya et al. 2022). The chemical make-up of quinoa is constituted of 60.0–69.0% carbohydrates, 13.0–20.0% protein, 9.1–12.6% moisture, and 4.0–10.0% lipids, respectively.

Quinoa is also a vial source of vitamins and minerals. The fatty acid composition of the quinoa is unique. However, due to its low glycaemic index, it has a lot of health benefits, especially for those with diabetes (Nowak et al. 2016). Quinoa is a great product for celiac disease patients since it is free of gluten and has the greatest degree of food quality (Anaya et al. 2022). Earlier studies have shown the development of bread, pasta, pancakes, biscuits, noodles, cakes, and crackers using the quinoa seeds (Adhikari et al. 2017). Due to its good nutritional profile, quinoa has grown in popularity; nevertheless, it also has certain disadvantages and contains some anti-nutritional substances, such as saponins, which give flour a bitter flavour (Nath et al. 2022). The anti-nutrients can be removed by washing or dehulling. Saponin is present in the seed coat, and after washing, the fluid is used as a biopesticide. Extrusion and roasting processes further decrease the saponin contents and bitter flavour (Samtiya et al. 2020). The quantity of phytates, a mineral inhibitor, is decreased in quinoa by several procedures, including boiling, washing, germination, and fermentation.

6.2.2 Roots and Tubers

(a) Taro

Taro (Colocasia esculenta L., Schott.) is an herbaceous, monocotyledonous, perennial root crop which is being cultivated in the tropical areas of Southeast Asia and Pacific Islands. Taro, also known as Arvi in Pakistan, is a root vegetable that is widely cultivated in many parts of the country (Aditika et al. 2022). Taro cultivation in Pakistan typically begins in late winter or early spring and lasts until the end of summer. The crop requires warm temperatures and plenty of water and is usually grown in well-drained soils that are rich in organic matter. The crop's most valuable parts in terms of economic value are the corm and corbels (Mitharwal et al. 2022). In tropical and subtropical areas, taro tubers are common diet and a significant source of energy. The primary factor driving its cultivation is the underground corms' starch content, which varies from 70% to 80%. Taro's corm has a modest protein contents (i.e., 1.5%) and fat contents (i.e., 0.2%) (Shah et al. 2022). It contains higher magnitudes of starch (i.e., 70–80% d.w.b.), 0.8% dietary fibres, and 1.2% ash. Taro contains appreciable concentrations of vitamins consisting of riboflavin, thiamine, vitamin B_6, vitamin C, and niacin. The taro roots also contains remarkable amounts of the mineral elements including the potassium, copper, manganese, phosphorus, iron, and zinc. Taro consumption could be viable against numerous health maladies such as diabetes, digestive health, constipation and cardiovascular disorders (Murthy and Paek 2020) (Fig. 6.1).

Taro is used to make starch and alcohol in addition to being used as food and animal feed. Taro possess a variety of positive traits with a wide range of useful applications. Taro starch is used in a wide range of food-related products. Taro is a

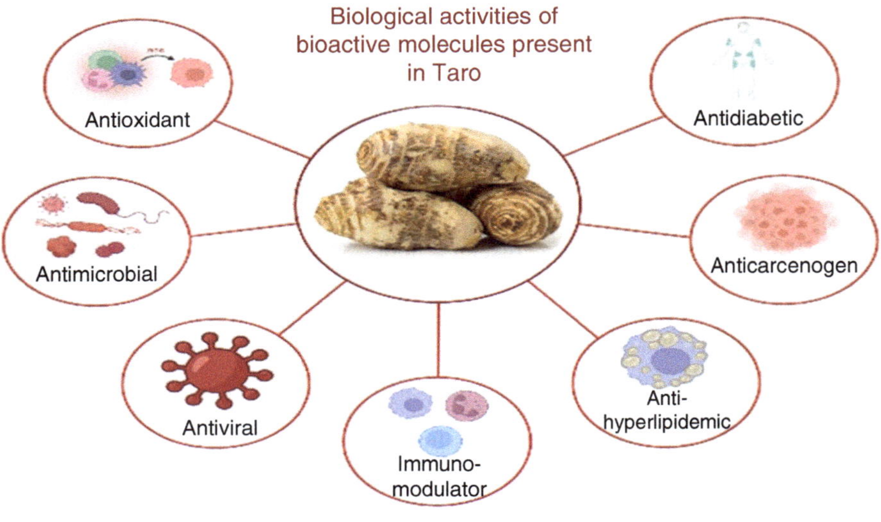

Fig. 6.1 Biological activities of taro (Shah et al. 2022)

special substance used in the creation of many different products as a stabilizers, emulsifiers, and prebiotic agent (Shah et al. 2022). Boiling, steaming, and oven baking are commonly used for the preparation of taro. A traditionally made taro product with healing properties is called poi. Taro is a staple ingredient for the developing of baby food products, biscuits, and noodles due to its easier digestibility (Adhikari et al. 2017).

Anti-nutrients in taro can restrict the applicability of taro in various food products are listed below. When sliced into pieces, the exposed sides of undercooked taro corms release mucilage droplets. A large amount of mucilage is either prevented or quickly released once the corm slices or pieces are rinsed or submerged in potable water (Samtiya et al. 2020). Oxalates, which give taro meals an unpleasant flavour or irritate the mouth when eaten raw or undercooked, are a significant barrier to their use. To reduce their oxalate-salt content, taro corms can be cooked for 30 minutes at 90 °C or for 24 hours at 30 °C. Additionally, the RDA recommends taking oxalate not more than 50–60 mg per day (Tosif et al. 2022). Due to presence of protease inhibitors, acridity, which is present in the corms, can appear as a milder form of external skin irritation or itching, as well as a more intense version that causes swelling in the mouth and throat. Variety of techniques like washing with sodium bicarbonate or an acidic ingredient can be used to reduce it (Samtiya et al. 2020). Human salivary and pancreatic amylases may become inactive due to the presence of alpha amylase inhibitors in taro corms. However, alpha-amylase inhibitors are not heat-resistant; after 30 minutes of boiling, they lose their effectiveness (Samtiya et al. 2020).

6.2.3 Pulses and Nuts

(a) Pigeon pea

Pigeon pea (*Cajanus cajan*) is categorized under legumes which is cultivated in the tropical and subtropical regions of the world. Pigeon peas, also known as "Toor daal" in Pakistan, are a widespread legume crop that are farmed extensively over the bulk of the country. It is a popular crop among farmers because of its ability to fix nitrogen in the soil and boost soil fertility (Murthy and Paek 2020). Pigeon pea farming typically begins in Pakistan in the summer and lasts until the monsoon season. The crop must grow in hot, sunny climates with high temperatures, and it usually grows in organically rich, well-drained soils. Pigeon peas contain significant amounts of proteins (i.e., 15.5–28.8%), starch content (i.e., 39.0–59%), and fat content (i.e., 0.6–3.8%) (Oluwole et al. 2022). The legumes also known to contain good magnitudes of vitamins and minerals. The mineral elements of the pigeon peas consisting of calcium, phosphorus, magnesium, iron, and potassium. Most of the indigestible carbohydrates, calcium and iron are found in seed coat which is primary nutrient source for consumers (Murthy and Paek 2020).

In Asia, de-husked splits are the predominant form in which pigeon pea is processed and used. The splits, entire (mature) seeds, or green (immature) seeds are commonly boiled in water with seasonings (Adhikari et al. 2017). The green seeds are typically preserved in brine or with meat in Africa. The green seeds are frequently included in “mixed veggies” dishes (Majili et al. 2020). The splits are either served with boiling rice or used to make traditional Indian dishes like sambar and khichri by being cooked in water. Recently, efforts to create a composite flour from pigeon pea seed have been made. Due to its high levels of protein, iron, and phosphorous pigeon pea flour has been referred as acceptable for eating and in development of various value-added foods such as breads, biscuits, and chapattis (Oluwole et al. 2022).

Pigeon pea seeds contain several anti-nutritional factors such as phenolics, phytates, cyanogenic compounds, lectins, saponins, subtilisin inhibitors, amylase inhibitors, and protease inhibitors (trypsin, chymotrypsin) (López-Moreno et al. 2022). The polyphenol content of the dark red to light red seeds of pigeon pea was higher than that of the white seeds (Samtiya et al. 2020). It is widely acknowledged that wet heat can render many proteases inhibitor actions inactive. Soaking the seeds in water results in reduction of the phytates, cyanogenic substances, and polyphenols. Dehusking also results in removal of the toxic tannins. When chickpea and pigeon pea were soaked in water overnight, about 50% of the polyphenolic chemicals were lost, and when germination was prolonged for 48 hours, an additional 10% was noticed (Nath et al. 2022).

(b) Cow pea

Cowpea (*Vigna unguiculata*) is a widely cultivated food crop around the globe. The food crop is grown in the Africa, Asia, and America. Although cowpea is mainly farmed for its seeds, which are packed with protein, it is also a significant source of hay, and forage (Murthy and Paek 2020). Cowpeas are a considered as a resilient crop which may be cultivated in a variety of soil types and climatic settings. It is renowned for its resistance to dryness and ability to flourish in desert environments. Additionally, the crop is well known for its capacity to fix nitrogen in the soil, so enhancing the soil fertility and lowering the demand for synthetic fertilizers (Ayalew and Yoseph 2022). Cowpea is a leguminous plant that is commonly known to as “lobia” in Pakistan. Smallholder farmers are the primary producers of cowpeas in Pakistan (Zia-Ul-Haq et al. 2013). The crop is directly planted in the field or in nurseries, and it is transferred 3–4 weeks later. For optimal growth and yield, cowpea need warm temperatures and adequate rainfall (Murthy and Paek 2020). Cowpeas typically include 23–24% protein, 60–65% carbohydrates, 4–6% fiber, and 1–2% fat per 100 g of their weight in terms of nutrition. It also includes vitamins (B_1, B_2, B_3, B_5, B_6, and folate) and minerals (iron, magnesium, phosphorus, potassium, and zinc) (Abebe and Alemayehu 2022). Cowpeas are also known to contain the bioactive compounds such as flavonoids and polyphenols, which have the tendency to work as anti-inflammatory agent and antioxidant. Animal feed, human consumption, and soil improvement through nitrogen fixation are just a few of the domestic uses for cowpea. Cowpeas are also used in a variety of industrial processes, such as the production of flour, starch, and protein isolate, as well as the utilization of their fibers in the textile and paper industries (Osei-Asibey et al. 2022).

The anti-nutritional factors of cowpeas include oxalate, phytate, tannin, lectin, and stachyose, however these can be significantly reduced by soaking, heating, boiling, and pre-processing techniques (Nath et al. 2022). For instance, phytates and oxalates can damage the kidneys by decreasing mineral absorption and causing several diseases associated with mineral insufficiency, such as anemia (Samtiya et al. 2020). These substances are referred to as "allelochemicals" in contrast to nutrients.

(c) Hyacinth Bean

Hyacinth/lablab bean (*Lablab purpureus*), is planted in Pakistan as food and feed crop belonging to pulses. It is primarily grown in the provinces of Punjab, Sindh, and Khyber Pakhtunkhwa. The crop is sown during the monsoon season, and it is harvested in the autumn. Because they can tolerate drought and grow in unfavorable soil, hyacinth beans are an appropriate crop for marginal soils (Bhat et al. 2022). Hyacinth beans are a good source of protein (22–25%), dietary fibers (11–17%), carbohydrates (50–60%), and minerals including iron and calcium. However, they also include tannins, phytic acid, and anti-nutritional compounds such as protease inhibitors (Bhat et al. 2022). Hyacinth beans are used in soups and stews, while the seeds can be boiled and used as food or as animal feed. In contrast to the beans, which are used in industry as a viable source of starch. Seeds can be processed into flour, protein concentrate, or oil by using different methods of processing (Murthy and Paek 2020). Anti-nutritional factors in hyacinth beans are tannins, phytates, and protease inhibitors may prevent the absorption and use of micro-nutrients (Nath et al. 2022). The anti-nutritional factors affect the micro-elements in hyacinth beans. Soaking hyacinth beans in water for a while before cooking can lessen their anti-nutritional content. Boiling hyacinth beans for at least 30 minutes can significantly reduce their tannin content. For a few days, hyacinth beans can be fermented to increase nutrient availability and reduce anti-nutritional elements (Samtiya et al. 2020).

(d) Groundnut

Groundnut is an important cash crop in Pakistan, particularly in Baran, (dry) upper Punjab, and parts of the Northwest Frontier Province (NWFP). It is grown under irrigated conditions in Sindh. Approximately 85% of Pakistan's total groundnut area is in Punjab, 10% in NWFP, and 5% in Sindh. The harvest usually takes place in October and November after the crop has been sown in March and April (Nile and Park 2013). The optimal growth conditions for groundnuts are warm temperatures and sandy loam soil. It is cultivated largely for the oil that can be produced from it and used for cooking and the production of many different food items. Additionally, biodiesel, animal feed, and peanut butter are all made from groundnuts. Groundnut farming is a source of income for many farmers in Pakistan (Mehmood et al. 2021).Groundnuts contain good protein contents (i.e., 22–30%) while the oil content vary between 4% and 46%. Additionally, peanuts are a good source of vitamins including vitamin B, E, and K. The literature also shows the presence of appreciable magnitudes of minerals such as phosphorus, calcium, magnesium, and potassium (Toomer 2018). The nutritional value and storage properties of peanut oil depend on the ratio of saturated to unsaturated fatty acids (Nile and

Park 2013). Roasted peanuts include more antioxidants and eight (8) essential fatty acids, of which four – palmitic, stearic, oleic, and linoleic acids – account for more than 90% of the triacylglycerols in peanuts. The kernels are used to manufacture the peanut butter, oil, roasted snack peanuts, confections, and peanuts. Only a very small portion of peanut skins are often employed to extract polyphenol compounds or produce cow feed (Toomer 2020). The rest of peanut skins are discarded as waste products of the peanut processing industry. Peanut skins are considered as an enriched source of antioxidants. Peanut skins can be used as renewable raw material for antioxidant compounds (Adhikari et al. 2017).

Groundnuts include several anti-nutrients such as phytates, tannins, lectins, and protease inhibitors that can affect the nutrient absorption and results in impaired digestion. Processing techniques such as soaking, roasting, boiling and fermentation can improve the digestibility and bioavailability of minerals in groundnuts (Samtiya et al. 2020). Roasting can reduce lectin levels, whereas soaking groundnuts for several hours in water before cooking can reduce phytic acid levels. Boiling and fermentation are further methods for reducing the antinutritional components of groundnuts (Eltom et al. 2022).

(e) Coconut

Coconut production only takes place in the coastal areas of Sindh and Baluchistan. Because coconut trees require a warm, humid environment with plenty of rainfall, they are frequently grown in sandy soils close to the coast (Ollivier et al. 1999). The crop is frequently multiplied by seedlings, and for best growth, it requires effective irrigation and drainage systems. Growing coconuts may generate income for farmers while also boosting the regional economy (Abdul and Zafar Iqbal 2011). More study and development are needed to examine the potential benefits and applications of this underutilized species due to Pakistan's limited coconut cultivation and use. Coconut is a very nourishing food that is rich in fiber, vitamins, and minerals. The coconut's meat, or edible part, contains protein, iron, and zinc in addition to having a lot of saturated fat (Ollivier et al. 1999). Coconut water is a fantastic source of hydration because it is low in calories and high in electrolytes like potassium and sodium (Abdul and Zafar Iqbal 2011).

Milk, minced meat for recipes, and cooking oil are domestic uses for coconut. It is also used in cosmetics, and soap. Commercially, coconut is used to make lubricants, fuel, and raw materials for the manufacture of textiles, furniture, and building materials (Kaur et al. 2019). The bioavailability of various minerals may be lowered by anti-nutrient elements included in coconut are phytic acid, tannins, and trypsin inhibitors (Nath et al. 2022). To reduce these antinutritional components, treatment methods include soaking, fermenting, and boiling. By soaking in water or brine, phytic acid levels can be lowered, and fermentation can reduce phytic acid and tannins (Samtiya et al. 2020).

6.2.4 Horticultural Produce

(a) Chenopodium album

Alternative plants can produce vegetables for human use in place of conventional plants, offering a nutritious food source. Although these widely grown veggies are largely free of agricultural toxins, it is uncertain how they might impact human health. These plants may provide the nutrients that humans require. The nutrients are required for a healthy metabolism and physiology in the body (Zhang et al. 2022). However, a lot of these low-cost nutritive plants have not yet undergone enough research. The use of culinary herbs and wild vegetables is common in regions of the world where these species are regarded as the primary nutritional food (Murthy and Paek 2020). Consumption of *C. album* may provide the nutrients and can help in lowering the chance of contracting illnesses such as aging, cancer, diabetes, coronary heart disease, and neurological disorders. Due to their high concentration of phytochemicals including n-6 and n-3 essential fatty acids, vitamins E and C, and polyphenols, wild plants should be a staple of any diet for the prevention of age-related disorders and cancer (Koca et al. 2015). These vegetables can be used as a beneficial food source and alternative vegetable supply for human consumption (Adhikari et al. 2017).

C. album has long been used in traditional medicine and as a food source for both humans and animals. This plant has been used for many years as Ayurvedic medicine to treat a variety of ailments of digestive tract, respiratory tract, and skin (Zhang et al. 2022). Young shoots are pleasant with the slightly bitter-tasting leaves are characteristics of *C. album*. It could be used after roasting and/or salads preparation. It serves as a valuable source of food for livestock, especially sheep and goats (Adhikari et al. 2017). *C. album* has been investigated as a potential source of biofuel due to its high oil content. The plant also has natural hues that are employed in the textile industry. *C. album*, like many other plants, contains anti-nutritional elements that could hinder the body's ability to absorb nutrients (Samtiya et al. 2020). *C. album* contains tannins, phytic acid, and oxalic acid, among other antinutritive factors. Blanching, soaking, fermenting, and boiling can help in the reduction of antinutritional factors in *C. album* (Nath et al. 2022).

(b) Moringa

Moringa (*Moringa oleifera* L.) is one of the 14 species that make up the genus Moringa and belongs to family Moringaceae. In the sub-Himalayan region, it is indigenous to Afghanistan, Pakistan, Bangladesh, and Northwestern India (Chhikara et al. 2021). The species are grown in tropics and subtropics of the world. Drought-resistant, with roots, sturdy grey bark, delicious green leaves, fragrant white blooms in the spring, and delicate pods with 10–12 seeds each are the characteristics of small to medium-sized M. oleifera trees. It also develops swiftly, is not unduly depending on soil or climate (Wang et al. 2018). Numerous parts of the very nutritious plant moringa provides a variety of health benefits. The leaves include a variety of minerals and vitamins consisting of calcium, magnesium, iron, and vitamin C, vitamin A, and vitamin K (Chhikara et al. 2021). While moringa pods are a rich source of vitamins C, A, potassium, and fiber, moringa seeds are a good source of protein, fiber, and vitamins like vitamin E, magnesium, and zinc. The blooms are loaded with antioxidants, potassium, and calcium and have been used in traditional medicine (Saa et al. 2019).

M. oleifera can be fed to lambs, goats, and cattle. Additionally, used for aesthetic purposes is the oil that is derived from the seeds. The anti-fungal activity of the extract, the protective action against the toxic effects of arsenic element, and the usage of the fat from the seeds as vegetable oil for cooking that is resistant to oxidative degradation have all been studied. (Saa et al. 2019) In India, it is common practice to prepare various foods into various meals without considering their nutritional value because of a hatred of the flavor (Adhikari et al. 2017). Moringa contains several anti-nutritional factors such as phytates, oxalates, tannins, and alkaloids which could halt body from assimilation nutrients (Saa et al. 2019). Numerous techniques such as heat treatments, sprouting, soaking and fermentation are available to reduce these anti-nutrients to a permissible range (Samtiya et al. 2020).

6.3 Traditional Uses of Neglected Food Plants

The traditional uses of these neglected food plants including pulses, beans, legumes, root, and tubers are very widespread (Ebert 2014). Bioactive compounds in the neglected plant foods such as tubers and roots these are favored for their nutritional and medicinal benefit to treat various health disorders (Okigbo and Anyaegbu 2021). Pulses being rich source of proteins ranging between 20% and 40% are considered as an alternative against malnutrition and served as breakfast food and used in the development of soups and snacks (Luzardo-Ocampo et al. 2020). Drought tolerant ability of cereal crops contribute their major role in diet and served as excellent source of carbohydrates. Mostly used dishes of tubers such as taro, eddo or arvi in Pakistan are gravy based, arvi gosht being popular including mutton, beef, and lamb (Luzardo-Ocampo et al. 2020). Hummus, made from chickpeas, falafel, made from chickpeas or mung beans, and dahl, made from lentils, are all common foods made using pulses (Teye et al. 2020). Dry rice bean seeds can be crushed into flour, cooked with rice, or used as a substitute for rice. Dishes with "mixed vegetables" usually contain the green pigeon pea seeds. The seeds are either eaten with hot rice or boiled in water and added to sambar and khichri, two traditional Indian cuisines (Teye et al. 2020). However, further research is required to understand and disseminate the underlying mechanisms of the domestic food products to the commercial or industrial level.

6.4 Industrial Valorization of Neglected Food Plants

Pakistan is home to a wide variety of plant species that have historically been used as food and medicine. Through the industrial value of these underutilized food plants, which will also aid in biodiversity preservation and improve food security, it may be feasible to produce financial rewards for farmers and the public (Massawe et al. 2015). To promote the commercial use of these underutilized food plants, several activities can be performed. Research and development are foremost required to ascertain the potential applications of these neglected plant food species due to their nutritive potential, medicinal properties, and industrial food applications (Mayes et al. 2012). Consumers can be attracted due to their therapeutic and nutritional advantages by encouraging the development of novel food products and value-added services. A nexus between the growers, processors, and distributors may be required to create a sustainable supply chain for these plants. Likewise, promotion, preservation and cultivation of the abandoned food plants by setting up seed banks, nurseries, and farmer training programs could be a viable approach in uplifting the use of the neglected plant food species (Adhikari et al. 2017). This could help to ensure a steady supply of these plants as well as the preservation of biodiversity and the quality of the soil. Overall, Pakistan's industrialization of closed food factories has the promise of enhancing food security, fostering the preservation of biodiversity, and creating new employment opportunities for farmers and the public (Shweta and Yadav 2017). Collaborative work among academia and government could play their pivotal role in ensuring the neglected plant foods species for the revival and food applications and value addition.

6.5 Nutritional Significance and Opportunities

Underutilized plant foods of Pakistan have a huge nutritional significance because these plants are enriched sources of the essential micro and macronutrients. These plant species have historically been used as ingredient of choice, food, feed and/or medicine (Mayes et al. 2012). Consuming them, however, can significantly improve the nutritional quality of meals, particularly in rural areas where access to a range of foods may be limited (Shweta and Yadav 2017). It could be a viable approach to diversify the agricultural produce and enhance the sustainable farming practices using the underutilized plant foods. This can require setting up seed banks, nurseries, and farmer training programs to promote the growth of these plants and boost their yields (Massawe et al. 2015).

6.6 National Policies Towards Preserving Biodiversity and Promoting Neglected Food Plants Production

Food Security Policy of Pakistan (FAO, 2023)

Vision
A Food Secure Pakistan

Mission
To create a modern, efficient food production and distribution system that can, in terms of availability, access, utilization, and stability, most effectively support food security and nutrition.

Goals
More specifically food security policy aims to:

1. Reduce hunger and malnutrition.
2. Promote 4% annual growth rates in sustainable food production systems.
3. Boost agriculture's profitability, competitiveness, and climate resiliency

Major Challenges to Food Security

- The current level of food insecurity has also demonstrated that people are not going hungry because there is not enough food accessible on a global scale, but rather because they are unable to afford it or lack the resources to produce adequate food. Putting more of an emphasis on dietary diversification, nutrient dense nutrition, and healthy eating lifestyle for everybody is tough.
- Lowering the cost of nutritious food purchases for the society's lower income groups.
- Increased climate unpredictability and extreme weather events have a considerable impact on the stability of the food supply, food production, and the lives of farmers and vulnerable populations. The national average yields of almost all crops are poor, and productivity is gradually declining due to climate change. Therefore, reducing and adapting to the effects of climate change on agriculture and cattle is a significant problem.
- The conditions of food security and nutrition are impacted by critical concerns such as emergencies, natural disasters, and migration. Regional emergencies, internal population displacement brought on by security operations, floods, and droughts are the main causes of situations like this.
- Increasing the supply of agricultural inputs and their availability, timing, and quality
- Putting together the infrastructure and technologies required for post-harvest management and value addition.
- Accelerating the diffusion of technical innovation.
- Utilizing water and land resources wisely.
- Identifying qualified persons for food security and food systems analysis.

- Improving the ability of National Agricultural Research Systems (NARS) to handle and complete the complex research goals of modern agriculture.

Policy Imperatives
Sustainable agriculture not only influence the dietary diversification but also promotes the rural development by enhancing the livestock and crop productivity, affordable price ranges for consumers, and high farm income.

Policy Measure
It is advised to launch some significant measures to address the major challenges occurring by following these steps.

- Encouraging the development and consumption of pulses and oilseeds as crops that can replace imported goods
- Agricultural diversification driven by value chains will aid in nutritional variety.
- Increasing the output of crucial crops to free up natural resources for the cultivation of other high-value crops.
- To secure the agriculture sector's viability, production shortages must be filled, and farm profitability must be maintained.
- Impact assessment and adaptation strategy optimization for climate change.
- Establish a crop-livestock breeding program that enlists both domestic and international research organizations.
- Improving food and nutrition through community-based acute malnutrition management, targeted food distribution strategies, and programs to improve livelihoods.
- Increased media participation and educational initiatives to promote media literacy.
- The National Zero Hunger Program for evaluating school meals, nutrition, and family farming programs.
- Working with the National Disaster Management Authority (NDMA) to develop an emergency plan for the agriculture, livestock, and food security industries.
- Encouraging and assessing the impact of agricultural innovations, such as precision farming, nanotechnology, and next-generation biotechnology, from the perspective of sustainable food systems.
- Government assistance with the launch of exclusive agricultural channels promoting the dietary intake of the most neglected plant foods in collaboration with the researchers from the academia.

6.7 Toxicological Challenges and Possible Solutions

Anti-nutrients are the barriers for the consumers towards the use of these most neglected plant foods in our dietary routine. These anti-nutrients adversely affects the bioavailability of essential minerals such as zinc, iron and calcium. Anti-nutrients of poisonous nature associated with the neglected/under-rated plant foods includes phytates, oxalates, tannins, phenols, saponins, mucilage, amylase, and trypsin inhibitors (Adhikari et al. 2017). Antinutrients in plants influence whether they may be used as food by reducing nutrient use or interfering with nutrient absorption (Okigbo and Anyaegbu 2021). Antinutrient levels in the edible plant species are typically modest and below dangerous levels (Samtiya et al. 2020). To curtail the losses, the processing techniques (López-Moreno et al. 2022) including frying, soaking, blanching, and sodium bicarbonate washing are adopted. Different food preparation techniques have a variable effect on the presence of antinutritional elements in legumes (Teye et al. 2020). More significantly, after soaking for 12 hours, highly strong trypsin inhibitory activity in dry Indian beans decreased gradually by 51%, whereas tannins and phytic acids decreased to 46% and 36%, respectively (Teye et al. 2020). Therefore, it is crucial to emphasize that precooking procedures are essential when incorporating underused plant species into food (Okigbo and Anyaegbu 2021).

6.8 Conclusion

Literature insights on the neglected and underutilized species (NUS) of plants spotlights the hidden potential of these species for humans due to their immense nutritional significance pertaining to proteins, dietary fibers, essential fats, minerals and vitamins. Likewise, the bioactive compounds linked with these un-rated plant foods are the windows for freshness. In the backdrop of their promising food potential, neglected food plants and crops are encouraged to grow on the marginal lands, in rotation with other crops or intercropped with them, and using a range of farming methods. FAO created the Future Smart Food Initiative with the goal of identifying NUS with a high potential for integration into agricultural systems. But creating an environment that will promote NUS production, marketing, and consumption is crucial. The promotion of NUS production, the improvement of regulatory frameworks, the promotion of agricultural diversity, and the enhancement of market access infrastructure can all be greatly aided by the governments. Advertising NUS as a component of sustainable diets may further boost their uses. To boost the output of underutilized crops, governments can help scientists to present research and put cutting-edge technologies into practice in cultivation, post-harvest processing, marketing, and consumption. Incorporating neglected and underutilized species into agricultural systems could lead to a nutrient-dense, climate-resilient, and sustainable agriculture that supports food security, malnutrition and livelihoods of the vulnerable population groups.

Conflicts of Interest All authors read and approved the final manuscript and declare that they have no conflict of interest.

References

Abah CR, Ishiwu CN, Obiegbuna JE et al (2020) Sorghum grains: nutritional composition, functional properties and its food applications. Eur J Nutr Food Saf 12(5):101–111. https://doi.org/10.9734/ejnfs/2020/v12i530232

Abdul HS, Zafar Iqbal M (2011) Chemical composition of meat (kernel) and nut water of major coconut (Cocos nucifera L.) cultivars at coastal area of Pakistan. Pak J Bot 43(1):357–363

Abebaw G (2021) Review on structure, functional and nutritional composition of barley (Hordeum vulgare). Nutr Food Process 4(2):1–8. https://doi.org/10.31579/2637-8914/046

Abebe BK, Alemayehu MT (2022) A review of the nutritional use of cowpea (Vigna unguiculata L. Walp) for human and animal diets. J Agric Food Res 10:100383. https://doi.org/10.1016/j.jafr.2022.100383

Adebiyi JA, Obadina AO, Adebo OA et al (2018) Fermented and malted millet products in Africa: expedition from traditional/ethnic foods to industrial value-added products. Crit Rev Food Sci Nutr 58(3):463–474. https://doi.org/10.1080/10408398.2016.1188056

Adhikari L, Hussain A, Rasul G et al (2017) Tapping the potential of neglected and underutilized food crops for sustainable nutrition security in the mountains of Pakistan and Nepal. Sustainability 9(2):291. https://doi.org/10.3390/su9020291

Aditika, Kapoor B, Singh S et al (2022) Taro (Colocasia esculenta): zero wastage orphan food crop for food and nutritional security. S Afr J Bot 145:157–169. https://doi.org/10.1016/J.SAJB.2021.08.014

Amadou I, Gounga ME, Le GW et al (2013) Millets: nutritional composition, some health benefits and processing – a review. Emir J Food Agric 25(7):501–508. https://doi.org/10.9755/ejfa.v25i7.12045

Anaya RB, De La Cruz E, Muñoz-Centeno LM et al (2022) Food and medicinal uses of ancestral Andean grains in the districts of Quinua and Acos Vinchos (Ayacucho-Peru). Agronomy 12(5):1014. https://doi.org/10.3390/agronomy12051014

Angelov A, Yaneva-Marinova T, Gotcheva V et al (2018) Oats as a matrix of choice for developing fermented functional beverages. J Food Sci Technol 55:2351–2360

Ayalew T, Yoseph T (2022) Cowpea (Vigna unguiculata L. Walp.): a choice crop for sustainability during the climate change periods. J Appl Biol Biotechnol 10(3):154–162. https://doi.org/10.7324/JABB.2022.100320

Babu S, Yadav GS, Singh R et al (2018) Production technology and multifarious uses of buckwheat (Fagopyrum spp.): a review. Indian J Agron 63(4):415–427

Bhat SS, Shivamallu C, Prasad KS et al (2022) Biomedical importance of Lablab purpureus: a review. Med Plants Int J Phytomed Relat Ind 14(1):20–29. https://doi.org/10.5958/0975-6892.2022.00003.X

Chhikara N, Kaur A, Mann S et al (2021) Bioactive compounds, associated health benefits and safety considerations of Moringa oleifera L.: an updated review. Nutr Food Sci 51(2):255–277. https://doi.org/10.1108/NFS-03-2020-0087

Decker EA, Rose DJ, Stewart D (2014) Processing of oats and the impact of processing operations on nutrition and health benefits. Br J Nutr 112(2):58–64. https://doi.org/10.1017/S000711451400227X

Ebert AW (2014) Potential of underutilized traditional vegetables and legume crops to contribute to food and nutritional security, income and more sustainable production systems. Sustainability 6(1):319–335. https://doi.org/10.3390/su6010319

Eltom HAE, Sulieman MA, Hassan AB et al (2022) Effect of boiling and roasting processing on nutritional quality of two Sudanese peanuts (Arachis hypogaea) cultivars. Legume Sci 5(1):159. https://doi.org/10.1002/leg3.159

FAO (2023) National Food Security Policy, Ministry of National Food Security and Research Islamabad, Government of Pakistan. http://faolex.fao.org/docs/pdf/pak184963.pdf

Grote U, Fasse A, Nguyen TT et al (2021) Food security and the dynamics of wheat and maize value chains in Africa and Asia. Front Sustain Food Syst 4:617009. https://doi.org/10.3389/fsufs.2020.617009
Gunawan S, Dwitasari I, Rahmawati N et al (2022) Effect of process production on antinutritional, nutrition, and physicochemical properties of modified sorghum flour. Arab J Chem 15(10):104134. https://doi.org/10.1016/j.arabjc.2022.104134
Kaur K, Chhikara N, Sharma P et al (2019) Coconut meal: nutraceutical importance and food industry application. Foods Raw Mater 7(2):419–427. https://doi.org/10.21603/2308-4057-2019-2-419-427
Koca L, Hasbay I, Bostanci S et al (2015) Some wild edible plants and their dietary fiber contents. Pak J Nutr 14(4):188–194. https://doi.org/10.3923/pjn.2015.188.194
López-Moreno M, Garcés-Rimón M, Miguel M et al (2022) Antinutrients: lectins, goitrogens, phytates and oxalates, friends or foe. J Funct Foods 89:104938. https://doi.org/10.1016/j.jff.2022.104938
Lu L, Murphy K, Baik BK et al (2013) Genotypic variation in nutritional composition of buckwheat groats and husks. Cereal Chem 90(2):132–137. https://doi.org/10.1094/CCHEM-07-12-0090-R
Luzardo-Ocampo I, Cuellar-Nuñez ML, Oomah BD et al (2020) Pulse by-products. In: Campos-Vega R, Oomah BD, Vergara-Castañeda HA (eds) Food wastes and by-products: nutraceutical and health potential. Wiley, pp 59–92
Majili ZS, Nyaruhucha C, Kulwa K et al (2020) Preferences and consumption of pigeon peas among rural households as determinants for developing diversified products for sustainable health. Sustainability 12(15):6130. https://doi.org/10.3390/su12156130
Mao H, Xu M, Ji J et al (2022) The utilization of oat for the production of wholegrain foods: processing technology and products. Food Front 3(1):28–45. https://doi.org/10.1002/fft2.120
Massawe FJ, Mayes S, Cheng A et al (2015) The potential for underutilised crops to improve food security in the face of climate change. Procedia Environ Sci 29:140–141. https://doi.org/10.1016/j.proenv.2015.07.228
Matin A, Grubor M, Jurišić V et al (2018) Hulled barley nutritional composition after drying and cooking thermal treatment. Krmiva: Časopis o hranidbi životinja, proizvodnji i tehnologiji krme 59(2):51–60. https://doi.org/10.33128/k.59.2.1
Mayes S, Massawe FJ, Alderson PG et al (2012) The potential for underutilized crops to improve security of food production. J Exp Bot 63(3):1075–1079. https://doi.org/10.1093/jxb/err396
Mehmood K, Rehman A, Khan A et al (2021) Farmers' perceptions, awareness and adoption of improved groundnut varieties in Potwar Plateau of Pakistan. Sarhad J Agric 37(4):1364–1376. https://doi.org/10.17582/JOURNAL.SJA/2021/37.4.1364.1376
Mitharwal S, Kumar A, Chauhan K et al (2022) Nutritional, phytochemical composition and potential health benefits of taro (Colocasia esculenta L.) leaves: a review. Food Chem 383:132406. https://doi.org/10.1016/J.FOODCHEM.2022.132406
Murthy HN, Paek KY (2020) Health benefits of underutilized vegetables and legumes. Ref Ser Phytochem. https://doi.org/10.1007/978-3-030-44578-2_1-1
Naheed S, Raza I, Anwar MZ et al (2015) Forecasting area and production of barley in Punjab, Pakistan. Pak J Agric Res 28(3):304–309
Nath H, Samtiya M, Dhewa T et al (2022) Beneficial attributes and adverse effects of major plant-based foods anti-nutrients on health: a review. Human Nutr Metab 28:200147. https://doi.org/10.1016/J.HNM.2022.200147
Nile SH, Park SW (2013) Fatty acid composition and antioxidant activity of groundnut (Arachis hypogaea L.) products. Food Sci Technol Res 19(6):957–962. https://doi.org/10.3136/fstr.19.957
Nowak V, Du J, Charrondière UR et al (2016) Assessment of the nutritional composition of quinoa (Chenopodium quinoa Willd.). Food Chem 193:47–54. https://doi.org/10.1016/j.foodchem.2015.02.111
Okigbo RN, Anyaegbu CF (2021) Underutilized plants of Africa. J Biol Nat 13(2):34–49

Ollivier J, Akus W, Bonneau X et al (1999) Coconut nutrition in Papua New Guinea. CORD: Coconut Res Dev 15(2):76–105. https://doi.org/10.37833/cord.v15i02.329

Oluwole OB, Nicholas-Okpara VAN, Elemo G et al (2022) Medicinal uses, nutraceutical potentials and traditional farm production of Bambara beans and pigeon pea. Glob J Epidemiol Public Health 6:41–50. https://doi.org/10.12974/2313-0946.2021.06.01.3

Osei-Asibey RK, Wireko-Manu FD, Aidoo R et al (2022) Farmers' perception of the use and benefits of cowpea storage methods in Northern Ghana. Sustainability 14(9):5129. https://doi.org/10.3390/su14095129

Rodríguez JP, Rahman H, Thushar S et al (2020) Healthy and resilient cereals and pseudo-cereals for marginal agriculture: molecular advances for improving nutrient bioavailability. Front Genet 11:49. https://doi.org/10.3389/fgene.2020.00049

Saa RW, Fombang EN, Ndjantou EB et al (2019) Treatments and uses of Moringa oleifera seeds in human nutrition: a review. Food Sci Nutr 7(6):1911–1919. https://doi.org/10.1002/fsn3.1057

Samtiya M, Aluko RE, Dhewa T et al (2020) Plant food anti-nutritional factors and their reduction strategies: an overview. Food Prod Process Nutr 2:1–14

Shah YA, Saeed F, Afzaal M et al (2022) Industrial applications of taro (Colocasia esculenta) as a novel food ingredient: a review. J Food Process Preserv 46(11):e16951. https://doi.org/10.1111/jfpp.16951

Shoaib M, Khan MN, Akhtar N et al (2018) Productivity of oat (Avena sativa L.)-berseem (Trifolium alexandrinum L.) forage mixture in irrigated plain of Pakistan. Pak J Agric Sci 55(2):302–312. https://doi.org/10.21162/PAKJAS/18.4947

Shweta, Yadav VK (2017) Exploiting potentials of neglected and underutilized crops for food security under climate change scenarios in India. In: Greener agriculture and environment through convergence of technologies. Proceedings of the international symposium on agriculture and environment – ISAE 2017, University of Ruhuna, Sri Lanka, 19th January 2017

Syed A, Raza T, Bhatti TT et al (2022) Climate impacts on the agricultural sector of Pakistan: risks and solutions. Environ Chall 6:100433. https://doi.org/10.1016/J.ENVC.2021.100433

Teye E, Deha CI, Dadzie R et al (2020) Delivering the nutritional needs by food to food fortification of staples using underutilized plant species in Africa. Int J Food Sci 2020. https://doi.org/10.1155/2020/8826693

Toomer OT (2018) Nutritional chemistry of the peanut (*Arachis hypogaea*). Crit Rev Food Sci Nutr 58(17):3042–3053. https://doi.org/10.1080/10408398.2017.1339015

Toomer OT (2020) A comprehensive review of the value-added uses of peanut (Arachis hypogaea) skins and by-products. Crit Rev Food Sci Nutr 60(2):341–350. https://doi.org/10.1080/10408398.2018.1538101

Tosif MM, Najda A, Klepacka J et al (2022) A concise review on taro mucilage: extraction techniques, chemical composition, characterization, applications, and health attributes. Polymers 14(6):1163. https://doi.org/10.3390/polym14061163

Ullah A, Ahmad A, Khaliq T et al (2017) Recognizing production options for pearl millet in Pakistan under changing climate scenarios. J Integr Agric 16(4):762–773. https://doi.org/10.1016/S2095-3119(16)61450-8

Wang P, Chen T, Sun J et al (2018) Research progress in the application of Moringa oleifera in animal feed. Chin J Anim Nutr 30(7):2488–2495

Zhang Z, Zhang J, Lu P et al (2022) Six underutilized grain crops for food and nutrition in China. Plan Theory 11(19):2451. https://doi.org/10.3390/plants11192451

Zia-Ul-Haq M, Ahmad S, Amarowicz R et al (2013) Antioxidant activity of the extracts of some cowpea (Vigna unguiculata (L) Walp.) cultivars commonly consumed in Pakistan. Molecules 18(2):2005–2017. https://doi.org/10.3390/molecules18022005

Chapter 7
Neglected Plant Foods of Bangladesh

M. Khairul Alam

7.1 Introduction

Bangladesh is an agriculture-based country, and agriculture produces around 90% of its food needs, including cereals and vegetables. According to a report from the Bangladesh National Nutrition Council, the country achieved self-sufficiency for rice crops in the fiscal year 1996–7 (BNNC 2020). The report indicates a gradual change in dietary behavior among households, with a shift from cereals to non-cereal foods such as fruits, vegetables, and animal products. The internationally accepted standard for cereals' contribution to daily caloric supplies is 54–55%. Contrarily, cereals dominate the dietary patterns of the people of Bangladesh, accounting for a daily caloric share of 74–76%. Nearly one-fifth of the total diet of people from rural backgrounds comes from vegetables. Due to the mounting caloric requirements of the ever-increasing population in Bangladesh, high-yielding cultivars of rice have dominated food crop production, while traditional varieties of the crop supply only around 10% of the total rice requirements. The same is true for other food crops, and indigenous crops are gaining the least interest from farmers due to their comparatively low productivity.

The absence of quality protein and micronutrients is evident in regular diets in Bangladesh. So, one of the Strategies of the National Nutrition Policy 2015 is to increase food diversity. In the last few years, the country has witnessed a substantial increase in the production of aquatic organisms and plants along with the constant increase in rice growth. Similarly, a considerable development in various food commodities and livestock has also been observed (BNNC 2020).

Neglected plant species provide a greater opportunity to overcome dietary need because these plants are rich in nutrients, provides a good margin for profit, and are climatically adaptable and regionally produced (Li et al. 2020). The common

M. Khairul Alam (✉)
Bangladesh Forest Research Institute, Chattogram, Bangladesh

T. Ismail et al. (eds.), *Neglected Plant Foods of South Asia*,
https://doi.org/10.1007/978-3-031-37077-9_7

practice of taking an imbalanced diet which includes excessive cereal consumption and inadequate intake of foods with good micronutrients can cause multiple health risks (Wright et al. 2022). The contemporary world's food system lacks sustainability. Currently, this system is based upon the consumption of staple crops and has ignored the potential of neglected plants, though the latter can provide a better substitution for the existing food system (Wright et al. 2022). Neglected crops are mostly grown at the domestic level and are considered wild. The recent advancement in farm practices has promoted high-yield and profitable crops which has considerably decreased the consumption of these conventional plants (Li et al. 2020). These plants are regionally grown and easily available, so can help people in achieving economic, nutritional, and food security.

There are wide diversities of underutilized food plants (Islam et al. 2012; Khatun et al. 2013; Alam 2020). Ethnic communities from the Chittagong Hill Tracts practice a traditional slash-and-burn system of agriculture, locally known as jhums. About 50 crops and other food plants are grown in jhums (Khisa and Mohiuddin 2015; Alam 2020). These Jhums serve as a source for various local food plants (Alam 1998, 2002; Miah and Chowdhury 2004; Chowdhury et al. 2009; Miah et al. 2011; Alam and Mohiuddin 2009, 2010; Khisa and Mohiuddin 2015; Paul et al. 2016). In addition to that, culinary herbs make up the majority of local food plants, and among them approximately 60 species of plants are consiered edible (Lewin 1869; Alam and Mohiuddin 2010).

Like a diverse number of plant species, also the parts used are diverse. But plant foods are limited in numbers. Few of them are used as scarce foods or substitutes for staple foods but a wide number of species are cooked as vegetables. Many traditional fruits are used for making pickles and are preserved in the form of jams, jellies, candid foods, etc. Most of the traditional food products are traditional innovations that vary a little from area to area. Knowledge about the food plants is women-centered. This knowledge system is perpetuated from generation-to-generation. Some prepared foods and pickles are being sold and have good trading potential. Most of the food plants are harvested from natural habitats, and some come from cultivations. Arbitrary harvesting is causing a decline in resources. Moreover, trading and commercialization also exert pressure on resources. So, there are scopes for domestication, cultivation, and conservation of agro-biodiversity of these underutilized food plants.

This chapter attempts to discuss neglected food plants and traditional plant food from Bangladesh. It also tries to throw light on their economic prospects, nutritional significance, pertinent development agenda, and national policies, toxicological and safety issues. Traditional knowledge and gender role about traditional food items has been discussed. Brief notes on distribution patterns and the domestication potentials of neglected food plants are also given.

The methodology adopted in developing this chapter mainly includes a review of pertinent literature and the author's personal experience gathered during the documentation of ethnobotanical information and traditional knowledge during his professional career over four decades.

7.2 Neglected Food Plants

In Bangladesh, the most consumed food crop is rice followed by wheat and maize consumption at second and third positions respectively. People living in remote areas fulfill their protein dietary needs through the intake of pulse crops. Along with animal proteins or alone different vegetables are cooked as dishes. Vegetables grow almost all the time in this country and a good number of vegetables are harvested from wild states. Farmers also grow fruits in their homesteads and orchards not only to meet their domestic demand but also for sale to generate cash. An extensive list of neglected food plants and wild fruits is available in the literature as mentioned in the Sect. 7.1. A list was published that classified food items into 14 types Islam et al. (2012). Figure 7.1 displays harvest of non-traditional food plants by a tribal woman from natural habitat in Chittagong Hill Tracts, Bangladesh.

Biswas et al. (2021) mentioned Panicum miliaceum, and Setaria italica, as minor cereal crops. Instead of rice or wheat people sometimes depend on millet, bamboo shoots, yams, and cassava as staple food substitutes. Often, they depend on cooked vegetables. Thus, underutilized food plants and non-traditional plant foods are linked with food security. Most of the wild fruits are taken independently and many of them are preserved. Within the edible plant parts, leaves and fruits contribute to the maximum, the remainder being edible tubers, flowers, and seeds. Wild and neglected food plants play an important role in the food system and dietary requirements of the local people's diet. In remote hill areas and poverty-pruned areas poor people sometimes depend on wild food plants to meet their dietary requirements. Figure 7.2 depicts Surplus harvest for sale in the street market by tribal women in Chittagong Hill Tracts, Bangladesh.

Fig. 7.1 Harvest of non-traditional food plants by a tribal woman from natural habitat in Chittagong Hill Tracts, Bangladesh

Fig. 7.2 Surplus harvest for sale in the street market by tribal women in Chittagong Hill Tracts, Bangladesh

Plant foods are mostly traditional items developed through local innovations and are deeply rooted in local cultures. The traditional food system is also less explored in Bangladesh. Neglected food plants have been traditionally used in various ways based on local practices, which can be categorized into the following major groups of uses. These staple food substitutes are cooked as vegetables; eaten raw as fruit; juice; sap; prepared as pickles; prepared as jams and jellies; prepared as candid (murabba), halwa (a paste fried in oil, butter oil, and sugar) and other food products, and used as condiments and spice. Based on local uses, the following major groups of uses could be identified as traditional uses of neglected food plants.

7.2.1 Cereals and Pseudo Cereals

Cereals are commonly used in the food industry because of their wide applications. A few examples of local fruits being used as cereals in various regions of Bangladesh include *Panicum miliaceum, Setariai talica, Zea mays*. They are extensively used as ingredients in bread, snacks, and flour production.

7.2.2 Roots and Tubers

Roots and tubers are used as a source of food in multiple cuisines around the world. These roots are also used for thickening sauces and soups. Some common examples of roots and tubers used in foods in Bangladesh include Colocasia esculenta,

C. gigantea, Dioscorea alata, D. bulbifera, D. esculenta, D. hamiltonii, D. pentaphylla, D. pubera, Homalomena aromatica, Manihot esculenta, Maranta arundinacea, Melocanna baccifera, Nymphaea nouchali, Pachyrhizus erosus.

7.2.3 *Pulses and Nuts*

Indigenous pulses and nuts of Bangladesh including *Arachis hypogaea, Cajanus cajan, Canavalia ensiformis, Firmiana colorata, Hodgsonia macrocarpa, and Macrotyloma uniflorum* are a plentiful source of plant proteins and are intended for making curries and soups, while nuts are used as high-value snacks.

7.2.4 *Horticultural Produce*

7.2.4.1 Fruits

Functional Foods

Functional foods play an imperative role in disease prevention. They are effective against multiple diseases and inflammations. In Bangladesh, these fruits play an important role in improving the health of individuals. Few examples of these fruits include *A. latifolia, Ananas comosus, A. montanum, Artocarpus chama, A. heterophyllus, A. lacucha, Averrhoa bilimbi, A. carambola, Bridelia retusa, M. B. stipularis, Buchanania cochinchinensis, Capparis zeylanica, Carallia brachiata, Carica papaya, C. hystrix, C. maxima, Dimocarpus longan, Diospyros blancoi, Elaeocarpus floribundus, E. tectorius, Elaeagnus latifolia, Emblica officinalis, Garcinia cowa, G. lanceifolia, G. pedunculata, Limonia acidissima, Manilkara zapota, Mangifera indica, M. sylvatica, Melastoma malabathricum, Morus indica, Nypa fruticans, Passiflora foetida, Phoenix paludosa, P. sylvestris, Pithecellobium dulce, Psidium guajava, Psidium guineense, Salacia chinensis, Sonneratia apetala, S. caseolaris, Spondias dulcis, Syzygium cumini, S. cymosum, S. malaccense, S. fruticosum, S. nervosum, Tamarindus indica, Tamilandia uliginosa, Trapa bispinosa, Willughbeia edulis, Xerospermum noronhianum, Ziziphus oenoplia, and Z. rugosa.* All of these local food products have high antioxidant and antibacterial activities, and their nutritional and therapeutic significance is being further investigated.

Confectionary and Beverages

Traditional fruits have multiple applications in the confectionery industry. They are used in making fruit-flavored candies, jellies, jams, and fruit beverages including drinks, juices, nectars, purees. Some of them are commonly used for their natural

Fig. 7.3 Non-traditional seasonal fruits in a street market at Narayanganj, Bangladesh

sweetness and flavor. *Antidesma acidum, A. bunius, A. ghaesembilla, Baccaurea ramiflora, S. pinnata, Stixix suaveolens, Potentilla indica, Protium serratum, Z. mauritiana, Haematocarpus validus, L. senegalensis, Limonia acidissima. Morus indica, Allophylus cobbe, Alphonsea ventricosa, Ampelocissus barbata, Baccaurea ramiflora. Borssus flabellifer, Bouea oppositifolia, D. malabarica, Ficus auriculata, F. semicordata, F. racemosa F. jangomas,* Indian *Grewia asiatica, Paramignya citrifolia, Terminalia bellirica* are few examples of fruits which are often used in the production of fruit fillings and coatings for chocolates and other confectionery products. Figure 7.3 shows Non-traditional seasonal fruits in a street market at Narayanganj, Bangladesh.

7.2.4.2 Vegetables

Herbal and Pharmaceutical

Local vegetables in Bangladesh encompass *Alternanthera philoxeroides, A. sessilis, Amaranthus spinosus, A. viridis, Ampelygonum chinense, Bacopa monneri, B. tulda, Begonia barbata, Benicasa hispida, Blumea lacera, Boerhavia diffusa, Bombax ceiba, Brassica alba, B. campestris, B. napus, Carica papaya, C. ambrosioides, Cissus adnata, C. repens, C. gigantea, Cucumis hystrix, Dillenia indica, D. pentaphylla, Dioscorea alata, D. bulbifera, D. esculenta, D. hamiltonii, D. pentaphylla, D. pubera, Diplazium esculentum, D. polypoides, Ficus racemosa, F. semicordata, Glinus oppositifolius, Ipomea aquatica, Ipomea batatas, Leucas aspera, Ludwigia adscendens, Malva verticillata, Marsilea minuta, Momordica cochinchinensis,*

Oroxylum indicum, Oxalis corniculata, Pachyrrhizus erosus, Pisum sativum, P. quadrifolia, Premna esculenta, Rumex dentatus, Senegalia rugata, Senna tora, Sesbania grandiflora, Typhonium trilobatum, and Xanthosoma sagittifolium. All of these vegetables plants have potential applications in herbal and pharmaceutical industries due to their medicinal properties. Socioeconomic wellbeing of a large number of Bangladesh farmers depends on herbs cultivation.

Curries and Soups

In Bangladesh, the following group of vegetables find its applications in curries, soups, salads, cuisines, and spices; *Bambusa balcooa, B. nutans, Citrus hystrix, C. diffusa, Corchorus capsularis, Coccinia grandis, Enhydra fluctuans, Glebionis coronaria, Nymphaea nouchchali, Paederia foetida, Parkia timorensis, Portulaca oleracea, Solanum indicum, S. lasiocarpum, S. nigrum, Sonneratia apetala, S. caseolaris, Sphenoclea zeylanica, Zehneria scabra, and Zingiber officinale.* Moreover, *Curcuma longa, Dendrocalamus giganteus, Lasia spinosa, and Mangifera indica* are commonly used for pickling.

Beverages and Puddings

Angiopteris evecta, Artocarpus heterophyllus, Chenopodium album, Ficus auriculata, Hibiscus sabdariffa, Melocanna baccifera, Spodias dulcis, and S. pinnata are used as essential ingredients in making canned slices, juices, jams, jellies. *Oxalis corniculata, Pachyrrhizus erosus, R. maritimus,* and *Canavalia ensiformis* are popularly used for making custard, porridge, and puddings.

Functional Foods

Vegetables that include *Acmella calva, Alocasia macrorrhizos, Alpinia nigra, Centella asiatica, Colocacia esculenta, Moringa oleifera,* have functional properties and are used as functional foods. *Amorphophalus bulbifer, Commelina benghalensis, Xanthium strumarium* are commonly used in bread making.

7.2.5 *Others*

7.2.5.1 Use as Spice and Condiments

Ajuga macrosperma, Allium sativum, Etlingera linguiformis, Eryngium foetidum, Ocimum americanum, O. basilicum, O. basilicum var. *purpurescene, O. gratissimum, Oenanthe javanica, Pandanus amaryllifolius, Sesamum indicum, Zanthoxylum rhetsa* are some common examples of condiments plants of Bangladesh which support livelihood of the farming communities.

7.2.5.2 Juice

Pulp of the fruits including *Ananas comosus, Citrus aurantium, C. maxima, Mangifera indica* are used at household to industrial level for production of juices.

7.2.5.3 Sap

Sap of *Borassus flabellifer, Nypa fruiticans, Phoenix sylvestris* is used as refreshing beverage of for the production of value added products, and is considered as promising source of rural income in various parts of Bangladesh.

7.2.5.4 Jams

Ananas comosus, Flacourtia indica, F. inermis, F. jangomas, Hibiscus sabdariffa, Morus indica, Psidium guajva, Sonneratia apetala, S. caseolaris, Spondias dulcis, S. pinnata are a few amongst most exploited native fruits of Bangladesh traditionally or commercially exploited for the production of jams.

7.2.5.5 Drinks

Aegle marmelos, Averrhoab bilimbi, A. carambola, Cocos nucifera (green coconut), *Mangifera indica* (green), *Tamarindus indica* are preferred foods of plant origin to develop value added refreshing drinks in different seasons and different cultures of the country.

7.2.5.6 Pickles/Leathers/Food Preserves

Allium sativum, Averrhoa bilimbi, A. carambola, Bambusa balcooa, B. nutans, B. tulda, Carissa carandas, Citrus assamensis, Citrus hystri, Dendrocalamus giganteus, D. hamiltonii, Dillenia indica, D. pentagyna, Elaeocarpus floribundus, E. tectorious, Emblica officinalis, Flacourtia indica, F. inermis, F. jangomas, Garcinia cowa, G. lanceifolia, G. pedunculata, G. xanthochymus, Mangifera indica (green), *Melocanna baccifera, Spondias dulcis, S. pinnata, Tamarindus indica, Zehneria scabra, Ziziphus mauritiana* (local variety) are extensively used to develop an array of salted and sweetened products such as pickles, fruits leathers and preserves in various cultures of urban and rural Bangladesh. Figure 7.4 shows Pickles and leather from non-traditional fruits in a street market at Narayanganj, Bangladesh.

Fig. 7.4 Pickles and leather from non-traditional fruits in a street market at Narayanganj, Bangladesh

7.2.5.7 Sauce

Brassica campestris, Tamarindus indica are traditionally used in Bangladesh for making a variety of sauces.

7.2.5.8 Murabba/Candy

Aegle marmelos, Ananas comosus, Arachis hypogea, Benicasa hispida, Cocos nucifera, Emblica offinalis, Mangifera indica, Sesamum indicum are used for making candy or *murabba* – a sugar dense preserve. Figure 7.5 elaborates Murabba from non-traditional fruits in a street market, Bangladesh.

7.2.5.9 Halwa

Artocarpus heterophyllus, Carica papaya, Cucurbita máxima are conventionally used for making sweet puddings in alone or in combination with semolina or other edible grains flours.

Wild, neglected and lesser used edible plants associated to different categories of uses are given in Table 7.1.

Fig. 7.5 Murabba from non-traditional fruits in a street market, Bangladesh

Table 7.1 Neglected food plants from Bangladesh associated to different use categories

Botanical name	Bangla name (s)	Edible parts	Traditional food /other uses	Citations
Acmella calva	Ojoni shag	Tender leaves	Cooked as vegetables. Inflorescences paste is used tooth ache.	Khatun et al. (2013)
Acronychia pedunculata	Bon jamir	Fruits	Ripe fruit is eaten.	Pasha and Uddin (2019)
Aegle marmelos	Bel	Fruits	*Murabba* is made from unripe fruit pulp, ripe fruit pulp is used to make drink, and also eaten. It has medicinal uses.	Begum (1987)
Ajuga macrosperma	Ajuga	Leaves	Leaves used as condiment in curries. Also have medicinal uses.	Khatun et al. (2013)
Alangium salvifolium	Ankarkanta	Fruits	Ripe fruit is eaten.	Das (1987)
Allium sativum	Roshun	Bulbs	Vegetables, spices, pickles. Also have medicinal uses.	Khatun et al. (2013)
Allophylus cobbe	Chumigola gach	Fruits	Ripe fruits are eaten.	Pasha and Uddin (2019)
Alocasia macrorrhizos	Mankachu	Rhizomes	Cooked in curries. Also have medicinal uses.	Rahman and Khatun (2020)

(continued)

Table 7.1 (continued)

Botanical name	Bangla name (s)	Edible parts	Traditional food /other uses	Citations
Alphonsea ventricosa	Ramkola	Fruits	Ripe fruits are eaten.	Pasha and Uddin (2019)
Alpinia nigra	Tara	Tender stem	Cooked in curries. Also has medicinal value.	Islam et al. (2012)
Alternanthera philoxeroides	Haicha shak	Leaves	Cooked as vegetables in curries.	Khatun et al. (2013)
Alternanthera sessilis	Hechi shak	Leaves	Cooked as vegetables in curries.	Khatun et al. (2013)
Amaranthus spinosus	Kanta notey	Stems and leaves	Cooked as vegetables. Also has medicinal value.	Khatun et al. (2013)
Amaranthus viridis	Notey shak	Stems and leaves	Cooked as vegetables. Also has medicinal value.	Khatun et al. (2013)
Amorphophalus bulbifer	Bon ool	Rhizomes and petioles	Cooked as vegetables. Also has medicinal value.	Alam (2020)
Amorphophalus paeonifolius var. *campanulatus*	Ool kachu	Rhizomes	Cooked as vegetables. Also has medicinal value.	Alam (2020)
Ampelocissus barbata	Jaria lahiri	Tender shoots and leaves. Fruits	Cooked as vegetables and ripe fruits eaten fresh. Also has medicinal value.	Khatun et al. (2013)
Ampelocissus latifolia	Pahar angur	Fruits	Ripe fruits are edible.	Author's observation
Ampelygonum chinense	Mohicharan sak	Tender shoots and leaves	Cooked as vegetables. Acidic in taste.	Khatun et al. (2013)
Ananas comosus	Anarash	Fruits, tender base of young leaves	Ripe fruits are eaten fresh. Drink, jam and *murabba* are made from fruits. Leaf-base juice is used as vermifuge.	Begum (1987)
Angiopteris evecta	Raj denkhi shag	Tender fronds	Cooked as vegetables.	Khatun et al. (2013)
Annona muricata	Kanta ata	Fruits	Ripe fruits are eaten.	Pasha and Uddin (2019)
Annona reticulata	Nona, ataphal	Fruits	Ripe fruits are eaten.	Pasha and Uddin (2019)
Annona squamosa	Sharifa	Fruits	Ripe fruits are eaten.	Pasha and Uddin (2019)
Antidesma acidum	Chutki	Fruits	Ripe fruits are eaten.	Pasha and Uddin (2019)

(continued)

Table 7.1 (continued)

Botanical name	Bangla name (s)	Edible parts	Traditional food /other uses	Citations
Antidesma bunius	Bon shiyalboka	Fruits	Ripe fruits are eaten.	Pasha and Uddin (2019)
Antidesma ghaesembilla	Elena	Fruits	Ripe fruits are eaten.	Das (1987)
Antidesma montanum	Shiyalbuka	Fruits	Ripe fruits are eaten.	Pasha and Uddin (2019)
Arachis hypogea	Chinabadam	Fruits / nuts	Roasted seeds are eaten, made into candies. Edible oil is extracted from seeds.	Begum (1987)
Artocarpus chama	Cham, Chapalish	Fruits	Ripe fruit pulp is edible.	Das (1987)
Articarpus heterophyllus	Kaanthal	Fruits, seeds	Juicy flesh around the seed is eaten, unripe fruits and seeds are cooked as vegetables, seeds are roasted and taken as nuts and also made as smash, seeds are also boiled and made into *halwa*.	Begum (1987)
Artocarpus lacucha	Barta, Dewa	Fruits	Ripe fruit pulp is edible.	Alam (1998)
Averrhoa bilimbi	Bilimbu	Fruits	Used to prepare drink, pickles and jam.	Pasha and Uddin (2019)
Averrhoa carambola	Kamranga	Fruits	Used to prepare drink, pickles and jam.	Pasha and Uddin (2019)
Baccaurea ramiflora	Bhubi, Lotkon	Fruits	Ripe fruit pulps are edible.	Alam (1998)
Bacopa monneri	Brahmishak	Leaves	Leaves cooked as vegetables. Also have medicinal uses.	Khatun et al. (2013)
Bambusa balcooa	Borak bans	Young shoots, culms	Young shoots are cooked as vegetables, and also used to prepare pickles. Culms in construction and wicker work.	Banik (1998)
Bambusa nutans	Mahal bans	Young shoots, culms	Young shoots are cooked as vegetables, and also used to prepare pickles. Culms in construction and wicker work.	Banik (1998)

(continued)

Table 7.1 (continued)

Botanical name	Bangla name (s)	Edible parts	Traditional food /other uses	Citations
Bambusa tulda	Talla bans	Young shoots, culms	Young shoots are cooked as vegetables, and also used to prepare pickles. Culms in construction and wicker work.	Banik (1998)
Begonia barbata	Sil tetoi / Tokpata	Petioles	Vegetables, acidic in taste. Also eaten raw.	Khatun et al. (2013)
Benicasa hispida	Chalkumra	Tender shoots, leaves, green fruits. Ripe fruits.	Vegetables. *Murabbas* are prepared from mature ripe fruits.	Khatun et al. (2013)
Blumea lacera	Kukurshinga	Tender shoots and leaves	Cooked as vegetables.	Khatun et al. (2013)
Boerhavia diffusa	Punornova	Leaves	Vegetables and medicinal in dropsy.	Khatun et al. (2013)
Bombax ceiba	Shimul	Stamen bundles	Cooked as vegetables.	Alam (2002)
Borassus flabellifer	Tal	Sap from male inflorescence, pulp juice	Fresh sap as drink and prepared molasses by boiling and thick juice syrup is used as dessert.	Islam et al. (2012)
Bouea oppositifolia	Mailla aam	Fruits	Fruit is eaten, acidic. Medicinal use of fruits in dog bites.	Das (1987)
Brassica alba	Sada sharisha	Tender leaves and seeds	Leaves as vegetables, seed paste as spices in curries.	Islam et al. (2012)
Brassica campestris	Sharisha	Leaves, tender shoots and seeds	Leaves and tender shoots are cooked as vegetables. Sauce from seed paste. Edible oil extracted from seeds.	Author's observation.
Brassica napus	Maghi sharisha	Seeds	Seed paste as spices in curries.	Khatun et al. (2013)
Bridelia retusa	Kantakosoi	Fruits	Fruits are eaten.	Das (1987)
Bridelia stipularis	Koombi lodhi	Fruits	Fruits are eaten.	Pasha and Uddin (2019)
Buchanania cochinchinensis	Nala amshi	Fruits	Fruits are eaten.	Das (1987)
Cajanus cajan	Arhar	Green pods, seeds, leaves	Green pods are boiled and eaten as a substitute of main dish; seeds used as pulses. Leaves are used in jaundice.	Alam (2020)

(continued)

Table 7.1 (continued)

Botanical name	Bangla name (s)	Edible parts	Traditional food /other uses	Citations
Canavalia ensiformis	Makhan shim	Green pods, seeds	Vegetables. Seeds cooked as pulses	Alam (2020)
Capparis zeylanica	Kalookra	Ripe fruits	Edible.	Pasha and Uddin (2019)
Carallia brachiata	Roshoko, Kiabong	Ripe fruits	Edible.	Das (1987)
Carica papaya	Pepe	Green and ripe fruits	Green fruits are cooked as vegetables. *Halwa* is made from boiled green fruit smeshs. Ripe fruits are eaten.	Begum (1987)
Carissa carandas	Karamcha	Fruits	Cooked in curries, also pickled.	Pasha and Uddin (2019)
Catunaregam spinosa	Monkanta	Fruits	Ripe fruits are eaten.	Das (1987)
Centella asiatica	Thankuni	Leaves	Green leaves are taken salad, paste as curry substitute. Also has medicinal uses.	Khatun et al. (2013)
Centrostachys aquatica	Panitaki / Thuash	Tender shoots	Vegetables.	Author's observation
Cheilocostus speciosus	Keomul	Tender stems	Vegetables.	Author's observation
Chenopodium album	Batuashak	Leaves	Vegetables. Medicinal – vermiguge.	Khatun et al. (2013)
Chenopodium ambrosioides	Ghandha batua	Leaves	Vegetables. Medicinal – vermiguge.	Rahman and Khatun (2020)
Cissus adnata	Vatialota, Shogor amila	Tender shoots and leaves	Vegetables.	Khatun et al. (2013)
Cissus repens	Marmairalota	Tender shoots and leaves	Vegetables.	Khatun et al. (2013)
Citrus assamensis	Adajamir	Fruit s	Drink, pickles.	Pasha and Uddin (2019)
Citrus aurantium	Komola	Fruits	Fruits are eaten.	Pasha and Uddin (2019)
Citrus hystrix	Shatkora	Fruits	Rind in beef and other curries. Pickled.	Pasha and Uddin (2019)

(continued)

Table 7.1 (continued)

Botanical name	Bangla name (s)	Edible parts	Traditional food /other uses	Citations
Citrus maxima	Jambura	Fruits	Mesocarps/ pulps are edible, processed in different ways. Peeled rind is used in making marmalades.	Begum (1987)
Coccinia grandis	Telakucha	Leaves and green fruits	Vegetables. Leaf juice is reported to be affective in diabetes.	Khatun et al. (2013)
Cocos nucifera	Daab, Narikel	Fruits	Fruit water as drink, kernels used many ways: milk extracted from kernel paste is cooked in different curries, kernel scarps used in sweetmeats and candies.	Begum (1987)
Colocacia esculenta	Kochu	Petioles, leaves, stolons, rhizomes	Cooked as vegetables.	Alam (2020)
Colocacia gigantea	Boro kachu, Salad kachu	Petioles, leaves, rhizomes	Cooked as vegetables.	Khatun et al. (2013)
Commelina benghalensis	Kanshira	Tender leaves	Cooked as vegetables.	Khatun et al. (2013)
Commelina diffusa	Botboitta shag	Tender leaves	Cooked as vegetables.	
Corchorus capsularis	Mara shak	Tender leaves	Cooked as vegetables	Khatun et al. (2013)
Cordia dichotoma	Bohal	Fruits	Ripe fruits are eaten.	Pasha and Uddin (2019)
Cucumis hystrix	Bandar marfa	Fruits	Green fruits are cooked as vegetables.	Author's observation
Cucurbita máxima	Mishti kumra	Fruits	Both green and matured ripe fruits are cooked as vegetables. Ripe fruits are also made in to Halwas.	Khatun et al. (2013)
Curcuma longa	Halud	Rhizomes and inflorescence	Fresh rhizome paste is fried and taken as curry. Inflorescence is also cooked in curries. Has a wide range of uses from spice to medicinal and cultural.	Khatun et al. (2013)
Dendrocalamus giganteus	Budhum bans	Young shoots	Cooked as vegetables, and also made into pickles.	Banik (1998)

(continued)

Table 7.1 (continued)

Botanical name	Bangla name (s)	Edible parts	Traditional food /other uses	Citations
Dendrocalamus hamiltonii	Pencha bans	Young shoots	Cooked as vegetables, and also made into pickles.	Banik (1998)
Dillenia indica	Chalta	Green fruits	Cooked in curries and made into pickles.	Das (1987)
Dillenia pentagyna	Ajugi	Green fruits	Cooked in curries and made into pickles.	Das (1987)
Dimocarpus longan	Ashphal	Fruits	Ripe fruits are eaten.	Pasha and Uddin (2019)
Dioscorea alata	Chupri alu	Root tubers	Cooked as vegetables and also eaten as scarcity food.	Alam (2020)
Dioscorea bulbifera	Chamalu	Root tubers and bulbils	Cooked as vegetables and also eaten as scarcity food.	Alam (2020)
Dioscorea esculenta	Mou alu	Root tubers	Cooked as vegetables and also eaten as scarcity food.	Alam (2020)
Dioscorea hamiltonii	Bonalu	Root tubers	Cooked as vegetables and also eaten as scarcity food.	Alam (2020)
Dioscorea pentaphylla	Suara alu	Root tubers	Cooked as vegetables and also eaten as scarcity food.	Alam (2020)
Dioscorea pubera	Kukur alu	Root tubers	Cooked as vegetables and also eaten as scarcity food.	Alam (2020)
Diospyros blancoi	Beelati gab	Fruits	Ripe fruits are eaten.	Pasha and Uddin (2019)
Diospyros malabarica	Deshi gab	Fruits	Ripe fruits are eaten. Tannins squeezed from fresh fruits were once used in country fishing net culling.	Das (1987)
Diospyros toposia	Katgula	Fruits	Ripe fruits are eaten.	Das (1987)
Diplazium esculentum	Denki shak	Tender fronds	Cooked as vegetables	Khatun et al. (2013)
Diplazium polypoides	Denki shak	Tender fronds	Cooked as vegetables	Khatun et al. (2013)
Elaeagnus latifolia	Amjam	Fruits	Ripe fruits are eaten.	

(continued)

Table 7.1 (continued)

Botanical name	Bangla name (s)	Edible parts	Traditional food /other uses	Citations
Elaeocarpus floribundus	Jalpai	Fruits	Eaten fresh, cooked in curries and pickled.	Das (1987)
Elaeocarpus tectorious	Belfoi	Fruits	Eaten fresh, cooked in curries and pickled.	Pasha and Uddin (2019)
Emblica officinalis	Amloki	Fruits	Eaten fresh, pickled and made into *murabba*.	Alam (1998)
Enhydra fluctuans	Helencha	Tender shoots and stems	Cooked as vegetables.	Khatun et al. (2013)
Eryngium foetidum	Bilati dhonia	Leaves	As a flavoring agent and condiment in curries.	Alam (2020)
Etlingera linguiformis	Chengi tara	Rhizomes	Used as a spice and condiment in meat curries by tribal people in the Hill Tracts.	Author's observation
Ficus auriculata	Bara dumur	Figs	Fresh figs are cooked in curries and ripe ones are eaten.	Alam (1998)
Ficus racemosa	Joygga dumur	Figs	Fresh figs are cooked in curries and ripe ones are eaten. It has also medicinal uses in diabetics.	Alam (1998)
Ficus semicordata	Sadimadi dumur	Figs	Fresh figs are cooked in curries and ripe ones are eaten.	Alam (1998)
Firmiana colorata	Faisaudal	Seeds	Eaten roasted.	Das (1987)
Flacourtia indica	Baichani	Fruits	Ripe fruits eaten, and also made into jam and pickles.	Das (1987)
Flacourtia inermis	Loai	Fruits	Ripe fruits eaten, and also made into jam and pickles.	Das (1987)
Flacourtia jangomas	Paninnya mala, Lukluki	Fruits	Ripe fruits eaten, made into jam and pickles.	Alam (1998)
Garcinia cowa	Kao phal	Fruits	Ripe fruits eaten, green fruits are cooked in curries and pickled.	Alam (1998)
Garcinia lanceifolia	Bandairi phal, Thisuru	Fruits	Ripe fruits eaten, green fruits are cooked in curries and pickled.	Pasha and Uddin (2019)
Garcinia pedunculata	Taikor	Fruits	Ripe fruits eaten, green fruits are cooked in curries and pickled.	Pasha and Uddin (2019)

(continued)

Table 7.1 (continued)

Botanical name	Bangla name (s)	Edible parts	Traditional food /other uses	Citations
Garcinia xanthochymus	Dephal	Fruits	Ripe fruits eaten, green fruits are cooked in curries and pickled.	Alam (1998)
Glebionis coronaria	Baburi shak, Chandramallika	Leaves	Cooked as vegetables.	Author's observation
Glinus oppositifolius	Gima sak	Leaves	Cooked as vegetables.	
Grewia asiatica	Phalsa	Fruits	Ripe fruits are eaten.	Alam (1998)
Haematocarpus validus	Agnipholi, Rosoko	Fruits	Ripe fruits are eaten.	Author's observation
Hibiscus sabdariffa	Chukai	Leaves, fruits	Cooked in curries, and also made into jam.	Alam (2020)
Hodgosia macrocarpa	Pilagula	Matured seeds	Roasted and eaten.	Author's observation
Homalomena aromatica	Ghandabi kachu	Petioles and rhizomes	Cooked as vegetables.	Khatun et al. (2013)
Ipomea aquatic	Kolmi sak	Young shoots and leaves	Cooked as vegetables.	Khatun et al. (2013)
Ipomea batatas	Mishti alu	Young leaves and root tubers	Cooked as vegetables, boiled tubers are taken as food substitutes and made into *halawas*.	Khatun et al. (2013)
Lasia spinosa	Kanta kachu	Petioles and rhizomes	Cooked as vegetables. Also have medicinal values.	Khatun et al. (2013)
Lepisanthes rubigonosum	Bon harina	Fruits	Ripe fruits are eaten.	Das (1987)
Lepisanthes senegalensis	Aamjoon	Fruits	Ripe fruits are eaten.	Pasha and Uddin (2019)
Leucas aspera	Shetodron	Tender leaves	Cooked as vegetables. Also have medicinal uses.	Khatun et al. (2013)
Limonia acidissima	Kodbel	Fruits	Ripe fruit pulp is eaten.	Pasha and Uddin (2019)
Ludwigia adscendens	Kasardam	Tender leaves and shoots	Cooked as vegetables.	Khatun et al. (2013)
Macrotyloma uniflorum	Kurti kolai	Seeds	Cooked as pulses.	Author's observation
Malva verticillata	Napashak	Leaves	Cooked as vegetables.	Khatun et al. (2013)

(continued)

Table 7.1 (continued)

Botanical name	Bangla name (s)	Edible parts	Traditional food /other uses	Citations
Mangifera indica	Aam	Fruits	Green fruits cooked in curries, pickled, and made into drinks, *murabbas*. Ripe fruits are eaten, and made juice.	Author's observation
Mangifera sylvatica	Uriaam	Fruits	Ripe fruits eaten.	Alam (1998)
Manihot esculenta	Shimulalu	Root tubers	Boiled and taken as main food substitute, powdered into arrowroots.	Alam (2020)
Manilkara zapota	Sofeda	Fruits	Ripe fruits are eaten.	Pasha and Uddin (2019)
Maranta arundinacea	Bans alu	Root tubers	Boiled and taken as main food substitute, powdered into arrowroots	Author's observation
Marsilea minuta	Sushnisak	Leaves	Cooked as vegetables	Khatun et al. (2013)
Melastoma malabathricum	Datranga	Fruits	Ripe fruit is eaten.	Das (1987)
Melocanna baccifera	Mulibans	Young shoots	Young shoots cooked as vegetables, sometimes taken as a substitute of staple food. Also pickled.	Khatun et al. (2013)
Mollugo spergula	Khetpapara	Leaves	Cooked as vegetables.	Khatun et al. (2013)
Momordica cochinchinensis	Bonkakrol	Green fruits	Vegetables. Also have medicinal uses.	
Moringa oleifera	Sajna	Leaves, green fruits	Vegetables. Mature seed powder has antibacterial properties.	Khatun et al. (2013)
Morus indica	Tunt	Fruits	Fresh fruits are eaten, also made into jams.	Pasha and Uddin (2019)
Musa paradisica	Kancha kola	Pseudostem piths	Cooked as vegetables.	Alam (2020)
Nymphaea nouchchali	Shapla, Shaluk	Petioles, rhizomes	Petioles are cooked as vegetables. Rhizomes are boiled or burned and taken as a starch sustitute.	Begum (1987)
Nypa fruiticans	Golpata	Fruits and young inflorescence	Sap by cutting young inflorescence is taken as drink, used to prepare molasses and sweet porridges, endosperm of fruits are eaten.	Das (1987)

(continued)

Table 7.1 (continued)

Botanical name	Bangla name (s)	Edible parts	Traditional food /other uses	Citations
Ocimum americanum	Bontulshi	Leaves	As spice and condiment in curries.	Alam (2020)
Ocimum basilicum	Babuitulshi	Leaves	As spice and condiment in curries.	Alam (2020)
Ocimum basilicum var. *purpurescene*	Laltulsi	Leaves	As spice and condiment in curries.	Alam (2020)
Ocimum gratissimum	Ramtulshi	Leaves	As spice and condiment in curries.	Alam (2020)
Oenanthe javanica	Painya adani, Pan turashi	Leaves and tender shoots	Cooked in curries as condiment.	Khatun et al. (2013)
Oroxylum indicum	Khona	Tender fruits, stem bark	Cooked as vegetables. Tender stem bark and fruits are reported to be used to cure jaundice.	Khatun et al. (2013)
Oxalis corniculata	Chuka amrul	Petioles and leaves	Cooked in curries.	Khatun et al. (2013)
Pachyrhizus erosus	Shak alu	Root tubers	Eaten raw and also cooked in vegetables. Root decoction is reported to be diuretic.	Alam (2020)
Paederia foetida	Gondho bhadali	Leaves	Cooked as vegetables.	Khatun et al. (2013)
Pandanus amaryllifolius	Polao pata	Leaves	Used in cooking fried rice and other desserts for fragrance.	Author's observation
Panicum miliaceum	Cheena	Grains	Substitute of rice.	Biswas et al. (2021)
Paramignya citrifolia	Bonjamir	Fruits	Fruit is eaten.	Das (1987)
Parkia timorensis	Kukitetui	Young pods	Cooked as vegetables.	Pasha and Uddin (2019)
Passiflora foetida	Jumkolota	Fruits	Fruits are eaten fresh.	Alam (2002)
Phoenix paludosa	Hantal	Fruits	Kernels and endosperms are eaten.	Das (1987)
Phoenix sylvestris	Khejur	Sap, fruits	Sap is taken fresh, also made into brown sugars. Ripe fruits are eaten.	Alam (1998)
Pisum sativum	Motorshuti	Leaves and seeds	Leaves cooked as vegetables. Seed a traditional pulse.	Khatun et al. (2013)
Pithecellobium dulce	Khiya babla, Jilapi	Fruits	Ripe fruits are eaten.	Pasha and Uddin (2019)

(continued)

Table 7.1 (continued)

Botanical name	Bangla name (s)	Edible parts	Traditional food /other uses	Citations
Portulaca oleracea	Nunia shak	Leaves	Cooked as vegetables.	Khatun et al. (2013)
Portulaca quadrifolia	Boro nunia	Leaves	Cooked as vegetables.	Khatun et al. (2013)
Potentilla indica	Indian strawberry	Fruits	Ripe fruits are eaten.	Author's observation
Premna esculenta	Lalong shak	Tender leaves	Cooked as vegetables.	Khatun et al. (2013)
Protium serratum	Gutguttya	Fruits	Eaten raw and cooked in curries.	Alam (1998)
Psidium guajava	Peyara	Fruits	Ripe fruits are eaten, and used in making jam.	Begum (1987)
Psidium guineensis	Bon peyara	Fruits	Ripe fruits are eaten, and used in making jam.	Author's observation
Rumex dentatus	Bon-palong	Young leaves	Cooked as vegetables.	Khatun et al. (2013)
Rumex maritimus	Ganga-palong	Young leaves	Cooked as vegetables.	Khatun et al. (2013)
Salacia chinensis	Madhuphal	Fruits	Ripe fruits edible.	Pasha and Uddin (2019)
Senegalia rugata	Kuchilot, Leronpata	Young leaves, fruits	Young roots cooked in curries. Dry fruits used for hair wash.	Alam (1998)
Senna tora	Chakunda	Young leaves	Cooked as vegetables.	Khatun et al. (2013)
Sesamum indicum	Til	Seeds	Spices, candies and oil extraction.	Alam (2020)
Sesbania grandiflora	Bakful	Flowers	Cooked as vegetables.	Khatun et al. (2013)
Setaria itálica	Kaon	Grains	Used as rice substitute.	Biswas et al. (2021)
Solanum indicum	Bon begun	Young fruits	Cooked as vegetables.	Rahman and Khatun (2020)
Solanum lasiocarpum	Tomato begun	Young fruits	Cooked as vegetables.	Author's observation
Solanum nigrum	Tit begun	Young fruits	Cooked as vegetables.	Rahman and Khatun (2020)
Sonneratia apetala	Keora	Fruits	Cooked in curries and made into jams.	Das (1987)
Sonneratia caseolaris	Soila	Fruits	Cooked in curries and made into jams.	Das (1987)
Sphenoclea zeylanica	Zilmorich	Tender leaves	Cooked as vegetables.	Khatun et al. (2013)

(continued)

Table 7.1 (continued)

Botanical name	Bangla name (s)	Edible parts	Traditional food /other uses	Citations
Spondias dulcis	Bilati aamra	Fruits	Eaten fresh, cooked in curries, pickled and made into jam.	Das (1987)
Spondias pinnata	Deshi aamra	Fruits	Eaten fresh, cooked in curries, pickled and made into jam.	Das (1987)
Sterculia foetida	Udal	Seeds	Roasted and eaten.	Alam (1998)
Stixix suaveolens	Madhumalati	Fruits	Ripe fruits edible.	Das (1987)
Syzygium cumini	Kalojam	Fruits	Ripe fruits are eaten.	Islam et al. (2012)
Syzygium cymosum	Khudijam	Fruits	Ripe fruits are eaten.	Pasha and Uddin (2019)
Syzygium fruticosum	Titijam	Fruits	Ripe fruits are eaten.	Das (1987)
Syzygium malaccense	Jamrul	Fruits	Ripe fruits are eaten.	Pasha and Uddin (2019)
Syzygium nervosum	Daphajam	Fruits	Ripe fruits are eaten.	Das (1987)
Tamarindus indica	Tentul	Fruit pulps	Made to drinks, sauce and pickled. Have other medicinal values.	Begum (1987)
Tamilandia uliginosa	Piralu	Fruits	Fresh fruits are cooked in curries and ripe ones are edible.	Das (1987)
Terminalia bellirica	Bohera	Fruits	Green fruits are eaten. Also is a component of myribolans.	Das (1987)
Trapa bispinosa	Paniphal	Fruits	Eaten raw.	Begum (1987)
Typhonium trilobatum	Ghetkochu	Petioles	Cooked as vegetables.	Khatun et al. (2013)
Willughbeia edulis	Lota aam	Fruits	Fruits are eaten.	Das (1987)
Xanthium strumarium	Ghagra	Tender shoots	Cooked as vegetables.	Khatun et al. (2013)
Xanthosoma sagittifolium	Dood kachu	Petioles and rhizomes	Cooked as vegetables.	Khatun et al. (2013)
Xanthosoma violaceum	Surma kachu	Petioles and rhizomes	Cooked as vegetables.	Khatun et al. (2013)
Xerospermum noronhianum	Bon lichu	Fruits	Ripe fruits are edible.	Pasha and Uddin (2019)

(continued)

Table 7.1 (continued)

Botanical name	Bangla name (s)	Edible parts	Traditional food /other uses	Citations
Zanthoxylum rhetsa	Bajna	Leaves	Cooked in curries and used as a condiment.	Alam (1998)
Zea mays	Bhutta	Seeds	Young combs burned or boiled and eaten as a substitute of rice.	Alam (2020)
Zehneria scabra	Rakhal shosha	Fruits	Consumed fresh and sometimes pickled. Also have medicinal properties.	Khatun et al. (2013)
Zingiber officinale	Ada	Inflorescence, rhizomes	Inflorescence cooked in curries as vegetables, rhizomes used as spices. Rhizome has multiple medicinal uses.	Khatun et al. (2013)
Ziziphus mauritiana (local variety)	Boroi, Kul	Fruits	Consumed fresh and pickled.	Begum (1987)
Ziziphus oenoplia	Anor, bon boroi	Fruits	Consumed fresh.	Das (1987)
Ziziphus rugosa	Bon boroi	Fruits	Consumed fresh.	Das (1987)

7.3 Industrial Valorization of Neglected Food Plants

People who are dependent on wild food plants mostly harvest food plants from wild states. The survival of various residents of Chittagong Hill is, directly and indirectly, dependent on these food plants (Miah and Chowdhury 2004; Miah et al. 2011). People who have adapted to consuming wild plant food but now live in areas outside of forest habitats often experience a sense of nostalgia for these unique products.

With limited quantities of gathered fresh food plants, no formal markets have developed. Traditional neglected food plants have some sort of informal market. These are mostly marketed in the local markets directly by the harvesters. Sometimes some street vendors or small shop owners buy from harvesters and sell them. Thus, there is a one or two-step marketing chain for neglected food plants in the country. But the market size is very small and confined to limited areas. As the traditional food plants are consumed by different communities from different areas, it creates a market demand for niche products and thus becomes commercialized. It also creates opportunities for domestication. Market access to underutilized food plants is, thus, an opportunity both for reducing poverty and domestication of potential crops. This will encourage in-situ conservation of local agro-biodiversity.

In Bangladesh plant food processing, marketing and export is under the purview of agro-processing industries. There are a few sporadic reports on the value chain analysis of some selected crops like mango, tomato, chili, turmeric, potato, mung bean, maize, and mustard oil seed (Katalyst 2016). Some value-added plant food products or preserved foods like pickles, jams, and food candied have some valorization. These have entered into both informal and formal markets but little information is available about these crops. The formal market systems of plant food products are included under the agro-processing industries sector. Value-added processed food items from mango, pine apple, jujube, and other fruits are prepared and marketed by some food manufacturing companies. These are shelved in departmental stores, grocers' shops and sold to consumers. Even some products are exported. This value chain process put some value to food plants. This sector has enormous potential for development through the forward and backward linkages. The major stakeholders of the informal markets are the producers, gatherers, and street vendors. The primary products are homemade pickles. The street vendors or hawkers procure fresh fruits from markets, prepare pickles, and sell those by themselves. Generally, they sell pickles in front of schools.

Bangladesh Agro-Processing Association (BAPA) is the Association for the food processing industries in Bangladesh. BAPA member industries produce limited trade-oriented products and concentrate mostly on export-oriented products. Bangladesh's government is implementing a wide cross-sector program on Employment that includes the agro-processing sector too (https://seipfd.gov.bd/publications/seip-at-a-glance-2022/ on 8 November 2022).

7.4 Nutritional Significance and Opportunities

Nutrition plays an imperative role in human health as it provides the energy necessary for human survival. The role of food commodity in meeting the nutritional requirements of the body increases the importance of that food product. This value is determined by the presence of essential nutrients in food.

In Bangladesh generally a dish is prepared either by cooking animal foods with plant recipes, or food plants alone. During food processing little nutrients are lost, mixing several plant parts comprising of various species makes this food more nutritious. Islam et al. (2012) prepared food composition tables and a database for 481 food items that included some selected ethnic food plants.

Reportedly, the local food plants in Bangladesh carry curative effects and some vegetables have high healing properties (Islam et al. 2013). When the fiber content of these food plants was analyzed the results showed that these traditional plants carried relatively large amounts of dietary fiber in comparison to the frequently used commercial food crops (Alam 2022). Pickles or *aachars* had been a traditional innovation in the food system for thousands of years. Recent studies indicate that this product has many health benefits.

7.5 Development Agenda, Constitutional Commitments and National Policies Toward Preserving Biodiversity and Promoting Neglected Food Plants Production

7.5.1 National Food Policy 2006

The centerpiece of the food policy targets the availability of adequate food for everyone with the following objectives (Ministry of Food and Disaster Management 2006):

1. To maintain ample, healthy, and wholesome food for individuals.
2. To strengthen an individual's buying capacity towards food availableness.
3. To guarantee the desired level of nutrients in the population.

7.5.2 National Nutrition Policy 2015

This policy targets to combat the challenges of nutritional deficiencies in less nutritionally privileged masses which include women, young girls, and children. Moreover, this policy also focuses on ameliorating the living conditions of the individuals in the state.

In addition, it will play a substantial role in formulating new plans and will also help in improving the already available measures to confront the challenges of malnutrition. These steps will pave a path toward enhancing the nutritional profile of the people in Bangladesh. The pivotal reasons behind malnutrition must be highlighted to guarantee proper nutrition.

7.5.3 The Industrial Policy 2016

The primary aim of the Industrial Policy of 2016 is to foster industrialization by facilitating collaboration between public and private entities. This collaboration seeks to create more opportunities and growth in the industrial sector. The policy's primary focus is on expediting the growth of the industrial sector, promoting entrepreneurship, and fostering the development of Small and Medium Enterprises (SMEs). The policy aims to establish and boost production, improve quality, and enhance the marketing abilities of industrial items. The agriculture and food processing industry has been designated as a high-priority sector due to its potential for rapid expansion, significant job creation, and ability to generate substantial export revenue.

7.5.4 The National Environment Policy 2018

This Policy envisions sustainable development through the preservation of the environment, and focus on the reduction of the detrimental impact of climate change. A certain section of the policy focuses on ensuring the safety of food and water throughout the entire processing.

7.5.5 The Second National Plan of Action for Nutrition (NPAN2) 2016–2025

The resurgence of the Bangladesh National Nutrition Council in 2015, led by the respected Prime Minister, has stimulated the creation of the second National Plan of Action for Nutrition (NPAN2). The plan is intended to implement the National Nutrition Policy 2015 for the next decade (2016–2025). Primarily, the second National Plan of Action for Nutrition (NPAN2) marked the first instance of genuine cross-sectoral collaboration and prioritization for nutrition in Bangladesh. The plan presented tangible measures to ensure coordination across multiple sectors and highlighted the significance of localized planning. Moreover, the plan has strengthened the country's existing commitment to foster a nutrition program that encompasses multiple sectors and encourages coordination among them. This approach aligns with the objectives set out in the National Nutrition Policy (NNP) of 2015 and moves Bangladesh closer to realizing its vision for improved nutrition.

7.5.6 The Bangladesh Food Safety Act, 2013

This legislation provides for the creation of a proficient, productive, and scientifically grounded Bangladesh Food Safety Authority. It aims to work in collaboration with other departments to ensure the regulation of food processing and distribution for both domestic consumption and international market. Additionally, the law intends to guarantee the populace's right to access safe food.

7.5.7 Bangladesh Food Safety Authority (BFSA)

A detailed review of the food safety system in Bangladesh was given by FISL (2021). The Food Safety Act of 2013 provided the base for the formation of the Food Safety Authority (BFSA) in 2015, whose work came under the ambit of the Food Ministry. The primary purpose of this authority is to regularize and observe food-related processes. This also focuses on the adequate availability of food using scientific techniques to deal with the challenge of rising food insecurity. Furthermore, food safety has been given due importance as BFSA regulates food safety rules in the state. This is the main body that defines the limits of the contaminants, heavy metals, toxins, additives and also determines permissible limits for pesticides and drugs.

The products being imported from other states came under the observation of the Bangladesh Standards and Testing Institution (BSTI) which verifies the label of compliance. The above-mentioned two authorities determine the safety of food samples at the domestic level. However, the absence of scientific technology has diminished the accomplishment of this task. Resultantly, the evaluation of local food plants and their products remains undermined.

7.6 Toxicological Challenges and Possible Solutions

Plants naturally contain toxins that can provide health benefits when consumed in specific quantities. Indigenous people have developed century-old recipes to prepare and consume these plants, making them toxin-free and wholesome. However, consuming certain plants in excessive amounts can lead to food poisoning. For example, in Bangladesh, numerous people were affected by food poisoning due to the excessive consumption of Xanthium strumarium (Khatun et al. 2013).

Raw consumption of Yams and taros can cause allergies, but native people have developed traditional methods to make these plants edible. To process Yams, the plant is first cut into small pieces and placed in baskets, which are then left overnight in flowing water. This process removes any undesirable material, and the plant becomes ready for cooking. Yam chunks are often soaked and boiled before their use in recipes to ensure they are fully cooked and safe to consume. Similarly, Taros are not consumed raw and must be boiled first to break down the calcium oxalate crystals present in them. In Taro curries, tamarind pulp is commonly used as an ingredient, as it is known to break down oxalate crystals (Saravanan et al. 2016). In Bangladesh, excessive use of chemical additives coupled with heavy adulteration and severe contamination is imposing health risks on the population. Certain chemicals are being used in food processing to avoid contamination. These chemicals have imposed a formidable challenge to food safety in the country (Mohiuddin 2019). Moreover, water and food are highly contaminated with Arsenic, and drinking this contaminated water has compromised the health of many people in various regions of the state (Rahaman et al. 2022).

Food becomes vulnerable to microbial contamination during pre-harvest and post-harvest processing (Noor and Feroz 2016). Reportedly, cooked food from the streets of Bangladesh is heavily contaminated with aerobic and coliform bacteria (Banik et al. 2019). The country is also confronting a serious challenge of food contamination owing to the excessive presence of metals and metalloids in food and water (Islam et al. 2018).

7.7 Traditional Knowledge and Gender Role

In any region, conventional methods and easily accessible food resources play a major role in the determination of the food system. The newly devised methods in any food system once become accepted by the locals move from generation to generation. Human's interaction with each other and with their surrounding is central to this socialization. Moreover, conventional knowledge has an edifice of adaptability and experience. Thus, the continual evolution of food commodities and consumption of neglected plants is primarily related to the availability of regional resources because the indigenous people are familiar with the nature of these plants and resources (Alam 2022).

The toxicity of mushrooms in Bangladesh is determined by several traditional methods. The development of clumps on the soil, when mushrooms are produced on bamboo, is an indication of the non-toxic mushrooms, which are considered safe for human consumption, while the absence of these clumps determines the presence of toxicity in mushrooms. Another technique for the identification of toxicity in mushrooms is the lime application. The blackening of this vegetable after treatment with lime shows the toxic nature of the mushrooms. Tanchangyaa, a tribe in Bangladesh, believes mushrooms with ants interaction is consumable for humans (Mohiuddin 2009).

Sun drying is an ancient method used for the preservation of fruits and vegetables. In Bangladesh, people living on the hill tops usually dry seasonal commodities to use them for the rest of the year (Alam 2022). Jujube is a common fruit dried for various purposes. This preservation technique provides a source of income for the locals and also makes the region food secure. There exist considerable variations in the harvesting of plants and people prefer to harvest multiple varieties instead of focusing on a single plant specie. The information about the local plants is gender-based in Bangladesh as women have better knowledge about the processing and preservation of these food plants (Alam 1998, 2002; Alam and Khisa 2000; Mohiuddin 2009; Alam 2022).

Reports from hilly parts of Bangladesh depicted that females gather approximately twice the amount of non-timber resources and fulfill their dietary needs in comparison to their male counterparts Kar (2010). Similarly, two tribes living in the hilly areas of Bangladesh were familiar with the knowledge of local plants and women had a better understanding of these local species. This is the very reason for the increase involvement of women in minor food business in these areas Mohiuddin (2009).

7.8 Ecology, Distribution Pattern, Genetic Resource Base and Domestication of Neglected Food Plants

In Bangladesh, it is common practice to deliberately grow underutilized plants that are native to the area. Local food plants have been grown in farmlands since ancient times (Alam et al. 1996; Alam 1997, 2020). People living in hilly areas also rely on numerous local food plants obtained from indigenous sources (Kar 2010; Alam 1998, 2002, 2022). For the growth of root crops and vegetables, limited shade is essential. These vegetables are typically cultivated near water streams (Alam 2022).

The growth of edible indigenous fruits primarily occurs in woody trees and in various weather conditions. The people use multiple sites and growing techniques throughout the different seasons to preserve biodiversity. These local plants, which are wild and have various uses, have a broad genetic foundation that contributes to their sustainability (Alam 1998, 2002, 2020, 2022). Multiple ecosystems favor the growth of non-timber food plants including neglected food plants with various species (Kar 2010; Alam 2022).

There exist genetic variations in crops although these variations have not been documented Moreover, unselective growing of certain species can lead to the gradual elimination of these resources. The cultivators can gain financial benefits by exploiting certain local products through business activities. This will increase the growth of these plants at domestic and commercial levels which will result in the sustainability and conservation of these resources (Alam 1998, 2002).

Limited data is available about the local food plants and their cultivation at domestic levels in Bangladesh. Although a wide variety of underutilized fruits and vegetables are being grown at a small scale on farmlands by the locals (Alam et al. 1996; Alam and Khisa 2000). Thus, the preservation of resources can only be possible by wise consumption, usage, and growth of these food plants.

7.9 Conclusion

This chapter explains a picture of the existing status of the various species of both wild and harvested food plants in Bangladesh. Data shows their limited consumption and provides a large list of underutilized plants available in the region. Unfortunately, these local plants do not find appropriate applications in the food items. The focus of this study is to highlight the lack of proper research and evidence on the consumption of these local plant foods. The idea of the consumption of these products is the creation of ancient people and this traditional consumption is being followed from one generation to another. However, the scientific investigation about the health benefits of these plants is still left uninvestigated by scientists in Bangladesh. The consumption of these neglected plants fulfills the dietary demands of less privileged people and plays a substantial role in ensuring food security for various segments of society. Intake of these local edible food plants has also been reported to enhance the nutritional status of individuals. One of the examples of these plants include pickle which possesses multiple health benefits and is effective against certain diseases.

In addition to that, the role of gender is important while discussing the consumption of these traditional plants because the knowledge about these plants is mostly gender-based. Females are more equipped with information relevant to edible local plants, their presence in various regions, and their mode of processing and preservation. In Bangladesh, food preparation and processing frequently comes in the female domain and it has encouraged women's role in small food business. The introduction of local food plants in the market will help to alleviate the issue of poverty and will also contribute to production increase of these crops at domestic levels. After meeting the domestic need these plants are brought into the market and are traded to earn money. Thus, these conventional plants are fulfilling the financial needs of poor people, along with satisfying their needs for nutrition and food security. This will act as a leading step towards biodiversity conservation in the region.

There is a huge formal and informal market for these local food plants and various innovations in the consumption of these plants can enhance their market value. The markets of these local products in limited in nature and involves few shareholders. There is a dire need to devise various other methods to enhance the usage of these local plants as their production is still in the novice stage.

Conflicts of Interest Author read and approved the final manuscript and declares no conflict of interest.

References

Alam MK (1997) Knowledge of ethnobotany towards socioeconomic development of agroforestry systems. In: Sukwong S et al (eds) Tropical forestry in the 21st century, vol 7. Community Forestry/Agroforestry. Faculty of Forestry, Kaesatsart University, Bangkok, pp 114–119

Alam MK (1998) Role of ethnobotany in sustainable development of hill farming system. In: Banik RL, Alam MK, Pei S, Rastogi A (eds) Applied ethnobotany. Bangladesh Forest Research Institute, Chittagong, pp 76–82

Alam MK (2002) Ethnobotanical knowledge and indigenous non-timber food crops for sustainable development of upland farming systems in CHT. In: Khan NA, Alam MK, Khisa SK, Millat-e-Mustafa (eds) Farming practices and sustainable development. CHTB- VFFP-IC, pp 155–163

Alam MK (2020) Traditional food plants and agrobiodiversity in Chittagong Hill Tracts, Bangladesh: food security, nutrition and conservation strategy. J Bangladesh Agric 10:7–21

Alam MK (2022) Traditional ecological knowledge on non-wood forest products management and biodiversity conservation: a focus on Chittagong Hill Tracts (CHTs), Bangladesh. In: Non-wood forest products of Asia: knowledge, conservation and livelihood. Springer International Publishing, Cham, pp 41–69

Alam MK, Khisa SK (2000) The perception of ethno-botany in Chittagong Hill Tracts and its linkage with biodiversity. In: Of popular wisdom: indigenous knowledge and practices in Bangladesh. BARCIK/IARD, Dhaka, pp 55–70

Alam MK, Mohiuddin M (2009) Shifting cultivation (*jhum*) agro-biodiversity at stake: Bangladesh situation. Acta Hortic 806:709–715

Alam MK, Mohiuddin M (2010) Scope of non-timber forest products (NTFPs) on economic development of Chittagong Hill Tracts (CHT), Bangladesh. Consultancy Report, submitted to INBAR (mimeographed)

Alam MK, Mohiuddin M, Basak SR (1996) Village trees of Bangladesh diversity and economic aspects. Bangladesh J For Sci 25:21–36

Bangladesh National Nutrition Council (BNNC) (2020) Bangladesh Nutrition Council: towards improved nutrition. Special publication on the Mujib Year. Bangladesh National Nutrition Council (BNNC), Health Service Division, Ministry of Health and Family Welfare, p 27

Banik RL (1998) Ethnobotany of bamboo and rattan and their indigenous management and utilization in Chittagong Hill Tracts. In: Banik RL, Alam MK, Pei S, Rastogi A (eds) Applied ethnobotany. Bangladesh Forest Research Institute, Chittagong, pp 93–102

Banik A, Abony M, Datta S, Towhid ST (2019) Microbiological quality of ready-to-eat food from Dhaka, Bangladesh. Curr Res Nutr Food Sci 7:161–168

Begum M (1987) In: Khan MS (ed) Edible fruits of Bangladesh. Hassan Book House, Dhaka, p 124

Biswas JC, Islam MR, Haque MM, Hamid A (2021) Minor cereal crop production and their future prospects in Bangladesh. Asian Soil Res J 5:48–56

Chowdhury MSH, Halim MA, Muhammed N, Koike M, Biswas S (2009) Indigenous knowledge in natural resource management by the hill people: a case of *Mro*tribe in Bangladesh. For Trees Livelihoods 19:129–151

Das DK (1987) Edible fruits of Bangladesh. Bulletin 3, Plant taxonomy series. Forest Research Institute, Chittagong, p 19

Feed the Future Innovation Lab for Food Safety (FISL) (2021) Food safety system in Bangladesh: current status of food safety, scientific capacity and industry preparedness. USAID, p 50

Islam SN, Khan MNI, Akhtaruzzaman M (2012) Food composition tables and database for Bangladesh with special reference to selected ethnic foods. Institute of Nutrition and Food Science, University of Dhaka, Bangladesh, p 138

Islam MMT, Talukder A, Rahman T, Islam JMM, Hossain S, Shekhar HU (2013) Food functionality of popular and commonly consumed indigenous vegetables and fruits from Bangladesh. Food Nutr Sci 4:741–747

Islam MM, Karim MR, Zhen X, Li X (2018) Heavy metal and metalloid pollution of soil, water and foods in Bangladesh: a critical review. Int J Environ Res Public Health 15:2825

Kar SP (2010) Non-timber forest product (NTFP) utilization and livelihood development in Bangladesh. Ph.D. Dissertation, School of Forest Resources, The Graduate School, The Pennsylvania State University

Katalyst (2016) Study on the roles and opportunities for private sector in agro-food processing industry of Bangladesh. Submitted by: Innovation Consulting Private Limited, p 114

Khatun M, Hassan MA, Islam SN, Rahman MO (2013) Taxonomy of the leafy vegetables in Bangladesh. Bangladesh J Plant Taxon 20:95–123

Khisa SK, Mohiuddin M (2015) Shrinking and changing livelihoods in the Chittagong Hill Tracts of Bangladesh. In: Erni C (ed) Shifting cultivation, livelihood and food security: new and old challenges for indigenous people in Asia. FAO, IWGIA and AIPP, Bangkok

Lewin TH Capt (1869) The Hill Tracts of Chittagong and the dwellers there in; with comparative vocabularies of the hill dialects. Bengal Printing Company, Limited. Calcutta, 151 p

Li X, Yadav R, Siddique KHM (2020) Neglected and underutilized crop species: the key to improving dietary diversity and fighting hunger and malnutrition in Asia and the Pacific. Front Nutr 7:593711

Miah M, Chowdhury MSH (2004) Traditional forest utilization practice by the *Mro* tribe in the Bandarban region, Bangladesh. Schweiz Z Forstwes 155:65–70

Miah MD, Chakma S, Koike M, Muhammed N (2011) Contribution of forests to the livelihood of the Chakma community in the Chittagong Hill Tracts of Bangladesh. J For Res 17:449–457

Ministry of Food and Disaster Management (2006) National Food Policy 2006. Ministry of Food and Disaster Management, Government of Bangladesh, 14 August 2006

Mohiuddin M (2009) Traditional knowledge of plant uses and conservation prospects by the hill communities in Bandarban Hill District, Bangladesh. Ph.D. thesis. Submitted to the Institute of Forestry and Environmental Sciences, University of Chittagong 4331, p 245

Mohiuddin A (2019) The mysterious domination of food contaminants and adulterants in Bangladesh. Int J Environ Sci Nat Res 16:555941

Noor R, Feroz F (2016) Food safety in Bangladesh: a microbiological perspective. Stamford J Microbiol 6:1–6

Pasha MK, Uddin SB (2019) Minor edible fruits of Bangladesh. Bangladesh J Plant Taxon 26:299–313

Paul AK, Alam S, Rahman MA, Alam AHMJ (2016) Wild edible vegetables used by the Chakma community of Rangamati hill district, Bangladesh. Eco-friendly Agric J 9:22–26

Rahaman MS, Mise N, Ichihara S (2022) Arsenic contamination in food chain in Bangladesh: a review on health hazards, socioeconomic impacts and implications. Hyg Environ Health Adv 2:100004

Rahman AHMM, Khatun MM (2020) Leafy vegetables in Chapai Nawabganj district of Bangladesh focusing on medicinal value. Bangladesh J Plant Taxon 27:359–375

Saravanan D, Thirumalai DA, Sivakumar A, Asharani IV (2016) A systematic review of natural toxins in food plants. Int J Res Ayurveda Pharm 7:52–57

Wright OR, Sivakumar D, Sultanbawa Y, Netzel ME (2022) Food and nutrition security: underutilized plant and animal-based foods. Front Nutr 8:821388

Chapter 8
Exploring Social-Ecological Systems for Mainstreaming Neglected and Underutilised Plant Foods: Local Solutions to Food Security Challenges in Sri Lanka

Sujith Ratnayake, Michael Reid, Danny Hunter, Nicolette Larder, Renuka Silva, Harsha Kadupitiya, Gamini Pushpakumara, Teresa Borelli, Sharon Mendonce, Athula Liyanage, Gamini Samarasinghe, Thushanthi Perera, and Champika Kariyawasam

8.1 Introduction

Sri Lanka is a country where links between biodiversity, food and culture have historically been strong (Bandara 2009). It has very high degree of biodiversity relative to other Asian countries (Myers et al. 2000) and is home to around 4500 edible plant species, of which 900 are endemic (Beltrame et al. 2019). Four pillars of diversity—biodiversity, agrobiodiversity, biocultural diversity, and climatic diversity are central to ensuring the food diversity of the country. The wide variability in climate, topography and soils create a wide diversity of natural and

S. Ratnayake (✉)
School of Humanities, Arts and Social Sciences, University of New England, Armidale, NSW, Australia

Applied Agricultural Remote Sensing Centre, School of Science and Technology, University of New England, Armidale, NSW, Australia

Ministry of Environment, Battaramulla, Sri Lanka
e-mail: rratnay2@myune.edu.au

M. Reid · N. Larder
School of Humanities, Arts and Social Sciences, University of New England, Armidale, NSW, Australia

D. Hunter · T. Borelli · S. Mendonce
Alliance of Bioversity International and CIAT, Rome, Italy

T. Ismail et al. (eds.), *Neglected Plant Foods of South Asia*,
https://doi.org/10.1007/978-3-031-37077-9_8

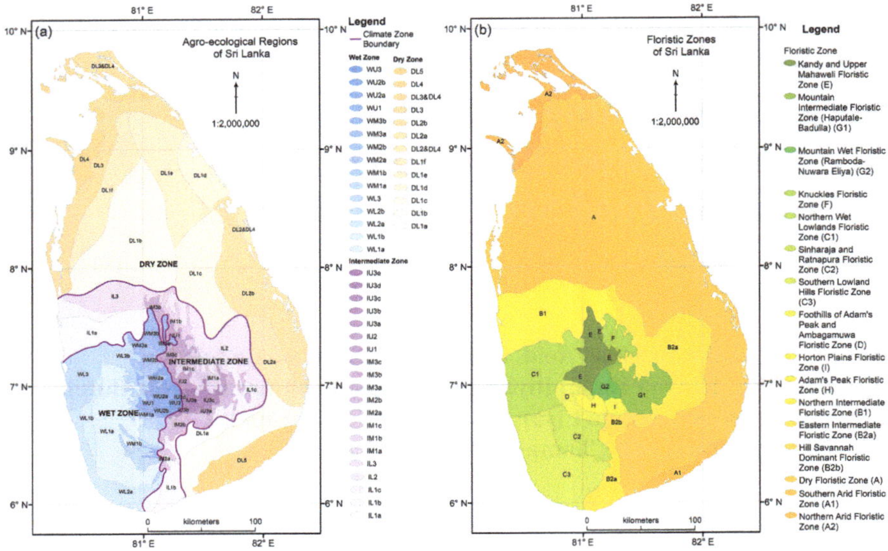

Fig. 8.1 Agroclimatic (**a**) and bioclimatic (**b**) diversity of Sri Lanka. (Data sources: (**a**) Kadupitiya et al. (2022); (**b**) Wikramanayake et al. (2022))

agroecosystems—sixteen floristic zones (Wikramanayake et al. 2022) and forty-six agroecological regions (Punyawardena 2020) that harbour cultivated agrobiodiversity and wild food plant species (Fig. 8.1). Over 792 plant species support the agrobiodiversity of cultivated crops in Sri Lanka, of which 6%, 44% and 50% of species are considered endemic, native and introduced species, respectively (Puspakumara et al. 2016; Gee et al. 2020). Wet evergreen (wet zone), dry mixed (dry zone) and montane (up-country) forests are important for providing non-cultivated diversity (forest foods)—wild edible food plants and tree resources. Sri Lanka also recorded 410 species of wild relatives of edible food crops, belonging to 47 families and 122

R. Silva · T. Perera
Faculty of Livestock, Fisheries & Nutrition, Wayamba University of Sri Lanka, Makandura, Gonawila, Sri Lanka

H. Kadupitiya
Natural Resources Management Centre, Department of Agriculture, Peradeniya, Sri Lanka

G. Pushpakumara
Faculty of Agriculture, University of Peradeniya, Peradeniya, Sri Lanka

A. Liyanage · G. Samarasinghe
Plant Genetic Resources Centre, Department of Agriculture, Gannoruwa, Peradeniya, Sri Lanka

C. Kariyawasam
School of Environmental and Rural Science, University of New England, Armidale, NSW, Australia

genera. Of this 15%, 74% and 11% of species are considered endemic, native and naturalised exotic species of wild relatives of food crops, respectively (Gee et al. 2020; Liyanage 2010; Ratnayake et al. 2021).

Social-Ecological Systems (SESs) in Sri Lanka have been using and thus maintaining, a wide range of cultivated crop plants and non-cultivated wild food plants over several thousand of years to fulfill household food supply, meet nutritional requirements of rural farming communities and support national food security (Punyawardena 2008; Ganashan et al. 1996; Mihiranie et al. 2020; Aheeyar 2013; MOE 2016). Wild foods from surrounding ecosystems were primary components of traditional food systems in SESs ensuring food security and resilience (Ahmed et al. 2022). In fact, SESs in Sri Lanka contribute substantially to maintaining and conserving wild, neglected and underutilised genetic resources for food (Bandara 2009; Liyanage 2010; Ranil et al. 2020; IUCN 2015a; Goonatilake et al. 2015; Goonatilake and Ekanayake 2016; Ekanayake et al. 2020; Pushpakumara et al. 2016; Godamulla 2018; Alwis et al. 2018). Three distinct terrestrial SESs are found in three different climatic zones of Sri Lanka; namely Kandyan Home Gardens (KHGs), Village Tank Cascade Systems (VTCSs) and Owita Systems, which are primarily maintained as smallholder farming systems. Distinct climate, edaphic and geomorphological factors in each of the three SESs, provide key ecosystem services that contribute to wild and neglected food environments in association with biocultural resources to ensure food security by providing dietary diversity to the local communities (Goonatilake and Ekanayake 2016; Pushpakumara et al. 2016; Dayananda et al. 2021; WFP 2015; Mijatović et al. 2018; Borelli et al. 2020a; SCBD 2010; Marambe et al. 2012a; DeClerck 2013). The distribution of SES landscapes in Sri Lanka is shown in Fig. 8.2.

Despite the significant amount of plant diversity, most of the country's edible plant diversity remains neglected or underutilised (Bandula et al. 2016; Bandula and Nath 2020; Peduruhewa et al. 2021). Sri Lanka's food supply is largely dependent on a few cultivated crops—mainly rice, maize, pulses and a few selected vegetables and fruits. This has resulted in less diversification of local food systems, which has contributed to the rise in national food and nutrition insecurity (Ratnayake et al. 2020). Potential climate change impacts and an increasing population in the Asian region has further worsened the current food insecurity crisis in Sri Lanka (Hemathilake and Gunathilake 2022; Esham and Garforth 2013; IPCC 2022).

Diversifying local food systems through Neglected and Underutilised Plant Foods (NUPFs) could be a promising solution for the sustainability of future food and nutrition security (Kumar et al. 2022). NUPFs ensure environmentally sustainable food production despite adverse environmental conditions and low input availability (Mihiranie et al. 2020; Bandula et al. 2016; Ratnayake et al. 2020). They also contribute substantially to dietary diversification—adding different tastes as well as various macro and micronutrients to the diet, particularly of rural communities (Gee et al. 2020; Borelli et al. 2020b; Li et al. 2020; Talabi et al. 2022). Furthermore, they help make agricultural systems more climate resilient (Padulosi et al. 2013). However, the future sustenance of NUPFs in Sri Lanka is challenging and their roles in food and nutritional security needs to be realised. While most of these species

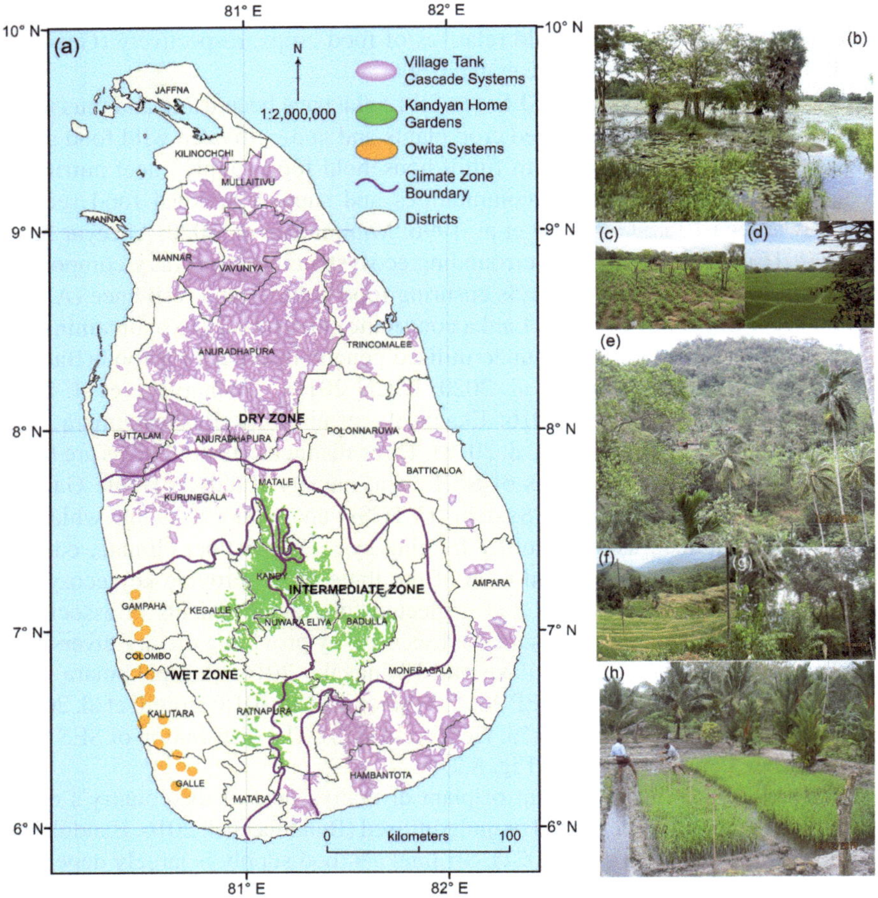

Fig. 8.2 Social-Ecological Systems (SESs) in Sri Lanka. (**a**) spatial distribution of SESs, (**b**) a tank of a Village Tank Cascade System (VTCS), (**c**) traditional Chena cultivation in VTCS, (**d**) command area paddy cultivation in VTCS, (**e**) multi-story tree canopies in Kandyan Home Garden (KHG), (**f**) paddy cultivation in KHG system, (**g**) multi-purpose horticultural tress in KHG, and (**h**) traditional rice cultivation in Owita system. (Data sources: (**a**) Department of Agrarian Development (DAD) and Natural Resources Management Centre (NRMC), Sri Lanka. Photo sources: Biodiversity for Adaptation to Climate Change (BACC) project)

were commonly used in ancient times, they have since lost their genetic diversity due to many reasons, i.e., gradually declining natural ecosystems and the cultivated agroecosystems.

Challenges facing social-ecological traditional food systems are aggravated by global environmental change especially changing climate and intensive commercialisation of traditional farming lands (IPCC 2022). Studies have indicated that several external and internal environmental and socio-economic factors, such as changing climate and agricultural land use intensification, negatively influenced the future sustenance and survival of wild, neglected and underutilised food plants in the SESs (Ahmed et al. 2022; Ratnayake et al. 2019; Wickramasinghe et al. 2023; Bharucha and Pretty 2010). Additionally, with the introduction of high-yielding improved crop varieties, the demand for NUPFs was depleted and they became disused (Li et al. 2020). However, there is still immense potential and plenty of opportunities, such as traditional knowledge and Sri Lankan food culture, to help improve the maintenance, cultivation and use of NUPFs in Sri Lanka (Malkanthi et al. 2014). With the present economic crisis in Sri Lanka amid increased food prices, NUPFs are becoming popular in home gardens, urban agriculture, as well as household and commercial-scale food production, with the support of government incentives and innovative commercialization initiatives (Borelli et al. 2020a; Malkanthi 2017; Hunter et al. 2019).

The outstanding potential of NUPFs for enhancing food security and human nutrition (health benefits) has neither been studied nor exploited. Information is scattered among stakeholders and not properly analysed to support their mainstreaming into national policies and planning strategies (Padulosi et al. 2011; Hunter et al. 2017). Thus, the objective of this study is to explore neglected and underutilised food plants/crops found in Sri Lanka (with special reference to the SESs), their spatial distribution, nutritional significance and future challenges to promote sustainable utilisation. The neglected and underutilised food plants were identified and documented based on six plant food groups—cereals and pseudo cereals, roots and tubers, pulses and legumes, vegetables and leafy vegetables, wild mushrooms and fruits, in three distinct types of terrestrial SESs (VTCSs, KHGs and Owita) located in three different climatic zones in Sri Lanka. Field explorations were conducted in selected SESs during 2019–2022 to collect data on plant species and to examine the utilisation of NUPFs in these systems. Initially plant species were identified through field observation and interviews with farmers, traditional medical practitioners and the community. Further, species were identified and validated in consultation with village-wise key informants, plant taxonomists, and also using records of passport data and reference specimens housed at the Plant Genetic Resources Centre (PGRC) and National Herbarium, during the period of June 2022 to March 2023. In addition to the field exploration data, the study examined data from past studies and assessments (revised handbook of Ceylon Flora of Ceylon (volumes I–XIV) and National Red List), and ecogeographic and baseline surveys conducted by different researchers, projects and organisations.

8.2 Neglected and Underutilised Food Plants of Sri Lanka

In Sri Lanka, neglected food plants mainly include non-commercially cultivated traditional crop plants, semi-wild and wild edible plants as well as tree resources that can be used whole or partly, which have food and nutritional value. In addition to food and nutritional value, some of these plants possess pharmacological properties and biocultural importance based on their traditional medicinal applications and use as in cultural food preparation. Neglected and wild edible plants, some of which may be related to crops (i.e., crop wild relatives) tend to occupy special niches in the local ecologies, natural or semi-natural ecosystems (Bandula et al. 2016; Peduruhewa et al. 2021; Malkanthi 2017; Pushpakumara et al. 2007; Ratnayake et al. 2019; Li and Siddique 2018; Mudau et al. 2022). According to traditional food recipes, different parts of these food plants—shoots, leaves, fruits, pods, buds, seeds, barks, stem, tubers, flowers, the stalk of flowers, mushrooms, roots, and latex are used for food preparations (Mihiranie et al. 2020; Abeywardhana and Perera 2016). However, deciding whether a particular plant or crop is neglected or underutilised is sometimes challenging due to the degree of their utilisation, which may be based on the social, ecological, economic, cultural, climatic and farming characteristics of a particular area. Thus, the authors considered the attributes listed in Table 8.1 to specify if a particular plant or crop species was underutilised or neglected.

Table 8.1 Attributes used to specify the underutilised or neglected food plant/crop

	Underutilised	Neglected
1	Commercially cultivated to small extents in related agroecosystems	Not commercially cultivated, semi-domesticated crop species are present in the local subsistence farming systems or wild/semi-wild ecosystems
2	Have a fully or partially known food, nutrition and associated cultural value of the whole or any part of the crop/plant	Very little is known about the crop/plant's food, nutrition and associated cultural value
3	Diversity within crop species (Intra-species diversity) studied in some crops	Intra-species diversity is rarely studied
4	Availability of many known indigenous varieties	Availability of indigenous varieties relatively less known
5	Partially mainstreamed into a formal national seed system	Absence in the formal national seed system
6	Availability of planting materials via local nurseries of government and private sector	Non-availability of planting materials via proper channel
7	Presence in community seed banks and community seed exchange/participatory breeding programs	Rarely maintained in community seed banks and community seed exchange/participatory breeding programs
8	Marginalised by research and development organisations	Hardly any emphasis in terms of any national research, extension programs and policies
9	Inadequate agronomic/horticultural information	Agronomic/horticultural information is not available or is very poorly documented
10	Partially unexploited potential for food and nutrition values	Limited or unexploited potential for food and nutrition values
11	Germplasm and morphological characterisation limited to a few crops	The minimum level of germplasm and morphological characterisation
12	Limited availability of germplasm at Plant Genetic Resources Centre (PGRC) or field gene banks	Unavailability of germplasm at PGRC or field gene banks
13	Inadequate awareness among the general public about the food, nutrition and medicinal value	Minimum or no awareness among the general public about the food, nutrition and medicinal value

(continued)

Table 8.1 (continued)

	Underutilised	Neglected
14	Local people can identify these plants from their surrounding environment	Most local people do not know about the presence of these plants in their surrounding environment
15	Availability of rural markets	Limited availability of rural markets
16	Food recipes available from various sources including electronic media such as You-Tube but not well promoted	No or limited availability of recipes
17	The percentage of consumption by rural or urban people is not at a satisfactory level	A very low percentage of consumption by the rural or urban people
18	Present in the rural or urban food basket but not at a satisfactory level	Rarely present in the rural food basket compared to past decades
19	Present in daily dietary intake (food plate) but not at a recommended level	Very rarely present in daily dietary intake (food plate); if present then not at the recommended level
20	Inadequate post-harvest technological development (processing, storage)	Post-harvest technological development not in place (processing, storage)
21	Evidence of common use by the ancestors of local communities but currently, limited use	Very few people have knowledge of traditional and ethnobotanical use
22	Used in rural and urban areas but not mainstreamed into their food systems systematically	Not considered in the local food systems
23	Inadequate (basic) market value chains development process	Not considered in the market value chains development process
24	Sometimes specific agroecological and climatic adaptation practices are required for sustainable production	Mostly found in niche environments and fit best into marginal/harsh environments with specific agroecological and climatic adaptations
25	Associated local knowledge partially documented	Associated local knowledge is with current users

Source: Based on author's knowledge and experience

In Sri Lanka, a diverse range of neglected food plants are found in varied landscapes and a considerable portion of these food plants are semi or fully domesticated (Goonatilake and Ekanayake 2016; Ekanayake et al. 2020; Malkanthi 2017; Pushpakumara et al. 2007). Some of these food plants are traditionally grown by farmers as indigenous varieties or landraces where they are still important for the subsistence of local communities (Bandula et al. 2016; Peduruhewa et al. 2021; Malkanthi 2017; Pushpakumara et al. 2007; Ratnayake et al. 2019; Li and Siddique 2018). In SESs the food plants are mostly grown together with other dominant commercial crops such as paddy, maize, cassava, pepper, banana, green chillies, coconut, papaya, in various types of farming systems, including shifting cultivation (Chena), home gardens, and off-season paddy farms (Bandula and Nath 2020). In addition, these food plants grow naturally; in forests, wetlands, aquatic systems; in semi-natural marginal environments such as scrub lands, and open forests; and around existing marginal agricultural lands as wild edibles and wild relatives of

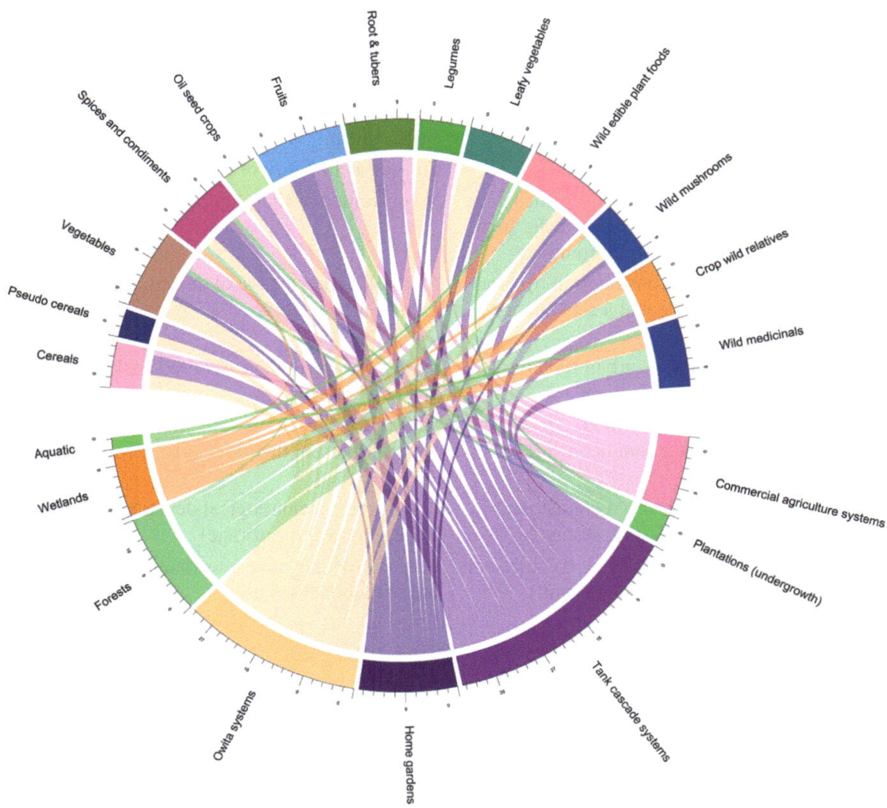

Fig. 8.3 Neglected and underutilised plant food groups and their landscapes in Sri Lanka. (The wider the links, the more food groups are found in that landscape). (Data sources: Ranil et al. (2020), Goonatilake and Ekanayake (2016), Pushpakumara et al. (2007), HLP-BLA (2020), Wasala et al. (2014), DOA (2005), and PGRC (1999))

these crops (Dahanayake 2015). According to the past NUPFs assessment and eco-geographic survey data, the three SES landscapes recorded the highest number of neglected and underutilised plant food groups relative to other ecosystems of Sri Lanka (Fig. 8.3).

8.2.1 Cereals and Pseudo Cereals

(a) Cereals

Many of the cereals used in the past are no longer in use today due to changes in food habits, difficulties in processing and preparation, obsolescence of old-fashioned cooking methods and ignorance of the present generation. Rice (*Oryza sativa* L.), the staple cereal cultivated in Sri Lanka, provides a significant contribution to the dietary energy, protein and fat intake of the population, some neglected traditional varieties also have medicinal value (Weerasekara et al. 2020). Paddy exhibits a very high intra-species genetic diversity including wild relatives, landraces and ancestor cultivars (Gee et al. 2020; Liyanage 2010). The existence of more than 2000 traditional rice accessions—identified as indigenous traditional rice varieties or landraces had been reported (Weerakoon and Somaratne 2021; Munasinghe and Price 2016; Samarasinghe et al. 2015; Dharmasena 2010; Kennedy and Burlingame 2003). Among these neglected traditional rice varieties, some varieties are most prominently cultivated in the traditional farming systems of SESs because of their health benefits, adaptability to soil conditions and resistance to different abiotic stress conditions (Dharmasena 2010; Abeysekera and Premakumara 2016). They are Pachchaperumal (Lord Buddha's rice), Suwendel (Fragrant rice), Suduheenati (White rice), Madathawalu (Red rice), Kaluheenati (Dark rice), Kuruluthudu-wee (Red rice), Rath suwendel (Red fragrant rice) Pokkali, Mudukiriel, Batapola wee, Sudu-wee, Muthumanikkan and Murungakayan (Fig. 8.4) (Ekanayake et al. 2020; Hunter et al. 2019; Munasinghe and Price 2016; IUCN 2015b; Rambukwella and Priyankara 2016). Traditional rice varieties are consumed as cooked rice and porridge while rice flour is used to make variety of rice flour-based foods such as string hoppers, hoppers, pittu and porridge mostly in combination with a coconut milk curry or sambol. Traditional rice varieties such as Maa-wee varieties were given to pregnant women, Heenati-wee varieties were given to children and convalescents (Rambukwella and Priyankara 2016; Silva and Chandrasekara 2020). Among some culturally recognised varieties having different properties each appealing to specific circumstances related to climate, soil, disease resistance, individual preference, and medicinal properties, many of them are fast disappearing from the SESs due to various reasons and it is high time to salvage those varieties and use them to current and future food security needs (IUCN 2015b).

Fig. 8.4 Nutritionally significant neglected traditional rice varieties (*Oriza sativa* L.) found in social-ecological systems of Sri Lanka. (**a**) suwadel, (**b**) kaluheenati, (**c**) kruluthuda, (**d**) madathavalu, (**e**) pachchaperumal, (**f**) pokkali, (**g**) sunduheenati, and (**h**) suduru-samba. (Photo source: Biodiversity for Food and Nutrition (BFN) project)

There are four neglected millet varieties found in SESs in Sri Lanka (Table 8.2). Ancient communities used millets in their daily diet by preparing them in various forms. Finger millet — locally known as Kurakkan (*Eleusine coracana* (L.) Gaertn) is the third most important cereal cultivated in Sri Lanka next to rice and maize. Large-scale Finger millet cultivations are found in the dry and intermediate zones mostly as rainfed upland cultivations, therefore, Finger millet cannot be considered a neglected cereal in Sri Lanka. However, their nutritional and potential health benefits have not yet been fully explored among traditionally cultivated varieties (Jayawardana et al. 2019), mostly cultivated in SESs (Fig. 8.5). Some articles mention Pearl millet—locally known as Kambu (*Cenchrus americanus* (L.) Morrone) as a neglected or underutilised cereal in Sri Lanka (Peduruhewa et al. 2021; Tyagi et al. 2018); however, a very limited extent of cultivation existence in SESs and only one accession record being found in the Plant Genetic Resources Centre (PGRC) (PGRC 1999). Sorghum— locally known as Idal irigu (*Sorghum bicolor* (L.) moench) is another cereal traditionally cultivated mainly in the uplands of dry and intermediate zones SESs, however, no indigenous varieties are found in Sri Lanka and its potential as food or feed crop has not been fully exploited (Kaluthanthri and Dasanayaka 2019).

(b) Pseudo cereals

The pseudo cereals group consists mainly of crops such as Quinoa, Amaranth, Buckwheat, Chia seeds, Wattle seed and Kañiwa. The incorporation of these seeds into the diets of coeliac patients should help lessen the deficit in fiber consumption. They can be included into commonly consumed foods such as bread, biscuits, cakes and pasta (Usman et al. 2022; Upasana and Yadav 2022). Among above mentioned species, only Amaranth is popular and cultivated to some extent in Sri Lanka (PGRC 1999).

Amaranthus (Local name: Thampala)

Amaranthus is one of the few non-grasses with the potential of becoming a cereal-like grain crop. However, in Sri Lanka, *Amaranthus* spp. are used as leafy vegetables and there are no reports of its seeds being used as food. Number of wild relatives of this plant are found in home gardens of SESs, along roadsides and marginal bare lands being freely available as a natural forage for foraging animals and other livestock in Sri Lanka (Leukebandara et al. 2015). About 72 accessions belonging to five species of *Amaranthus* are conserved in a seed gene bank at PGRC. These species are *A. hybridus* sub sp. *cruentus* (L.) Thellung, *A. dubius* Mart., *A. hybridus* L., *A. lividus* L. and *A. tricolor* L. However, there are several local names for Amaranthus and they are not specific to a particular species; for example, the local name Rathu (red) Tampala is sometimes used to refer to both *A. hybidus* and *A. tricolor.* In addition to the above five species, *Amaranthus spinosus* L. (Katu Thampala), *Amaranthus hypochondriacus* L. (Rana Thampala) and *Amaranthus viridis* L. (Koora Thampala) are reported in Sri Lanka (PGRC 1999). *Amaranthus cruentus* L. (Locally known as Landesi) is an ancestral variety a good source of carbohydrates, proteins, lipids, vitamins, minerals, and bioactive compounds (Rajapaksha 1998; Wolosik and Markowska

Table 8.2 Neglected millet crop species found in SESs of Sri Lanka

No	Local name	Common name/s	Family	Botanical name	Local availability, agroecology, conservation status and edible use	References/sources
1	Maha Meneri	Proso millet	Poaceae	*Panicum millaceum* L.	Drought-tolerant local, minor, millet cultivations found in rainfed low-input traditional farming systems—Village Tank Cascade Systems in Sri Lanka (VTCSs). Proso millet is the principal crop during the minor cultivation season (Yala) in the VTCSs. It is found in the Chena (shifting) cultivation systems, grown as a rainfed crop where minimum tillage conditions and low inputs are used during the drier part of the year. Fourteen out of twenty-four local Proso millet accessions are conserved in the seed gene bank at the Plant Genetic Resources Centre (PGRC) of Sri Lanka. Soaked and ground Proso millet flour can be used in many food preparations as rice flour.	Bandula and Nath (2020), Peduruhewa et al. (2021), Wasala et al. (2014), PGRC (1999), Tyagi et al. (2018), Wimalasiri et al. (2017), Rajapaksha (1998), and Dassanayake and Clayton (1997)
2	Heen Meneri	Little millet	Poaceae	*Panicum sumatrense* Roth	Similar to Maha Meneri, Little millet is drought tolerant and can be grown in a short period. The consumption rate of Little millet is very low. Apart from food value, the Little millet has medicinal value and provides several health benefits to consumers—reducing the risk of developing diabetes mellitus, obesity, heart diseases and constipation. Food preparation is similar to Maha Meneri. Three out of eight local accessions of little millets are conserved in the seed gene bank at the PGRC.	Peduruhewa et al. (2021), Malkanthi (2017), Wasala et al. (2014), PGRC (1999), Tyagi et al. (2018), Menaka (2019), and Dassanayake and Fosberg (1981a, b)
3	Thanahal	Foxtail millet, Italian millet	Poaceae	*Setaria italica* (L.) P. Beauv.	In the past, Foxtail millet was cultivated in the dry areas of Sri Lanka under Chena (shifting) cultivation during the Maha season. It is now treated as a less important crop both economically and as a food due to poor yields and less popularity as food or feed. Morphologically eight different varieties of Foxtail millet are found in Sri Lanka. According to the colour of the seeds, three varieties (yellow, gold and black) of Foxtail millets can be distinguished in the country. Parent variety names are represented in the local varieties collection at the PGRC. These names are Thanahal, Balathana, Kaluthana, Kirigothana, Mas-thanahal, Raththana, Thanakurahan and Thanawee. Like rice, Foxtail millet is boiled and used in many preparations.	Bandula and Nath (2020), Peduruhewa et al. (2021), Malkanthi (2017), Wasala et al. (2014), PGRC (1999), Tyagi et al. (2018), Rajapaksha (1998), and Dassanayake and Fosberg (1983)
4	Amu	Kodo millet	Poaceae	*Paspalum scrobiculatum* L.	Exceptionally hardy drought-resistant crop. The extent of Kodo millet cultivation in the country is very small. However, it is still grown in the dry and intermediate zone as small pockets as a Chena (shifting) cultivation crop in the Maha cultivation season. Grains are boiled and eaten as rice and milk-rice.	Peduruhewa et al. (2021), Wasala et al. (2014), PGRC (1999), Tyagi et al. (2018), Rajapaksha (1998), and Dassanayake and Fosberg (1985)

Fig. 8.5 Traditional finger millet (*Eleusine coracana* (L.) Gaertn) cultivation (Local name: Bala-kurakkan) in a Village Tank Cascade System landscape. (Photo source: Biodiversity for Adaptation to Climate Change (BACC) project)

2019) has totally disappeared from VTCSs areas (Ekanayake et al. 2020). According to the verbal description of plant features by village elders of North and North-central VTCSs, the crop was widely used as a subsistence food crop in 1970's and prior to that, in Chena lands of VTCSs. Currently, there is an increasing interest in their utilisation for the formulation of healthy and nutritious food products. In the future, Amaranthus varieties can be re-introduced to VTCSs for cultivation considering their nutrition value, low water consumption, climate change adaptation and significance as a health food (Ekanayake et al. 2020).

8.2.2 Roots and Tubers

Roots and tubers are important alternative food sources among Sri Lankans. The most wildly used local root and tuber crops in Sri Lanka are cassava (*Manihot esculenta* Crantz), sweet potato (*Ipomoea batatas* (L.) Lam.), yams (*Dioscorea* spp.) and edible aroids (*Colocasia esculenta* (L.) Schott and *Xanthosoma* spp.). Cassava and sweet potato were introduced to the country in the colonial period and are presently as popular as potatoes (Godamulla 2018; Alwis et al. 2018; Peduruhewa et al. 2021). Meanwhile, the consumption and cultivation of local root and tuber crops are neglected and some of these crops are threatened with extinction in Sri Lanka. According to the literature, there are about 60 neglected and underutilized indigenous varieties of roots and tuber crops in the country, however, a majority of consumers are aware of and use about 15–20 varieties due to the preference of their food-taste (Godamulla 2018; Sajiwanie and Rathnayaka 2016). More than 40 local names have been recorded in literature for indigenous edible root and tuber crops in Sri Lanka, however, there is considerable variation in the local names based on growing localities, cultural and traditional use and characteristics of the yam (i.e., color, shape, taste, nutritional value, habit). Some varieties that were introduced and naturalized in the country, grow in almost all parts of it and can resist harsh climate conditions; however, most of their nutritional properties and easily cultivable characteristics to address food and nutrition security have not been investigated yet (Godamulla 2018; Sajiwanie and Rathnayaka 2016; Alwis et al. 2007; Rupasingha et al. 2020). Neglected and underutilised indigenous root and tuber crop varieties recorded in SESs of Sri Lanka are listed in Table 8.3. Nutritionally significant indigenous root and tuber crop varieties selected from three SESs for promotion and value chain development under the Biodiversity for Food and Nutrition (BFN) project are shown in Fig. 8.6.

Dioscorea species are seasonal crops mainly cultivated in home gardens in the wet and intermediate zones but can also be cultivated in SESs of dry zone areas with good soil moisture content. Nine out of ten *Dioscorea* spp. reported in Sri Lanka are edible species. The most popular species are *Dioscorea alata* L. with different names such as water yam, white yam and purple yam (purple color can be seen in any part of the crop). Other species are *D. esculenta, D. rotundata, D. bulbifera, D. pentaphyla, D. tomentosa, D. trimenii, D. koyamae and D. oppositifolia. Dioscorea koyamae* Jayas. and *Dioscorea trimenii* Prain & Burkill are endemic to Sri Lanka and *Dioscorea alata* L. and *Dioscorea esculenta* (Lour.) Burkill were introduced and cultivated (Jayasuriya 1990). According to the analysis of food composition value of selected Sri Lankan traditional yam varieties, *Amorphophallus paeoniifolius* (Dennst.) Nicolson (Elephant foot yam) (Kidaran) and *Dioscorea bulbifera* L. (Aerial yam) (Udala) recorded the highest nutritional properties among tested samples (Sajiwanie and Rathnayaka 2016; Rupasingha et al. 2020). *Brachystelma lankana* Dassan. & Jayas (Pathan-ala) is an endemic variety with medicinal value and is only recorded from the Knuckles mountainous region (Pitawala Pathana and Laggala in the Central-Matale district of Sri Lanka) and the Balangoda area of Ratnapura district of Sri Lanka (Dassanayake and Jayasuriya 1974; MOE 2012). It is known to be critically endangered wild yam species due to unsustainable harvesting and the degradation of forest habitats, which is also affected by climate change and other environmental factors (MOE 2012).

Root and tuber crops provide supplementary food in rural areas, adding variety to the diet. Neglected yams are used for both savoury and sweet dishes because of their flavour and delicacy (Fig. 8.7). In addition to their main contribution as dietary energy, they provide important nutritional and health benefits such as antioxidative, hypoglycemic, hypocholesterolemic, antimicrobial, and immunomodulatory activities. Boiled or steamed roots and tubers are consumed with scrapped coconut, as chips and as curries (by adding coconut milk and spices). Chopped and dried tubers are also ground into flour and mixed with wheat flour to make rotti, bread and sweets. The young leaves of some of them are eaten as vegetables (Alwis et al. 2018; Rajapaksha 1998; Sajiwanie and Rathnayaka 2016; Rupasingha et al. 2020).

Table 8.3 List of neglected and underutilised root and tuber species and varieties found in SESs and wild ecosystems of Sri Lanka

No	Local name/s	Common name/s	Family	Botanical name	Habit	Storage structure
1	Artichoke/Kukul-buthsarana	Artichoke yam/Guinea arrowroot	Marantaceae	*Calathea allouia* (Aubl.) Lindl	Herb	Tuber
2	Buthsarana	Indian shot/Canna	Cannaceae	*Canna indica* L.	Rhizomatous herb	Rhizome
3	Dandila/Dandina	Purple yam	Dioscoreaceae	*Dioscorea alata* L.	Twiner	Tuber
4	Desiala/Kiri habarala/ Kiriala	Cocoyam/Tannia	Araceae	*Xanthosoma sagittifolium* (L.) Schott	Tuberous subshrub	Corm
5	Gahala	Taro	Araceae	*Colocasia esculenta* (L.) Schott	Tuberous herb	Corm
6	Habarala	Giant taro/Elephant ears	Araceae	*Alocasia macrorrhizos* (L.) G. Don	Herb	Corm
7	Angiliala	–	Dioscoreaceae	*Dioscorea alata* L.	Twiner	Tuber
8	Hingurala	–	Dioscoreaceae	*Dioscorea alata L.*	Twiner	Tuber
9	Hirithala	–	Dioscoreaceae	*Dioscorea oppositifolia* L.	Twiner	Tuber
10	Hondala	–	Passifloraceae	*Adenia hondala* (Gaertn.) W.J.de Wilde	Climber	Tuber
11	Hulankeeriya/Araluk	Arrowroot	Marantaceae	*Maranta arundinacea* L.	Herb	Rhizome
12	Innala	Country potato	Lamiaceae	*Coleus rotundifolius* (Poir.) A. Chev. & Perrot	Herb	Tuber
13	Kahataala	–	Dioscoreaceae	*Dioscorea alata* L.	Twiner	Tuber
14	Katuala	–	Dioscoreaceae	*Dioscorea pentaphylla* L.	Twiner	Tuber

15	Kidaran	Elephant foot yam	Dioscoreaceae	*Amorphophallus paeoniifolius* (Dennst.) Nicolson	Herb	Corm
16	Kiri kodol/Kodol	–	Dioscoreaceae	*Dioscorea alata* L.	Twiner	Tuber
17	Kukulala	Asiatic yam	Dioscoreaceae	*Dioscorea esculenta* (Lour.) Burkill	Twiner	Tuber
18	Maiokka	Cassava	Euphorbiaceae	*Manihot esculenta* Crantz	Tuberous shrub or tree	Tuber
19	Mawala[a]	–	Dioscoreaceae	*Dioscorea* sp.	Herb	Corm
20	Raja-ala	Greater yam/Water yam	Dioscoreaceae	*Dioscorea alata* L.	Twiner	Tuber
21	Jaffna rasawalli[a]	Greater yam/ Water yam	Dioscoreaceae	*Dioscorea alata* L.	Twiner	Tuber
22	Sewelala[a]	–	Araceae	*Alocasia* sp.	Herb	Corm
23	Udala	Aerial yam	Dioscoreaceae	*Dioscorea bulbifera* L.	Twiner	Tuber
24	Wel-ala kola	–	Araceae	*Colocasia esculenta* (L.) Schott	Tuberous herb	Corm
25	Relburulla[a]	–	Araceae	*Alocasia* sp.	Subshrub	Corm
26	Nattala[a]	–	Dioscoreaceae	*Dioscorea alata* L.	Twiner	Tuber
27	Sawkenda	–	Zingiberaceae	*Kaempferia rotunda* L.	Rhizomatous herb	Rhizome
28	Iniala	–	Dioscoreaceae	*Dioscorea alata* L.	Twiner	Tuber
29	Isuru desa-ala/Rathu kiriala[a]	–	Araceae	*Colocasia* sp.	Tuberous herb	Corm

(continued)

Table 8.3 (continued)

No	Local name/s	Common name/s	Family	Botanical name	Habit	Storage structure
30	Gonala/Kahata-gonala/Kiri-gonala/Hiritala	Wild yam	Dioscoreaceae	*Dioscorea koyamae* Jayas.[b] *Dioscorea oppositifolia* L. *Dioscorea spicata* Roth	Twiner	Tuber
31	Geriala	–	Dioscoreaceae	*Dioscorea alata* L.	Twiner	Tuber
32	Jamburala	–	Dioscoreaceae	*Dioscorea alata* L.	Twiner	Tuber
33	Kahata-angala	–	Dioscoreaceae	*Dioscorea alata* L.	Twiner	Tuber
34	Rata-ala	–	Dioscoreaceae	*Dioscorea alata* L.	Twiner	Tuber
35	Pathanala	Wild yam	Apocynaceae	*Ceropegia lankana* (Dassan. & Jayas.) Bruyns[b] *Brachystelma lankana* Dassan. & Jayas[b]	Herb	Tuber
36	Uyala	–	Dioscoreaceae	*Dioscorea tomentosa* J. Koenig ex Spreng.	Twiner	Tuber
37	Dahiya-ala	–	Dioscoreaceae	*Dioscorea trimenii* Prain & Burkill[b]	Twiner	Tuber
38	Panuala	–	*Araceae*	*Typhonium trilobatum* schott	Herb	Corm

Data sources: Mihiranie et al. (2020), Goonatilake et al. (2015), Ekanayake et al. (2020), Godamulla (2018), Alwis et al. (2007, 2018), Bandula and Nath (2020), PGRC (1999), IUCN (2015b), Silva and Chandrasekara (2020), Rajapaksha (1998), Dassanayake and Clayton (1996, 1997, 1998, 1999, 2000), Dassanayake and Fosberg (1980, 1981a, b, 1983, 1985, 1987, 1991, 1994, 1995), Sajiwanie and Rathnayaka (2016), Rupasingha et al. (2020), Jayasuriya (1990), and Weerasekara et al. (2018)

[a]Exact local and botanical names not found in national checklists

[b]Endemic

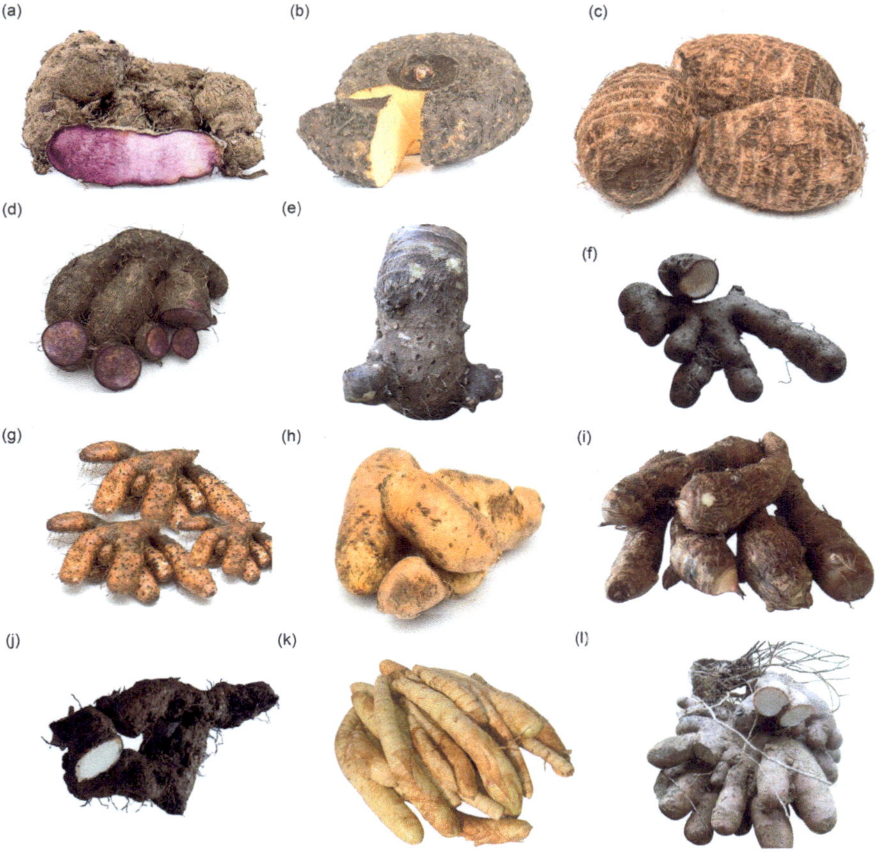

Fig. 8.6 Selected nutritionally significant neglected root and tubers from SESs in Sri Lanka for promotion and reintroduction through Biodiversity for Food and Nutrition (BFN) project. (**a**) raja-ala, (**b**) kidaran, (**c**) kiriala, (**d**) dandila, (**e**) mawala, (**f**) angiliala, (**g**) hingurala, (**h**) kukulala, (**i**) Isuru desa-ala, (**j**) katuala, (**k**) hulankeeriya, and (**l**) iniala. (Photo sources: Biodiversity for Food and Nutrition (BFN) project; https://lankancart.com/index.php?route=product/product&product_id=2561; http://science.sjp.ac.lk/backtonature/traditional-foods-in-sri-lanka-yams-root-tubers; https://www.ceylone.lk/product-category/local-yam-for-power-full-life-desheeya-ala-warga/)

Fig. 8.7 Value-added sweetness and delicacy of local yams. (**a**) kiriala watalapan (*Xanthosoma sagittifolium* (L.) Schott), (**b**) sevelala cake (*Alocasia* spp.), (**c**) kiriala chips (*Xanthosoma sagittifolium* (L.) Schott), (**d**) dandila dodoal (*Dioscorea alata* L.), and (**e**) keriala toffee (*Xanthosoma sagittifolium* (L.) Schott). (Photo source: Biodiversity for Food and Nutrition (BFN) project)

8.2.3 Pulses and Legumes

Pulses are rich in protein and a source of supplementary protein in the daily diets of Sri Lankans. The most commonly cultivated pulses and nuts in Sri Lanka are; common bean, yard long-bean, cowpea, soy bean, winged bean, ground nut, pigeon pea, lentils, black gram and green gram. Neglected pulses are mostly found in the traditional farming systems in SESs, especially in the VTCSs as ancestral cultivars (IUCN 2015a, b; Ekanayake et al. 2020). Diversity was observed in yard long-bean especially in the VTCSs. Most farmers prefer to cultivate traditional cultivars of these crops; locally known as polon-ma, kotu-ma, bushita, patta-ma and mas-ma due to their adaptation for shifting cultivation (Chena) conditions. Out of the traditional varieties, locally called kotu-ma and polon-ma, rathu-ma, hawari-ma are popular as they give more yield, taste good and are quite pest and disease tolerant. These traditional varieties of yard long-bean maintained by local farmers in their farms as family field gene banks and need to be characterised for better understanding of the available varieties for future value chain development and utilisation (Wasala et al. 2014) (Fig. 8.8 and Table 8.4).

8.2.4 Horticultural Crops

The neglected horticultural crops in Sri Lanka are mostly indigenous varieties or traditional cultivars, which are still cultivated and consumed to a lesser extent by rural communities as well as some peri-urban and urban communities. However, neglected horticultural crops have unexplored potential to further contribute to the rural and urban food systems to achieve national food and nutrition security (Ranil et al. 2020; Peduruhewa et al. 2021; Pushpakumara et al. 2007; Ratnayake et al. 2019; Borelli et al. 2021). These species can be mainly divided into neglected

Fig. 8.8 Traditional pulses and yard long-bean varieties maintained by local farmers of social-ecological systems in family field gene banks. (**a**) bala-cowpea seeds, (**b**) mas-ma seeds, (**c**) dambala seeds, (**d**) polon-ma seeds, (**e**) bushita seeds, (**f**) polon-ma pod (farmer's field gene bank), (**g**) polon-ma pods, (**h**) rathu-ma pods, and (**i**) kotu-ma pods. (Photo source: Biodiversity for Food and Nutrition (BFN) and Biodiversity for Adaptation to Climate Change (BACC) projects)

Table 8.4 Neglected pulse and legume species/varieties found in SESs of Sri Lanka

No.	Local name/s	Common name	Family	Botanical name	Habit
1	Kollu	Horse gram	Fabaceae	*Macrotyloma uniflorum* (Lam.) Verdc.	Twining, climbing or partially erect vines
2	Wanduru-ma	Velvet bean	Fabaceae	*Mucuna pruriens* (L.) DC	Twining, climbing vines
3	Pothu-dambala/ dambala/dara-dambala	Lima bean	Fabaceae	*Phaseolus lunatus* L.	Small climbers or erect herbs
4	Boo-ma	Rice bean	Fabaceae	*Vigna umbellate* (Thunb.) Ohwi & H. Ohashi	Small vines/ climbers
5	Polonma[a]	Yard-long bean/Snake bean	Fabaceae	*Vigna unguiculate* subsp. *sesquipedalis* (L.) Verdc.	Trailing or climbing vines
6	Koth-averay/Seeni bonchi[b]	Cluster bean	Fabaceae	*Cyamopsis tetragonoloba* (L.) Taub.	Erect, bushy herb
7	Gas-awra/Bu-awara/ Wal-awara	Jack bean	Fabaceae	*Canavalia ensiformis* (L.) DC.	Bushy or climbing
8	Awara	Sword bean	Fabaceae	*Canavalia gladiate* (Jacq.) DC.	Vines, climbing or trailing
9	Ho-dambala/Kos-ata-dambala/Kiri-dambala/ Ratu-peti-dambala/ Sudu-peti-dambala	Lablab bean/ Hyacinth bean	Fabaceae	*Lablab purpureus* (L.) Sweet	Twining/climbing/ trailing or upright herb
10	Ala-bonchi[b]	Yam bean	Fabaceae	*Pachyrhizus erosus* (L.) Urban	High-climbing vines
11	Wal-kollu	Peanut grass (Wild relative of Pigeon pea)	Fabaceae	*Cajanus scarabaeoides* (L.) Thouars	Small, twining vines

Data sources: Goonatilake and Ekanayake (2016), Bandula and Nath (2020), Peduruhewa et al. (2021), Malkanthi (2017), PGRC (1999), Weerasekara et al. (2020), IUCN (2015b), Tyagi et al. (2018), Rajapaksha (1998), Dassanayake and Clayton (1996, 1997, 1998, 1999, 2000), Dassanayake and Fosberg (1980, 1981a, b, 1983, 1985, 1987, 1991, 1994, 1995), MOE (2012), Joshi et al. (2020), and Saxena et al. (1990)

[a]Exact local name and botanical name not given in national checklists

[b]Exact local name not found in national checklists

vegetables and fruits that can be found in varied agroecosystems and wilderness areas in Sri Lanka including SESs. However, potentials of these vegetables and fruits have not been properly realised thus importance of these species has not properly surfaced. Further, comprehensive assessments of the distribution of these species have not been completed (Dahanayake 2015). Based on the past ecogeographic survey data, high-value areas of neglected and underutilized vegetables and fruits and their wild relatives are shown in Fig. 8.9.

(a) Neglected vegetables

The taxonomic diversity of indigenous vegetables and leafy vegetables in Sri Lanka comprises over 105 species from 47 families (Ranil et al. 2020). Many of these vegetable crop species are neglected or underutilised (Table 8.5). Currently, vegetable production in the country is heavily dependent on around 25–40 species (Ranil et al. 2020, 2021; Chamara et al. 2021). Neglected and underutilised vegetables are grown throughout the country and most of them are indigenous varieties mostly found in wet zone law and mid-country KHGs and peri-urban Owita Systems. In the VTCSs, neglected vegetables are cultivated mainly as shifting cultivation (Chena). While the rural population in Sri Lanka use these vegetables in various traditional food preparations to diversify their diet and ensure food security in their households, new initiatives have begun to incorporate these varieties into urban food systems and the urban model vegetable gardens through the GEF-funded Biodiversity for Food and Nutrition (BFN) project (Esham and Wijeratne 2021). Ultimately, the demand for indigenous neglected vegetable varieties is increasing due to the realisation of their nutritional and medicinal value. Some therapeutic properties of neglected leafy greens have received scientific attention for their use for some chronic disease conditions such as cardiovascular diseases, cancer, diabetes mellitus, Alzheimer's disease, cataracts, and age-related functional decline, cognitive function and to resolve skin and digestive afflictions (Gee et al. 2020; Nadeeshani et al. 2018). In addition, several herbal greens collected from wild and semi-wild areas are utilised for culinary and traditional medicinal purposes (Ranil et al. 2020, 2021; IUCN 2015b; Nadeeshani et al. 2018).

(b) Wild mushrooms

Apart from indigenous vegetable varieties, a number of edible mushroom varieties are found in the SESs. Mushrooms are grown in the KHGs, Owita Systems and bear lands of the VTCSs and are consumed by the local community living there. These mushroom varieties are locally known as; Nerihathu, Rathkevilla, Weerahathu, Kukulubadawel, Kirikuttanhathu, Netibimmal, Uruhathu, Metihathu, Pidurubimmal, Humbasbimmal, Heenweliya, Mahaweliya, Lokapahurubimmal, Welibimmal, Lenahathu, Indaloluhathu. However, their botanical names are unknown due to a lack of taxonomic research. The nutritional composition of Sri Lankan wild edible mushrooms has not been studied in detail (Ekanayake et al. 2020; HLP-BLA 2020).

(c) Neglected and underutilised fruits

The diversity of fruit species in the country, in natural and cultivated systems, has been recorded to be over 200 species (including 18% endemic and 41% indigenous)

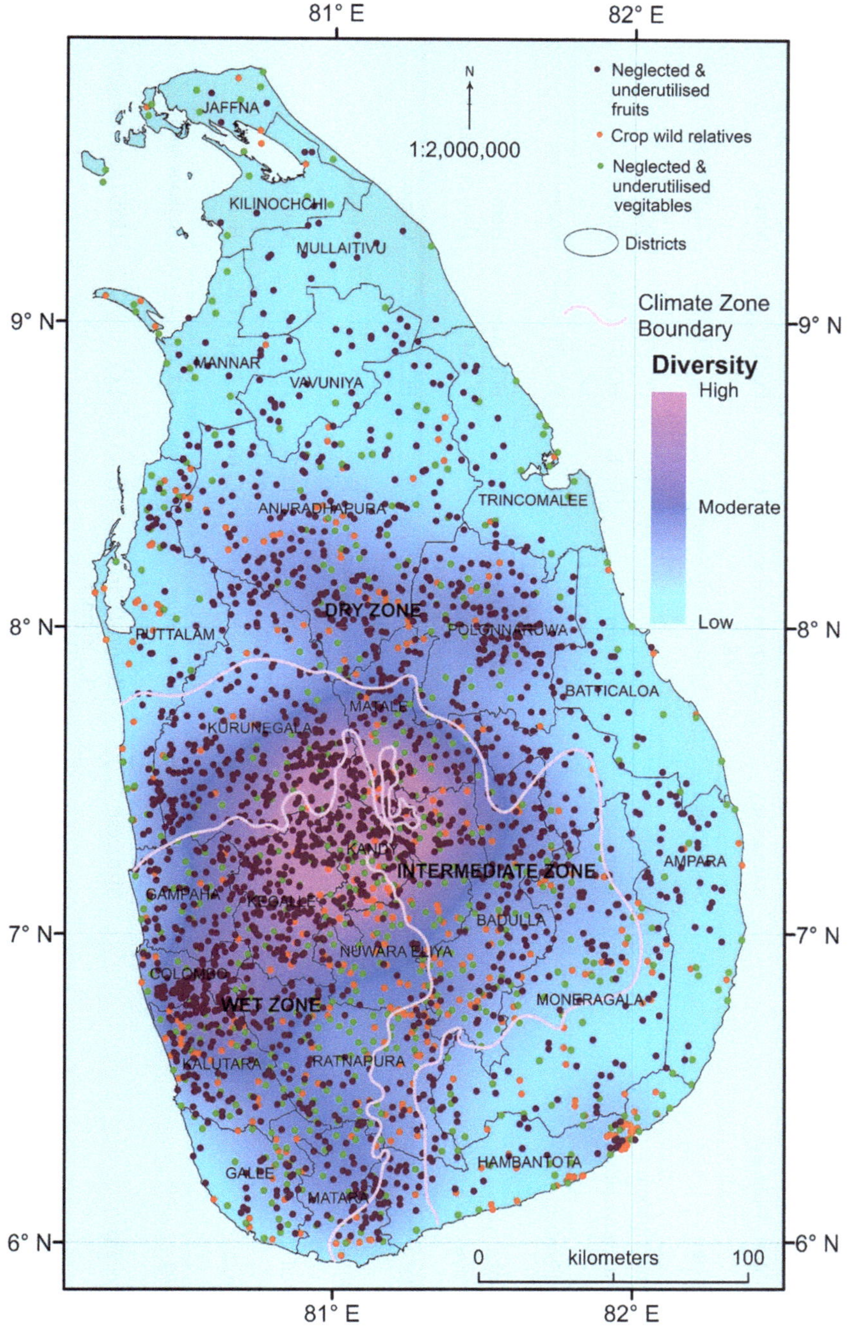

Fig. 8.9 Sri Lanka's high-value areas of occurrence of neglected and underutilised fruit and vegetable species as well as their wild relatives based on the spatial distribution of species occurrence data. (Data sources: Liyanage (2010), Ranil et al. (2020), Ratnayake et al. (2019), and DOA (2005))

Table 8.5 Neglected vegetable species and varieties found in SESs of Sri Lanka

No	Local name/s	Family	Botanical name	Edible part/s	Food preparations	Habit
1	Thittatibbatu/Thibbatu	Solanaceae	*Solanum violaceum* Orteg	Fr	FC, FT	Shrub
2	Welthibbatu	Solanaceae	*Solanum trilobatum* L.	Fr	FC, FT	Creeping shrub
3	Gonabatu/Thibbatu	Solanaceae	*Solanum torvum* Sw.	Fr	FC, FT	Shrub
4	Ahaswambatu/Elabatu	Solanaceae	*Solanum macrocarpon* L.	Fr, L	FC, FT, LM, LP	Shrub
5	Delbatu/Naibatu	Solanaceae	*Solanum capsicoides* All	Fr	FC, FT	Shrub
6	Kalukanberiya	Solanaceae	*Solanum nigrum* L.	Fr, L	FC, LM	Herb
7	Kalukanmeriya/Kalukanweriya	Solanaceae	*Solanum americanum* Mill.	Fr, L	F, LM	Herb or small shrub
8	Diyalabu	Cucurbitaceae	*Lagenaria siceraria* (Molina) Standl.	Fr	FC, FB, LC, LT, C	Climber
9	Yakkommadu	Cucurbitaceae	*Citrullus colocynthis* (L.) Schrad.	Fr	FC	Climber
10	Alupuhul	Cucurbitaceae	*Benincasa hispida* (Thunb.) Cogn.	Fr, L, Se, Flb	F, FC, SC	Climber
11	Lunu-kola	Amaryllidaceae	*Allium hookeri* Thwaites	L, Bu	LT, SF, FCS	Herb
12	Wal-lunu	Amaryllidaceae	*Pancratium zeylanicum* L.	L	LT, SF	Bulbous herb
13	Maha-sarana	Aizoaceae	*Zaleya decandra* (L.) Burm. f.	L	LC	Herb (prostrate stem)
14	Polpala	Amaranthaceae	*Ouret lanata* (L.) Kuntze	L+St	LM, LB	Herb

15	Katuthampala/Katukeera	Amaranthaceae	*Amaranthus spinosus* L.	L+St	LC, SF	Herb
16	Kirihenda	Amaranthaceae	*Celosia argentea* L.	L+St	LM, LC, LP	Herb
17	Anguna/Anukkola/Kirianguna/Thitha-anguna	Apocynaceae	*Wattakaka volubilis* (L.f.) Stapf	L	LM	Woody twiner
18	Kurinnan	Apocynaceae	*Gymnema lactiferum* (L.) R.Br. ex Schult.	L	LM	Twining or straggling woody plant
19	Welkohila	Araceae	*Nephthytis triphylla* Nash	L	LC, LT	Climber
20	Monarakudumbiya	Asteraceae	*Cyanthillium cinereum* (L.) H. Rob.	L+St	LM, LP, LB	Herb
21	Lolu	Boraginaceae	*Cordia dichotoma* G. Forst	Flb, Fr, L	LM, FCST, LC	Tree
22	Welabakola[a]	Brassicaceae	*Cardamine hirsuta* L.	L+Flb	LC, LM	Herb
23	Abakola	Brassicaceae	*Brassica juncea* (L.) Czern.	L	LS, LM	Herb
24	Walabakola	Cleomaceae	*Cleome viscosa* L.	L	LM	Herb
25	Barukoku/Pathra-werealla	Blechnaceae	*Blechnopsis orientalis* (L.) C. Presl	L	SF, LM, LC	Herb (Pteridophytes)
26	Giritilla	Convolvulaceae	*Argyreia kleiniana* (Schult.) Raizada.[b]	L, Fr	LM, FC, FT	Liana/Woody creeper
27	Gon-kekiri/Heen-kekiri/Len-kekiri	Cucurbitaceae	*Cucumis maderaspatanus* L.	Fr	F, C	Climber

(continued)

Table 8.5 (continued)

No	Local name/s	Family	Botanical name	Edible part/s	Food preparations	Habit
28	Kawudu-kekiri/Telaberiya	Cucurbitaceae	*Solena umbellata* (J.G. Klein ex Willd.) W.J.de Wilde & Duyfjes	Fr	F, FC, FB, FS	Climber
29	Madu	Cycadaceae	*Cycas nathorstii* J. Schust.	Se, Lsh	SC, LSC	Tree
30	Godapara	Dilleniaceae	*Dillenia retusa* Thunb.[b]	Fr	FC	Tree
31	Getatumba	Lamiaceae	*Leucas zeylanica* (L.) W.T. Aiton	L	LC, LS	Herb
32	Kahata	Lecythidaceae	*Careya arborea* Roxb.	Fr	FC	Tree
33	Kapu-kinissa	Malvaceae	*Abelmoschus angulosus* Wall. ex Wight & Arn.	Fr	FC	Herb
34	Heen-bovitiya	Melastomataceae	*Osbeckia octandra* (L.) DC.	L+St	LC, LP	Subshrub
35	Heenpala/Heenala	Molluginaceae	*Glinus oppositifolius* (L.) Aug.DC	L	LP	Creeping herb
36	Attikka	Moraceae	*Ficus racemosa* L.	Fr	FC	Tree
37	Wathabanga/Lechchakotta	Nyctaginaceae	*Ceodes grandis* (R.Br.) D.Q. Lu	L	LM	Small tree
38	Kebella	Phyllanthaceae	*Aporosa cardiosperma* (Gaertn.) Merr. *Aporosa lindleyana* (Wight) Baill.	L	LM, LC, LT	Tree

39	Sweet-leaf/Singappuru-leaf/Japan batu/ Mella-dumkola	Phyllanthaceae	*Sauropus androgynus* (L.) Merr.	Fr, L, Lsh, Flb	FC, LS, SF	Shrub
40	Lunuwila	Plantaginaceae	*Bacopa monnieri* (L.) Wettst.	L+St	LC, LS, LB	Herb
41	Keran-koku; Miyena-dalu	Pteridaceae; Athyriaceae	*Acrostichum aureum* L.; *Diplazium esculentum* (Retz.) Sw.	L	LC, LS, SF	Herb (Pteredophytes)
42	Diyahabarela/Jabara	Pontederiaceae	*Pontederia hastata* L.	L, Flb	LC, LT	Aquatic herb
43	Gendakola	Portulacaceae	*Portulaca oleracea* L.	L+St	LC	Herb
44	Penelawel/Welpenela	Sapindaceae	*Cardiospermum microcarpum Kunth*	L	LS, LT, LP	Climbing or sub scandent herb
45	Gasnivithi	Talinaceae	*Talinum paniculatum* (Jacq) Gaertn	L+St	LC, LT	Subshrub
46	Kowkka	Cucurbitaceae	*Coccinia grandis* (L.) Voigt	L, Fr	FC, LS, LP	Climber
47	Pothupala	Convolvulaceae	*Aniseia martinicensis* (Jacq.) Choisy	L	LS, LT	Short creeper
48	Mella	Olacaceae	*Olax zeylanica* L.	L	LC, LS	Shrub or small tree

(continued)

Table 8.5 (continued)

No	Local name/s	Family	Botanical name	Edible part/s	Food preparations	Habit
49	Ikiriya/Neermulliya/Katuikiriya	Acanthaceae	*Hygrophila auriculata* (Schumach.) Heine	L+St	LS, LC	Herb
50	Batukarawila/Karawila	Cucurbitaceae	*Momordica charantia* L.	Fr	FC, FT	Climbing annual vine
51	Weda-mukunuwenna	Amaranthaceae	*Alternanthera sessilis* L.	L+St	LS, LM, LT, LP	Herb
52	Thebu/koltan	Costaceae	*Hellenia speciosa* (J.Koenig) S.R.Dutta	L	LS, LM, LT, LC, LP	Rhizomatous herb

Data sources: Bandara (2009), Ranil et al. (2020, 2021), IUCN (2015a, b), Goonatilake and Ekanayake (2016), Peduruhewa et al. (2021), Mudau et al. (2022), HLP-BLA (2020), Weerasekara et al. (2018, 2020), Rajapaksha (1998), Dassanayake and Clayton (1996, 1997, 1998, 1999, 2000), Dassanayake and Fosberg (1980, 1981a, b, 1983, 1985, 1987, 1991, 1994, 1995), Alwis et al. (2007), Nadeeshani et al. (2018, 2021), Liyanaratne (2003), and Thamilini et al. (2019)

Edible parts: *L* Leaf, *Fr.* Fruit, *Lsh.* Leaf shoot, *Se* Seed, *St.* Stem, *Flb.* Flower/bud, *P* Pod, *Bu* Bulb

Food preparations: *F* Fresh, *FC* Fruit Curry, *LC* Leaf Curry, *LS* Leaf Salad/Samble, *SC* Seed Curry, *FS* Fruit Salad, *FT* Fruit Tempered, *LM* Leaf Mallum, *LP* Leaf Porridge, *LT* Leaf Tempered, *FB* Fruit Beverage, *LB* Leaf Beverage-Herbal drink, *SF* Stir-Fry, *FCST* Flower Curry/Salad/Tempered, *C* Chutney, *LSC* Leaf Shoot Curry

[a]Exact local name is not found in national checklists

[b]Endemic

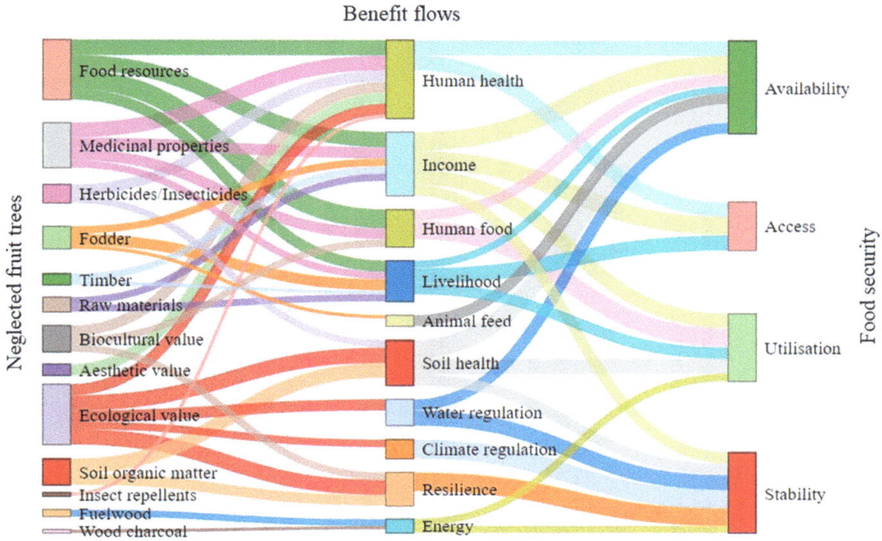

Fig. 8.10 Traditional uses of neglected fruit tree resources in social-ecological systems and link to ecosystem services and food security. The thicker the link, the higher the benefits and importance for food security. (Data source: Author's calculation from food security perception survey data of the upper Nachchaduwa VTCS complex of Anuradhapura, Sri Lanka (2023))

representing 56 plant families. However, more than 60 varieties of fruit species come under the neglected and underutilised category (Pushpakumara et al. 2016; Ratnayake et al. 2019; DOA 2005); Among the SESs, KHGs recorded the highest neglected and underutilised fruit species diversity (Pushpakumara et al. 2016; DOA 2005; Thamilini et al. 2019). More than 50% of the neglected fruits in the country are found as wild/forested food tree resources (DOA 2005; Ickowitz et al. 2022). A majority of fruit trees have multiple utility values and communities use these fruit and tree resources for various purposes (IUCN 2015a; Goonatilake et al. 2015) (Fig. 8.10). In addition to the nutritional value most neglected fruit species contain medicinal properties, with various parts of plants used for the preparation of traditional medicines by ayurvedic practitioners and the local community (Ratnayake et al. 2019). Neglected and underutilised fruit species recorded in SESs from past literature and ecogeographic surveys are listed in Table 8.6.

8.3 Food Security Challenges in SESs and Nutritional Significance of NUPFs

Although no systematic documented historical data is available regarding food consumption and nutritional quality of traditional diets, a plant-based diet with a diversity of foods had been available for consumption in the past, especially in the traditional farming communities in SESs. These diets offered various health and

Table 8.6 The richness of neglected and underutilised fruit species found in SESs of Sri Lanka

No	Local name/s	Common name/s	Botanical name	Family	Habit	Occurrence
1	Katu-annoda/Katu-attha/ Rata-anoda	Soursop	*Annona muricata* L.	Annonaceae	Tree	Cultivated
2	Weli-anoda/Weli-attha	Bullock's heart	*Annona reticulata* L.	Annonaceae	Tree	Cultivated
3	Sini-anoda/Seeni-attha	Custard apple	*Annona squmosa* L.	Annonaceae	Shrub or tree	Cultivated
4	Cherimoya	Cherimoyer	*Annona cherimola* Miller	Annonaceae	Shrub or tree	Cultivated
5	Beli	Bael fruit	*Aegle marmelos* (L.) Corrêa	Rutaceae	Tree	Cultivated
6	Cheena-pera	Chinese guava	*Psidium catleianum*	Myrtaceae	Shrub or small tree	Cultivated
7	Madan	Black plum	*Syzygium cumini* (L.) Skeels	Myrtaceae	Tree	Native (wild/forest)
8	Heen-dan	Indian blackberry	*Syzygium caryophyllatum* (L.) Alston	Myrtaceae	Shrub or small tree	Native (wild/forest)
9	Balu-dan	Coralberry	*Ardisia elliptica* Thunb	Primulaceae	Shrub	Native (wild/forest)
10	Gaduguda	Rambai	*Lansium domesticum* Corrêa	Meliaceae	Tree	Cultivated
11	Pera	Guava	*Psidium guajava* L.	Myrtaceae	Tree	Cultivated
12	Jam-pera	Strawberry guava	*Psidium cattleianum* Afzel. ex Sabine	Myrtaceae	Shrub or small tree	Cultivated
13	Pini-jambu	Water apple	*Syzygium samarangense* (Blume) Merr. & L.M. Perry	Myrtaceae	Tree	Cultivated

14	Seeni-jambu/Veli-jambu	Rose apple	*Syzygium jambos* (L.) *Alston*	Myrtaceae	Tree	Cultivated
15	Kamaranga	Star fruit	*Averrhoa carambola* L.	Oxalidaceae	Small tree	Cultivated
16	Heen-karamba	Conkerberry/bush plum	*Carissa spinarium* L.	Apocynaceae	Shrub	Native (wild/forest)
17	Desi-puhul/Tun-tun	Giant granadilla	*Passiflora quadrangularis* L.	Passifloraceae	Climber	Naturalized exotic
18	Biling	Bilimbi	*Averrhoa bilimbi* L.	Oxalidaceae	Small tree	Cultivated
19	Wal-del/Bedi-del/Hingala-del	Ceylon breadfruit	*Artocarpus nobilis* Thwaites[a]	Moraceae	Tree	Wet tropical forest
20	Maha-karamba	Karonda, Jamson	*Carissa carandas* L.	Apocynaceae	Shrub or small tree	Native (wild/forest)
21	Koon	Ceylon oak	*Schleichera oleosa* (Lour.) Oken	Sapindaceae	Tree	Native (wild/forest)
22	Rata-lawulu	Canistel, yellow sapote	*Pouteria campechiana* (Kunth) Baehni	Sapotaceae	Tree	Cultivated
23	Lovi	Batoko plum	*Flacourtia inermis* Roxb.	Salicaceae	Tree	Cultivated
24	Telabu	Wild almond	*Sterculia foetida* L.	Malvaceae	Tree	Native (wild/forest)
25	Jam	Jamaica cherry	*Muntingia calabura L.*	Muntingiaceae	Tree	Naturalized exotic
26	Daminiya/Damunu	Salvia leaved crossberry	*Grewia tiliifolia* Vhal[a]	Malvaceae	Medium sized tree	Native (wild/forest)
27	Heen-naran	Mandarin	*Citrus × aurantium f. deliciosa* (Ten.) M. Hiroe	Rutaceae	Shrub or small tree	Cultivated
28	Masan/Debara/Maha-debara	Indian jujube	*Ziziphus mauritiana* var. mauritiana Lam.	Rhamnaceae	Small thorny tree	Native (wild/forest)

(continued)

Table 8.6 (continued)

No	Local name/s	Common name/s	Botanical name	Family	Habit	Occurrence
29	Mora/Penni-mora/Rasa-mora	Longan/Dragon eye	*Dimocarpus longan* Lour	Sapindaceae	Tree	Native (wild/forest)
30	Heen-himbutuwel	Lolly berry	*Salacia chinensis* L.	Celastraceae	Large woody climber	Native (wild/forest)
31	Gal-karanda/Karanda/Magul-karanda	Indian beech	*Pongamia pinnata* (L.) Pierre	Fabaceae	Tree	Native (wild/forest)
32	Nam-nam	Nam nam	*Cynometra cauliflora* L.	Fabaceae	Tree	Cultivated
33	Siyabala	Tamarind	*Tamarindus indica* L.	Fabaceae	Tree	Cultivated
34	Beheth-nelli	Amla	*Phyllanthus emblica* L.	Phyllanthaceae	Small tree	Native (wild/forest)
35	Rata-nelli	Star gooseberry	*Phyllanthus acidus* (L.) Skeels	Phyllanthaceae	Tree	Cultivated
36	Palu	Palu	*Manilkara hexandra* (Roxb.) Dubard	Sapotaceae	Tree	Native (wild/forest)
37	Jambola	Pummelo	*Citrus maxima* (Burm.) Merr.	Rutaceae	Tree	Cultivated
38	Sapodilla	Sapodilla	*Manilkara zapota* (L.) P. Royen	Sapotaceae	Tree	Cultivated
39	Uguressa	Governor's plum	*Flacourtia indica* (Burm.f.) Merr.	Salicaceae	Shrub or small tree	Native (wild/forest)
40	Ambul-dodam	Sour orange	*Citrus* × *aurantium* L.	Rutaceae	Small tree	Cultivated
41	Pani-dodam	Sweet orange	*Citrus sinensis* (L.) Osbeck	Rutaceae	Shrub or tree	Cultivated
42	Mahaeraminiya	Wild Jujube	*Ziziphus rugosa* Lam.	Rhamnaceae	Spiny, shrub or small tree	Native (wild/forest)
43	Embul pera	Sour guava	*Psidium guineense* Sw.	Myrtaceae	Shrub or small tree	Cultivated

44	Ketambilla	Ceylon gooseberry	*Dovyalis hebecarpa* (Gardner) Warb.[a]	Salicaceae	Shrub or small tree	Wet tropical wild/forest areas
45	Galsiyabala	Velvet tamarind	*Dialium ovoideum* Thw.[a]	Fabaceae	Tree	Dry tropical wild/forest areas
46	Weralu	Wild olive	*Elaeocarpus serratus* L.	Elaeocarpaceae	Tree	Native (wild/forest)
47	Batadamba	Damba	*Syzygium nervosum* A. Cunn. ex DC.	Myrtaceae	Tree	Native (wild/forest)
48	Weera	Weera	*Drypetes sepiaria* (Wight & Arn.) Pax & K. Hoffm.	Putranjivaceae	Tree	Native (wild/forest)
49	Wel embilla/Walangasal	False black pepper	*Embelia ribes* subsp. ribes Burm.f.	Primulaceae	Climber	Native (wild/forest)
50	Divul	Wood-apple	*Limonia acidissima* L.	Rutaceae	Tree	Native (wild/forest)
51	Etamba/Wal-amba	Wild-mango	*Mangifera zeylanica* (Blume) Hook.f.[a]	Anacardiaceae	Tree	Wet tropical wild/forest areas
52	Mee-amba	Honey-mango	*Mangifera indica* L.	Anacardiaceae	Tree	Native (wild/forest)
53	Karawala-kebella	Bignay/Wild cherry	*Antidesma bunius* (L.) Spreng.	Phyllanthaceae	Tree	Native (wild/forest)
54	Kara	Coromandel boxwood	*Canthium coromandelicum* (Burm.f.) Alston	Rubiaceae	Shrub	Native (wild/forest)
55	Heen-embilla	Ceylon bignay	*Antidesma alexiteria* L.	Phyllanthaceae	Shrub or small tree	Native (wild/forest)
56	Katu-embilla/Wel-embilla	Bastard oleaster	*Elaeagnus latifolia* L.	Elaeagnaceae	Scrambling shrub	Native (wild/forest)
57	Kotamba/Kottan	Indian almond	*Terminalia catappa* L.	Combretaceae	Tree	Naturalized exotics
58	Amberella	Jew plum/Golden apple	*Spondias dulcis* Sol. ex Parkinson	Anacardaceae	Tree	Cultivated

(continued)

No	Local name/s	Common name/s	Botanical name	Family	Habit	Occurrence
59	Galmora/Bulu-mora, Na-imbul	Fijian longan	*Pometia pinnata* J.R. Forst. & G. Forst.	Sapindaceae	Tree	Native (wild/forest)
60	Kothala himbutu/Himbutu-wel	–	*Salacia reticulata* Wight	Celastraceae	Large woody climber (Liana)	Native (wild/forest)
61	Indi/Wal-indi	Ceylon date palm	*Phoenix farinifera* Roxb.	Arecaceae	Palm tree	Native (wild/forest)
62	Peni-gedi	Miracle-fruit berry	*Synsepalum dulcificum* (Schumach. & Thonn.) Daniell	Sapotaceae	Large shrub	Native (wild/forest)
63	Kaju-puhulan	Cashew apple (fruit)	*Anacardium occidentale* L.	Anacardiaceae	Tree	Cultivated
64	Local-delum	Pomegranate (local variety)	*Punica granatum* L.	Punicaceae	Large shrub or small tree	Cultivated
65	Gin-pol[b]	Nipa palm fruit/ Water coconut	*Nypa fruticans* Wurmb	Arecaceae	Mangrove palm	Native (wild/ mangrove forest)

Data sources: Goonatilake et al. (2015), Ekanayake et al. (2020), Pushpakumara et al. (2016), Marambe et al. (2012a), Bandula and Nath (2020), Malkanthi (2017), Ratnayake et al. (2019), Dahanayake (2015), DOA (2005), PGRC (1999), Weerasekara et al. (2018, 2020), IUCN (2015b), Dassanayake and Clayton (1996, 1997, 1998, 1999, 2000), Dassanayake and Fosberg (1980, 1981a, b, 1983, 1985, 1987, 1991, 1994, 1995), Thamilini et al. (2019), Abeysuriya et al. (2020), and Rajapakse and Wijesekera (2019)

[a]Endemic

[b]Found in coastal social-ecological systems

nutritional benefits, including protection from non-communicable diseases (NCDs) (Ekanayake et al. 2020; Silva and Chandrasekara 2020). However, phenomenal change in dietary behaviours as evident from neglecting foods derived from traditional crops has been witnessed that may be linked to colonisation and westernisation of the lifestyle (Weerasekara et al. 2018). According to the food production and utilisation survey among households in the upper Nachchaduwa VTCS complex Anuradhapura, Sri Lanka found that community, within which relatively low amounts of fruits and leafy vegetables are consumed. This may lead to micronutrient deficiencies among the VTCS community (Thamilini et al. 2019; Malkanthi et al. 2007). Further, overall variety of foods (dietary diversity) in the VTCSs are also unsatisfactory. On average, only 6–7 out of 13 food groups and around 9 individual food items were consumed by the respondents on a normal day (Fig. 8.11).

The communities of the surveyed VTCS areas also perceived that their local food system has the potential to provide nutritious and healthy NUPFs, but they admitted that it is not utilised optimally for their household diets due to low income, lack of food literacy, nutrition transition, food preferences and issues related to their agricultural practices (HLP-BLA 2020; Silva and Chandrasekara 2020; Thamilini et al. 2019; Malkanthi et al. 2007). Although the local food systems in the studied SESs are vulnerable due to some current socioeconomic reasons and climate change effects, the majority of home garden systems still show potential to enhance the supply of fruits and vegetables with systematic agricultural practices (Pushpakumara et al. 2016; Marambe et al. 2012b; Weerahewa et al. 2012). The supply of wild plant foods from village forests and semi-natural areas in the VTCSs has also become limited to a few vegetables and fruit varieties due to changes in land use pattern and structure—driven principally by land clearance to accommodate a growing population and commercial agriculture. In reality, most of the indigenous crop varieties and wild food resources consumed by the village community in their childhood are no longer available (Silva and Chandrasekara 2020).

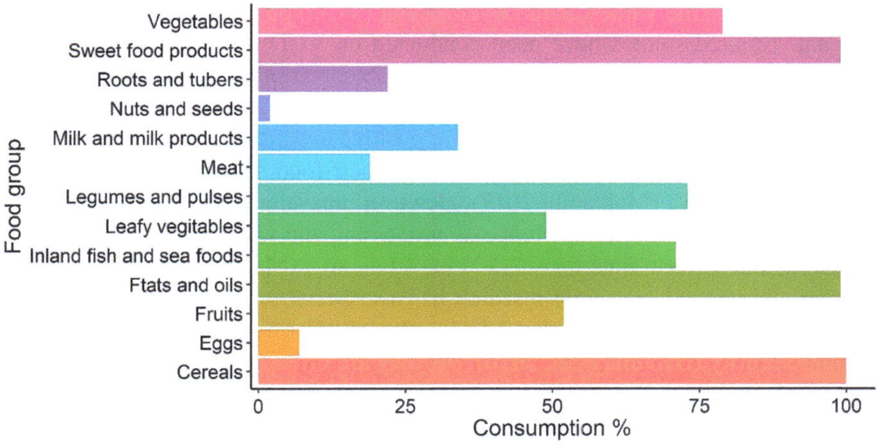

Fig. 8.11 Percentage of participants who consumed different type of food group on the previous day of the food consumption survey in upper Nachchaduwa VTCS complex located in Anuradhapura district of Sri Lanka

The recently revised food-based dietary guidelines for Sri Lankans encourages the consumption of local indigenous food varieties to achieve optimal nutritional status among the population (MOH 2022). Many of the neglected and underutilised food crops found in Sri Lanka and elsewhere are not only rich sources of fibre, vitamins, and minerals but also contain bioactive plant compounds having antioxidant, anti-cancer, and antimicrobial properties (Peduruhewa et al. 2021; Hunter et al. 2019; Hettiarachchi and Gunathilake 2020). However, only a few studies have been conducted that use local samples of neglected food plant varieties to determine the phytochemical and nutritional composition of NUPFs in Sri Lanka (Table 8.7).

The varietal level variation in nutritional composition has also been largely unexplored (Hunter et al. 2019). Though research data on nutritional composition of traditional food crops of Sri Lanka are scant, the available information cited in the Table 8.7 demonstrates superiority of traditional varieties of locally important species (cereals, yams, pulses, vegetables and fruits) over high demanding commonly consumed varieties. Micronutrients such as β-carotene, vitamin C, Ca and K, fibre and other beneficial phytochemical contents are high in several neglected food crops (Hettiarachchi and Gunathilake 2020). Traditional food crops may be considered cost effective sources of dietary energy and proteins. The generation of food composition data in NUPFs is one of the strategies for popularising the consumption of these. Compositional data play a critical role in advocacy, nutritional planning, food processing, promotion, marketing and crop improvement (Hunter et al. 2019; Oliveira et al. 2018).

8.4 Value Chain Development of NUPFs: Constraints and Opportunities in Sri Lanka

Though much attention has been drawn to NUPFs globally, in light of nutrition and health concerns, the Value Chain Development (VCD) on these plants in Sri Lanka is scant. Limited studies have been conducted on VCD of underutilised crops—fruits, traditional rice varieties and few vegetables varieties (Bandula et al. 2016; Bandula and Nath 2020; Barry 2008; Arampath et al. 2021; Weerasinghe et al. 2020). One study found that farmers were positive towards the cultivation of underutilised crops because these crops enhanced their food security, income and had low production cost (Malkanthi 2017). However, unavailability of seeds and planting materials, and lack of knowledge on processing and value additions were highlighted as main constraints for VCD in the study area (Malkanthi and Silva 2014). Limited market and consumer research has also been conducted on NUPFs value chains. This research showed that NUPFs play an important role in rural food and income security (Bandula et al. 2016), but they are often overlooked by urban consumers. A study conducted in urban areas of Colombo district in Sri Lanka found that their residents are not consuming neglected and underutilised fruits at a satisfactory level, although sixteen underutilised fruit varieties are grown in these areas (Rangani et al. 2017). It is possible that gender-sensitive approach to VCD of these

Table 8.7 Nutritional significance of selected neglected and underutilised plant foods based on locally collected samples

Food group	Botanical name	Local name/s	Nutritional significance	References
Cereals	*Oryza sativa* L. (Traditional rice varieties)	Suwandel	Vitamin E was found only in traditional rice varieties, Kaluheenati, Madathavalu and Suduheenati but not in other traditional varieties and improved varieties studied. Kaluheenati contained the highest amount of vitamin E (6.34 ± 0.01 mg/100 g), B1 (1.58 ± 0.06 mg/100 g) and B2 (2.11 ± 0.03 mg/100 g) and Pokkali had the highest vitamin K (7.55 ± 0.02 mg/100 g) among the studied rice varieties. The second highest vitamin B1 content was found in Suduheenati (1.04 ± 0.01 mg/100 g). Pachchaperumal and Pokkali contained similar amounts of vitamin B2, second only to Kaluheenati among the traditional varieties. Kaluheenati and Suduheenati recorded the highest iron and fibre content.	Hunter et al. (2019) and Kulasinghe et al. (2019a)
		Kaluheenati		
		Kuruluthuda		
		Madathavalu		
		Pachchaperumal		
		Pokkali		
		Suduheenati		
	Setaria italica (L.) P. Beauv.	Finger millet (Bala kurakkan)	Traditional Sri Lankan millet varieties (Kurakkan and Thanahal) tested for Zn, Fe, Ca, K, Na, and Mg. Finger millet varieties recorded comparatively high Ca and K content. Foxtail millet varieties showed comparatively high levels of Zn and Na content. Both finger millet and foxtail millet varieties showed a high level of Mg content.	Hunter et al. (2019)
		Finger millet (Wadimal kurakkan)		
		Foxtail millet (Golden) (Thanahal-Ran)		
		Foxtail millet (Yellow) (Than hal- Kaha)		
		Foxtail millet (Black) (Than hal -Kalu)		

(continued)

Table 8.7 (continued)

Food group	Botanical name	Local name/s	Nutritional significance	References
Root and tubers	*Dioscorea alata* L.	Raja-ala	Among the roots and tubers explored for their nutritional significance, Raja-ala and Hingurala recorded higher protein content compared with other local cultivars of tubers. Elephant foot yam (Kidaran) showed the highest calclum content (15.87 mg/100 g). Yams contained fat-soluble vitamins like vitamin A and K. Highest levels of vitamin K, vitamin B1 and crude proteins were recorded in Raja-ala. Antioxidant capacity and phenolic contents were significantly higher in both Udala and Kidaran yams. However, Dandila, Guruala, Hingurala, Rasawalli and Kahata-ala also recorded high amount of antioxidant capacity. In raw form, Dandila, Guruala and Kahata-ala recorded high phenolic content; Raja-ala and Udala recorded comparatively high ascorbic acid content.	Sajiwanie and Rathnayaka (2016), Rupasingha et al. (2020), Weerasekara et al. (2018), and Kulasinghe et al. (2019a)
	Dioscorea alata L.	Kahata angala		
	Dioscorea alata L.	Dandila		
	Dioscorea alata L.	Guruala		
	Dioscorea alata L.	Hingurala		
	Dioscorea alata L.	Jaffna rasawalli		
	Dioscorea alata L.	Kahata-ala		
	Amorphophallus paeoniifolius (Dennst.) Nicolson	Kidaran		
	Dioscorea bulbifera L.	Udala		
	Dioscorea esculenta (Lour.) Burkill	Kukulala	These cultivars contain a significant quantity of crude fibre and minerals. Kukulala contained the highest crude fat content among the studied yams. Kiriala recorded moderate antioxidant capacity and a comparatively higher ascorbic acid content (12 mg/100 mg) in yam. Potassium (K), Mg, Na, Zn and Fe were the most abundant mineral elements in tested yams, where potassium showed the higher availability.	Rupasingha et al. (2020), Kulasinghe et al. (2018), and Chandrasekara and Josheph Kumar (2016)
	Xanthosoma sagittifolium (L.) Schott	Kiriala		
	Colocasia spp.	Isuru desa-ala, Rathu-kiriala	Highest ascorbic acid content (12 mg/100 mg) in raw form.	Rupasingha et al. (2020)
	Dioscorea alata L.	Angiliala	Angiliala contained the highest ash percentage and also contained vitamin K. Vitamin A was found only in Katuala, which also had the highest vitamin B1 content.	Kulasinghe et al. (2019a)
	Dioscorea pentaphylla L.	Katuala		

Vegetables and green leafy vegitables	*Solanum torvum* Sw.	Thibbatu (wild eggplant, turkey berry)	The three *Solanum* species studied were *S. torvum, S. melongena* L. (Wambatu) and *S. melongena* var. insanum L. (Elabatu). Linoleic, α-linolenic and oleic acids were more abundant than most other unsaturated fatty acids in the Solanum species studied. *S. torvum* is the best source of minerals among the three with K 526 mg/100 g; Ca 329 mg/100 g. It also contains considerable amounts of vitamin E (90 μg/100 g). The overall nutritional value of *S. torvum* along with *S. melongena* (Elabatu) were in general superior to that of *S. melongena* (Wambatu).	Nadeeshani et al. (2021)
	Alternanthera sessilis L.	Weda-mukunuwenna	These green leafy vegetables are good sources of minerals.	Nadeeshani et al. (2018)
	Wattakaka volubilis (L.f.) Stapf	Aguna-kola		
	Coccinia grandis (L.) Voigt	Kowakka		
	Hellenia speciosa (J.Koenig) S.R.Dutta	Thebu-kola		

(continued)

Table 8.7 (continued)

Food group	Botanical name	Local name/s	Nutritional significance	References
Pulses	*Vigna radiata* (L.) R. Wilczek	Loku-mung Kaha-mung Heen-mung	Among the analysed traditional varieties of pulses, the highest crude fat content was found in Kollu (brown) (2.88 g/100 g), followed by White-cowpea (1.34 g/100 g) and Kadala-cowpea (1.29 g/100 g). The highest oleic acid content was found in Kaha-mung (742.84 mg/100 g) and essential fatty acids linoleic and α-linolenic was highest in Kadala-cowpea (352.26 mg/100 g) and Kollu (brown) (348.26 mg/100 g). Heen-mung and Kaha-mung had the highest crude protein content (20.72 g/100 g and 20.32 g/100 g, respectively), followed by Kadala-cowpea (11.82 g/100 g), Kollu (back) (9.62 g/100 g) and White-cowpea (3.26 g/100 g). Total carbohydrate contents of all the pulses ranged between 52.15 g/100 g (White-cowpea) and 41.71 g/100 g Kollu (black). Dietary fiber levels were rich in selected pulses where all the levels were above 20%. Kollu (black) contained the highest dietary fiber (33.71 g/100 g) while the lowest was found in Heen-mung (20.45 g/100 g). Red-cowpea and Kadala-cowpea contained the highest K content (11.76 m g/100 g and 12.35 mg/100 g, respectively). The highest Ca content was found in Kollu (black) (3.12 mg/100 g). Heen-mung had the highest iron content (130.59 μg/100 g). The highest Zn content was found in Red-cowpea, Kadala-cowpea and white-cowpea varieties (47.94 μg/100 g, 46.74 μg/100 g and 50.86 μg/100 g, respectively). The highest vitamin A and vitamin K were found in Loku-mung (606.26 μg/100 g) and Kadala-cowpea (26.96 μg/100 g), respectively. The highest vitamin B1 and B2 were found in White-cowpea (8.8 μg/100 g and 30.21 μg/100 g, respectively).	Kulasinghe et al. (2019b) and Kulasinghe (2018)
	Macrotyloma uniflorum (Lam.) Verdc.	Kollu (black) Kollu (brown)		
	Vigna unguiculata (L.) Walp.	Kadala-cowpea Red-cowpea White-cowpea		

Fruits	*Phyllanthus emblica* L.	Beheth-nelli (Amla)	Among the studied fruits, *P. emblica* showed the highest content of vitamin C (ascorbic acid equivalents/AAE): 136.8 ± 0.2 mg AAE/100 g and 96.5 ± 0.2 mg AAE/100 g. Additionally, it showed high antioxidant activities. *P. emblica* also reported high ascorbic acid which is of significance as it has the ability to enhance the non-heme iron absorption.	Abeysuriya et al. (2020) and Silva and Sirasa (2018)
	Averrhoa bilimbi L.	Biling/Bilimbi	Vitamin C 9.4 mg AAE/100 g (AAE – ascorbic acid equivalent)	
	Flacourtia indica (Burm.f.) Merr	Uguressa	Vitamin C 12.2 mg AAE/100 g	
	Phoenix farinifera Roxb.	Indi /Wal-indi	Vitamin C 50.4 mg AAE/100 g	
	Synsepalum dulcificum (Schumach. & Thonn.) Daniell	Peni-gedi (Miracle fruit)	Vitamin C 37.4 mg AAE/100 g	
	Annona muricata L.	Katu-anoda	Vitamin C 12.6 mg AAE/100 g	
	Annona reticulata L.	Weli-anaoda	Vitamin C 9.5 mg AAE/100 g	
	Annona squmosa L.	Sini-anoda	Vitamin C 34.7 mg AAE/100 g	
	Citrus × aurantium f. deliciosa (Ten.) M.Hiroe	Heen-naran	Vitamin C 17.3 mg AAE/100 g	
	Spondias dulcis L.	Amberella	Vitamin C 53.5 mg AAE/100 g	
	Elaeocarpus serratus L.	Weralu	Vitamin C 37.3 mg AAE/100 g	
	Aegle marmelos (L.) Corrêa	Beli/Bael fruit	Vitamin C 31.1 mg AAE/100 g; It also showed high antioxidant properties.	
	Anacardium occidentale L.	Kaju-puhulan	Both yellow (189.2 mg AAE/100 g) and red (188.3 mg AAE/100 g) varieties of *A. occidentale* showed high values of vitamin C (ascorbic acid) and antioxidant levels.	
	Punica granatum L.	Local-delum	Local variety of *P. granatum, P. emblica, A. occidentale* and *D. ovoideum* reported the highest iron content (0.9 mg/100 g).	
	Dialium ovoideum Thw.	Galsiyambala		

plants may increase their consumption while also improving the livelihoods of women across the value chain. This potential is shown by recent value chain analysis conducted with different farming systems which found a higher contribution of females in the marketing of NUPFs (Liyanagamage et al. 2017).

Current Sri Lankan NUPFs value chains are very basic maintaining a typical traditional supply chain and very rarely linked to supply and demand-driven strategies. Most of NUPFs in Sri Lanka are seasonal and thus cannot react to market changes accordingly (Barry 2008; Wimalaweera and Samarasinghe 2013). Therefore, careful investigation of opportunities, constraints and innovations along the value chains of neglected food crops needs to be evaluated. If the outstanding barriers at each level of the value chain are not clearly identified, it is difficult to determine the specific measures needed to address, promote and disseminate the crops among the population. Based on the literature compiled from other counties and South Asian regions, the most common barriers to VCD of neglected crops are not exceptional to the Sri Lankan situation and can be attributed to (i) marginalised economic competitiveness due to other commercial crops; (ii) the constraints in producing, storing, and processing and value addition; (iii) poor knowledge of nutritional and health benefits along with nutraceutical and therapeutical applications; (iv) underdeveloped and unstructured value chains; and (v) undesirable associations with a poor rural lifestyle and low social status (Malkanthi and Silva 2014; Raneri et al. 2019). Actors and drivers identified in value chains of Sri Lankan neglected and underutilised fruits and vegetables are shown in Fig. 8.12.

With proper technology and industry involvement, NUPFs products can be tailor-made to cater to certain nutritionally vulnerable populations in the community such as pre-schoolers and school children (through school feeding programmes), as

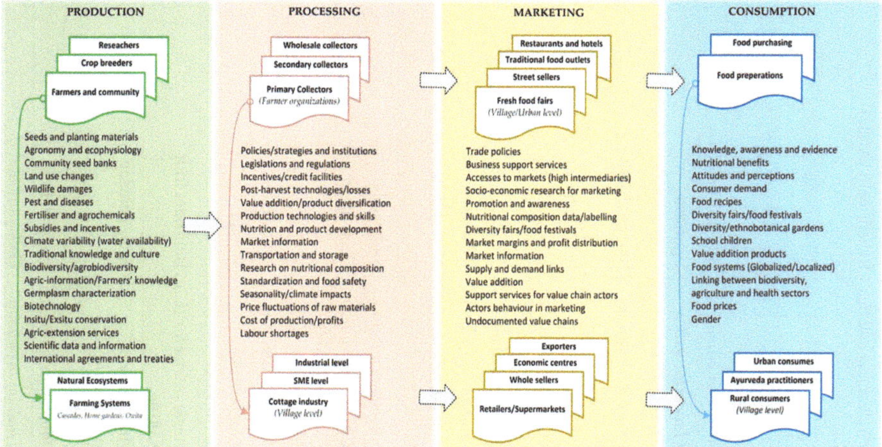

Fig. 8.12 Potential actors and drivers identified in future value chain development for neglected and underutilised fruits and vegetables in Sri Lanka. (Sources: Gee et al. (2020), Liyanage (2010), Bandula et al. (2016), Ratnayake et al. (2019), Mudau et al. (2022), Dahanayake (2015), DOA (2005), Barry (2008), and Wimalaweera and Samarasinghe (2013))

well as pregnant and lactating mothers. Such initiatives will widen and diversify consumer segments; helping to create enhanced and continued demand for NUPFs. Industrial valorisation applications can also help minimise some of the toxicological effects concerning the anti-nutritional constituents in these neglected food crops and the potential to extract such compounds to create potent nutraceutical and therapeutical applications. Some of the important and ancestor cultivars, historically present but currently rare and neglected in VTCS landscapes such as of Henwattakka and Pitiwattakka (*Cucurbita maxima* Duchesne), Rathkehel (*Musa* × *paradisaca*), Landesi (*Amaranthus cruentus* L.), Kirisorgam (*Sorghum bicolor* (L.) Moench) and several traditional paddy varieties (*Oryza sativa* L.) are also some potential crops for super food development (Ekanayake et al. 2020).

However, currently, little is known about where neglected food plant species come from, how they taste, how to process the plants, recipes, their nutritional properties, health benefits, and so on. Production and marketing systems that promote the consumption of the value-added products coming from the NUPFs are vital (Fig. 8.13). Strong public relations and marketing methods, through mass and social media platforms, about these product innovations are necessary to enhance consumer acceptance. Showcasing ancestral crop cultivars in their original farming habitats by establishing model ethnobotanical gardens in SESs can

Fig. 8.13 Creative value-added dishes with restaurant quality introduced at traditional food outlets (Hela Bojun Centres) of the Biodiversity for Food and Nutrition (BFN) project using NUPFs. (**a**) braised cabbage and tofu with olu-rice, (**b**) sweet and sour vegetable with traditional rice, (**c**) vegetable ratatouille with indigenous bean and sweet potato mash, and (**d**) assorted indigenous pluses salad. (Photo source: BFN project (at national food festival 2017 commemorating the World Food Day at Plant Genetic Resources Centre))

Fig. 8.14 Model crop garden established as a pilot site to demonstrate local root and tuber crops at Community Development Centre, Aranayake, Sri Lanka. (Photo source: Biodiversity for Food and Nutrition (BFN) project)

Fig. 8.15 Community seed bank established to store seeds of traditional neglected and underutlised crops at one of the Village Tanka Cascade System landscapes. (Photo source: Biodiversity for Adapatation to Climate Change (BACC) project)

also raise awareness among communities (Fig. 8.14). Establishment of community seed banks is one of the best practices to maintain diversity of neglected and underutilised food crops that help availability of seeds and planting materials for farmers (Fig. 8.15).

8.5 Biodiversity for Food and Nutrition (BFN) Approach for Mainstreaming NUPFs for Improved Nutrition and Livelihoods in Sri Lanka

Mainstreaming the use of neglected food plants for human nutrition and health is an important step towards strengthening markets and, value chains, for value addition, creating more linkages to new employment opportunities in local and rural communities, and supporting overall economic productivity. The recent GEF-supported Biodiversity for Food and Nutrition (BFN) project, implemented in three pilot sites of SESs in Sri Lanka, highlights that despite the abundance and nutritional significance of neglected and underutilised food plants, they are often under-valued, underused and rarely considered in local food systems. Consequently, there is limited research and development investment, therefore policy decisions related to nutrition and health often fail to take these NUPFs into account or recognize the contribution NUPFs can make to improving diets. This situation is beginning to change with the benefits of consuming a nutrient-rich biodiverse food, as part of a healthy, sustainable diet, being increasingly recognised as an important health and economic goal for more and more countries (Fanzo et al. 2017; HLPE 2019; Hunter et al. 2020).

The BFN approach, a cross-sectoral initiative spearheaded by Sri Lanka (together with Brazil, Kenya and Turkey) set out to research and initially promote local food biodiversity of SESs using an interdisciplinary mainstreaming approach. Mainstreaming biodiversity for improving diets and nutrition involves the integration of nutrient-rich biodiversity including neglected and underutilised species into all relevant policies, strategies, plans and programs addressing malnutrition in all its forms contributing to health and nutrition (FAO 2016). It is this mainstreaming local biodiversity through SESs for improving diets and nutrition of the people that the BFN approach aligns with closely.

The BFN approach represents a unique approach to link NUPFs, diets and nutrition through three main interlinked strategies: (i) strengthening evidence for the nutritional value and social-ecological importance of NUPFs, (ii) targeting policies for promoting NUPFs and emerging markets, and (iii) raising awareness of NUPFs among the consumers as alternative food options. This approach is non-prescriptive and can be readily adapted to the context of most countries. It can help identify entry points for NUPFs and offers numerous insights into how to best mainstream biodiversity into local to wider national food systems (Fig. 8.16).

8.6 Constraints to Mainstreaming and Better Use of NUPFs in Sri Lanka

Despite increasing awareness of diet-related health problems and the benefits of NUPFs contributing to varied, healthy foods (Hunter et al. 2020), there still are multiple obstacles to mainstreaming NUPFs use for improved human nutrition and wellbeing. These barriers consist of the growing commercialisation of agriculture

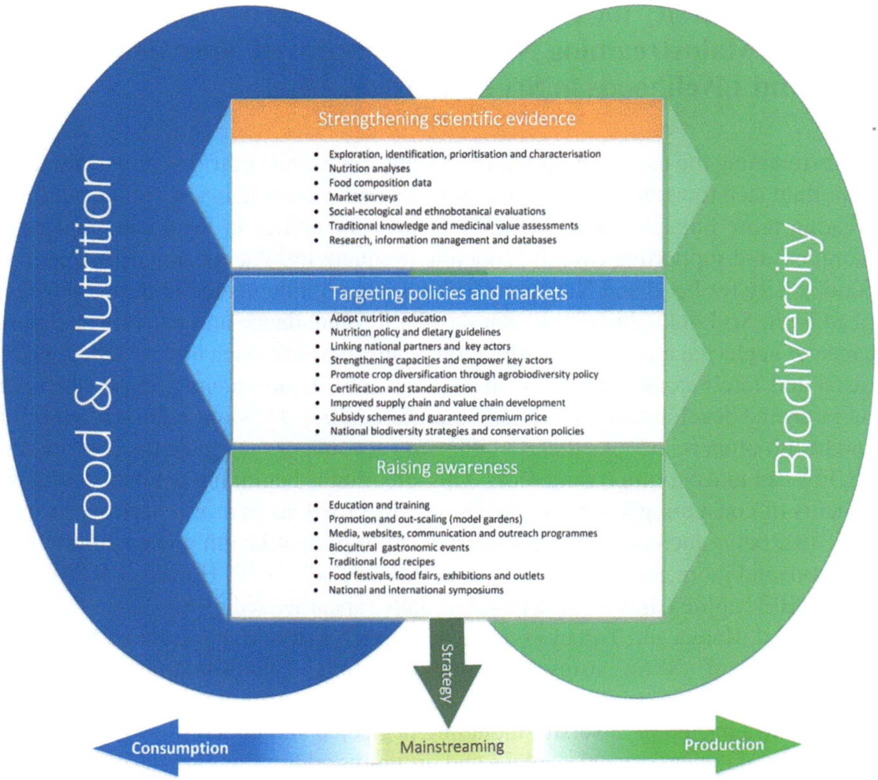

Fig. 8.16 The Biodiversity for Food and Nutrition (BFN) approach for mainstreaming Neglected and Underutilised Plant Foods (NUPFs) into local food systems from production to consumption. (Sources: Beltrame et al. (2019), Borelli et al. (2020a), and Chacha et al. (2022))

production systems leading to the narrowing down of certain species or some varieties. In many cases, NUPFs are not able to compete with these more commercialised species due to several socio-economic or technical challenges. The limitation of incentives to conserve and grow NUPFs with greater nutrition together with unsuitable land use practices, habitat destruction and over-harvesting has led to the erosion of much of these NUPFs genetic diversity.

Furthermore, this natural diversity of NUPFs, and the traditional/ethnobotanical knowledge and biocultural values involved with it, is being undermined in many countries including Sri Lanka. Globalisation of trade, urbanisation and changing food and dietary habits have had a considerable influence on the awareness and utilisation of NUPFs and the intergenerational transmission of biodiverse-food knowledge. For example, increased urbanisation and changing food habits have contributed to the demise of many NUPFs among modern generations in Sri Lanka. There is no organised cultivation of these crop species and most of these neglected plant species can now only be found in the wild or grown in home gardens (Pushpakumara 2011). With the modern lifestyle, communities are becoming less dependent on local resources and traditional food cultures, resulting in the depletion or limited transmission of traditional knowledge. Thus, these valuable resources are weakened down and is being deprived of a purpose, threatening local management systems.

Limited availability of scientific evidence such as food composition data and agronomic and horticultural research information, low priority on variety improvement and technology development are major challenges for promotion and awareness of NUPFs, in local and international markets. Though there are some databases developed for major commodity crops, a lack of systematic databases or information systems for neglected and underutilised plants/crops is one of the major constraints to increasing utilisation and conservation (Nizar et al. 2021). Also, inadequate financial and non-financial incentives to create a conducive environment, for the promotion of research and development, production and marketing along with a limited scarcity of planting materials are other contributing factors for the limited availability of NUPFs products in the markets. Further, inadequate guidelines for the production or collection of seeds or planting materials to guide the NUPFs growers and lack of awareness of recipes and culinary practices are also reported as barriers for the promotion of NUPFs (Weerahewa and Pushpakumara 2019).

National diet-related nutrition and health issues can be resolved by providing access to a diversity of nutrients required for healthy growth and living; yet, there are a number of barriers to their production and consumption. At present, indigenous fruits and vegetables are considered 'inferior', mainly due to limited awareness of consumers and the lack of aesthetic standards among local products. More technical, political and financial support, as well as subsidies and public awareness, are imperative for farmers to conserve and grow these foods and for consumers to find them desirable and create a demand (Rathnasinghe et al. 2019).

The barriers to the promotion and mainstreaming of NUPFs are also considerable. Past and ongoing research reveals that the major causes for this often are related to several factors. These include ineffective policies, lack of skills and institutional capacity, poor economic competitiveness with commodity cereal crops, a lack of crop improvement, poor agronomic practices, inefficient processing and value adding, disorganised or non-existent market chains, lack of data linking NUPFs to dietary diversity and nutrition outcomes, and poor perception of these foods (Hunter and Borelli 2013; IFAD and Alliance of Bioversity International and CIAT 2021) (Fig. 8.17).

8.7 Conclusion

Sri Lankan SESs are hotspots of neglected and underutilised food plants and a significant proportion of the neglected plant food diversity identified contributes to the local food basket. Diversifying local food systems through greater mainstreaming of NUPFs could be a promising solution for the sustainability of future food

Fig. 8.17 Constraints to mainstream and better use of Neglected and Underutilised Plant Foods (NUPFs) in Sri Lanka

security and nutrition in Sri Lanka. As this study highlights neglected and underutilised food plants can ensure environmentally sustainable food production despite adverse environmental conditions and low input availability. NUPFs substantially contribute to generating sustainable ecosystem services and promoting dietary diversification—adding different tastes as well as various macro and micronutrients to the diet, particularly for rural communities. Furthermore, they contribute to greater climate-resilience and climate-friendly agri-food systems especially in the SESs. The study assessment revealed that Sri Lankan SESs such as Kandyan Home Gardens, Village Tank Cascade Systems and Owita Systems have great potential to maintain and conserve majority of neglected and underutilise food plants found in the country. However, these much-needed resources are still a long way from being fully realised and undertaken in the broader national context.

Despite growing recognition of the benefits of NUPFs contributing to human nutrition and wellbeing, there are many barriers and obstacles to mainstreaming NUPFs into food systems still remaining. These barriers mainly include ineffective policies, disconnected collaboration between key national partners of NUPF environments, lack of capacity and inadequate promotion and awareness and weak conservation strategies and plans that limit the utilisation of NUPF diversity in the country. In addition, it is important to understand NUPF environments in the context of projected climate change and intensive commercialisation of traditional farming lands (land use intensification) in the SESs. Further, the study highlighted support for the multi-pronged approaches for mainstreaming biodiversity for food and nutrition. Scaling up production of NUPFs in line with the recommendations proposed in the light of lessons learned would go a long way to improving NUPF conservation and their contribution to local food security and nutrition.

Acknowledgments The first author thankfully acknowledges the GEF funded Biodiversity for Adaptation to Climate Change (BACC), Biodiversity for Food and Nutrition (BFN), Crop Wild Relatives (CWR) and Healthy Landscapes (HL) projects for facilitation in field data collection and key informant interviews, and providing secondary data and information.

Conflicts of interest All authors read and approved the final manuscript and declare that they have no conflict of interest.

Funding This research was funded by an Australian Government Research Training Program (RTP) and Destination Australia Program (DAP) Scholarships through the University of New England (UNE), Australia to the first author.

Ethical aspects The key informant interviews undertaken in this study were carried out in accordance with the relevant guidelines and approvals provided by the Human Research Ethics Committee of the University of New England, Australia (approval no: HE22-030, dated 3rd May 2022).

References

Abeysekera WKSM, Premakumara GAS (2016) Health food properties of traditional rice in Sri Lanka. Lambert Academic Publishing, Germany

Abeysuriya HI, Bulugahapitiya VP, Loku Pulukkuttige J (2020) Total vitamin C, ascorbic acid, dehydroascorbic acid, antioxidant properties, and iron content of underutilised and commonly consumed fruits in Sri Lanka. Int J Food Sci 2020:1–13

Abeywardhana R, Perera A (2016) Identification of traditional Food Recipe of Sri Lanka. AGRIEAST: J Agric Sci 8:1–10

Aheeyar MMM (2013) Alternative approaches to small tank/cascade rehabilitation: socio-economic and institutional perspective. Hector Kobbekaduwa Agrarian Research and Training Institute (HARTI), Colombo

Ahmed S, Warne T, Stewart A, Byker Shanks C, Dupuis V (2022) Role of wild food environments for cultural identity, food security, and dietary quality in a rural American state, vol 6. Front Sustain Food Syst, p 153

Alwis SD, Fonseka H, Senanayake G (2007) Lesser known indiginous food crops in Sri Lanka. Biodiveristy Secretariat (BDS), Ministry of Environment Sri Lanka, Colombo

Alwis LMHR, Dissanayakea MUK, Rathnayakea RMWMC (2018) Species diversity of local root and tuber crops in agricultural habitats in Uva region. Agric Value Addit 1:47–61

Arampath P, Dekker M, Dharmasena D (2021) Occurrence of bioactive health promoting compounds in commercial products of Mango, Pineapple and Wood apple in Sri Lanka. Trop Agric Res 32:243–255

Bandara M (2009) Exploring the link between culture and biodiversity in Sri Lanka. SANSAI: Environ J Glob Community 4:1–23

Bandula A, Nath TK (2020) Underutilised crops in the agricultural farms of Southeastern Sri Lanka: farmers' knowledge, preference, and contribution to household economy. Econ Bot 74:126–139

Bandula A, Jayaweera C, De Silva A, Oreiley P, Karunarathne A, Malkanthi S (2016) Role of underutilised crop value chains in rural food and income security in Sri Lanka. Proc Food Sci 6:267–270

Barry IN (2008) A value chain analysis for the Sri Lankan Beli and Woodapple Subsectors: Research Report No. 4. International Centre for Underutilised Crops (ICUC), Colombo

Beltrame D, Eliot G, Güner B, Lauridsen NO, Samarasinghe WG, Wasike VW, Hunter D, Borelli T (2019) Mainstreaming biodiversity for food and nutrition into policies and practices: methodologies and lessons learned from four countries. AARI 29:25–38

Bharucha Z, Pretty J (2010) The roles and values of wild foods in agricultural systems. Philos Trans R Soc B Biol Sci 365(1554):2913–2926

Borelli T, Hunter D, Padulosi S, Amaya N, Meldrum G, de Oliveira Beltrame DM, Samarasinghe G, Wasike VW, Güner B, Tan A (2020a) Local solutions for sustainable food systems: the contribution of orphan crops and wild edible species. Agronomy 10:231–256

Borelli T, Hunter D, Powell B, Ulian T, Mattana E, Termote C, Pawera L, Beltrame D, Penafiel D, Tan A (2020b) Born to eat wild: an integrated conservation approach to secure wild food plants for food security and nutrition. Plan Theory 9:1–78

Borelli T, Beltrame D, Wasike VW, Samarasinghe G, Tan A, Hunter D (2021) The BFN mainstreaming toolkit: a roadmap to using neglected and underutilised species for food system change. In: Padulosi SED et al (eds) Orphan crops for sustainable food and nutrition security. Routledge, Oxon/New York, pp 163–173

Chacha JS, Ofoedu CE, Suleiman RA, Jumbe TJ, Kulwa KB (2022) Underutilised fruits: challenges and constraints for domestication. In: Future foods. Elsevier, pp 133–150

Chamara R, Wijerathna M, Herath I, Rajapaksha R (2021) Knowledge, attitudes and practices (KAP) of rural women in the dry zone of Sri Lanka on underutilized vegetable crops. Ethnobot Res Appl 21:1–25

Chandrasekara A, Josheph Kumar T (2016) Roots and tuber crops as functional foods: a review on phytochemical constituents and their potential health benefits. Int J Food Sci 2016:1–16

Dahanayake N (2015) Some neglected and underutilised fruit-crops in Sri Lanka. Int J Sci Res Publ 5:1–7

Dassanayake MD, Clayton WD (eds) (1996) A revised handbook to the Flora of Ceylon. Nord. J. Bot., vol X. Oxford IBH, New Delhi

Dassanayake MD, Clayton WD (eds) (1997) A revised handbook to the Flora of Ceylon, vol XI. Oxford IBH, New Delhi

Dassanayake MD, Clayton WD (eds) (1998) A revised handbook to the Flora of Ceylon, vol XII. Oxford IBH, New Delhi

Dassanayake MD, Clayton WD (eds) (1999) A revised handbook to the Flora of Ceylon, vol XIII. Oxford IBH, New Delhi

Dassanayake MD, Clayton WD (eds) (2000) A revised handbook to the Flora of Ceylon, vol XIV. Oxford IBH, New Delhi

Dassanayake MD, Fosberg FR (eds) (1980) A revised handbook to the Flora of Ceylon, vol I. Oxford IBH, New Delhi

Dassanayake MD, Fosberg FR (eds) (1981a) A revised handbook to the Flora of Ceylon, vol II. Oxford IBH, New Delhi

Dassanayake MD, Fosberg FR (eds) (1981b) A revised handbook to the Flora of Ceylon, vol III. Oxford IBH, New Delhi

Dassanayake MD, Fosberg FR (eds) (1983) A revised handbook to the Flora of Ceylon, vol IV. Oxford IBH, New Delhi
Dassanayake MD, Fosberg FR (eds) (1985) A revised handbook to the Flora of Ceylon, vol V. Oxford IBH, New Delhi
Dassanayake MD, Fosberg FR (eds) (1987) A revised handbook to the Flora of Ceylon, vol VI. Oxford IBH, New Delhi
Dassanayake MD, Fosberg FR (eds) (1991) A revised handbook to the Flora of Ceylon, vol VII. Oxford IBH, New Delhi
Dassanayake MD, Fosberg FR (eds) (1994) A revised handbook to the Flora of Ceylon, vol VIII. Oxford IBH, New Delhi
Dassanayake MD, Fosberg FR (eds) (1995) A revised handbook to the Flora of Ceylon, vol IX. Oxford IBH, New Delhi
Dassanayake M, Jayasuriya A (1974) A new species of Brachystelma (Asclepiadaceae) from Sri Lanka. Ceylon J Sci Biol Sci 11:39–41
Dayananda D, Weerahewa J, Weerasooriya S (2021) Water availability, crop choices and profitability of farming: a case study of Mahakanumulla Tank Village. Trop Agric Res 32:81–94
DeClerck F (2013) Harnessing biodiversity: from diets to landscapes. In: Diversifying food and diets. Routledge, pp 49–66
Dharmasena P (2010) Assessment of traditional rice farming. A case study from Moneragala District of Sri Lanka, COMPAS Project Future in Our Hands Development Fund Badulla
DOA (2005) Exploration, collection, conservation and characterisation of underutilised fruits: SL-USA cooperate germplasm conservation project-annual report 2005. Horticultural Crops Research and Development Institute (HORDI), Department of Agriculture, Peradeniya
Ekanayake SP, Ratnayake SS, Sugathadasa S, Rajapakse RM (2020) Chapter Three—Biodiversity baseline assessment. In: Baseline assessment of the UNEP-GEF managing agricultural landscapes in socio-ecologically sensitive areas to promote food security, well-being and ecosystem health— healthy landscapes project. SACEP—South Asia Co-operative Environment Programme, Colombo
Esham M, Garforth C (2013) Climate change and agricultural adaptation in Sri Lanka: a review. Clim Dev 5:66–76
Esham M, Wijeratne W (2021) Risk and vulnerability assessment for resilience building in the colombo city region food system (CRFS). International Water Management Institute (IWMI), Colombo
Fanzo J, Arabi M, Burlingame B, Haddad L, Kimenju S, Miller G, Nie F, Recine E, Serra-Majem L, Sinha D (2017) Nutrition and food systems: a report by the High Level Panel of Experts on Food Security and Nutrition of the Committee on World Food Security. Food and Agriculture Organisation of the United Nations (FAO), Rome
FAO (2016) Voluntary guidelines for mainstreaming biodiversity into policies, programmes and national and regional plans of action on nutrition. Secretariat of the Commission on Genetic Resources for Food and Agriculture, Rome
Ganashan P, Balendira S, Dassanayake MD (1996) Sri Lanka country report to the FAO international technical conference on plant genetic resource. World Food and Agriculture Organisation (FAO), Leipzig
Gee E, Borelli T, de Oliveira Beltrame DM, Oliveira CNS, Coradin L, Wasike V, Ndiwa AM, Samarasinghe G, Güner B, Tan A (2020) The ABC of mainstreaming biodiversity for food and nutrition: concepts, theory and practice. In: Hunter D, Borelli T, Gee E (eds) Biodiversity, food and nutrition: a new agenda for sustainable food systems. Routledge, Oxon/New York, pp 85–186
Godamulla D (2018) Indigenous tuber crops that brings health and well being. Biodiversity for Food and Nutrition (BFN) Project, Plant Genetic Resources Center (PGRC), Gannoruwa, Peradenya
Goonatilake S, Ekanayake SP (2016) Traditional knowledge use in Sri Lankan agroecosystems for livelihoods and adaptation to climate change. Department of Agriculture, Sri Lanka, Peradeniya

Goonatilake S, Ekanayake SP, Perera N, Wijenayake T, Wadugodapitiya A (2015) Biodiversity and ethnobiology of the Kapiriggama Small Tank Cascade System in Sri Lanka. IUCN programme on Restoring Traditional Cascading Tank Systems Technical Note No. 2. International Union for Conservation of Nature (IUCN), Colombo

Hemathilake DMKS, Gunathilake DMCC (2022) Chapter 31 – Agricultural productivity and food supply to meet increased demands. In: Bhat R (ed) Future foods. Academic Press, pp 539–553

Hettiarachchi H, Gunathilake K (2020) Bioactives and bioactivity of selected underutilised fruits, vegetables and legumes grown in Sri Lanka: a review. J Med Plants Stud 8:34–44

HLP-BLA (2020) Baseline assessment report of the Healthy Landscapes Project (UNEP-GEF): managing agricultural landscapes in socio-ecologically sensitive areas to promote food security, well-being and ecosystem health. South Asia Co-operative Environment Programme (SACEP) and Alliance of Bioversity International and CIAT, Colombo

HLPE (2019) Agroecological and other innovative approaches for sustainable agriculture and food systems that enhance food security and nutrition: a report by the High Level Panel of Experts on Food Security and Nutrition of the Committee on World Food Security. Food and Agriculture Organisation of the United Nations, Rome

Hunter D, Borelli T (2013) Introduction. In: Fanzo J et al (eds) Diversifying food and diets: using agricultural biodiversity to improve nutrition and health. Routledge, Oxon/New York, pp 3–14

Hunter D, Guarino L, Spillane C, McKeown PC (2017) Routledge handbook of agricultural biodiversity. Routledge, Oxon/New York

Hunter D, Borelli T, Beltrame DM, Oliveira CN, Coradin L, Wasike VW, Wasilwa L, Mwai J, Manjella A, Samarasinghe GW (2019) The potential of neglected and underutilised species for improving diets and nutrition. Planta 250(3):709–729

Hunter D, Gee E, Borelli T (2020) Nourishing people, nurturing the environment: biodiversity for food systems transformation and healthier diets. In: Hunter D, Teresa B, Gee E (eds) Biodiversity, food and nutrition: a new agenda for sustainable food systems. Routledge, Oxon/New York, pp 3–20

Ickowitz A, McMullin S, Rosenstock T, Dawson I, Rowland D, Powell B, Mausch K, Djoudi H, Sunderland T, Nurhasan M (2022) Transforming food systems with trees and forests. Lancet Planet Health 6:e632–e639

IFAD and Alliance of Bioversity International and CIAT (2021) How to do mainstreaming neglected and underutilised species in national policy for nutrition outcomes. In: International Fund for Agricultural Dvelopment (IFAD), Rome

IPCC (2022) Climate change 2022. In: Pörtner HO et al (eds) Summary for policymakers, impacts, adaptation, and vulnerability. Contribution of Working Group II to the Sixth Assessment Report of the Intergovernmental Panel on Climate Change, Intergovernmental Panel on Climate Change (IPPC). Cambridge University Press

IUCN (2015a) Tank ecosystem restoration of the Kapiriggama Small Tank Cascade System. IUCN programme on Restoring Traditional Cascading Tank Systems Technical Note No.3. International Union for Conservation of Nature (IUCN), Colombo

IUCN (2015b) IUCN's engagement in the Mainstreaming Biodiversity Conservation and use in Sri Lankan Agro-ecosystems for Livelihoods and Adaptation to Climate Change (BACC) Project. International Union for Conservation of Nature (IUCN), Colombo

Jayasuriya AM (1990) Dioscorea koyamae (Dioscoreaceae), a new edible yam from Sri Lanka. Brittonia:142–146

Jayawardana SAS, Samarasekera JKRR, Hettiarachchi GHCM, Gooneratne J, Mazumdar SD, Banerjee R (2019) Dietary fibers, starch fractions and nutritional composition of finger millet varieties cultivated in Sri Lanka. J Food Compos Anal 82:103249

Joshi BK, Shrestha R, Gauchan D, Shrestha A (2020) Neglected, underutilized, and future smart crop species in Nepal. J Crop Improv 34:291–313

Kadupitiya HK, Madushan RN, Gunawardhane D, Sirisena D, Rathnayake U, Dissanayaka D, Ariyaratne M, Marambe B, Suriyagoda L (2022) Mapping productivity-related spatial characteristics in rice-based cropping systems in Sri Lanka. J Geovis Spat Anal 6:1–10

Kaluthanthri DVS, Dasanayaka PN (2019) Sun loving sorghum – a better option for the sustainable production (production and use of sorghum (*Sorghum bicolor*) and its adaptability to Sri Lanka: a review). Int J Prog Sci Technol:213–219

Kennedy G, Burlingame B (2003) Analysis of food composition data on rice from a plant genetic resources perspective. Food Chem 80(4):589–596

Kulasinghe WMAA (2018) Evaluation and comparison of the nutritional composition of selected traditional cereals, pulses and yams grown in Sri Lanka. MPhil thesis, Postgraduate Institute of Agriculture (PGIA), Peradeniya, Sri Lanka, University of Peradeniya, Sri Lanka

Kulasinghe W, Wimalasiri K, Samarasinghe G, Silva R, Madhujith T (2018) Macronutrient composition of selected traditional yams grown in Sri Lanka. Trop Agric Res 29:113–122

Kulasinghe W, Wimalasiri S, Samarasinghe G, Silva R, Madhujith T (2019a) Evaluation and comparison of vitamin profiles of selected traditional rice and yams grown in Sri Lanka. Trop Agric Res 30:23–31

Kulasinghe WMAA, Wimalasiri KMS, Samarasinghe WLG, Silva R, Madhujith T (2019b) A nutritional evaluation of selected traditional underutilised pulses grown in Sri Lanka. In: Symposium on agrobiodiversity for climate change adaptation, food and nutrition, Plant genetic Resources Center, Peradeniya, Sri Lanka, Mainstreaming Agrobiodiversity Conservation and use in Sri Lankan Agro-ecosystems for Livelihoods and Adaptation to Climate Change (BACC) project, pp 23–24

Kumar L, Chhogyel N, Gopalakrishnan T, Hasan MK, Jayasinghe SL, Kariyawasam CS, Kogo BK, Ratnayake S (2022) Climate change and future of agri-food production. In: Bhat R (ed) Future foods: global trends, opportunities, and sustainability challenges. Elsevier, pp 49–79

Leukebandara IK, Premaratne S, Peiris BL (2015) Nutritive quality of Thampala (*Amaranthus* spp.) as a forage crop in Sri Lanka. Trop Agric Res 26(4):624–631

Li X, Siddique KHM (eds) (2018) Future smart food-Rediscovering hidden treasures of neglected and underutilised species for zero hunger in Asia. Food & Agriculture Organisation of the United Nations, Bangkok, pp 51–59

Li X, Yadav R, Siddique KH (2020) Neglected and underutilised crop species: the key to improving dietary diversity and fighting hunger and malnutrition in Asia and the Pacific. Front Nutr 254:1–28

Liyanagamage T, De Silva D, Karunaratne A, O'Reilly P, Nilmalgoda H (2017) Value chain analysis of underutilised crops of different farming systems in Sri Lanka. In: Beneragama CK, Kumara GDK, Ariyaratne WMTP (eds) Fourth international conference on postharvest and quality management of horticultural products of interest for tropical regions, Kandy, Sri Lanka. International Society for Horticultural Science (ISHS)-Acta Horticulturae, pp 6–8

Liyanage ASU (2010) Eco-geographic survey of crop wild relatives. In situ conservation of crop wild relatives through the enhanced information management and field application (CWR) project. Plant Genetic Resources Centre (PGRC), Peradeniya

Liyanaratne J (2003) Some Sri Lankan common pot-herbs. Anc Sci Life 23:71

Malkanthi S (2017) Importance of underutilised crops in Thanamalwila Divisional Secretariat Division in Monaragala District in Sri Lanka. J Agric Sci Sri Lanka 12(3):197–206

Malkanthi S, Silva P (2014) Opportunities and challenges in cultivating underutilised field crops in Moneragala district of Sri Lanka. Asian J Agric Rural Dev 4:96–105

Malkanthi R, Silva K, Chandrasekera G, Jayasinghe J (2007) High prevalence of malnutrition and household food insecurity in the rural subsistence paddy farming sector

Malkanthi SHP, Karunaratne AS, Amuwala SD, Silva P (2014) Opportunities and challenges in cultivating underutilised field crops in Moneragala district of Sri Lanka. Asian J Agric Rural Dev 4:96–105

Marambe B, Pushpakumara G, Silva P (2012a) Biodiversity and agrobiodiversity in Sri Lanka: village tank systems. In: The biodiversity observation network in the Asia-Pacific region. Springer, pp 403–430

Marambe B, Weerahewa H, Pushpakumara D, Silva G, Punyawardena B, Premalal S, Miah G, Roy J (2012b) Vulnerability of home garden systems to climate change and its impacts on food

security in south asia: project report. Asia-Pacific Network for Global Change Research (APN), APN, Bangkok
Menaka S (2019) Potential uses of Little millet (Panicum sumatrense) a valuable cereal neglected in contemporary world. In: Proceeding of the Open University Research Sessions (OURS 2019), Colombo, Sri Lanka, Open University of Sri Lanka
Mihiranie S, Jayasinghe JK, Jayasinghe CV, Wanasundara JP (2020) Indigenous and traditional foods of Sri Lanka. J Ethn Foods 7:1–19
Mijatović D, Sakalian M, Hodgkin T (2018) Mainstreaming biodiversity in production landscapes, United Nation Environment Programme (UNEP), New Delhi
MOE (2012) The national red list 2012 of Sri Lanka; conservation status of the fauna and flora. Ministry of Environment, Colombo
MOE (2016) National biodiversity strategic action plan 2016–2022. Biodiversity Secretariat, Ministry of Mahaweli Development and Environment, Colombo
MOH (2022) Food based dietary guidelines for sri lankans:training aid. Nutrition Division, Ministry of Health, Colombo
Mudau FN, Chimonyo VGP, Modi AT, Mabhaudhi T (2022) Neglected and underutilised crops: a systematic review of their potential as food and herbal medicinal crops in South Africa. Front Pharmacol 12:4054
Munasinghe M, Price AH (2016) Genetic and root phenotype diversity in Sri Lankan rice landraces may be related to drought resistance. Rice 9:1–13
Myers N, Mittermeier RA, Mittermeier CG, Da Fonseca GAB, Kent J (2000) Biodiversity hotspots for conservation priorities. Nature 403:853–858
Nadeeshani H, Wimalasiri K, Samarasinghe G, Silva R, Madhujith T (2018) Evaluation of the nutritional value of selected leafy vegetables grown in Sri Lanka. Trop Agric Res 29:255–267
Nadeeshani H, Samarasinghe G, Silva R, Hunter D, Madhujith T (2021) Proximate composition, fatty acid profile, vitamin and mineral content of selected banana varieties grown in Sri Lanka. J Food Compos Anal 100:103887
Nizar NMM, Jahanshiri E, Tharmandram AS, Salama A, Sinin SSM, Abdullah NJ, Zolkepli H, Wimalasiri EM, Suhairi TASTM, Hussin H (2021) Underutilised crops database for supporting agricultural diversification. Comput Electron Agric 180:1–13
Oliveira BDM, Oliveira CNS, Borelli T, Santiago RAC, Coradin L, Hunter D (2018) Brazilian underutilised species to promote dietary diversity, local food procurement, and biodiversity conservation: a food composition gap analysis. Lancet Planet Health Meet Abstr 2:22
Padulosi S, Heywood V, Hunter D, Jarvis A (2011) Underutilised species and climate change: current status and outlook. In: Yadav SS, Redden RJ, Hatfield JL (eds) Crop adaptation to climate change. Blackwell Publishing Ltd, Oxford, pp 507–521
Padulosi S, Thompson J, Rudebjer P (2013) Fighting poverty, hunger and malnutrition with neglected and underutilised species (NUS): needs, challenges and the way forward. Bioversity International, Rome
Peduruhewa P, Jayathunge K, Liyanage R (2021) Potential of underutilised wild edible plants as the food for the future–a review. J Food Secur 9:136–147
PGRC (1999) Plant genetic resources catalogue passport information, vol II. Plant Genetic Resources Centre, Department of Agriculture, Gannaoruwa, Peradeniya
Punyawardena BVR (2008) Rainfall of Sri Lanka and agroecological zones. Department of Agriculture, Peradeniaya
Punyawardena BVR (2020) Climate. In: Mapa RB (ed) The soils of Sri Lanka. Springer Nature Switzerland AG, Cham, pp 13–22
Pushpakumara DKNG (2011) Analysis of the state of agrobiodiversity and traditional knowledge in selected ecosystems in Sri Lanka including threats to them: baseline Study reported prepared for the GEF-funded Biodiversity for Food and Nutrition (BFN) project. Bioversity International, Rome

Pushpakumara DKNG, Gunasena HPM, Singh VP (eds) (2007) Underutilised fruit trees in Sri Lanka, vol 1. World Agroforestry Centre, South Asia Office; Sri Lanka Council for Agriculture Research Policy; Asian Centre for Underutilised Crops in Sri Lanka, New Delhi

Pushpakumara D, Heenkenda H, Marambe B (2016) Kandyan home gardens: a time-tested good practice from Sri Lanka for conserving tropical fruit tree diversity. In: Tropical fruit tree diversity. Routledge, pp 171–190

Puspakumara DKNG, Liyanage ASU, Kodagoda MM (2016) Sri Lanka's country report for the state of world's biodiversity for food and agriculture (SoWBFA). Department of Agriculture, Peradeniya

Rajapakse K, Wijesekera I (2019) Proximate analysis and developlemt of cookies from underutilized nipa palm kernel. In: Symposium on agrobiodiversity for climate change adaptation, food and nutrition. Plant genetic Resources Center, Peradeniya, Sri Lanka, Mainstreaming Agrobiodiversity Conservation and use in Sri Lankan Agro-ecosystems for Livelihoods and Adaptation to Climate Change (BACC) project, pp 23–24

Rajapaksha UKRMH (1998) Traditional food plants in Sri Lanka. Hector Kobbekaduwa Agrarian Research and Training Institute, Colombo

Rambukwella R, Priyankara EAC (2016) Production and marketing of traditional rice varieties in selected districts in Sri Lanka: present status and future prospects. Hector Kobbekaduwa Agrarian Research and Training Institute (HARTI), HARTI, Colombo

Raneri JE, Padulosi S, Meldrum G, King OI (2019) Promoting neglected and underutilised species to boost nutrition in LMICs. UNSCN Nutr 44:10–25

Rangani G, Mahaliyanaarachchi R, Sivashankar P, Hettiarachchi I, Bandara B (2017) A study on consumption pattern and awareness of underutilised fruits among urbanites within Maharagana DS division. J Appl Microbiol 119:859–863

Ranil RHG, Pushpakumara G, Fonseka RM, Fonseka H, Bandaranayake PCG, Weerakkody WAP, Ariyaratne WMTP, De Silva AN, Gunawardena NPT (2020) Utilising neglected crop genetic resources for food and nutritional security: special reference to indigenous vegetables of Sri Lanka. In: Silva RP et al (eds) Agricultural research for sustainable food systems in Sri Lanka, vol 2: A pursuit for advancements. Springer, Singapore, pp 39–66

Ranil R, Chamara R, Pushpakumara D, Bussmann RW (2021) Exploration, conservation, and utilisation of ethnobotanical knowledge: Sri Lankan perspective. In: Abbasi AM, Bussmann RW (eds) Ethnobiology of mountain communities in Asia. Springer, Cham, pp 409–431

Rathnasinghe D, Samarasinghe G, Silva R, Hunter D (2019) 'True Sri Lankan taste' food outlets: promoting indigenous foods for healthier diets. Nutrition Exchange South Asia, Emergency Nutrition Network (ENN) (12). www.ennonline.net/nex/12/truesrilankantaste

Ratnayake SS, Kumar L, Kariyawasam CS (2019) Neglected and underutilised fruit species in Sri Lanka: prioritisation and understanding the potential distribution under climate change. Agronomy 10(1):34

Ratnayake SS, Kariyawasam CS, Kumar L, Hunter D, Liyanage A (2021) Potential distribution of crop wild relatives under climate change in Sri Lanka: implications for conservation of agricultural biodiversity. Curr Res Environ Sustain 3:1–9

Rupasingha P, Ratnayake K, Senarathne A (2020) Antioxidant capacity of indigenous root and tuber crops of Sri Lanka as affected by simple processing. Asian Food Sci 19:18–17

Sajiwanie JWA, Rathnayaka RMUSK (2016) Investigation of nutritional value of selected Sri Lankan traditional root and tubers. Scholars Acad J Biosci 4:656–660

Samarasinghe WLG, Dalshini K, Sumanasinghe VA, Ubesekara NMU, Hettiarrachchi K (2015) Characterisation of suduru samba rice (*Oryza sativa* L.) accessions using genomic wide SSR poly-morphism and seed morphology. Trop Agric Ann Sri Lanka Dep Agric 163:36–39

Saxena K, Singh L, Reddy M, Singh U, Lateef S, Sharma S, Remanandan P (1990) Intra species variation in *Atylosia scarabaeoides* (L.) Benth., a wild relative of pigeonpea (*Cajanus cajan* (L.) Millsp.). Euphytica 49:185–191

SCBD (2010) In: Belair C et al (eds) Sustainable use of biological diversity in socioecological production landscapes: background to the 'Satoyama initiative for the benefit of biodiversity

and human well-being': Technical Series No. 52. Secretariat of the Convention on Biological Diversity (SCBD), Montreal
Silva R, Chandrasekara A (2020) Food security and human health: baseline assessment report of the Healthy Landscapes Project (UNEP-GEF): managing agricultural landscapes in socio-ecologically sensitive areas to promote food security, well-being and ecosystem health. South Asia Co-operative Environment Programme (SACEP) and Alliance of Bioversity International and CIAT, Colombo
Silva K, Sirasa M (2018) Antioxidant properties of selected fruit cultivars grown in Sri Lanka. Food Chem 238:203–208
Talabi AO, Vikram P, Thushar S, Rahman H, Ahmadzai H, Nhamo N, Shahid M, Singh RK (2022) Orphan crops: a best fit for dietary enrichment and diversification in highly deteriorated marginal environments. Fronts Plant Sci 13
Thamilini J, Wekumbura C, Mohotti AJ, Kumara AP, Kudagammana ST, Silva KRR, Frossard E (2019) Organised homegardens contribute to micronutrient intakes and dietary diversity of rural households in Sri Lanka. Front Sustain Food Sys 3:94
Tyagi RK, Pandey A, Agrawal A, Varaprasad KS, Paroda RS, Khetarpal RK (2018) Regional expert consultation on underutilised crops for food and nutritional security in Asia and the Pacific – Thematic, strategic papers and country status reports. Asia-Pacific Association for Agricultural Research Institutions (APAARI), Bangkok
Upasana, Yadav L (2022) Pseudocereals: a novel path towards healthy eating. In: Waisundara VY (ed) Pseudocereals. IntechOpen, India, pp 1–14
Usman M, Patil PJ, Mehmood A, Rehman A, Shah H, Haider J, Xu K, Zhang C, Li X (2022) Comparative evaluation of pseudocereals peptides: a review of their nutritional contribution. Trends Food Sci Technol 122:287–313
Wasala WMD, Danwatta ASJ, Kondasinghe KMCL, Abeyrathna PWMDT (2014) Agro bio diversity survey in kandyan home garden, owita system and village tank system. Plant Genetic Resources Centre, Department of Agriculture, Peradeniya
Weerahewa J, Pushpakumara G (2019) Towards updating of the national action plan for conservation and sustainable utilisation of agrobiodiversity in Sri Lanka. In: Proceedings of symposium on agrobiodiversity for climate change adaptation, food and nutrition, Plant Genetic Resources Center, Gannoruwa, Sri Lanka, 23–24th May 2019
Weerahewa J, Pushpakumara G, Silva P, Daulagala C, Punyawardena B, Premalal S, Miah M, Roy J, Jana S, Marambe B (2012) Are homegarden ecosystems resilient to climate change? An analysis of adaptation strategies of homegardeners in Sri Lanka. APN Sci Bull 2:22–27
Weerakoon S, Somaratne S (2021) Development of a core collection from Sri Lankan traditional rice (Oryza sativa) varieties for phenotypic and genetic diversity. Nusantara Biosci 13(1)
Weerasekara PC, Withanachchi CR, Ginigaddara G, Ploeger A (2018) Nutrition transition and traditional food cultural changes in Sri Lanka during colonisation and post-colonisation. Foods 7:111
Weerasekara PC, Withanachchi CR, Ginigaddara G, Ploeger A (2020) Understanding dietary diversity, dietary practices and changes in food patterns in marginalised societies in Sri Lanka. Foods 9:1659
Weerasinghe PASS, Kopiyawattage KPP, Abeyrathne AHMSWB (2020) Effectiveness of Wood apple budded sweet orange cultivation as a coping strategy to climate change: value chain analysis in Nawagaththegama, Sri Lanka. In: 12th annual research symposium proceedings, Rajarata University of Sri Lanka, Mihintale, Rajarata University of Sri Lanka, pp 20–21
WFP (2015) Sri Lanka food security atlas: livelihoods, food security and resilience. World Food Programme, Colombo
Wickramasinghe MRCP, Dayawansa NDK, Jayasiri MMJGCN, Silva RPD (2023) A study on external pressures of an ancient irrigation cascade system in Sri Lanka. Agric Syst 205(C):103593
Wikramanayake E, Fernando S, Wickramaratne C, Akbarally Z, Wikramanayake S (2022) Sri Lankan moist forests ecoregion: an imperiled rsland rainforest. In: Reference module in earth systems and environmental sciences, pp 37–52

Wimalasiri E, Ashfold M, Walker S, Nissanka S, Karunaratne A (2017) The relationship between rainfall characteristics and Proso millet (*Panicum miliaceum* L.) cultivation in low country dry zone, Sri Lanka
Wimalaweera GNM, Samarasinghe HMUSR (2013) Sri lankan vegetable and fruit supply chain: a comprehensive overview. Young business technology and entrepreneurial researchers (YBTER) University of Kelaniya, Sri Lanka, Young Business Technology and Entrepreneurial Researchers
Wolosik K, Markowska A (2019) Amaranthus Cruentus taxonomy, botanical description, and review of its seed chemical composition. Nat Prod Commun 14: 1934578X19844141

Chapter 9
Neglected Food Plants of Nepal

Bal Krishna Joshi

9.1 Introduction

Nepal is one of the south Asian countries, popularly known as Mount Everest country. Agriculture is the basis of life for the majority of people. Among 24,300 total biological species in Nepal, 28% are agricultural genetic resources (i.e. agrobiodiversity) (Joshi et al. 2020b). Agrobiodiversity includes six components namely crops, forages, livestock, insects, microorganisms and aquatic species. Each of these components has four sub-components (domesticated, semi-domesticated, wild relative and wild edible). Among these components and sub-components, food plant species include only crops (cultivated plant species) and wild edible plant species (Fig. 9.1). Food plant species are those species whose parts or extract or processed items are orally taken. In Nepal, there are 996 such species (Fig. 9.1) of which 252 are under cultivation (Joshi et al. 2020a). Crop cultivation ranges from 60 to 4700 masl for which many agricultural technologies have been developed and recommended through agricultural research and development (Joshi 2017).

Agricultural research was started formally in 1950 in Nepal. Due to the varied climates, agro-eco-zones and ethnicity, many different kinds of indigenous food plant species were cultivated across the country (Upadhyay and Joshi 2003). However, research, development and education have always given priority to only a few exotic food plant species since the beginning of agricultural research and development. Due to this strategy, a large number of food plant species and landraces are neglected. All wild food plant species (which include semi-domesticated, wild relative and wild edible) are neglected. At the species level, 133 food crop species (65% of total food crops) are neglected whereas, at the landrace level, 99.7% are under the neglected group (Joshi et al. 2017a, 2019b). Only 38 local landraces of 19 crops have been used to develop 41 varieties (Joshi 2017). Under the formal system, a

B. K. Joshi (✉)
Nepal Agricultural Research Council, Kathmandu, Nepal

T. Ismail et al. (eds.), *Neglected Plant Foods of South Asia*,
https://doi.org/10.1007/978-3-031-37077-9_9

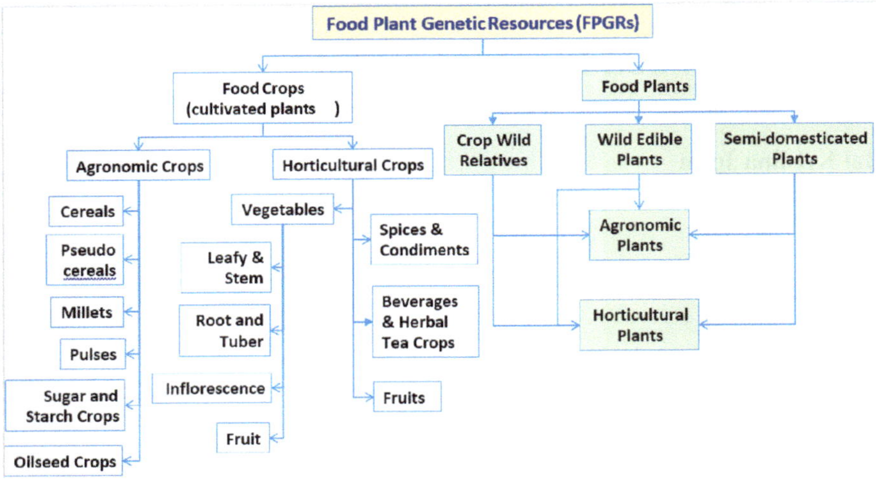

Fig. 9.1 Groups of food plant genetic resources in Nepal

total of 623 varieties of 65 crops have been registered. Among these varieties, 544 (83.3%) are exotic and 79 (12.7%) are native (Joshi et al. 2017a). Research history indicates that agricultural research depends 95% on foreign germplasm in Nepal (Joshi 2017).

Even with the introduction of a large number of genetic resources, food insecurity is increasing year by year and remains a big challenge (Bista et al. 2013; ADS 2015; MoAD 2016; Acharya et al. 2018; Joshi and Shrestha 2018). One of the strategies for food security could be of considering neglected native food plant species and crop landraces in research, development and education targeting to develop site-specific improved varieties and diversify the site-specific staple crops. Nepal possesses a large number of neglected and underutilized species and some of them are defined as future smart food crops (Li and Siddique 2018; Joshi et al. 2019b). This chapter describes only parts of neglected food crops which are under cultivation and some semi-domesticated food plant species. Here the crop is defined as domesticated species under kingdom Plantae and plant (though include crops but in a narrow sense) is non-domesticated species under kingdom Plantae.

9.2 Neglected Food Plants of Nepal

In Nepal, one farming family consumes on an average 122 different plant species; grows, maintains and harvests 200 plant species each year and among them, 100 species are food plants (Govinda Sharma, Personal Communication, 2022). Farming household grows deliberately 40 food plant species. Within these species, there may be many more cultivars (varieties and landraces), but the majority of the native landraces are being wiped out. Most Nepali people consume about 16 plant species in each meal (Joshi et al. 2020a). These plant diversities have contributed to secure annual family food and nutritional needs among many of the farming communities along with income generation. For sustainable food and nutrition security, all these cultivars need to be improved genetically. However, most of them are neglected. They are cultivated in three different agro-ecozones of Nepal, namely High-Hill, Mid-Hill and Tarai (Upadhyay and Joshi 2003).

Neglected and underutilized species (NUS) are those plant species that are not included in formal research, education, and development (Padulosi et al. 2013; Joshi et al. 2019a, b; Shrestha et al. 2019), if included, only for few years. Under this group, neglected food plant species (NFPs) are the most important for food and nutrition security along with environmental security. NFPs have immense capacity to aid in food security, nutritional enhancement, dietary and culinary variation, as well as income generation. They are also capable of thriving in challenging climatic conditions and even on soils that are considered less fertile (Kunwar et al. 2012; Dangol et al. 2014, 2017; Joshi et al. 2019b; Pudasaini 2013). NFPS are very easy to cultivate and perform well even in poor soil and under ordinary management. Based on the knowledge of agriculturists and policy makers on these species, NFPS can be divided into four groups (Fig. 9.2). Examples in each group are given in Fig. 9.3 and Table 9.1. The word neglected applies to both species and landraces, which means neglected food plant species and neglected food crop landraces. There are many neglected landraces of even major staple crops (Table 9.1). Both neglected species and landraces have a greater role in sustaining the main smallholder

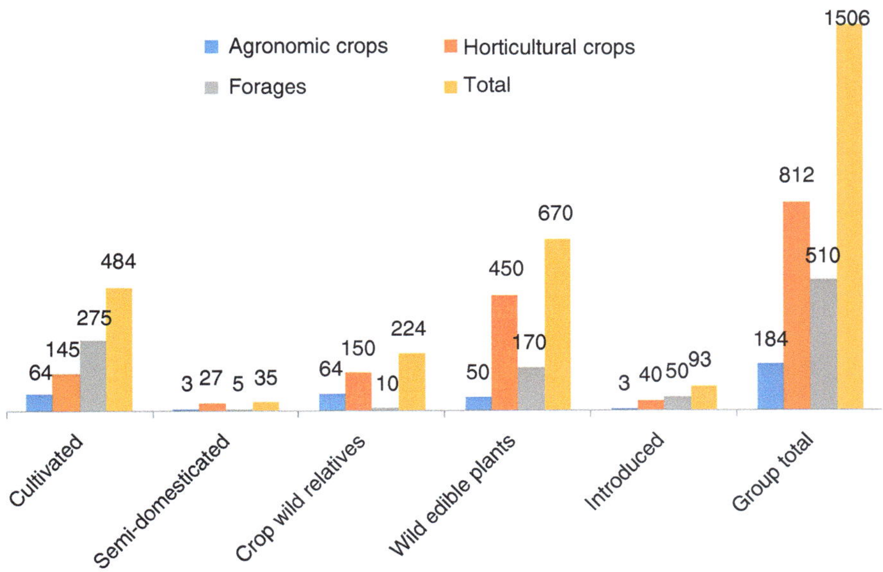

Fig. 9.2 Total agricultural plant species available in Nepal. (Adapted from: Joshi et al. 2017b)

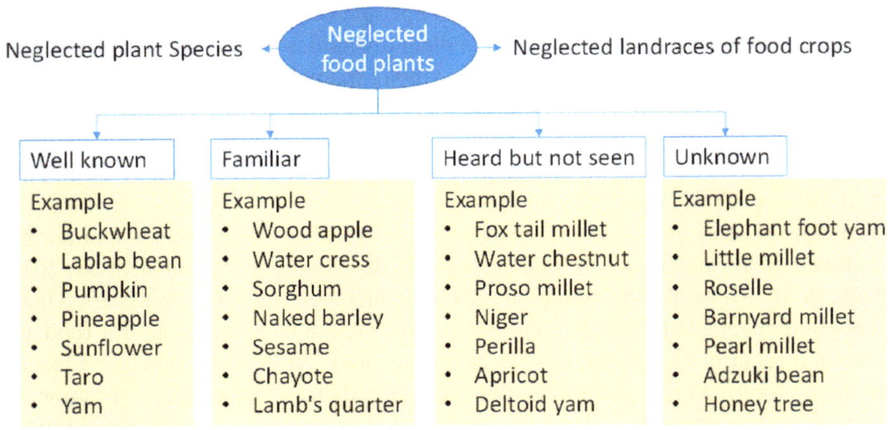

Fig. 9.3 Groups of neglected food plants (species and landraces) based on the knowledge of researchers, policy makers, academicians and agriculturists. (Modified from: Joshi et al. 2019b)

Table 9.1 Some neglected landraces of food crops in Nepal

SN	Common name	Nepali name	Scientific name	Neglect group
A.	**Cereals, pseudo cereals and millets**			
1.	Barnyard millet, Japanese millet	Saamaa	Echinochloa frumentacea Link	Unknown
2.	Foxtail amaranth, Grain amaranth	Jhule latte, Maarse, Raamdaanaa	Amaranthus caudatus L.	Heard but not seen
3.	Foxtail millet	Kaaguno	Setaria italica (L.) Beauv.	Heard but not seen
4.	Grain amaranth	Suntale latte, Maarse, Raamdaanaa	Amaranthus cruentus L.	Heard but not seen
5.	Kodo millet	Kodee	Paspalum scrobiculatum L.	Unknown
6.	Little millet	Dhaankodo, Sujeekodo, Kutkee	Panicum sumatrense Roth.	Unknown
7.	Naked barley	Uwaa, Karu	Hordeum vulgare L. var. nudum Hook.f.	Familiar
8.	Pearl millet	Ghoge, Baajraa	Pennisetum glaucum (L.) R.Br.	Unknown
9.	Prince's feather, Grain amaranth	Pwaankh latte, Maarse	Amaranthus hypochondriacus L.	Familiar
10.	Proso millet	Chino	Panicum miliaceum L.	Heard but not seen
11.	Sorghum	Junelo	Sorghum bicolor (L.) Moench.	Familiar
12.	Tartary buckwheat	Tite faapar	Fagopyrum tataricum Gaertn.	Well known

(continued)

Table 9.1 (continued)

SN	Common name	Nepali name	Scientific name	Neglect group
B.	**Roots and tubers**			
13.	American taro	Lahure pindaalu, Dudhe pindaalu	Xanthosoma sagittifolium (L.) Schott.	Unknown
14.	Cassava, Tapioca	Simal tarul	Manihot esculenta Crantz.	Heard but not seen
15.	Deltoid yam, Nepal yam	Vyaakur	Dioscorea deltoidea Wall. ex Griseb	Heard but not seen
16.	Edible canna, Achira	Pushtakaree, Fool tarul	Canna edulis Ker Gawl.	Unknown
17.	Elephant foot yam, Stink lily	Ol	Amorphophallus paeoniifolius (Dennst.) Nicolson	Unknown
18.	Giant taro	Ghyaampe pindaalu	Alocasia macrorrhizos L. G. Don	Familiar
19.	Sugar beet, Beet root	Chukandar	Beta vulgaris L.	Well known
20.	Sweet potato	Sakharkhanda, Suthunee	Ipomoea batatas (L.) Lam.	Well known
21.	Taro	Pindaalu	Colocasia esculenta Schott.	Well known
22.	White yam, Greater yam	Seto tarul, Ghar tarul	Dioscorea alata L.	Well known
C.	**Pulses and nuts**			
23.	Adzuki bean	Maaslaharee, Guraans, Raatomaas	Vigna angularis (Willd.) Ohwi & H. Ohashi	Unknown
24.	Areca nut	Supaaree	Areca catechu L.	Well known
25.	Cluster bean	Jhuppe simee	Cyamopsis tetragonoloba (L.) Taub.	Unknown
26.	Double bean	Asare simee	Phaseolus vulgaris L.	Familiar
27.	Faba bean	Bakullaa	Vicia faba L.	Well known
28.	Field pea	Saano keraau, Kalaa	Pisum sativum var. arvense L.	Familiar
29.	Garden vetch	Kutil kosaa	Vicia sativa L.	Familiar
30.	Horse gram	Gahat	Macrotyloma uniflorum (Lam.) Verdc.	Well known
31.	Lablab bean, Hyacinth bean	Hiunde simee, Taate simee, Maachhe simee	Lablab purpureus (L.) Sweet.	Well known
32.	Rice bean	Masyaang, Jhilunge, Siltung	Vigna umbellata (Thunb.) Ohwi & Ohashi	Familiar
33.	Sword bean	Tarbaare simee	Canavalia ensiformis (L.) DC.	Heard but not seen

(continued)

Table 9.1 (continued)

SN	Common name	Nepali name	Scientific name	Neglect group
D.	**Fruits and vegetables**			
34.	Apricot	Khurpaanee	Prunus armeniaca L.	Heard but not seen
35.	Ash gourd, Wax gourd	Kubhindo	Benincasa hispida (Thunb.) Cogn.	Familiar
36.	Balsam apple	Barelaa, Chuchche karelaa	Momordica balsamina L.	Heard but not seen
37.	Bread fruit	Raamphal	Artocarpus altilis Fosb.	Heard but not seen
38.	Butter tree, Nepal butter fruit	Chiuree	Aesandra butyracea (Roxb.) Bachni.	Heard but not seen
39.	Chayote	Iskus	Sechium edule (Jacq.) Sw.	Familiar
40.	Citron	Bimiro	Citrus medica L.	Familiar
41.	Coconut	Nariwal	Cocos nucifera L.	Well known
42.	Custard apple	Sitaafal, Sarifaa	Annona squamosa L.	Familiar
43.	Drumstick	Sajiwan, Sitalchinee	Moringa oleifera Lam.	Heard but not seen
44.	Eureka lemon	Nibu Kaagatee	Citrus limon (L.) Eureka	Heard but not seen
45.	Garlic pear, Nepal caper	Sipleegaan	Crateva unilocularis Buch.-Ham.	Unknown
46.	Grape fruit	Sankhatro	Citrus paradisi Macfad.	Heard but not seen
47.	Hamilton's bamboo	Taamaa baans	Dendrocalamus hamiltonii Nees Am. ex Munro	Well known
48.	Honey tree	Mauwaa	Madhuca longifolia (J. Konig) J.F.Macbr.	Unknown
49.	Indian gooseberry, Myrobalan	Amalaa	Emblica officinalis Gaertn.	Familiar
50.	Jackfruit	Katahar	Artocarpus heterophyllus Lam.	Well known
51.	Java Plum	Jaamun	Syzygium cumini (L.) Skeels	Heard but not seen
52.	Jujube	Bayar	Ziziphus jujuba Mill.	Familiar
53.	Lamb's quarter	Bethe	Chenopodium album L.	Familiar
54.	Leaf amaranth	Saag latte, Chaulaai saag	Amaranthus dubius Mart. ex Thell.	Familiar
55.	Lemon	Nibuwaa	Citrus limon (L.) Burm.f.	Well known
56.	Morning glory, Water spinach	Karmee saag, Kaangkun	Ipomea aquatica Forsk	Heard but not seen
57.	Mulberry	Kimbu, Kiunkaafal	Morus alba L.	Familiar
58.	Musk melon	Kharbujaa	Cucumis melo L.	Heard but not seen
59.	Nepalese hog plum	Lapsi	Choerospondias axillaris (Roxb.) B.L.	Familiar

(continued)

Table 9.1 (continued)

SN	Common name	Nepali name	Scientific name	Neglect group
60.	Olive	Jaitun	Olea europea var. cuspidate (Wall.ex G.Don) Cif.	Unknown
61.	Papaya	Mewaa	Carica papaya L.	Well known
62.	Passion fruit	Lahare aanp	Passiflora edulis Sims	Heard but not seen
63.	Pecan nut	Chuchche okhar	Carya illinoinensis (Wangenh.) K.Koch	Heard but not seen
64.	Persimmon	Haluwaabed	Dyospyros virginiana L.	Familiar
65.	Pineapple	Bhuin katahar	Ananas comosus (L.) Merr.	Well known
66.	Plum	Aalu bakhada	Prunus domestica L.	Familiar
67.	Pummelo, Shaddock	Bhogate	Citrus grandis Osbeck.	Familiar
68.	Pumpkin	Farsee	Cucurbita maxima Duchesne.	Well known
69.	Ridge gourd	Tirai, Paate ghiraunlaa	Luffa acutangula (L.) Roxb.	Heard but not seen
70.	Rough lemon	Jyaamir, Kaathe jyaamir	Citrus junos Tanaka	Heard but not seen
71.	Snake gourd	Chichindo	Trichosanthes anguina L.	Familiar
72.	Sour orange	Kaalee jyaamir	Citrus aurantium L. ssp. aurantium	Heard but no seen
73.	Spiny gourd	Chattel, Jhuse karelaa	Momordica dioica Roxb.	Familiar
74.	Sweet belladonna, Pokeweed	Jaringo	Phytolacca acinosa Roxb.	Unknown
75.	Sweet lime	Chaaksee	Citrus limettoides Tanaka	Heard but not seen
76.	Tamarillo, Tree tomato	Rukh tamaatar	Solanum betaceum Cav.	Familiar
77.	Tamarind	Imlee, Titree	Tamarindus indica L.	Heard but not seen
78.	Tamilnadia	Pidaar, Pedaar	Xeromphis uliginosa (Retz.) Maheshw.	Unknown
79.	Water chestnut	Singadaa	Trapa natans var. bispinosa (Roxb.) Makino.	Heard but not seen
80.	Water melon	Tarbujaa	Citrullus lanatus (Thunb.) Matsum & Nakai	Familiar
81.	Watercress	Sim saag, Khole saag, Sim raayo	Nasturtium microphyllum Boenn. ex Rchb.	Familiar
82.	Wood apple	Bel	Aegle marmelos L.	Familiar
E.	**Others**			
83.	Aniseed, Anise	Sop	Pimpinella anisum L.	Familiar
84.	Black cumin	Himaalee jiraa, Kaalo jiraa,	Bunium persicum (Bioss.) B. Fedtsch.	Heard but not seen
85.	Black mustard	Kaaltoree	Brassica nigra L.	Heard but not seen

(continued)

Table 9.1 (continued)

SN	Common name	Nepali name	Scientific name	Neglect group
86.	Black pepper, Hot pepper	Marich	Piper nigrum L.	Familiar
87.	Cardamom	Sukumel	Elettaria cardamomum L.	Familiar
88.	Catnip, Catmint	Ngetaa	Nepeta cataria L.	Unknown
89.	Chinese leek, Fragrant onion	Dundu	Allium tuberosum Rottler. ex Spreng.	Heard but not seen
90.	Cinnamon tree	Daalchinee, Sinkaulee	Cinnamomum verum J.Presi	Familiar
91.	Clove	Lwaang	Syzgium aromaticum (L.) Merr. & Perry	Familiar
92.	Cumin	Jeeraa	Cuminum cyminum L.	Well known
93.	Felwort	Chiraaito	Swertia chirata Roxb.	Heard but not seen
94.	Giloy	Gurjo	Tinospora sinensis (Lour.) Merr.	Heard but not seen
95.	Hemp	Bhaango, Bhaang	Cannabis sativa L.	Familiar
96.	Holy basil, Sacred basil	Tulsaee	Ocimum sanctum L.	Well known
97.	Lovage, Caraway, Ajwain	Jwaano	Trachyspermum ammi (L.) Sprague ex Turrill	Familiar
98.	Marijuana	Gaanjaa	Cannabis indica Lam.	Familiar
99.	Mint	Pudinaa	Mentha spicata L.	Well known
100.	Mint, Garden mint	Pudinaa	Mentha longifolia (L.) Huds.	Well known
101.	Nepal aromatic garlic	Jimbu	Allium hypsistum Stearn	Heard but not seen
102.	Nepal pepper	Timur	Zanthoxylum armatum DC.	Familiar
103.	Niger	Jhuse til, Filinge	Guizotia abyssinica (L.f.) Cass.	Heard but not seen
104.	Perilla	Silaam	Perilla frutescens Brittoon	Heard but not seen
105.	Roselle	Belchanaa, Chukaa, Amilee bhindee	Hibiscus sabdariffa L.	Unknown
106.	Sacred basil, Holy basil	Tulsee	Ocimum sanctum L.	Familiar
107.	Sarson, Yellow mustard	Sarsyun	Brassica campestris L. var. sarson Prain	Familiar
108.	Serpentina	Sarpagandhaa	Rauvolfia serpentine (L.) Benth. ex Kurz.	Unknown
109.	Sesame	Til	Sesamum indicum L.	Familiar
110.	Spicewood	Siltimur	Lindera neesiana (Wall. ex Nees) Kurz	Heard but not seen
111.	Sunflower	Suryamukhee	Helianthus annus L.	Well known
112.	Sweet flag	Bojho	Acorus calamus L.	Familiar
113.	Utrasum bead tree	Rudraakchya	Elaeocarpus sphaericus (Gaertn.) K. Schum.	Heard but not seen
114.	Wild onion	Jimmu	Allium hyspsistum Stearn	Heard but not seen

farmers. Almost 80% of farming households are directly associated with neglected species and landraces for their livelihoods. Species and landraces may get excluded from the neglected category over time and location as there are many potentials to include such species in formal research and education.

Farmers have diverse types of land, agri-ecosystem, production environment and requirements. There are many site-specific, season-specific and even household-specific landraces and species which have a high level of genetic diversity that suit their diverse growing environment and meet their requirements, including religious and cultural demands. Seeds of almost all NFPs are produced by farmers themselves due to which agricultural systems in NFPs are completely self-dependent with zero new environmental shock to growing NFPs. Some NFPS (HCRP 1996; Upadhyay and Joshi 2003; Limbu and Thapa 2011; Kunwar et al. 2012; Shah 2014; Gurung et al. 2019; Joshi et al. 2019b) are described below.

9.2.1 *Cereals, Pseudo Cereals and Millets*

Foxtail millet It is a staple crop among low-income farmers and has a long-standing tradition of cultivation in the arid regions of high-altitude landscapes. It also thrives in mid-hill areas and can be utilized in various ways, such as being cooked as a rice substitute, ground into a substantial porridge, or employed in the production of alcoholic drinks. Additionally, it holds cultural significance and has been historically used as an offering to deities worshipped in the area. Regular use of foxtail millet leads to a significant fall in blood glucose, making it a useful food to help manage and prevent diabetes. Foxtail millet cooked with milk and sugar (kheer) as a pudding makes an excellent and upscale dessert.

Grain amaranth For more than eight millennia, humans have cultivated amaranth, a plant that the ancients considered to be a symbol of eternal life. This versatile crop produces seeds that can be consumed as a gluten-free grain substitute, leaves are eaten as fresh vegetables, and stalks as animal feed, resulting in minimal waste. In mountainous regions, amaranth is a popular source of nutritious grain and leafy vegetables rich in iron. It is also a staple food for religious fasting days, often prepared as a thick porridge from flour, or popped like corn. Additionally, amaranth is a rich source of protein and manganese, both of which are essential for maintaining healthy bones, and studies have indicated that it can help to lower cholesterol levels. Furthermore, amaranth is a valuable source of lysine, an essential amino acid that is especially important for vegetarians. It can be part of a nutritious breakfast for busy urbanities.

Little millet Little millet is a traditional crop grown in Nepal, but in relatively small quantities. It is a type of millet that is closely related to proso millet and is known for having the smallest grains, which cook much faster than those of other millets. Farmers favor little millet for its ability to withstand unfavorable agro-climatic conditions and its early maturity. This crop can endure both drought and water logging. Additionally, little millet is a suitable food for individuals of all ages due to its high nutritional content, gluten-free nature, and non-sticky, non-acidic grains.

Naked barley Naked barley, which has been cultivated for over 8000 years, is among the earliest grains to have been grown by humans. It is grown in Nepal at elevations of up to 3800 m, which is the highest altitude where it is cultivated anywhere in the world. Naked barley is an abundant source of whole grains and complex carbohydrates that provide various health benefits such as reducing cholesterol levels and the risk of developing type-2 diabetes. Historically, Buddhist monks have regarded barley as a nutritious breakfast food that helps to conserve energy in cold weather.

Proso millet Proso millet is a highly resilient crop that can be grown at elevations of up to 3500 m and requires very little water compared to other cereal crops. This crop has been widely cultivated for thousands of years around the globe. Proso millet is completely gluten-free and rich in essential minerals, particularly potassium, which supports the health of the nervous system. Furthermore, it may serve as a substitute for corn in the production of biofuels, which could potentially position proso millet as a key crop in the green-energy industry in the future. Proso millet is traditionally consumed in the mountain communities as a main food, prepared same as rice, where access to rice is still limited. It is used to prepare indigenous alcoholic drink 'Chhyang' and distilled liquor (rakshi). 'Rakshi' have cultural importance in many different communities.

Tartary buckwheat Tartary buckwheat is a pseudo-cereal that serves as an excellent gluten-free alternative and also as a leafy vegetable. It is primarily grown by small and marginal farmers who reside in remote, hilly eco-zones and often belong to the lower income brackets. Tartary buckwheat grain contains almost twice as much protein as traditional rice and corn, and it is also an abundant source of fiber, magnesium, copper, and manganese. These minerals, combined with the high rutin content in buckwheat, contribute to cardiovascular health and may potentially help prevent diabetes. Additionally, honey made from buckwheat flour has a distinct flavor and is commonly used as a remedy for coughs. Mountain communities in Nepal have used Tartary buckwheat as a treatment for bleeding gums.

9.2.2 Roots and Tubers

Deltoid yam This is a very useful perennial climber and tasty tuber crop. Earlier it was collected from the forest, but nowadays, many farmers grow it on their land. Even in some cities, it is an important component of urban agriculture. This crop is typically found in the mid-hill regions of Nepal, ranging from 600 to 1800 m above sea level. It is known to require minimal water and can be cultivated on marginal lands that have low fertility and are prone to drought. The underground and aerial tubers are consumed after boiling directly or as a vegetable.

Edible canna It is grown mainly in mid and high hills of East Nepal as an annual crop, though it is a perennial in nature. It grows up to the height of 2 m. It is also grown as an ornamental plant as it produces colorful flowers. A huge diversity can be observed in flower colors. The tubers are a staple crop in some areas and it also helps to purify water. Tubers and leaves are also used for livestock.

Elephant foot yam It is a perennial tuber crop and much diversity has been reported both in wild and cultivated species. Tubers can be harvested annually. It is found in the Tarai of Nepal at elevations up to 1000 masl and prefers a hot climate. It requires medium water, is heat tolerant and very easy to cultivate. Insect pests and diseases are not a problem. It is generally consumed in Maghe Sankranti (a religious festival). It is boiled and cooked as curry or pickle.

Greater Yam This is an annual tuber climber and grown in Tarai and Mid Hill (tropical to temperate zone). It loosens the soil, covers soil surface areas and therefore helps to suppress weeds and protect the soil. It requires less water and is grown in summer and harvested in winter. It has religious and cultural importance and is a very common and widely grown crop. A large number of different landraces are seen in the market during Manghe Sankranti (a religious festival).

Sweet potato It is a perennial vine crop but grown as an annual in tropical and subtropical areas. Its tubers are sweet and lectin-free. All age's people prefer sweet potato. It is consumed raw or after a boil. It has many religious and cultural values and is commonly consumed in the winter season.

Taro Taro is a very common tuber crop with diverse uses. It is found mostly in Mid Hill, controls evaporation, suppresses weeds, and is grown in summer. It is heat tolerant but flood susceptible herb. Tuber and leaf are used as vegetables. Tubers are also consumed directly after boiling as a staple food. Manghe Sankranti, a religious festival, exhibits a significant amount of diversity.

9.2.3 Pulses and Nuts

Adzuki bean It is an herb commonly grown in West Nepal. It is an annual vine, similar to rice bean. Many different types of adzuki bean are available. Pulses are used as a vegetable.

Coconut It is a tall woody herb palm. It is cultivated extensively in tropical fruits. This palm is with slender ringed trunk and featherlike giant leaves. The coconut fruit can be consumed in various forms such as fresh, dried, or processed into coconut milk or oil. The liquid found inside the coconut fruit, known as coconut water, is a popular beverage. The coconut flesh is rich in fat and emits a pleasantly sweet aroma.

Faba bean It is a winter leguminous vegetable crop, distributed from Tarai to Mid Hill. It improves soil and performs well even in dense planting. It is generally grown in a kitchen garden organically and young shoots, young pods as well as grains are used as a vegetable. It is a drought and flood-susceptible crop.

Horse gram It is a nitrogen-fixing rainfed crop that is easy to grow. It performs well even in marginalized land. Generally, it is grown by mixing with finger millet in Mid Hill (subtropical to warm temperate zones). It also helps in losing weight, improves soil, and grows as cover crop for soil surface. Its straw is useful for livestock. Horse gram is a medicinal crop that is tolerant to drought and is known for its ability to cure stones. In Nepal, it is customary for people to consume horse gram soup at least once during the winter season.

Rice bean It is a nitrogen-fixing crop, grown in tropical to warm temperate regions (600–2200 msl). This crop has the ability to adapt to marginal lands and has characteristics that make it tolerant to drought. It is versatile and can be cultivated in a variety of soil types, including steep and sloping areas. The straw of this crop is commonly used as cattle feed. It holds cultural and religious significance in the Newars and Magars communities. The seeds are cooked and consumed as soup or pulses (Daal), while the grains from young pods are used as a vegetable. Popular local dishes made using this crop include Batuk, Baara, Biraunla, Masyaura, Kwanti, Rot, Furaula, and Khichadi. The Newar community has a tradition of preparing soup (called Kwanti) from a mixture of nine-grain legumes during the festival of Janai-Purnima and rice bean constitutes one of these grain mixtures.

9.2.4 Fruits and Vegetables

Apricot It is a small fruit tree grown in mountain areas. It is similar to peach and eaten fresh. Seeds are also edible. Fruits can be kept long after drying and dried apricots are good iron. Apricots are a good source of vitamin A. It is a self-pollinated crop with white flowers.

Ash gourd It is a vine summer fruit vegetable crop grown for its large round or oblong fruit. The size and color of the fruit are similar to watermelon. The fruit is covered by very fine hairs when young and later during maturity, it is covered by white powder (wax coat). This wax (ash) coat helps fruit keep longer. It has yellow flowers and broad leaves. It is a medicinally and religiously valued crop. People offer their fruit to God. The fruits are sliced into small, circular pieces and dehydrated for preservation and later consumption. Dehydrated vegetables are also delicious in taste. Juice from mature fruits is considered very important for health and has a cooling effect.

Butter tree The tree known as Chiuri is found in Nepal. In the Chepang community, these trees are considered a valuable resource and are often gifted to daughters during their wedding. Due to the tree's versatility and usefulness, it is not commonly cut down or removed. The Chepang people are renowned for their extensive knowledge of the collection and preparation of Chiuri products. They have a special relationship with Chiuri trees as they have a custom of giving Chiuri trees to their daughters as a gift during the marriage. Hence, it is regarded as a private resource. About 35–40% of oil can be extracted from fully ripe dried seeds based on the amount. In the Chepang community, Chiuri ghee is a primary source of cooking oil for vegetables and roti. The juice of the Chiuri fruit is also consumed as a thirst quencher. Additionally, the Chiuri tree provides ecological advantages by protecting against soil erosion. This tree is cultivated at an altitude range of 400–1400 m above sea level, specifically in the sub-Himalayan region on steep and challenging slopes. It flowers between November and January and produces fruit from April to June. The propagation of this tree is done through its seeds.

Chayote It is a climbing vegetable crop. Both shoots and fruits (immature and mature) are used as vegetables. It is used to prepare multipurpose food, such as vegetables, snacks, etc. It is grown organically and covers large areas producing many shots and fruits. Commonly it is grown in sloppy or terraces land of mid-hill in a wide range of well-drained soil to cover the soil surface and suppress weeds. It tolerates salinity. Many livestock prefer chayote vines and leaves.

Drumstick This tree is a member of the legume family and is commonly cultivated on marginal lands, particularly along the periphery of fields. It is perennial and commonly grown in a wide range of soil types in Tarai areas. It is a natural water purifier and shoots, leaves and pods can also be used for livestock. It is an expensive vegetable and there is high demand in the market. It tolerates both drought and heat.

Honey tree The tree commonly known as *Mahua* is a large deciduous tree indigenous to Nepal. Its branches produce clusters of white flowers that emit a musky scent. The flowers are delicious and nutritious and are eaten as unprocessed or cooked. Collected flowers are also used for making alcohol and have very popular medicinal value. Seeds are also processed for oil extraction and the product is locally named as *Mahua* butter.

Indian gooseberry This is a versatile fruit tree that can be consumed fresh, dried, or pickled. It is native to tropical Asia and can be grown at altitudes of up to 1500 m. The tree is able to withstand both low and high temperatures, and can grow in soils with pH levels ranging from 6 to 10. It thrives in areas with plenty of light and can tolerate drought and forest fires. The tree is well-suited for subtropical regions and requires warm temperatures during the formation of flower buds. Dry periods during fruiting can lead to premature fruit dropping. This tree is beneficial in carbon trading efforts, as it helps reduce carbon emissions and prevents soil erosion. The fruit is culturally accepted as having medicinal properties and is used in pickling. It has an acidic taste and is often used for flavoring and seasoning in cooked dishes.

Jackfruit This fruit tree is typically grown in tropical to subtropical regions, and its fruits are commonly available in the market from April to July. The tender fruits are often used for cooking vegetables, while the ripe fruits are typically consumed directly as a table fruit. Seeds are very commonly used to make tasty vegetables as well can be eaten after roasting. It contains almost every vitamin and mineral. It creates good shade, controls evaporation, and houses many birds. This tree species has ecological benefits such as reducing carbon emissions, mitigating climate change, and preventing soil erosion.

Java plum It is a dark and evergreen tree that produces juicy and delicious fruits. It has medicinal properties and is grown in Tarai and mid-hill. It is drought and heat tolerant and keeps the soil moist due to the blocking of sunlight. It grows well in well-drained light soil, with sandy loam. The species has ecological benefits such as reducing carbon emissions and preventing soil erosion, and also have wood that is strong and durable. The wood from this tree is often used by local people to make plowing machines that are pulled by bullocks.

Tamarind It is a leguminous perennial tree that produces edible and very sour fruit. Fruit is full of fleshy, juicy acidic pulp which is used to prepare pau as well as to prepare cuisines. It performs well in acidic soil and is drought tolerant. The flesh color is brown or reddish brown when the fruit matured. Diversity can be noticed in the size of fruits, some with long pods and some with short to medium length pods.

Watercress It is an aquatic perennial herb that produces white flowers. It is generally found in wet areas of Mid hill. A dark green leaf is round with long petiole. It is a leafy vegetable and commonly found during the summer season. It tolerates a cold climate, and grows submerged, floating on the water. It forms bushy colonies in the water and root freely from the stems.

Wood apple It is a religious fruit tree and can be grown in marginal lands. It provides a house for many birds. It is a drought and heat-tolerant crop and grows well in sandy loam soil. Fruits are nutritious and have medicinal properties. Wood apple is considered sacred by Hindus, as its leaves are used in worship of Goddess *Shiva/Pashupati*. The plant's fruit is also popularly consumed for table purposes. In the Newari community, it is believed that girls should marry for the first time with this fruit. Additionally, the plant has ecological benefits such as reducing carbon emissions and preventing soil erosion which make it useful for carbon trading and environmental conservation efforts.

9.2.5 *Others*

Hemp It is an erect annual herb grown widely in different localities including in marginalized land. Hemp has compound leaves which are palmate in shape, meaning that they resemble the shape of a hand with fingers radiating outwards. The plant also has small flowers that are greenish-yellow in color. It is also grown for medicinal uses. Pickles made from the seeds of hemp are very tasty and popular.

Nepal pepper It is a shrub with thorns on leaves and stems. The plant looks attractive and produces seeds with a strong aroma. It is also called prickly ash and is used for pickling. Many Nepali prefer its flavor and taste. This plant is found in mid to high hills of Nepal at an altitude range of 1100–2900 msl. Seeds mature in Nov-Dec. It is drought tolerant and natural rainfall is enough. It works as an insect repellent and has medicinal value.

Niger A crop that is an erect, stout, branched annual herb primarily grown for its oil and seeds. The seeds are also used to make pickles. This crop is considered rare, and is grown in the Tarai and Mid Hill regions of Nepal. When it is in flower, the entire field has an attractive appearance.

Perilla It is an annual herb and very popular for making pickles. It can grow in different soil types and have an efficient water and nutrient utilization potential. Perilla can be cultivated in a wide range of environmental conditions and is mostly found in hilly areas of Nepal. It is suited to grow at an altitude of 1200–2400 msl (warm to cool temperate zone). The crop is generally ready to harvest in Oct-Nov and is produced organically with low inputs and care. Its pickle is commonly served in restaurants and hotels.

Roselle It is a perennial herb having alternate three leaves which are 3–6 inches long with five to eight lobes. The flowers are red due to red petals which are sour. These petals are used to make pickles. Leaves are also used to make vegetables.

Sesame It is an annual herb plant grown for its seeds. Seeds are used to make pickles as well as to extract oil. It is grown in Tarai and Mid Hill. It is popular for its aroma and taste.

9.3 Traditional Uses of Neglected Food Plants

Neglected food plants are everything for many smallholder farmers in Nepal. These plants are the main sources of supplying daily foods, generating income, feeding livestock, maintaining a balanced ecosystem, etc. Smallholder farmers are rich in knowledge on using, growing, and maintaining these neglected plants. Many neglected food plants are for food and nutrition security at household levels, considering them a staple food. They also have been considered as emergency crops which have been used traditionally even during natural calamities, famine, and to minimize the risk of survival. For example, grains of finger millet have been kept for more than 50 years by some farmers for future any kind of famine. Some of them are medically valued and used for special health problems, nutrition deficiency, sick people, etc. These are being used for the treatment of both internal and external medical conditions. The dosage form may be a paste, soup, cooked and mixed with special herb, as a tonic during weak, etc. Many plant species have multiple uses including medicinal values (Kunwar and Bussmann 2008; Pangeni et al. 2020).

In Nepal about 126 different ethnic groups have their own different culture and religious systems, for which these neglected food plants have a very significant role. Different forms and species are used during different cultural functions, e.g. weddings, from birth to death events, and different festivals (Uprety et al. 2012; Shrestha 2013; Khanal et al. 2014; Neupane and Poudel 2021). Hindu culture offers only native crop landraces to Gods and Goddesses. Processed and improved seeds and breeds are not acceptable for offering to Gods and Goddesses. Some of these species have a very special functional use and special dishes are prepared for various occasions. Special value has been assigned to some plant species which are only offered to special guests as specialty food.

Many requirements of household e.g. artistic items, pot, mat, bedding materials, storage items, cleaning items, etc. are prepared from such neglected food plants. Most of the livestock is completely reared on these species. NFPs have also economic values for the households who keep products of these neglected species in *hat bazar*, temporary roadside shops, and different kinds of fairs. Ecological services from such species are also well recognized and traditionally these plant species are being used for the improvement of the environment. For example, edible canna is used to purify the industrial waste water. Some specific uses along with the medicinal values of the specific crop (Joshi et al. 2019b; Paneru et al. 2020; Panthi 2021) are described below.

Tartary buckwheat can be utilized in the preparation of pancakes by blending flour with freshly ground ginger and garlic, and salt. The people living in high hill areas also favor porridge as a staple food. This food is gluten-free and helps in managing diabetes. Amaranth grains are popped and can be used in various ways, such as being mixed with tea, milk, or sugar to make sweet pudding (kheer), non-fermented bread (ground) and sweet balls (laddu) made by mixing it with sugar syrup, or flat round sweets made by mixing it with honey. Amaranth is beneficial in preventing hair loss and lowering cholesterol. Foxtail millet can be used in making pancakes, beverages (jaand), porridge (Dhindo), pudding, etc. It helps in preventing heart diseases and improving memory. Proso millet is an important source of food during periods of scarce and dry weather, and is typically consumed as a staple food, cooked like rice. It benefits the nervous system, lowers cholesterol, and possesses anti-aging properties.

The edible parts of taro include the leaves, petioles, corms, and cormels, which are utilized in preparing various food dishes. Taro helps to improve digestion, lower blood sugar levels, prevent certain types of cancers, protect the skin, and enhance vision health. Greater yam is a type of tuber that is boiled and fried to make snacks and curries. It also has benefits such as lowering blood pressure and improving brain cognition. Deltoid yam is boiled and served as a snack and can also be used as a potato substitute for vegetable dishes. It is known to aid in digestion, treat roundworms, and alleviate constipation. Elephant foot yam is boiled and used in curry dishes. It is a high-energy food and has medicinal properties that can help with bronchitis, asthma, abdominal pain, emesis, dysentery, and other similar ailments.

Grass pea is commonly used in the preparation of daal (soup), and its seeds are also used locally in homeopathic medicine. Horse gram seeds are cooked to make hot soup and curry, and they are particularly useful for treating diarrhea, stomach issues, hemorrhoids, and menstrual disorders. Horse gram has also been found to have benefits for kidney stones, weak liver, gall stones, hypertension, and excessive perspiration. The large seeds of the faba bean are roasted, while the small seeds are split and prepared as a soup. The green pods and seeds of the faba bean are also used as green vegetables, soups, and snacks. Faba bean has been found to help alleviate symptoms of Parkinson's disease, prevent birth defects, and contain immune-boosting nutrients. Rice bean is also good for soup and curry, and it has anti-inflammatory properties that can reduce swelling and relieve pain while also stopping bleeding. The succulent fruits of the drumstick plant can be cooked as curry or boiled and used for pickling. Soup made from drumstick increases bone density, reduces period cramps in women, and is beneficial for brain injury, improving digestion problems, and lowering diabetes levels.

The tender fruits and shoots of the chayote plant are commonly cooked as a vegetable in curries. Chayote is known for its ability to help lower high blood pressure, dissolve kidney stones, and treat atherosclerosis, or hardening of the arteries. The tender leaves of fenugreek are harvested and used as green vegetables during the winter months, and the seeds are often used as an additive in vegetable preparations. Fenugreek is beneficial for milk production in new mothers, regulating blood sugar levels, boosting testosterone and male libido, and treating inflammation. The young and tender leaves of grain amaranth can also be consumed as a vegetable. Grain amaranth has antioxidant properties and is known to reduce inflammation, which can help with weight loss efforts.

Indian gooseberry fruits can be consumed raw, or cut into pieces and dried to make pickles. They have high medicinal value and are commonly used as a remedy for coughs and colds. The young fruit can also be used as a vegetable, and the ripe fruit is eaten fresh. Mature seeds can be roasted and cooked as a vegetable. Jackfruit is known for preventing anemia and helping to alleviate mental stress and skin diseases. Ripe wood apple fruits are eaten as is or used to make semi-processed juice. The leaves have antibacterial properties and can be used to strengthen teeth and gums. The fruit and seeds are also used to treat diabetes, diarrhea, and ringworm. Ripe Java plum fruits are also eaten as is or used to make semi-processed juice, and are highly recommended for diabetic patients. The bark of the stem or trunk can be used to cure diarrhea and dysentery. Custard apple fruits are eaten fresh, with the flesh either scooped out or pressed through a sieve and added to milkshakes, custards, or ice cream. The leaf, young fruit, young tip, seed, and bark of the custard apple tree all have insecticidal properties, and the leaves and bark can be used for dye extraction as well.

Linseed is commonly used in pickling and is also known for its ability to help alleviate mental stress and skin diseases. Nepali butter tree is used to produce ghee, and its fresh fruits can be eaten. It also has pharmaceutical value and is believed to be effective for treating rheumatism. Perilla seeds are roasted and used in pickling, while the juice extracted from its leaves can be applied to cuts and wounds. Nepal pepper, also known as prickly ash, is a popular additive in pickles due to its flavor and taste. It is also known for its medicinal properties and is used to treat fever, coughs, and stomach problems.

9.4 Industrial Valorization of Neglected Food Plants

Industrial applications of neglected food plants are less known. Due to the subsistence farming and a low number of industries in the country, the industrial values of many NFPs were not explored and used. Based on the traditional knowledge and practices along with key informant surveys, many NFPs have industrial applications. Some food recipe has been made from some NFP and offered during large gatherings and big festivals indicating their industrial applications. Cereals and pseudo-cereals which are produced in Himalayan regions can be used to prepare biscuits, cookies, cakes, noodles, etc. with a very special taste and color. This product possesses the five-traits i.e. purity, quality, tasty, healthy and nutritious, and could be a choice of many people around the world. These foods called Himalayan super foods are lectin and gluten-free. Buckwheat is a very healthy product mainly due to rutin content in all parts of buckwheat. Rutin can be extracted for its multiple uses and sold, e.g. rutin tablets for many different kinds of health problems.

Cereals and pseudo cereals are very common and can be used at an industrial level to make various products like baked items and other ready to eat products. Millets are used to prepare different types of alcohol (e.g. *raksi, ailaa, jaad, tongbaa, chaang, tin paane, nigaar, arac, bhaatiko jaad*, etc). Alcohol and beer made from millets, and naked barley are very popular. Alcohol from the Mahuwa tree is considered medicinally very useful and this product might be very famous. Another industrial product might be the oil from NFP. Oil from coconut, sesame and niger are considered healthy and has a high market value. Traditionally, horse gram is excellent for stone problems in humans. Therefore, horse gram soup powder can be of industrial value.

Ash gourd is offered to Gods instead of animals in many temples. It has a very special cooling property. Its dried slice tastes like meat when cooked as a vegetable. It can be therefore used to prepare juice and dried vegetables on a larger scale. Similarly, the juice of wood apple could be of industrial application. Soup powder from stinging nettle might be marketed on at a large scale as it has good market demand. Roasted jack fruit seeds might be an alternative to ground peanut.

Another big potent is tamarind, Indian gooseberry and taro. Many different kinds of *titauraa* and *paau* (similar to candy) from Nepalese hog plum, tamarind and Indian gooseberry are very famous and have high market demand. They are sour, sweet, and spicy. There is a big market for *paau*, therefore, these NFPs could be utilized on a large scale. *Titauraa* is of two kinds and has a high market value. It is prepared by mixing different ingredients with the fruits of Indian gooseberry. It is ready-made sour food and is commonly used during travel. Another kind of *titauraa* is made up of taro mixed with other items e.g. black gram. This is a dried ball vegetable and can be used during the scarcity of vegetables. Taro is very easy to grow and has multiple uses and values. Starch can be extracted from taro. Its leaves are used to make *gaavaa* which is also a dried vegetable and can be used during the scarcity of vegetables.

Nepal is traditionally very rich in preparing diversified pickles. There are many factories for preparing pickles. Pickle from Nepal pepper and Niger test very unique and pickle industries can be established from these NFP. Nepalese butter tree is famous for butter (vegetable ghee). This butter has many potential uses i.e., raw material for confectionary and vegetable ghee manufacturing. Other industries like soap, pharmaceutical and candle manufacturers also can get benefit. Therefore, this tree can be used as an industrially valued NFP.

9.5 Nutritional Significance and Opportunities

NFP is grown organically mostly following the natural way of selection (allele selection, genes and genotypes selection, nutrient selection, primary and secondary metabolite selection, etc). They have balanced nutrients and are very important for the nutrient fulfillment of human beings who lived nearby the production areas (Adhikari et al. 2017; Pant 2018; Joshi et al. 2020c). Human forces may change the nutrition constitution, e.g., genetically improved crop varieties possess a higher amount of carbohydrates and lower amount of other nutrients as compared to landraces (Joshi et al. 2020c). Seasonal production of these NFPs from a well-familiar environment contains a diverse nutritional profile. The nutrition of these NFPs is essential for human beings, and other living organisms around the production sites. Growing environments, as well as stresses, also trigger NFP to produce diverse metabolites.

Nutritionally NFPs are very important and famous for a particular use (Gautam et al. 2009; Shah 2014; Li and Siddique 2018). General nutritional profiles are tabulated in Table 9.2. Detail nutrition profiling is needed to see the differences among the products of different NFPs grown from different areas and seasons. The general understanding is that NFPs are nutrition dense, work for a particular nutrient deficiency, and have a high market value because of naturally grown and tasty. Many of

Table 9.2 Some neglected landraces of food crops in Nepal

SN	Crop	Landraces
1.	Ash gourd	*Kuvindo*
2.	Banana	*Maalbhog, Rataulo, Fusre, Mirkute*
3.	Chayote	*Kaade Ishkush, Hariyo Ishkush, Kaalo Ishkush, Seto Ishkush*
4.	Finger millet	*Kaalo kodo, Smadhi kodo, Seto kodo, Lafre kodo, Kaatike kodo, Dudhe kodo*
5.	Indian gooseberry	*Amalaa*
6.	Lamb's quarter	*Saano bethe, Thulo bethe, Bethu, Dhule bethe*
7.	Papaya	*Lahare mewaa, Dalle mewaa, Thulo mewaa, Kankre mewaa*
8.	Proso millet	*Dhude chino, Hande chino*
9.	Rice	*Samundrafini, Tilki, Bhaati, Anghaa, Aapjhutte, Garue ghaiya, Praanpyuri, Ghiupuri*
10.	Sesame	*Kaalo til, Seto till*
11.	Soybean	*Saano batmaas, Kaalo batmaas, Seto batmaas*
12.	Taro	*Haatipaau, Ujarka, Khujure, Koili, Pancha Mukhe*
13.	Water cress	*Sim saag*
14.	Wheat	*Mudule gahun, Naaphal, Bhaabri gahun, Daabdi gahun, Raani gahun*
15.	Wood apple	*Bel*

them are used as a tonic and to boost the immune system. Farming communities offer different kinds of foods prepared from NFP to different conditional people, e.g. sick people, children, pregnant women, aged people, etc. due to their specific nutritional value. Their roles have also been recognized in enhanced ecological services including the improvement of soil fertility. Many of them are gluten-free, lectin-free, and high in polyphenols and therefore have a significant role in providing balanced human nutrition. One of the most important features of these NFPs is *aadilo*, which means after eating the products of some NFPs, it takes a long time to feel hungry. This is because of very dense and balanced nutrients in the products which digest slowly and release energy slowly.

NFPs are very important for health, nutrition, environment, food, income generation and others. There is a great opportunity to use NFPs for fulfilling the nutrient requirements of humans and livestock. Because of the high market value, smallholder farmers can generate income by selling nutrient-dense products. The trend of feeding artificial nutrient tablets, e.g. vitamin A and C tablets, and iron tablets is increasing over the years. These NFPs if utilized properly, such nutrient tablets may not be needed. In addition, some of their test and nutrition are associated with geographical areas, which indicate the potentiality of applying geographical indication right for market assurance. NFPs can play significant roles in improving the nutritional status of Nepalese people and meeting the targets of many different agreements and targets.

According to the World Bank report (World Bank 2011), undernutrition in Nepal costs 190 million USD annually which is critical to country economy, physical and intellectual health of the future generations. Many initiatives have been taken by the government to manage the risk of food security to coup up the future challenges of developmental abnormalities. The agricultural reforms 2015–2035 are the key steps to boost the economy of the people. This will resolve the issue of chronic malnutrition in women of reproductive age, lactating mothers and children below the age of five. In 2019, Nepal was ranked 142 nd among 189 UNDP observed countries with a 17.4% population suffering from multidimensional poverty. The Nepal Demographic and Health Survey (2016) (MoH 2017) have reported 36% of the children below five as stunted and 10% are wasted. Nearly 53% of children aged 6–59 months are micronutrient deficient and 41% females having an average age of 15–49 are anemic. Eleven percent of women aged 15–49 are short (less than 145 cm), and 17% are thin (BMI less than 18.5). Among men, 17% are thin, and 17% are overweight or obese. 86% of children aged 6–59 months received a vitamin A capsule.

The government has shown its commitment to the Sustainable Development Goals (SDGs) and the Goal 2 which aims at zero hunger by 2030. The goals were set and launched by Ministry of agriculture and livestock development on 19 December 2014 (MoAD 2016) to end all form malnutrition till 2025. The main focus was to improve the health status of wasted and stunted children, breastfeeding women and the adolescent girls. The policy was adopted as a national agenda to make Nepal a healthy nation and free from multidimensional food insecurity holding the vision of preparing a National Action Plan to eradicate hunger by 2025. The ZHC initiative was based on natural resource management and the use of crops that have the potential to grow under stress conditions and have efficient nutrient utilization potential with no wastage of any food highlighting the importance of interconnectedness various food systems that impact poverty, hunger and malnutrition. The strategic pillars and priorities envisaged by the ZHC initiative are given in Fig. 9.4 (Table 9.3).

Fig. 9.4 The envisioned strategic pillars by the ZHC initiative that emphasized on the improvement of agricultural and livestock systems to manage hunger, poverty and malnutrition. (Source: MoAD 2016)

Table 9.3 Nutritional value and medicinal application of neglected food plants

NFP	Nutritional value	Medicinal value
Adzuki bean	Rich in protein, fiber, iron and B vitamins. They contain polyphenols, anthocyanins, tannins, flavonoids, and saponins	Improves blood sugar control in diabetics, improves digestion, reduces cholesterol and triglyceride levels
Apricot	High amount of vitamins (A and beta carotene), flavonoids, and potassium, low in fat	Protect cells from damage, reducing your risk for obesity, diabetes, and heart disease
Ash gourd	A low-calorie vegetable with about 2.9 g of dietary fiber; low in calcium, magnesium, and phosphorus; has high antioxidant contents	Improves lung health, treats ulcers naturally, provides a cooling effect, helps digestion
Chayote	High carbohydrate and mineral	The chayote can lower blood pressure, treat atherosclerosis and dissolve kidney stones
Coconut	Contain high fat, several important minerals, rich in manganese, copper, iron and antioxidants	Help block the growth of certain bacteria strains, for blood sugar control, protect cells from oxidative damage
Custard apple	Mature fruits are used a food source and have medicinal value	All parts of the plant i.e., leaves, bark, stem, young fruits and bark powder have insecticidal properties and are also used dye particularly bark and leaves
Deltoid yam	Good source of carbohydrate	Locally different parts of the plants are used as treatment for constipation and as an Appetizer. Also, to treat roundworm.
Drumstick	High protein content	Soup of the plant improves bone mineral density and reduces cramping during periods, enhances memory, beneficial for neural injury and aids in digestion. It has antidiabetic properties.
Edible canna	Contains 70% carbohydrate, 22% fiber and 12% protein on dry basis	Used in various women health related issues and are also believed to be a diuretic, demulcent and diaphoretic
Elephant foot yam	Source of carbohydrate	Elephant foot yam is considered a divine plant that can cure multiple ailments i.e., asthma, bronchitis, abdominal pain, dysentery, emesis and other such diseases
Faba bean/Broad bean	High in protein	Faba beans are conventionally used to prevent birth defects, contain immune boosting nutrients and helps to reduce the symptoms of Parkinson disease.

(continued)

Table 9.3 (continued)

NFP	Nutritional value	Medicinal value
Fenugreek	It is the source of beta carotene, vitamin A, Vitamin C and other phytochemicals having antioxidant properties	Fenugreek is used to increase the milk production in lactating mothers and also to treat the inflammation. It reduces the blood sugar level, improves libido and enhance the testosterone in suffering males.
Foxtail millet	Contain nutrients like protein, fat, minerals and dietary higher than wheat and rice and is gluten free.	Foxtail millet is associated with improved memory functions and prevents cardiac ailments.
Grain amaranth	It is the source of protein, folic acid, minerals i.e., Ca, P, Mg and variety of vitamins and is considered as the high value food for lactating women's.	Helps to lower cholesterol and improve hair loss.
Grass pea	Rich source of protein	Grass pea seeds are locally used in used in homeopathic medicines
Greater yam/ White yam	Rich in minerals	It improves cognition and decrease blood pressure
Green pigweed, green amaranth	Its high nutritional value is the base of its popularity as it is rich in Vitamin C, Vitamin A, minerals and proteins.	Have high anti-oxidant potential and is used for weight loss and reduce inflammation.
Hemp	Contains all the essential amino acids and is rich in protein, a good source of Mg, P, K, Na, S, Ca, Fe, Zn and Vitamin E. Rich in omega-6 (linoleic acid) and omega-3 (alpha-linolenic acid) fatty acids.	Hemp lowers the risk of cardiac diseases, improve digestion, skin condition and reduce inflammation.
Honey tree	Highly rich in healthy fats, very nutritive and taken as a general tonic	Used in skin disease, rheumatism and headache. It is also traditionally used as astringent, tonic, cooling agent, and a treatment for bronchitis and chronic and acute tonsillitis
Horse gram	Rich in Calcium, Iron and protein and helps to lose weight	The general uses of the horse gram include preventing stomach hemorrhoids, diarrhea, and menstrual disturbances. Also found useful for gall stones, kidney stones, weak liver, excessive perspiration and hypertension.
Indiangooseberry	Fruits are a good source of vitamin C, minerals	High in medicinal value, used in chewing whole fruit while getting cough or cold
Jack fruit	Source of minerals and vitamins	Prevents skin diseases, anemia and helps in curing mental stress

(continued)

Table 9.3 (continued)

NFP	Nutritional value	Medicinal value
Java plum	The fruits of java plum are juicy, delicious and rich in Vitamin C. it has medicinal properties	Mostly used for the management of diabetes, stem and trunk barks are used against dysentery and diarrhea
Linseed	Rich in antioxidant, high omega 3 fatty acid	Helps in curing mental stress and skin diseases, used to shine hair, reduce cholesterol, and prevent diabetes. Good for weight management
Nepali butter tree	Dried seeds of the tree contain 35–40% of oil when fully ripe. Chiuri is a good source of fats and oils	Oil is effective against rheumatism and used to massage the persons suffering from headache.
Nepali pepper, Prickly ash	Contains essential oils, flavours, vitamin and sulphurous compounds	Multidimensional usage in homeopathic medicines i.e., to treat cough, fever, stomach trouble
Niger	Richness in various essential nutrients, a good source of important amino acids, high levels of omega-3 fatty acids and linoleic acid	Improve heart health, reduce inflammation, boost the immune system
Perilla	Good source of carbohydrates, Ca, Fe, Cu, Molybdenum, omega 3 fatty acids and vitamin	It has wound healing properties and leaves are applied to wounds and cuts. It serves as an additive for pickling.
Proso millet	Proso millets contains more fiber, proteins and iron contents than wheat and rice.	Its fiber make it cholesterol lowering, anti-aging and beneficial for the nervous system
Rice bean	High lysine, minerals and vitamins, including riboflavin, thiamine, ascorbic acid and niacin	Reducing swelling, inflammation, relieve pain and stop bleeding
Roselle	Rich in organic acids i.e., malic, citric, allo-hydroxycitric and tartaric acids. The plant is known for its vitamin C, Beta carotene, total sugars and protein contents	Used to treat different medical conditions including cancer, helmenthic disease and cardiovascular disorders. It also have antioxidant potential and is used to manage management
Sesame	Oil-rich, a good source of protein, fiber, thiamine, niacin, and vitamin B6; high in magnesium and antioxidants	Help reduce heart disease risk factors, help lower blood pressure, important for blood cell formation and function
Sweet potato	Rich in starch and fiber, an excellent source of beta carotene, vitamin C, and potassium	Improved blood sugar regulation and reduced oxidative damage
Tamarind	High in vitamin C and B vitamins as well as magnesium, potassium	Used to treatment constipation, sore throats and sunstroke
Taro	Rich in minerals	Lowers blood sugar levels, Improve digestion, prevent cancer, boost vision and protect the skin

(continued)

Table 9.3 (continued)

NFP	Nutritional value	Medicinal value
Tartary buckwheat	Rich source of lipids, dietary fiber, rutin, amino acid particularly lysine and minerals	As buckwheat is gluten-free it is used to manage diabetes, also reduce cholesterol due to high rutin contents
Watercress	Rich in folate, vitamin A, E and C and iron, calcium, and K, high in antioxidants	Prevent chronic diseases, such as diabetes, cancer and heart disease; benefits for heart health, including lowering blood pressure and cholesterol levels
Wood apple	The nutritious fruits have medicinal value	Leaves of wood apple have antibacterial properties and are used to strengthen teeth and gums. Seeds are used to treat ringworm, diarrhea and diabetes and bark is anthelminthic

9.6 National Policies Toward Preserving Biodiversity, Agrobiodiversity and Promoting Neglected Food Plants Production

Conservation in Nepal is mainly focused on non-agricultural genetic resources and big size species, e.g. tiger, elephant, *Shorea robusta*, etc. Non-agricultural genetic resources are conserved by allocating certain areas for which there are different acts and policies (Table 9.4). There are 12 National Parks, 1 Wildlife Reserve, 1 Hunting Reserve and 6 Conservation Areas in Nepal. NFPs are generally ignored in such protected areas, as policies do not explicitly mention these species. For agrobiodiversity, National Agriculture Genetic Resources Center (National Genebank) has been established under Nepal Agricultural Research Council. National Genebank has started conserving six components of agrobiodiversity, namely crops, forages, insects, microbes, livestock and aquatic agricultural genetic resources based on the Agrobiodiversity Policy. For prioritizing conservation, National Genebank has developed a method called five cell distribution analysis of landraces and species (Fig. 9.5). Genebank has developed and adopted 101 different good practices, and action plans for the conservation and utilization of agricultural genetic resources (Joshi et al. 2017b, 2020b). Objectives and gaps in relevant policies are given in Table 9.4. Many policies favor only a few common crops. However, about half of the total conserved germplasms in the National Genebank are NFPs. Because of adopting four conservation strategies i.e., ex-situ, on-farm, in-situ and conservation breeding, many NFPs are being conserved in different kinds of banks across the country.

To make awareness among the Nepalese people, Nepal Government has declared this year as National Agrobiodiversity Year 2022/23 and some of the focus areas in this declaration are promoting site-specific neglected and underutilized species.

Table 9.4 Policy provision and gaps concerning NFPs

SN	Policy	Provision	Gaps
1.	Agriculture Development Strategy 2015	Proposes to strengthen, collect, classify, and assess diverse bio-resources relevant to food and agriculture Initiate a system of registering agro-biodiversity Develop regulations for research, experimentation and evaluation of the biodiversity and genetic pool and regulations for the use of GMOs that have negative impact on biodiversity, gene pools and consumer's health. Promote community-based approaches to produce seeds at farm level which have greater survival for agro-biodiversity in the remote areas due to natural genetic variations Implement agro-biodiversity policy Improved resilience of farmers to climate change, Research on climate-resilient varieties Reorientation of investment in public research towards biosafety, mitigation of the effects of climatic change, conservation of the biodiversity and environment among others	Site-specific and NFPs focus activities lacking No specific activities for the interlinkages between agrobiodiversity conservation, nutrition and climate change
2.	Agro-biodiversity Policy (1st amendment 2014)	Management, conservation and sustainable use of conventional knowledge and agrobiodiversity to balance the ecosystem and creation of harmony with climate resilience of the production systems to overcome the effects of climate change. The regulatory authorities should ban imports of GMOs and if impossible must conduct research on their potential impacts on changing biodiversity, emerging risks and negative concerns to the environment. The government should protect and promote the rights, interests and economy of farmers, their traditional innovations, skills, technology and most importantly knowledge and practices. There must be provision of an equitable and judicious system for the even distribution of opportunities and benefits to the farming community both in term of material and genetic resources	Conservation priority not given. Many NFPs are endangered and their status and conservation strategy not mentioned. No clear linkage between nutrition and climate change-related policies
3.	Community Seed Bank Program Implementation Guidelines 2008	Seed multiplication of local varieties through the community-based seed production groups Exploration, identification, selection and production of local seeds	Focus on multiplication and supplying of seeds locally rather than conservation of local and NFPs Linkage with climate change and nutrition not mentioned

4.	Environment Protection Act (EPA) 2053 (1997)	Make legal systems to maintain a healthy and clean environment and minimize the possible adverse effects on nature, human beings, plants, wildlife and physical objects; and environmental protection with efficient use and management of natural assets Any area containing rare wild life, plants, biodiversity and places of historical and cultural significance must be delineated as an environmental protection area and should considered an area of highly significance from environmental protection point of view.	Contributions of agrobiodiversity including NFPs not recognized Pollinators for maintaining and creating genetic diversity not recognized
5.	Food and nutritional security plan of action 2013	Promote climate resilient and sustainable agriculture to reduce hunger, chronic malnutrition, and poverty and improve economy of small land holders among the poorest families for their better livelihoods. Nutrition education and training Promotion of locally available and indigenous foods and crops Promotion of mixed farming and IPM for enhanced environmental sustainability Livelihoods and farming systems-related interventions are more resilient to climate change and environmentally sustainable	Mobilization of location-specific NFPs, genetic enhancement of NFPs for better nutrition harvest Interlinkage between nutrition, agrobiodiversity and climate has been acknowledged to some extent but not reflected in proposed actions
6.	Forest Act 2049 BS (1992)	Impose restrictions on the forest products in term of their cutting, collection, use, transport, distribution, sale or export for the conservation of biodiversity and environment	Legal mechanisms for NFPs Repatriation for conservation Ecological services from agricultural genetic resources
7.	ITPGRFA-MLS Implementation Strategy and Action Plan (IMISAP) 2017	Devise a documentation structure of APGRs at national, regional and local levels to be placed in action to facilitate the accessions under MLS. NARC, MoALD, DoA, and many other similar organizations to have a yearly program to materialize the IMISAP. All relevant policies, acts and regulations at the local and national levels must be integrated to implement the IMISAP.	Action plans on NFPs Nutrition and climate change not explicitly taken into account

(continued)

Table 9.4 (continued)

SN	Policy	Provision	Gaps
8.	Multisectoral nutrition plan II 2018–22	Reducing the prevalence of malnutrition in women, children and adolescents through a multisector approach with the involvement of health, education, agriculture and livestock, water and sanitation, women, children and social welfare and local development Nutrition-specific and nutrition-sensitive interventions to improve nutritional status of adolescents, children under five and women's. Increase the quantity and improve the quality of institutions responsible for delivering services and scale up nutrition-sensitive as well as nutrition-specific services and build capacities at national, provincial and community levels.	Nutritional and medicinal values of NFPs, diversifying the sourcing of planting materials and foods Less emphasis on nutrition-sensitive agriculture No specific mention of the interlinkages between nutrition, climate change and agrobiodiversity
9.	National Adaptation Programs of Action (NAPA) 2010	Adaptation to Climate change Priority projects/programs related to biodiversity, agriculture, water resources, forest, infrastructure, health and calamities have been proposed Has defined coordination tools and execution modality for adaptation programs regarding climate change	Lacks specific programs on agro-biodiversity focusing on NFPs conservation and nutrition Interlinkage between climate change, agro-biodiversity conservation and nutrition largely missing
10.	National Biodiversity Strategy and Action Plan 2014–2020	Promotion of farming systems which are environment-friendly (minimum tillage, organic agriculture, etc.) and climate-oriented agriculture programs. Integrating climate vulnerability attitude in biodiversity management-related guidelines, programs/projects to ensure climate compatible development	Endangered NFPs Utilization of NFPs including value addition No clear linkage with nutrition
11.	National climate change policy 2019	Agriculture and food security-related policies, strategies and working policies for dealing with climate change issues Promotion of crop diversification, protection of agricultural biodiversity and organic farming	The capacity of NFPs to buffering climate changes and other stresses Linkage with nutrition non-existent
12.	National Environment policy 2019	Mainstreaming environment in development plans and policies, pollution control, waste management, nature/environment-friendly sustainable development	More focus on physical and mechanical means to environment protection and largely ignores interlinkage between environment, agro-biodiversity, climate change and nutrition

13.	National nutritional policy and strategy 2004	13 strategic nutrition approaches with long-term and short-term objectives to address the issue of malnutrition Strategies for lifestyle disease control Strategies for nutrition in exceptionally difficult circumstances	Location-specific seasonal nutritional value of NFPs
14.	National Parks and Wildlife Conservation Act 2029 BS (1973)	Wildlife conservation through habitat conservation	Habitat for agricultural genetic resources not perceived Agro plantation in protected areas
15.	National Seed Policy 1999	Conserving local crop varieties having distinct genetic traits and protecting the rights of the local community over seeds of such varieties Agro-biodiversity conservation and establishing varietal rights Regulation of GMO and transgenic plants coming from abroad Formulation and enforcement of bio-safety rules	Use values of NFPs not considered Making NFPs competent not given due priority
16.	National Seed Vision 2013–2025	Natural selection of regional landraces, exotic gene pool and their ancestry relatives to develop climate smart and nutrient-dense varieties. Promote the use of local seeds and conserve the agrobiodiversity. Promote VDC level seed bank to conserve agro-biodiversity Promote a coordination mechanism and establish a link between national and international varieties, regional genebanks and community seed stocks for the genetic diversity and exchange of materials to conserve biodiversity At local and regional levels designate zones, stations, farms and pockets to maintain variety and conserve biodiversity	The values and importance of NFPs Location-specific stable crop development covering NFPs No specific actions on climate change and nutrition in the Vision
17.	Seed Act 1988 (Amendment 2008)	Supply of quality seeds Regulating seeds Ownership rights to local varieties	Values of agrobiodiversity, including NFPs in the context of climate change and nutrition Promotion of local seeds in marketing and formal seed system Conservation and safeguarding of local germplasm through the use Marketing of seeds not linked with ownership rights

(continued)

Table 9.4 (continued)

SN	Policy	Provision	Gaps
18.	The right to food and food sovereignty 2018	Mutual coordination between three tiers of government for preventing and controlling starvation Establish food councils at national level throughout the country, provinces and regional levels to warrant that not a single person would starve. The provincial and local government of Nepal shall store the food necessary to fulfill the nutritional needs in any situation/disaster by mounting and updating their storage facilities with joint ventures Traditional and conventional foods produced at regional level shall be distributed on priority basis. The government must make necessary provisions for the mitigation of likely adverse impacts of climate change on food production	Role of NFPs in household food and nutrition security Interlinkage between agrobiodiversity, climate change, and nutrition recognized implicitly Role of agro-biodiversity for food and nutrition security not acknowledged explicitly
19.	Zero hunger challenge national action plan 2016–25	Focus on enhanced and quality food production with greater nutritional significance, increased resource utilization in agriculture, establishment of infrastructure, localization availability and development of agriculture based business which will increase employment opportunities while making agricultural occupations attractive for youths, support for smallholder and landless producers with access to productive resources, safety net support for vulnerable groups, and improved food governance Actions to promote climate-smart food crops with high nutritive value and comparative advantage in the market, Actively engaging farm families in linking the transactions of valuable indigenous crops to the markets	Role of location-specific NFPs on food, nutrition and environment security No provision of linkage between agrobiodiversity, nutrition and climate change

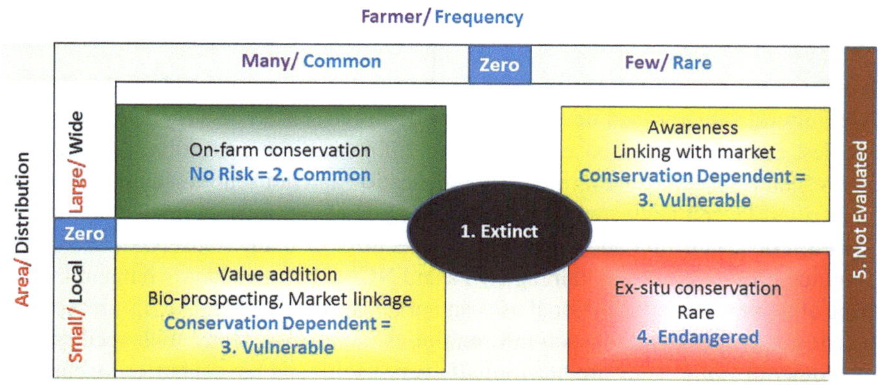

Fig. 9.5 Listing landraces and species based on their distribution under five cells to identify the conservation priority genotypes. (Adapted from: Joshi and Gauchan 2017)

Current policies favor the utilization of native crop genetic resources. However, promotional activities are not deliberately mentioned in policies focusing on NFPs. Nepal Agricultural Research Council started working on barley, buckwheat, amaranth, finger millet and proso millet through the NUS-related project in 1991 with support from IPGRI. Since then, few activities have been implemented in the country. Recently the provision of registration of landraces (farmer's varieties) has been created with the requirement of simple information. Any landraces including NFPs can be registered by farmers, communities and any other researchers and relevant organizations.

9.7 Toxicological Challenges and Possible Solutions

Toxicants of NFPs in Nepal are less studied. All information is based on farmers' knowledge and traditional practices. This knowledge needs further to verify. The young shoot of wild buckwheat is used as a vegetable during the lean period of green vegetables. If people consumed more, they may cause body parts lethargic and less sensitive. Sweet potatoes may increase the risk of kidney stone formation due to their oxalate content. Rice bean in traditional knowledge is ranked as a cold food, unsuitable for consumption during winter, causes flatulence and is believed to cause gastritis as well. Likewise, jack fruit is considered hot food; some people do not like the scent while it is ripened. Some landraces of taro cause throat problems. To lower the effects of toxicants, farmers eat less, boil, dry, cook well or mix with other items depending on the nature of harmful compounds in the NFPs.

9.8 Conclusion

A large number of food plant species and crop landraces are neglected in Nepal. Agricultural research and development are mainly focused on very few food plant species. Exotic varieties have significantly influenced the education, development and research systems, and farming areas. Location and household-specific contributions of NFPs and neglected crop landraces (NCLs) are underestimated and undervalued. Food, nutrition and environmental security of many households are due to continue growing and maintaining NFPs and NCLs. Farmers have firsthand knowledge of these species. Traditional uses and practices on NFPs and NCLs are rich and recently documentation of such information is increasing. Many such species have an industrial value. They are nutritionally very useful and have a buffering capacity for different kinds of stresses. Policies are lacking on activities for NFPs and NCLs. Policies should also focus on advancing site-specific NFPs and NCLs as staple varieties, which helps the conservation of many endangered NFPs. NFPs are further needed to explore their potential for commercialization. In addition, many neglected plant foods are prepared through many different processes and methods. Traditional plant foods never got priority in research, development and education. The research therefore should be focused on both neglected plant foods and neglected food plants for making agri-business profitable, diversified, unique and attractive.

Conflicts of Interest All authors read and approved the final manuscript and declare that they have no conflict of interest.

References

ADS (2015) Agriculture Development Strategy (ADS) 2015 to 2035. Ministry of Agricultural Development, Kathmandu. https://www.fao.org/faolex/results/details/en/c/LEX-FAOC171433/

Acharya AK, Paudel MP, Wasti PC, Sharma RD, Dhital S (2018) Status report on food and nutrition security in Nepal. Ministry of Agriculture, Land Management and Cooperatives, Kathmandu, pp 1–82

Adhikari L, Hussain A, Rasul G (2017) Tapping the potential of neglected and underutilized food crops for sustainable nutrition security in the mountains of Pakistan and Nepal. Sustainability 9:291

Bista DR, Amgain LP, Shrestha S (2013) Food security scenario, challenges, and agronomic research directions of Nepal. Agron J Nepal 3:42–52

Dangol DR, Gautam B, Oli BB (2014) Wetland plants and their local uses: observations from Rampur ghol, Chitwan. Nepal J Nat Hist Mus 28:142–159

Dangol DR, Maharjan KL, Maharjan SK, Acharya AK (2017) Wild edible plants in Nepal. Conservation and utilization of agricultural plant genetic resources of Nepal. NAGRC, Dhulikhel, pp 390–407

Gautam R, Sthapit B, Subedi A, Poudel D, Shrestha P, Eyzaguirre P (2009) Home gardens management of key species in Nepal: a way to maximize the use of useful diversity for the well-being of poor farmers. Plant Genet Resour 7:142–153

Gurung R, Dhakal R, Pudasaini N, Paneru PB, Pant S, Adhikari AR, Gautam S, Yadav RK, Ghimire KH, Joshi BK, Gauchan D, Shrestha S, Jarvis DI (2019) Catalog of traditional crop landraces of mountain agriculture in Nepal. NARC, LI-BIRD, Bioversity International, Pokhara. Available at: http://himalayancrops.org/project/catalogue-of-traditional-mountain-crop-landraces-in-nepal/

HCRP (1996) Annual report 1094/95. Hill Crops Research Program (HCRP), Nepal Agricultural Research Council. Agricultural Research Station, Kabre, Dolakha

Joshi BK (2017) Plant breeding in Nepal: past, present and future. J Agric For Univ 1:1–33

Joshi BK, Gauchan D (2017) Rebuilding local seed system of native crops in earthquake affected areas of Nepal, Proceedings of a National Sharingshop 18 Dec 2017. NAGRC; Bioversity International; Crop Trust, Kathmandu

Joshi BK, Shrestha R (2018) Nepal. In: Li X, Siddique KHM (eds) Future smart food – rediscovering hidden treasures of neglected and underutilized species for zero hunger in Asia. FAO, Bangkok, pp 161–178. Available at: http://www.fao.org/documents/card/en/c/I8907EN/

Joshi BK, Bhatta MR, Ghimire KH, Khanal M, Gurung SB, Dhakal R, Sthapit BR (2017a) Released and promising crop varieties of mountain agriculture in Nepal 1959-2016. Bioversity International, Pokhara

Joshi BK, Hari Bahadur KC, Acharya AK (2017b) Conservation and utilization of agricultural plant genetic resources in Nepal. In: Proceedings of 2nd National Workshop 22, p 23

Joshi BK, Pandey B, Adhikari RC (2019a) Progress and prospects of agricultural biotechnology and biosafety in Nepal: present status, challenges and way forward. Agricultural Biotechnology and Biosafety in South Asia: Progress and Prospects SAARC Agriculture Centre, p 59

Joshi BK, Shrestha R, Gautam IP, Poudel AP, Gotame TP (2019b) Neglected and Underutilized Species (NUS), and Future Smart Food (FSF) in Nepal. Government of Nepal, Nepal Agricultural Research Council, National Agriculture Genetic Resources Centre

Joshi BK, Shrestha R, Gauchan D, Shrestha A (2020a) Neglected, underutilized, and future smart crop species in Nepal. J Crop Improv 34:291–313

Joshi BK, Humagain R, Dhakal LK, Gauchan D (2020b) Integrated approach of national seed systems for assuring improved seeds to the smallholder farmers in Nepal. In: Shrestha RB, Penunia ME and Asim M (eds) Strengthening seed systems- Promoting Community Based Seed Systems for Biodiversity Conservation and Food & Nutrition Security in South Asia. SAARC Agriculture Center, Bangladesh; Asian Farmers' Association, the Philippines; and Pakistan Agricultural Research Council, Pakistan, p 181–194

Joshi BK, Ojha P, Gauchan D, Ghimire KH, Hari Bahadur KC (2020c) Nutritionally unique native crop landraces from mountain Nepal for geographical indication right. Traditional crop biodiversity for mountain food and nutrition security in Nepal. In: Gauchan D, Joshi BK, Bhandari B, Manandhar HK, Jarvis DI (eds) Tools and research results of UNEP GEF local crop project. NAGRC, LI-BIRD and the Alliance of Bioversity International and CIAT, Kathmandu

Khanal R, Timilsina A, Pokhrel P, Yadav RKP (2014) Documenting abundance and use of underutilized plant species in the Mid Hill Region of Nepal. Ecoprint: Int J Ecol 21:63–71

Kunwar RM, Bussmann RW (2008) Ethnobotany in the Nepal Himalaya. J Ethnobiol Ethnomed 4:1–8

Kunwar RM, Mahat L, Sharma LN, Shrestha KP, Kominee H, Bussmann RW (2012) Underutilized plant species in far west Nepal. J Mt Sci 9:589–600

Li X, Siddique KHM (2018) Future smart food - rediscovering hidden treasures of neglected and underutilized species for zero hunger in Asia. FAO, Bangkok. Available at: http://www.fao.org/documents/card/en/c/I8907EN/

Limbu P, Thapa K (2011) Chepang food culture: contribution to wild edible and neglected plant species. Local Initiatives for Biodiversity Research and Development, LIBIRD, Pokhara

MoAD (2016) Nepal: zero hunger challenge national action plan (2016–2025). Ministry of Agricultural Development, Kathmandu

MoH (2017) Nepal Demographic and Health Survey 2016. Ministry of Health, Nepal; New ERA; and ICF. Kathmandu, Nepal

Neupane B, Poudel S (2021) Documentation and on farm conservation of neglected and underutilized plant species in Lamjung district, Nepal. Heliyon 7:e05887. https://doi.org/10.1016/j.heliyon.2020.e05887

Padulosi S, Thompson J, Rudebjer P (2013) Fighting poverty, hunger and malnutrition with neglected and underutilized species (NUS): needs, challenges and the way forward. Bioversity International, Rome

Paneru S, Ghimirea S, Adhikarib S, Dhunganac S, Bharatia S (2020) Role of underutilized and neglected plant species to assure food security in Chepang community. Sustain Agric Res 1:6–9

Pangeni B, Bhattarai S, Paudyal H, Chaudhary RP (2020) Ethnobotanical study of Magar ethnic community of Palpa district of Nepal. Ethnobot Res Appl 20:1–17

Pant KK (2018) Neglected vegetables of Nepal: status and need for further research and utilization. Indian J Plant Sci 7:18–21

Panthi B (2021) Present status, importance and challenges of neglected and undeerutilized crops species (NUCS) in Nepal. Socio Econ Policy Stud 1:61–65. https://doi.org/10.26480/seps.02.2021.61.65

Pudasaini R (2013) The role of integrated home gardens and local, neglected and underutilized plant species in food security in Nepal and meeting the Millennium Development Goal 1 (MDG). In: Fanzo J et al (eds) Diversifying food and diets: using agricultural biodiversity to improve nutrition and health. Routledge, New York, pp 242–256

Shah SK (2014) Dietary contribution of underutilized minor crops and indigenous plants collected from uncultivated lands and forests in Nepal. In: Durst P, Bayasgalanbat N (eds) Promotion of underutilized indigenous food resources for food security and nutrition in Asia and the pacific. FAO, Bangkok, pp 46–59

Shrestha D (2013) Indigenous vegetables of Nepal for biodiversity and food security. Int J Biodivers 5:98–108

Shrestha R, Joshi BK, Shrestha J, Timalsina A (2019) Crop groups based on research priority: priority crops, Neglected and Underutilized (NUS), Future Smart Food (FSF). In: Joshi BK, Shrestha R (eds) Working groups of Agricultural Plant Genetic Resources (APGRs). Proceedings of National Workshop. NAGRC, NARC, Kathmandu, pp 65–69

Upadhyay M, Joshi BK (2003) Plant genetic resources in SAARC countries: their conservation and management: Nepal chapter. SAARC Agriculture Information Center. http://www.elibrary.icrisat.org/SATELib_files/SATELibNewsV4I3Mar2007.htm#NAB

Uprety Y, Poudel RC, Shrestha KK, Rajbhandary S, Tiwari NN, Shrestha UB, Asselin H (2012) Diversity of use and local knowledge of wild edible plant resources in Nepal. J Ethnobiol Ethnomed 8:1–15

World Bank (2011) Nepal – Nutrition at a glance. Washington DC; Worldbank, (also available at http://documents.worldbank.org/curated/en/398911468324019161/Nepal-Nutrition-at-a-glance)

Chapter 10
Grain Millet: Potential to Fill Nutrition Gaps in the Context of Food Security and Climate Change

Muhammad Amir, Saeed Akhtar, and Aneela Hameed

10.1 Introduction

Food security is a growing challenge of developing countries, especially for the South Asia and Africa which accounts highest number of undernourished populations. Moreover, climate change witnessed with higher frequency of catastrophic famines, droughts, and floods have significantly affected dietary patterns of peoples in wide geographical settings (Miyan 2015). The production of major cereals failed to meet the primary food needs of peoples in various parts of the world due to the issues of water shortages, frequent pest attacks, and fungal diseases. Consumers' dependency on primary staples such as wheat and rice has also drastically impacted the current agriculture production system and the food biodiversity (Hussain et al. 2020). Inadequate availability of food to ever increasing populations has thus resulted in higher prevalence rate of malnutrition and associated disorders. Such a situation necessitates incorporation of alternate or non-conventional food resources to existing food system (WHO 2020).

Millets are a group of small-seeded cereals with high nutritional value, and have the potential to overcome food and nutritional insecurity in devastating climatic changes (Lakew and Berhanu 2019). Millets are cultivated as minor crop in India, Niger, Nigeria, China, Pakistan, United States, Burkina Faso, Mali, Chad, Ethiopia, and other countries of subcontinent Asia and Africa (Eduru et al. 2021). In the past, millet remained a traditional staple for hundreds of thousands of people in a wide geographical region (Verma et al. 2020). Use of millets as dietary staples historically dates back to 7000 years in the sub-Saharan nomads. Millets are now

M. Amir · S. Akhtar (✉)
Department of Food Science & Technology, Faculty of Food Science & Nutrition, Bahauddin Zakariya University, Multan, Pakistan

A. Hameed
Department of Dairy Technology, Bahauddin Zakariya University, Multan, Pakistan

T. Ismail et al. (eds.), *Neglected Plant Foods of South Asia*,
https://doi.org/10.1007/978-3-031-37077-9_10

cultivated in most parts of the world but their production is steadily declining. It is estimated that 25.7% of the millets crop cultivated area dropped between 1961 and 2018 while per-hectare yield has increased from 575 to 900 kg during the same period (Meena et al. 2021; FAO 2022). The global production of millet is 29,563,000 MT of which 42% is produced in India (FAS USDA 2023). Major varieties of millet include pearl millet, finger millet, teff, proso millet, and foxtail millet while kodo millet, little millet, and barnyard millet are cultivated on a small scale. Millets are annual cereals with drought-resistant potential (Saini et al. 2021). Most of the varieties are adapted to arid and tropical climates and are resistant to diseases and pest attacks due to their unique genetic characteristics and crop physiology. When compared to other cereals such as wheat, rice, and corn, millets yield best under drought conditions and marginal soil fertility (Jayawardana et al. 2019).

The nutritional status of millets makes them potential candidates to overcome malnutrition due to their high carbohydrate and protein contents (Saini et al. 2021). The other unique features of millets are their high calcium, dietary fiber, and polyphenols distinguishing them from the rest of the cereals (Amadou et al. 2013). Millets contain 2–4% minerals, 12–20% dietary fiber, 6–19% protein, 1.5–5% fat, and 60–80% carbohydrates making them a nutrient-dense food (Mahajan et al. 2021; Sharma et al. 2021). The comparative study among millet varieties highlighted that the highest carbohydrates (79.8%) are present in finger millet (Ashwini et al. 2022) followed by proso, barnyard, pearl, kodo, and small millets i.e., 70.4%, 68.8%, 67.0%, 66.6%, and 65.6%, respectively, and lowest carbohydrates contents were found in foxtail millet (Bhatt et al. 2022a, b). Protein contents in different varieties of millet like proso and pearl millet were reported 12.5% and 11.8%, respectively (Anbukkani et al. 2017). Referred study report foxtail millet, barnyard millet, kodo millet, tiny and finger millet to contain 11.2%, 10.5%, 9.8% and 7.3% protein contents, respectively. Micronutrient profiling of indicated millets grains as appreciable carriers of vitamin B1 (5.5 mg/kg), riboflavin (20 mg/kg), niacin (51.1 mg/kg), and antioxidants (Sarita and Singh 2016). Total phenol contents reported in millet grains amounts ~19.7 mg GAE/g and the flavonoid contents ~11.1 mg CE/g. Millets are also carriers of phytosterols with an estimated amount of 58 mg/100 g. Millets are superior to wheat and rice due to their higher contents of essential amino acids and minerals (Sharma et al. 2021; IIMR Annual Report 2021). The mineral contents of millets vary due to their adaptability to diverse soil and climatic conditions. Among microelements reported in different cultivars of millet, the dominant ones are calcium, zinc, iron, phosphorus, sodium, potassium, magnesium, copper and manganese (Abioye et al. 2022). Millets due to their antioxidant, anti-carcinogenic, antimicrobial, and anti-ageing properties are effective against a variety of age-related changes and metabolic disorders such as diabetes mellitus, celiac disease, cardiovascular disease, arthritis, and gut associated disorders (Gupta and Mishra 2021).

Globally 850 million people are suffering from chronic hunger and over one billion are consuming insufficient food with poor supply of either macro or micronutrients. South Asia and central Africa are the most affected zones with 10–30% population suffering from micronutrient deficiency disorders and protein-energy malnutrition. An agreement exists among the world nations to tackle the present and forecasted food insecurity challenges aimed at ensuring the sustainability of effective food production system (Kondeti et al. 2022; Wakjira et al. 2022). Conventional food crops have the tendency to help mitigate the challenges to existing food production system. Year 2023 has therefore been declared as millet year and is intended to promote dietary diversity among urban and rural populations. Such an initiative is also expected to create awareness among the farmers, processors and the consumers on ecological, economical, nutritional and health promoting aspects of the millets (Shah et al. 2023)

10.2 Cultivation Practices of Millets

Agriculture in arid and semi-arid geographical settings is heavily dependent on weather conditions, and its the rainy season that decides the crop cultivation activities such as planting date, tillage practices, and method of sowing. Listed parameters do not merely affect crop yield (Alvar-Beltrán et al. 2020) but also ensure farmers' economy. The unpredictable onset of individual rainfall and huge variability in total rainfall is the major threat to the current agriculture production system and ultimately to food security when production is uncertain and insufficient (Mekonnen et al. 2021). Many sustainable solutions have been introduced to improve agriculture production system and to ensure the sufficient food supply in existing threats of environmental degradation and climate change. Cultivation of conventional or somewhat neglected food crops like millet is considered one amongst the many sustainable solutions to distressed food production system (UN FAO 2017). Millets are C4 plants, and their crop physiology enables them to grow under diverse soil and climatic conditions. These are the most resilient cereals that yield well even in severe abiotic stress i.e., high salinity, less fertile soils, droughts, erratic rainfalls, high temperature, and low soil pH which are the major limiting factors for crop production. Those regions which are highly affected by climate change are best suited for millet cultivation (Xie et al. 2020).

The planting season of millet begins with the start of monsoon rainfall usually in the month of May-July and their harvesting starts in the month of September with 60–90 days of short crop duration (Sood et al. 2020). In general, millet cultivation requires little farm input like manpower, machinery, and fertilizers. It needs minimal land preparation i.e., clearing of a field from shrubs, weeds, and grasses to achieve zero tillage cultivation. Farmers also till the soil using animal-drawn shine hoes, tractors, and power tillers. The method of sowing varies from region to region depending upon the soil and climatic conditions and availability of water. Usually, broadcasting is done after a single plough, but some farmers use rakes to incorporate the seeds into the soil (Singh et al. 2020; Mubiru et al. 2019). Fertilizers are not usually applied for millet cultivation, but some farmers use farmyard manure and handpick the tall grasses to clean the field. No other intermediary operations are needed for crop management until maturity is reached. The crop is harvested manually that is the most challenging aspect of its cultivation. Small-scale farmers thresh the crop manually while mechanical threshing is practiced at large scale (Sonko et al. 2022). The application of fertilizers (NPK) at the rate of 50 kg/ha as basal and 50 kg of urea after 4 weeks of planting as a top dressing has been reported to improve the crop yield to its maximum potential (Pradhan et al. 2017).

Depending upon varietal differences, millets cultivation varies from cultivar to cultivar. The cultivation of pearl millet starts at the onset of the monsoon season from mid-May to the first week of July with an optimum time of early June. The recommended seed rate for pearl millet is 4 kg per acre. The crop grows well in sandy and sandy loam soils (Sher et al. 2019). The seeding depth for pearl millet sowing should be half to 1 inch. Drill sowing of pearl millet creates additional problems as it is difficult to manage seed rates due to its small size and planter depth compared to corn. The crop is well adapted to those areas which are less fertile and characterized by droughts, alkaline and acidic soils (Jagdish 2021) where other cereals fail to produce. Summer crop of pearl millet is well suited for double cropping and rotations.

The sowing time of finger millet starts in May. The crop needs fine bed preparation for its cultivation. Approximately 8–10 tons per hectare of the farmyard is recommended for the soils to attain good moisture retention capacity. In some areas, a nursery is prepared for finger miller which is then transplanted to the field. The long-duration crop remains from May to September while short-duration crop is harvested during the month of December. Weeding is practiced twice after 2–3 weeks of emergence of the seeds and early trimming is recommended after 2 weeks. All types of soils except the coarse soils are suitable for the cultivation of finger millet (Gafoor et al. 2021; Taylor and Duodu 2019).

Foxtail millet is cultivated during the late spring and is a warm-season crop that matures within 75–90 days of cultivation to yield grains. Foxtail millet is a short lifespan crop and is water efficient in dry zones (Zhang et al. 2022). The kodo millet is propagated from seeds and plantation is done instead of broadcast sowing under diverse soil types. Proso millet is a relatively low-demanding crop with unknown diseases.

The diverse nature of millet in terms of its production technology, adaptability to climatic changes and resistance to pest activities make it suitable for cultivation in wide geographical settings. Hence, there is a need to develop a sustainable approach to promote millets cultivation as an environmental and farmers' friendly crop.

10.3 Production Statistics of Major and Minor Millets

Millets are mostly cultivated in dry zones of developing countries and data are scant to report true crop production statistics. However, available set of information on annual production of millets highlight importance of crop in future challenges to global food security. Pearl millet, finger millet, teff, proso millet, and foxtail millet are the majorly grown cultivars of economic and nutritional significance (Wilson and VanBuren 2022). Among all, pearl millet is cultivated on large scale in Africa and

Asia with an estimated production area of 14 and 12 million ha, respectively. In Asia, 10 million ha out of 12 is under millet cultivation in India and constitute almost 50% of the total millet cultivated area in the world. Asia have a major share of millet production in the world than Africa and Europe and other North American regions (Macauley 2015; Khairwal et al. 2007; Azare et al. 2020). Table 10.1 highlights the production statistics of deferent millet verities being cultivated in the region.

Table 10.1 Area under cultivation and production statistics of millets

Millet Type	Country	Year	Cultivation Area/ha	Total production Tons	References
Millet production	Total production	2022	–	30.48 Million Tons	Pavithra (2023)
	India	2022	12.45 m ha,	15.53 MT	Singh et al. (2022c)
	Niger	2020	>7 m ha	3.8 MT	Nourou et al. (2020)
	China	2021	–	2.7 MT	Mundus Agri (2022)
	Nigeria	2021	747,580 ha	2.0 MT	Azare et al. (2020)
	Sudan	2021	250 m ha	1.5 MT	USDA, FAS (2022)
	Senegal	2021	857,973	0.9 MT	Global Yield Gap Atlas (2022)
	Chad	2021	1,180,000 ha	0.689 MT	Index Mundi (2022)
	Russia	2021–22	264,000 ha	0.360 MT	Index Mundi (2022)
	Pakistan	2021	460,000 ha	0.315 MT	Index Mundi (2022)
	Nepal	2021	271,183 ha	0.314 MT	Luitel et al. (2017)
	Ukraine	2022	160,000 ha	0.275 MT	Index Mundi (2022)
Pearl millet	Africa	2020	14 million ha	11.36 MT	Azare et al. (2020)
	India	2019	7.48 m ha	9.26 MT	Malik and PAWAR (2019)
	Niger	2021	>7 million ha	3.808 MT	Nourou et al. (2020)
	Nigeria	2021	747,580 ha	2 MT	Doris (2022)
	Pakistan	2015	0.5 million	0.33 MT	Ullah et al. (2017)
	Bangladesh	2017–18	28.74	0.38 MT	Sarwar and Biswas (2021)
	Brazil	2018	5 million ha	0.8tons/ ha	Dias-Martins et al. (2018)
	Global	2019	29 million	31.723MT	Malik and Pawar 2019
Finger millet	India	2022	15.34 lakh ha	2.354 MT	Arya and Joshi (2022)
	Kenya	–	65,000 ha	84,500 T	Mgonja et al. (2007)
	Uganda	2021	175,173 ha	2.36484 MT	Ekwangu (2021)
	Nepal	2021	263,261 ha	0.237862 MT	Gairhe et al. (2021)
	Ethiopia	2021	455,581 ha	3.834021 MT	Gebreyohannes et al. (2021)
	Africa	2022	–	2 MT	Shinggu et al. (2022)
	Sri Lanka	2018	6000 ha	6946 T	Kumari et al. (2018)

(continued)

Table 10.1 (continued)

Millet Type	Country	Year	Cultivation Area/ha	Total production Tons	References
Paroso millet	United States	2021	267,901 ha	15,376 T	NASS (2022)
	India	2019	0.03 m ha	456,057 T	Vetriventhan et al. (2019)
	Bangladesh	2018–19	809 ha	1000 T	Sarwar and Biswas (2021)
	China	2019	0.32 m ha	–	Vetriventhan et al. (2019)
	Russia	2019	0.82 m ha	1.1 MT	Vetriventhan et al. (2019)
Kodo millet	India	2020	2.44 lakh ha	0.73 lakh tone	Gowri and Shivakumar (2020)
Foxtail millet	China	2021	2 million ha	6 MT	Ma et al. (2016)
	India	2021	0.98 lakh ha	0.56 lakh tone	Das et al. (2021)
	France	2017	–	40,604 T	Knoema 2022
Barnyard millet	India	2018	1.95 lakh ha	1.67 lakh tone	Sujatha (2018)

10.4 Millet's Potential as an Alternative to Conventional Cereals

Wheat, maize, oats, millet, barley, triticale, sorghum and rye are the major cereals of the world. Tremendously increasing population size and higher vulnerability of food crops production system to climate changes are extremely important intervention areas (Alexandratos and Bruinsma 2012). The abrupt variations in temperature, water shortage, and resistance in pest and microbial pathogenic species are responsible for reduced crop yield (Raza et al. 2019). Additionally, an exponential growth in world population that is expected to reach nine billion by 2050 may worsen food security situation especially in the developing economies (Khanal and Mishra 2017). Hence, there is a quest among the scientific community to find resilient crops that may better thrive present era and the future challenges to food crops production (Shah et al. 2021).

Millets are resilient crops that can yield best under diverse environmental conditions and are drought resistant. Small-seeded millets are resilient food crop and it's edible grains are high in dietary fiber, minerals, proteins, and bioactive compounds that may anticipate several health benefits and protect against modern lifestyle-associated chronic and degenerative diseases (Lenka et al. 2020). The bioactive and hypoglycemic profile of millets is important to develop traditional and functional recipes of health significance. Due to the superior functional and nutritional profile, millets based value-added products can boost the functional food market (Sharma et al. 2021). When compared with other cereals, carbohydrate contents of millets are lower but have more fiber. The average dietary fiber contents are 12–20% which is higher than wheat (12.2%), oats (11.8%), barley (14.5%), rye (16.8%), maize (8.1%), and rice (3.5%) (Joye 2020; Hamid et al. 2017). Fiber-rich foods are associated with reduced risks of inflammatory bowel diseases (Liu et al. 2015), cardiovascular complications (Kim and Je 2016), and depression (Miki et al. 2016).

Millets grains are carrier of high-value protein made with amino acids such as glutamic acid that constitute ~23% of the total amino acid pool (Hassan et al. 2021). This non-essential amino acid is the precursor of γ-aminobutyric acid, a neurotransmitter which reduces the excitability of the neurons and inhibits nerve transmission thus maintaining a balance for proper neural functions (de Leon and Tadi 2022). The supplementation of glutamic acid is an alternate therapy to manage estrogen deficiency and reduce the symptoms of menopause (Adebiyi et al. 2017). The contents of essential amino acids like leucine and isoleucine are also higher in millets when compared to oats, wheat, and rice (Wiedemair et al. 2020). The lipid contents of millet grains (6.4%) are more than wheat, maize, sorghum, rye, barley, triticale, and rice but lower than oats (7.6%). Major fatty acids of millets are linoleic (C18:2), oleic (C18:1) and palmitic acid (C16:0) with a share of 39–45%, 21–27% and 20–21%, respectively. Millets grains have unique fatty acids profile that may anticipate hypoglycemic properties (Annor et al. 2015; Taylor 2016) and protection against cardiac and gastrointestinal health disorders. Its fiber-bound chylomicrons promote the health of epithelial cells in the large intestine when fermented by the gut microbiota. The ash contents of millets are more than conventional cereals and like that of barley and oats, and vary depending upon the varietal differences and soil conditions. The minerals in millets are calcium, phosphorus, potassium, magnesium, iron, and zinc i.e., ~300, ~3.34, ~3.93, ~1.33, ~ 18, and ~43 mg/kg dry basis, respectively. The lower contents of iron and zinc are a major concern and limiting factor of millet utilization which can be managed by efforts on promoting fortification and biofortification of millet (Singh et al. 2022a; Gowda et al. 2022; Ullah et al. 2019; Taylor 2016). N'Dri et al. (2013) found that millets have comparatively higher phenolic acids (64.8 mg/kg) than wheat (1171 μg/g), maize (307.57 mg GAE/100 gds), sorghum (27.3 mg/kg), oats (50.5 mg/kg), barley (410 μg/g), and rice (120.13 mg GAE/100 gds). Similarly, the antioxidant capacity of millets is higher than other conventional cereals. Nani et al. (2015) found that the phytochemicals in whole-grain cereals have immunosuppressive potential and can be used as dietary supplements to treat autoimmune diseases.

In the above-discussed context, millets may be an alternative to other cereals due to their diverse nature and adaptability to less fertile soils, droughts, temperature fluctuations, and salinity (Saxena et al. 2018). These superior traits are linked to extensive root system of the millets enabling crop to grow and take nutrients from the deeper soil horizon (Comas et al. 2013). Additionally, millets have fewer chances of contamination by toxigenic mold, other fungal species and pest attacks (Ferrigo et al. 2016).

10.5 Nutritional Features and Food Potential of Different Millets

The nutritional profile of different millet varieties is superior in many aspects to that of conventional cereals. Millets are a source of dietary fiber, carbohydrates, proteins, and minerals and have diverse phytochemical constituents (Ramashia et al. 2021). Pearl millet is the main millet cultivar cultivated in tropical and subtropical regions of the world with 17.4% protein contents while contents of carbohydrates, fat, fiber and the ash have been reported upto 72.2%, 6.3%, 2.8%, and 2.2%, respectively (Fig. 10.1) (Krishnan and Meera 2018). The fatty acid profile of the millet shows linoleic acid (39–45%) is a major fatty acid followed by oleic acid (21–27%) and palmitic acid (20–21%) (Liang et al. 2010; Amare et al. 2021).

Pearl millet is comparatively rich in insoluble and soluble dietary fiber, resistant starches, minerals, and antioxidants. However, these contents vary depending on the soil conditions, climate, genotype and processing operations (Satyavathi et al. 2021). The nutritional composition of finger millet showed that total carbohydrates are present in the range of 71.9–81.5% while moisture, fat, protein, fiber, and ash are 7.5–11.8, 1.2–1.7, 5–9.6, 15–20, and 2.5–3.5%, respectively (Kuruburu et al. 2022: Nakarani et al. 2021). Finger millet is well known for it's nutritional, phytochemical, and low-glycemic index. Its hypoglycemic property is attributed to the fiber contents and complex carbohydrates which are digested and metabolized slowly to anticipate low postprandial serum glucose response. Micro-nutrient profiling of the millet highlights its edible grains as carriers of Ca, Zn, Fe, K, Mn, and P

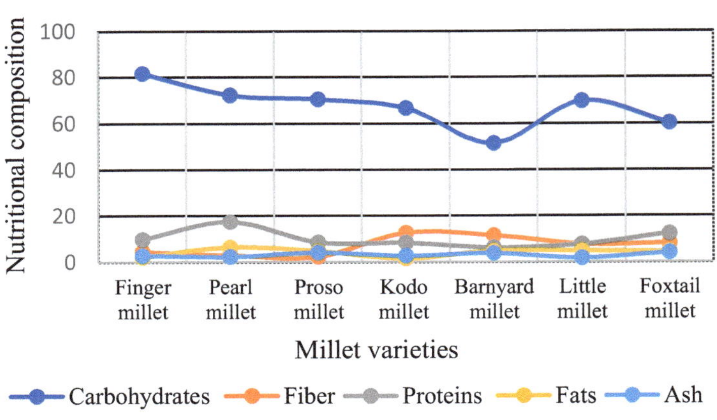

Fig. 10.1 Nutritional composition of millet varieties

at a rate of 2.4–3.4, 0.013–0.018, 0.041–0.048, 3.80–5.20, 0.051–0.062 and 1.60–2.90 mg/g, respectively. Calcium, the major mineral in the finger millet is 10 times greater than reported in other cereals and even 3 times higher than the calcium contents of the milk (Nakarani et al. 2021; Kumar et al. 2022; Muthamilarasan et al. 2016; Pucher et al. 2014).

Finger millet is rich in essential amino acids i.e., tryptophan, cystine, methionine, and aromatic amino acids when compared with other cereals (Maharajan et al. 2021). The crop is also a source of vitamins from the B group like thiamine, folic acid, niacin, and riboflavin (Jan et al. 2022). Barnyard millet is a minor cereal and contains more nutrients than conventional cereals (Shanmugapriya and Nazni 2020). The barnyard millet contains upto 10.5% protein while carbohydrates, fats and crude fiber contents have been reported in a range between 51.5–68%, 3.6–3.8% and 5.41–6.8%, respectively. Baryard millet essential fatty acids include oleic acid, linoleic acid and palmitic acid. The grain also compounds minerals like magnesium, calcium and iron (Ugare et al. 2014; Singh et al. 2022a, b). The nutritional profile of kodo millet indicates its edible grains as carriers of 66.6% carbohydrates, 8.3% protein, 1.4% fats, and 2.6% of inorganic residues. The dietary fiber contents of kodo millet vary from as low as 12.5% to as high 37–38% making it unique among all cereals. Despite of lower fat contents kodo millet is a good source of polyunsaturated fatty acids.

Proso millet is mostly cultivated in India and is the best source of carbohydrates (Bunkar et al. 2021). It's carbohydrate content varies from 70.4 to 76.7%. Other nutrients include fiber (2.2%), proteins (8.5–11.6%), fats (4.6%), and ash (4%) on dry weight basis. Levels of nutrients observed in Proso millet are comparable to the one reported in wheat and other cereals. The protein contents of proso millet are similar to that of wheat but rich in essential amino acids including methionine, isoleucine and leucine. Millets are therefore suitable for large-scale use in the production of food products like snack foods, baby foods, and dietary food because they contain all the necessary nutrients (Gowda et al. 2022). Additionally, diverse millet products have entered people's daily lives, such as millet nutrition powder, millet wine, and millet porridge which are developed either with the grain or its flour. Other promising nutritional and health attributes of the millet are that they are gluten-free, low glycemic index, and are promising carriers of high-quality protein and nutraceuticals. Little millet, another type of millet is conventionally fed to birds and animals as feed and fodder (Bhatt et al. 2022a, b). Nutritional composition of little millets indicates it's grains to hold carbohydrates, protein, fats and ash contents in a range between 69.70% to 78.53%, 6.87% to 7.26%, 4.64% to 4.70%, 4.74% to 5.75%, respectively. The utilization of little millet is limited due to the coarseness of edible grains, their low palatability and less diversified application in food preparations. In preview of the literature cited above on nutritional aspects of various types of millets, the conclusion can be drawn suggesting conventional and underutilized cultivars of millets as nutrient dense food crops which can help bridging food security gaps and ensure sustainability in food supply to the peoples at risk of food insecurity and malnutrition.

10.6 Resilience to Climate Change

The diversity in molecular, morpho-physiological and biochemical traits enables millets to better tolerate environmental stress than other cereals. Comparing with rice and wheat, millets have a short life cycle of 12–14 weeks. Plant characteristics such asthick cell wall, small leaf area, short stature, and deep root system enhance adoptability of millet to stress conditions (Saleem et al. 2021). The C4 photosynthetic mechanism of millets is advantageous to efficiently use CO_2 and inhibit photorespiration of ribulose 1,5-bisphosphate (RuBP) through ribulose-1,5-bisphosphate carboxylase/oxygenase (RuBisCO) pathway (Aubry et al. 2011). C4 mechanism increases the amount of CO_2 in the bundle sheath, reduces photorespiration by 80% (temperature dependent) and raises the catalytic activity of RuBisCO in plants (Sage et al. 2011).

Millets have improved photosynthetic rates in warm climates and confer immediate water and nitrogen use efficiency (NUE), which is 1.5–4 times higher than C3 photosynthesis due to RuBisCO of C4 plants working at elevated CO_2 levels (Sage and Zhu 2011). For instance, wheat and maize need 510 and 470 g of water to produce 1 g of dry biomass while foxtail millet only needs 257 g to produce similar amount of dry mass (Bandyopadhyay et al. 2017). Millets receive additional benefits from C4 photosynthesis in addition to WUE and NUE, such as greater environmental enactment in warm climates, more flexible biomass allocation patterns, and decreased hydraulic conductivity per unit leaf area (Sage and Zhu 2011).

Millets have developed a special set of adaptations to prevent, tolerate, survive and recover from stress caused by water deficiency. When compared to C3 cereals and other grain crops, sorghum, finger millet, and teff have water usage efficiencies (WUE) of 4.2–13.4 kg yield per hectare per millimetre of rain. In contrast to maize and winter wheat that need 500–900 mm of rainfall, millets merely need 200–500 mm of rainfall. High WUE helps maintain soil moisture content (Wilson and VanBuren 2022). Some millets such barnyard millet and fonio can mature in just 8 weeks and are completely drought resistant. Millets can escape unpredictable summer droughts and serve as a rescue crop when major cereals fail to yield. Millet has better anatomical and physiological adoptability, enabling it better withstand dehydration and low critical leaf water potential until the saturation reaches (Numan et al. 2021). Such adoptability factors render millets as next-generation crop attracting researchers to explore their resilient traitsto improve major cereals. One such intervention is made to engineer C4 features in rice using millet as models; nonetheless, it is essential for quick progress in creating crop species that can withstand climate change and have the potential of stress tolerance (Bandyopadhyay et al. 2017).

10.7 Millets in Malnutrition and Different Health Conditions

Millets due to their promising nutritional and phytochemical profile, and good dietary fiber contents, are considered Nutra cereals. Higher fiber contents of millets make them good laxatives and reduce the risks of gallstones development and alleviate issues including constipation, bloating, cramping and other issues of lower gastrointestinal tract. The fermentable dietary fibers of millet may also serve as prebiotic thus promoting probiotics growth and help eliminating pathogenic strains from the colon (Ambati and Sucharitha 2019). Oleic acid dominated millets' oil and their dietary fibers improve health of the epithelial cells of the colon. Unsaturated fatty acids like linoleic and oleic (50–60%) increase the level of HDL cholesterol and are associated with improved cardiovascular health. Glutamic acid a non-essential amino in millet is the precursor of gamma-aminobutyric acid and can prevent nerve excitability and reduced neural activity thus managing the health of the nervous system (Anitha et al. 2021).

The main polyphenols found in millet include a diversified range of phenolic acids and tannins which are believed to act as antioxidants and boost the immune system, and are considered effective in alleviating the risks of diabetes, cancer and heart diseases. Additionally, the polyphenols found in the seed coat of finger millet have antibacterial activity against *Bacillus cereus*. The enzymatic hydrolysis of complex carbohydrates can be inhibited by millet phenolics, which also inhibit malt amylase, glucosidase, and pancreatic amylase, all of which lower postprandial hyperglycemia. Millets as plausible carrier of plant calcium is considered beneficial for the bone health of growing young children, the elderlies, and pregnant women (Hassan et al. 2021). Millets grains consumption helps lactating women produce enough breast milk. Millet eases hip and knee discomfort and helps women's periods become more regular. Millets are good reserve of potassium that anticipate vasodilation properties and reduces blood pressure (SINHA et al. 2022).

Dietary intake of this non-conventional cereal protects against the risks of cardiovascular diseases and improves the circulatory system. Phytic acid and C-glycosyl flavones such apigenin, glucosylvitexin, glucosylorientin, and vitexin are the principal antinutrients in pearl millet (Taylor 2016). Most cereals include phytic acid, a natural antioxidant that prevents oxidation reactions in the seeds and helps the body retain inorganic phosphorus when ingested. However, due to their chelating activity with divalent mineral ions, heavy consumption of phytates may yield reduced bioavailability of zinc, calcium, manganese and other divalent ions (Gupta et al. 2017). Additionally, phytic acid may also attribute reduction in protein digestibility by limiting biological activity of proteases. The phytic acid concentration of millet grains is higher than that of oats, rice, barley, rye, and wheat, and it is comparable to that of cereals like maize and sorghum. Various authors claim that the phytic acid content of pearl millet can range from 588 mg/100 g to 1382 mg/100 g (Gabaza et al. 2018).

Dietary consumption of raw or unprocessed pearl millet grains may risk in reduced thyroid hormones production. This may relate to presence of large amounts of C-glycosyl flavones (1020 mg/kg), specifically glucosyl vitexin, glucosyl orientin, and vitexin and raw pearl millet grains. Scarcity of data on upper intake of raw or unprocessed millets grains or foods and its association with risks of reduced thyroid hormones production attracts researchers of the recent era to generate valid information on toxicological risks associated with millets consumption. Available literature also supports research on developing culturally acceptable solutions to mitigate millets' antinutrients and improve nutrient delivering potential of the crop (Babiker et al. 2020).

10.8 Recent Research on Mainstreaming Millets to Ensure Food Security in Pandemics

The root causes of food insecurity include but not limited to the ever-changing climate, pandemics like COVID-19 and conflicts which are critically influencing agriculture production system. A vast majority of the malnourished people in developing countries rely on agriculture for employment, income and food. Climate change and the COVID 19 pandemic has generated concerns on importance of integrating modern agricultural practices with low input and climate-resilient crops (Kakaei et al. 2022). During the recent era witnessing drastic climate changes and the pandemics like COVID 19, current staple food crops have shown their inadequacy to meet the global food security challenges. In their regions of cultivation, orphan millets are considered essential for the livelihood, food security, and nutrition of poor and malnourished communities.

Unconventional food sources like millets can effectively bridge dietary gaps linked with inadequate production of traditional staple crops, particularly in areas where food shortage is persistent. Better immunological health guarantees body's inherent capacity to counter microbiological infections. Millets, contrary to the conventional cereals provide more micronutrients which bear immune enhancing properties and are attributed as immune boosters (Muthamilarasan and Prasad 2021). Higher dietary fiber contents in the millets improve gut microbiota playing an essential role in maintaining a healthy gut microbiota profile and regulating immune functions (Rodríguez et al. 2020).

Millets are compounded with antioxidants such as curcumin, quercetin, ellagic acid, and catechins. Moreover, millet helps detoxification of the body from the factors like exogenous and indigenous free radicals, toxins, and neutralizing enzymes, and prevents risks of heart diseases and cancers. Antioxidants in millets have been reported significant in alleviating oxidative stress linked damages to the body (Liang and Liang 2019). Many of the COVID-19 symptoms including pneumonia, bronchitis, pulmonary inflammation, fibrosis, and fever are mediated by proinflammatory cytokines which can be mediated by phytochemicals in a better fashion (Makuvara 2022).

In addition to help ensure better food and nutrition security, millets production ensures fodder supplies to farm animals. Small millets in addition to their primary role of supplying food needs to the underprivileged farming, play an important role in supplying fodder for animals and increasing agro-biodiversity. With the facts reporting millets to supply 2000–3000 calories a day, tiny millets and other cultivars can help manage hunger hotspots (Muthamilarasan and Prasad 2021). UN identified "hunger hotspots" are geographically well suited for the cultivation of minor millets. Prioritizing millets cultivation in these regions under biodiversification model would not merely help achieving zero hunger score but also improve immunological and overall health indices of its consumers.

10.9 Challenges and Opportunities to Improve Millets Production in South Asia

In this recent era, ensuring adequate food supply to ~9 billion peoples of the world is the major challenge to our food production system. The 60% of the dietary requirements of the global population are being met with only a few crops such as wheat, rice, and maize which have shown tremendous vulnerability to the changing global environment. The next concern to climatic variations is the increased biodiversity and resistant in pests and microbial pathogens generating more than 30% per annum loss to the crop produced (Ali et al. 2017). Simultaneously, the worldwide increasing population has also increased pressure on plant breeders to meet increasing food demands by continuous improvement in yield of major food crops. These genetically modified food crops are unable to withstand environmental fluctuations and diseases. It further highlights suitability of non – conventional food crops like millets to the genetic improvement and bridging dietary gaps (Kavhiza et al. 2022).

The millets are highly nutritious and climate resilient crops (Bose 2018). Pearl millet is the main millet variety capable of producing grains under the extreme environments i.e., hottest climate, saline and acidic soils, erratic rainfalls, and low fertility rates. However, some constant threats such as biotic stress (blast and downy mildew), and abiotic stresses (heat stress at flowering and drought at seedling stage) still exist that can affect the yield of the pearl millet (Serba et al. 2020). The novel genes in germ plasm are being search out by the breeders to improve the biotic and abiotic stress tolerance in millets. Likewise, finger millet is major cereal crop in eastern Africa and southern Asia with extended storage period, extraordinary nutraceutical features, and exclusive capability to grow under semiarid and arid environments. Despite of these properties, the production rate of finger millet is very low, and it vary from region to region i.e., 300–695 kg/ha (Gupta et al. 2017). The general limiting factor to millet production is the infestation of Striga weed, less soil fertility, fungal diseases i.e., stem blast that can cause up to 50% yield loss. The blast disease instigated by the *Magnaporthe oryzae* (filamentous ascomycetous fungi) is considered as most damaging disease the stunt the growth at various stages and ultimately yield of finger millet in all millet producing areas. The evolving capability of virulent genes of different pathogens breakdown the blast resistance of finger millet by making it more susceptible and leads to yield instability in its all growing regions (Gebreyohannes et al. 2021). The current researches in South Asia and Africa solved the emerging problem of stem blast and have mitigated the production limitations of finger millets. These efficient and novel strategies can be implemented to gain durable and dynamic resistance against various virulent genes of pathogens and extensive ranges of agro-ecological zones ensuring the sustainable solution to production of food crops in the future (Mbinda and Masaki 2021).

Millet crops can provide dual benefits such as it has the potential to generates the higher capital for the farmers and can improve the health of the rising population. The scientific interventions can play a game changing role to overcome the limitation in acceptability and anti-nutritional components of the millets. The anti-nutritional contents of the different millet varieties can be reduced by using various processing techniques i.e., use of fermentations, germination, roasting and cooking (Kumar et al. 2018). Likewise, the acceptability rate of millet and millet-based products can be heightened by using the millet flour in different concentrations along with various other cereals having higher acceptability rates. Consequently, the utilization of millet and its flour in manufacturing of various packaged and commercial food products can inspire the farmers to sow millets and can also invigorate the famers.

Moreover, the various nutrients i.e., iron, calcium and protein deficiencies can be managed by introducing millet-based food products at state, national and international levels in developing nations. The planning and development of new strategies are needed to maximize the outputs to explore the opportunities, proper utilization and marketing of millets. The work on this opportunity can guarantee that millet can enhance the nutrition and food security in all over the world (Andreotti et al. 2022).

10.10 Conclusion

Millets have diverse environmental adoptability due to improved water and nitrogen usage efficiency, resistance to insect pests and pathogenic attacks, and climatic stresses. Millet crop is an ideal candidate for cultivation in UN-WFP-identified less fertile hotspots. Balanced nutritional profile and climate-resilient characteristics of millets have been widely explored highlighting their potential as the next generation staples to cope with the future risks of hunger. Global estimates on food requirements for future population load indicate the number of people experiencing acute food insecurity is expected to double necessitating urgent agricultural reforms to cope with the growing challenges of food insecurity and hunger. Maintaining nutritional security is also equally important to ensure health and nutritional well-being of the vulnerable populations. Mainstreaming small millets in conventional food crops production system of the regions at highest risk of food insecurity can help ensure sustainable food supply. Moreover, technical advancement is a necessity to alleviate the current production challenges of small millets that can significantly improve crop adoptability and marketing. Likewise, genetic improvement program may be placed in to reduce the intrinsic burden of phytates, tannins and other nutrients for better bodily utilization of millets grains. Using next-generation genomic methods including precise editing of pathway genes by the CRISPR/Cas9 approach also helps the production of superior millets cultivars. UN-FAO has declared the year 2023 as the International Year of Millets to improve crop production and wider consumer acceptability. The targets thus may be achieved by engagement of mainstream stakeholders including farmers, grain market, industry, government and the community organizations including non-governmental organizations.

Conflicts of Interest All authors read and approved the final manuscript and declare that they have no conflict of interest.

References

Abioye VF, Babarinde GO, Ogunlakin GO, Adejuyitan JA, Olatunde SJ, Abioye AO (2022) Varietal and processing influence on nutritional and phytochemical properties of finger millet: a review. Heliyon 8:1–15

Adebiyi JA, Obadina AO, Adebo OA, Kayitesi E (2017) Comparison of nutritional quality and sensory acceptability of biscuits obtained from native, fermented, and malted pearl millet (Pennisetum glaucum) flour. Food Chem 232:210–217

Agricultural Statistics by country, Agricultural production, supply and distribution. indexmundi.com (2022) https://www.indexmundi.com/agriculture/?country=ua&commodity=millet&graph=production

Alexandratos N, Bruinsma J (2012) World agriculture towards 2030/2050: the 2012 revision. https://www.fao.org/3/ap106e/ap106e.pdf

Ali S, Liu Y, Ishaq M, Shah T, Ilyas A, Din IU (2017) Climate change and its impact on the yield of major food crops: evidence from Pakistan. Foods 6:39

Alvar-Beltrán J, Dao A, Dalla AM, Heureux A, Sanou J, Orlandini S (2020) Farmers' perceptions of climate change and agricultural adaptation in Burkina Faso. Atmos 11:827

Amadou I, Gounga ME, Le GW (2013) Millets: nutritional composition, some health benefits and processing-a review. Emir J Food Agric 7:501–508

Amare E, Grigoletto L, Corich V, Giacomini A, Lante A (2021) Fatty acid profile, lipid quality and squalene content of teff (Eragrostis teff (Zucc.) Trotter) and Amaranth (Amaranthus caudatus L.) varieties from Ethiopia. Appl Sci 11:3590

Ambati K, Sucharitha KV (2019) Millets-review on nutritional profiles and health benefits. Int J Recent Sci Res 10:33943–33948

Anbukkani P, Balaji SJ, Nithyashree ML (2017) Production and consumption of minor millets in India-a structural break analysis. Ann Agric Res New Series 38:1–8

Andreotti F, Bazile D, Biaggi C, Callo-Concha D, Jacquet J, Jemal OM, King OI, Mbosso C, Padulosi S, Speelman EN, Noordwijk M (2022) When neglected species gain global interest: Lessons learned from quinoa's boom and bust for teff and minor millet. Glob Food Sec 32:100613

Anitha S, Botha R, Kane-Potaka J, Givens DI, Rajendran A, Tsusaka TW, Bhandari RK (2021) Can millet consumption help manage hyperlipidemia and obesity?: a systematic review and meta-analysis. Front Nutr 8:478

Annor GA, Marcone M, Corredig M, Bertoft E, Seetharaman K (2015) Effects of the amount and type of fatty acids present in millets on their in vitro starch digestibility and expected glycemic index (eGI). J Cereal Sci 64:76–81

Arya C, Joshi J (2022) Production of ragi and its prevalence aspects in Almora district (Uttarakhand). EPRA, ARER 10:1–7

Ashwini A, Prashanth SJ, Jammanakatti SV, Nayak S (2022) Quality attributes of composite flour of white finger millet (Eleusine coracana) and wheat (Triticum aestivum) for bread dough preparation. J Postharvest Technol 10:42–58

Aubry S, Brown NJ, Hibberd JM (2011) The role of proteins in C3 plants prior to their recruitment into the C4 pathway. J Exp Bot 62:3049–3059

Azare IM, Dantata IJ, Abdullahi MS, Adebayo AA, Aliyu M (2020) Effects of climate change on pearl millet (Pennisetum glaucum [LR Br.]) production in Nigeria. J Appl Sci Environ Manag 24:157–162

Babiker A, Alawi A, Al Atawi M, Al Alwan I (2020) The role of micronutrients in thyroid dysfunction. Sudan J Paediatr 20:13

Bandyopadhyay T, Muthamilarasan M, Prasad M (2017) Millets for next generation climate-smart agriculture. Front Plant Sci 8:1266

Bhatt D, Fairos M, Mazumdar A (2022a) Millets: nutritional composition, production and significance: a review. J Pharm Innov 11:1577–1582

Bhatt D, Rasane P, Singh J, Kaur S, Fairos M, Kaur J, Gunjal M, Dipendra M, Mehta CM, Avinashe H, Sharma N (2022b) Nutritional advantages of barnyard millet and opportunities for its processing as value-added foods. J Food Sci Technol:1–13. https://doi.org/10.1007/s13197-022-05602-1

Bose HK (2018) Farmers turn to millets as a climate-smart crop. Alternatives India. https://vikalp-sangam.org/article/farmers-turn-to-millets-as-a-climate-smart-crop/

Bunkar DS, Goyal SK, Meena KK, Kamalvanshi V (2021) Nutritional, functional role of kodo millet and its processing: a review. Int J Curr Microbiol App Sci 10:1972–1985

Comas LH, Becker SR, Cruz VMV, Byrne PF, Dierig DA (2013) Root traits contributing to plant productivity under drought. Front Plant Sci 4:442

Das IK, Krishnan S, Tonapi VA (2021) Current status and management of foxtail millet [Setaria italica (L.) Beauv.] blast disease. In: Blast disease of cereal crops. Springer, Cham, pp 201–209

de Leon AS, Tadi P (2022) Biochemistry, gamma aminobutyric acid. In: StatPearls. StatPearls Publishing. https://pubmed.ncbi.nlm.nih.gov/31869147/

Dias-Martins AM, Pessanha KLF, Pacheco S, Rodrigues JAS, Carvalho CWP (2018) Potential use of pearl millet (Pennisetum glaucum (L.) R. Br.) in Brazil: food security, processing, health benefits and nutritional products. Food Res Int 109:175–186

Doris DS (2022) Statista. Percentage change in production volume of millet in Nigeria from 2005 to 2022(in 1,000 metric tons). https://www.statista.com/statistics/1134509/percentage-change-in-production-of-millet-in-nigeria

Eduru A, Kamboj A, Reddy PM, Pal B, Vikas (2021) Nutritional and health benefits of millets, present status and future prospects: a review. Pharma Innov J 10:859–868

Ekwangu J (2021) Fertilizer micro dosing and timing of weeding for improved agronomic efficiency in finger millet production in eastern Uganda. Doctoral dissertation, Makerere University

FAO (2022) Food outlook biannual report on global food markets. https://www.fao.org/3/cb9427en/cb9427en.pdf

Ferrigo D, Raiola A, Causin R (2016) Fusarium toxins in cereals: occurrence, legislation, factors promoting the appearance and their management. Molecules 21:627

Food and Agriculture Organization of the United Nations (2017) The future of food and agriculture: trends and challenges. FAO. https://www.fao.org/3/i6583e/i6583e.pdf

Foreign Agriculture Service United State Department of Agriculture (2022) Ukraine millet production. https://ipad.fas.usda.gov/countrysummary/Default.aspx?id=UP&crop=Millet

Foreign Agricultural Services, United State Department of Agriculture (2023) Millet Explorer. https://ipad.fas.usda.gov/cropexplorer/cropview/commodityView.aspx?cropid=0459100

Gabaza M, Shumoy H, Muchuweti M, Vandamme P, Raes K (2018) Iron and zinc bioaccessibility of fermented maize, sorghum and millets from five locations in Zimbabwe. Food Res Int 103:361–370

Gafoor RA, Pillai S, Nishan MA (2021) Effect on integrated nutrient management on finger millet (Eleusine coracana (L.) Gaertn.): a review. J Pharmacogn Phytochem 10:292–298

Gairhe S, Gauchan D, Timsina KP (2021) Prospect and potentiality of finger millet in Nepal: nutritional security and trade perspective. J Agric Nat Resour 4:63–74

Gebreyohannes A, Shimelis H, Laing M, Mathe I, Odeny DA, Ojulong H (2021) Finger millet production in Ethiopia: opportunities, problem diagnosis, key challenges and recommendations for breeding. Sustainability 13:13463

Global Yield Gap Atlas (2022) https://www.yieldgap.org/senegal

Gowda NN, Siliveru K, Prasad PV, Bhatt Y, Netravati BP, Gurikar C (2022) Modern processing of Indian millets: a perspective on changes in nutritional properties. Foods 11:499

Gowri MU, Shivakumar KM (2020) Millet scenario in India. Econ Aff 65:363–370

Gupta E, Mishra P (2021) Functional food with some health benefits, so called superfood: a review. Curr Nutr Food Sci 17:144–166

Gupta SM, Arora S, Mirza N, Pande A, Lata C, Puranik S, Kumar J, Kumar A (2017) Finger millet: a "certain" crop for an "uncertain" future and a solution to food insecurity and hidden hunger under stressful environments. Front Plant Sci 8:643

Hamid A, Ilyas M, Kalsoom S, Bhatty N (2017) Effects of wheat bran diet and maize bran diet on the random blood glucose and weight of alloxan induced diabetic rats. J Anim Plant Sci 27:325–330

Hassan ZM, Sebola NA, Mabelebele M (2021) The nutritional use of millet grain for food and feed: a review. Agric Food Secur 10:1–14

Hussain M, Butt AR, Uzma F, Ahmed R, Irshad S, Rehman A, Yousaf B (2020) A comprehensive review of climate change impacts, adaptation, and mitigation on environmental and natural calamities in Pakistan. Environ Monit Assess 192:1–20

ICAR-Indian Institute of Millets Research (2021) Annual report 2021. Rajendranagar, Hyderabad, India. https://www.millets.res.in/annual_report/ar21-22.pdf

Jagdish (2021) Millet farming; millet cultivation practices. Agri farming. https://www.agrifarming.in/millets-farming-millets-cultivation-practices

Jan S, Kumar K, Yadav AN, Ahmed N, Thakur P, Chauhan D, Dhaliwal HS (2022) Effect of diverse fermentation treatments on nutritional composition, bioactive components, and anti-nutritional factors of finger millet (Eleusine coracana L.). J Appl Biol Biotechnol 10:4–2. https://doi.org/10.7324/JABB.2022.10s107

Jayawardana SAS, Samarasekera JKRR, Hettiarachchi GHCM, Gooneratne J, Mazumdar SD, Banerjee R (2019) Dietary fibers, starch fractions and nutritional composition of finger millet varieties cultivated in Sri Lanka. J Food Compos Anal 82:103249

Joye IJ (2020) Dietary fibre from whole grains and their benefits on metabolic health. Nutrients 12:3045

Kakaei H, Nourmoradi H, Bakhtiyari S, Jalilian M, Mirzaei A (2022) Effect of COVID-19 on food security, hunger, and food crisis. In: COVID-19 and the sustainable development goals. Elsevier, pp 3–29

Kavhiza NJ, Zargar M, Prikhodko SI, Pakina EN, Murtazova KMS, Nakhaev MR (2022) Improving crop productivity and ensuring food security through the adoption of genetically modified crops in sub-Saharan Africa. Agronomy 12:439

Khairwal IS, Rai KN, Diwakar B, Sharma YK, Rajpurohit BS, Nirwan B, Bhattacharjee R (2007) Pearl millet: crop management and seed production manual. International Crops Research Institute for the Semi-Arid Tropics, Patancheru, p 104

Khanal AR, Mishra AK (2017) Stock price reactions to stock dividend announcements: a case from a sluggish economic period. N Am J Econ Finance 42:338–345

Kim Y, Je Y (2016) Dietary fibre intake and mortality from cardiovascular disease and all cancers: a meta-analysis of prospective cohort studies. Arch Cardiovasc Dis 109:39–54

Knoema (2022) https://knoema.com/FAOPRDSC2020/production-statistics-crops-crops-processed?tsId=1125550
Kondeti AK, Ragimekula N, Ragavaiah RV (2022) Performance of improved pearl millet var. ABV-04 in Alfisols of Ananthapur district of Andhra Pradesh. Pharma Innov J 11:720–723
Krishnan R, Meera MS (2018) Pearl millet minerals: effect of processing on bioaccessibility. J Food Sci Technol 55:3362–3372
Kumar A, Tomer V, Kaur A, Kumar V, Gupta K (2018) Millets: a solution to agrarian and nutritional challenges. Agric Food Secur 7:1–15
Kumar RR, Singh SP, Rai GK, Krishnan V, Berwal M, Goswami S, Vinutha T, Mishra GP, Satyavathi CT, Singh B, Praveen S (2022) Iron and zinc at a cross-road: a trade-off between micronutrients and anti-nutritional factors in pearl millet flour for enhancing the bioavailability. J Food Compos Anal 111:104591
Kumar RR, Singh SP, Rai GK, Krishnan V, Berwal M, Goswami S, Vinutha T, Mishra GP, Satyavathi CT, Singh B, Praveen S (2022) Iron and zinc at a cross-road: a trade-off between micronutrients and anti-nutritional factors in pearl millet flour for enhancing the bioavailability. J Food Compos Anal 111:104591
Kumari WMR, Pushpakumara DKNG, Weerakoon WMW, Senanayake DMJB, Upadhyaya HD (2018) Morphological characterization of local and introduced finger millet (Elusine coracana (L.) Gaertn) germplasm in Sri Lanka. Trop Agric Res 29:167–183
Kuruburu MG, Bovilla VR, Naaz R, Leihang Z, Madhunapantula SV (2022) Variations in the anticancer activity of free and bound phenolics of finger millet (Eleusine coracana (L) Gaertn; variety KMR-301) seeds. Phytomedicine Plus 2:100276
Lakew A, Berhanu T (2019) Determination of intra and inter row spacing on the yield of pearl millet (Pennisetum glaucum L.) in the dry land areas of Wag Himra, eastern Amhara, Ethiopia. Arch. agric. environ. sci 4: 45–49.
Lenka B, Kulkarni GU, Moharana A, Singh AP, Pradhan GS, Muduli L (2020) Millets: promising crops for climate-smart agriculture. Int J Curr Microbiol App Sci 9:656–668
Liang S, Liang K (2019) Millet grain as a candidate antioxidant food resource: a review. Int J Food Prop 22:1652–1661
Liang S, Yang G, Ma Y (2010) Chemical characteristics and fatty acid profile of foxtail millet bran oil. J Am Oil Chem Soc 87:63–67
Liu X, Wu Y, Li F, Zhang D (2015) Dietary fiber intake reduces risk of inflammatory bowel disease: result from a meta-analysis. Nutr Res 35:753–758
Luitel DR, Siwakoti M, Jha PK, Jha AK, Krakauer N (2017) Climate, distribution and cropping pattern of finger millet in Nepal: a review. Int J Agric Environ Res 3:3854–3868
Ma Z, Yang X, Zhang C, Sun Y, Jia X (2016) Early millet use in West Liaohe area during early-middle Holocene. Sci China Earth Sci 59:1554–1561
Macauley H (2015) Feeding Africa: cereal crops: rice, maize, millet, sorghum, wheat. Abdou Diouf International Conference Centre, Senegal, p 74
Mahajan P, Bera MB, Panesar PS, Chauhan A (2021) Millet starch: a review. Int J Biol Macromol 180:61–79
Maharajan T, Ceasar AS, Krishna ATP, Ignacimuthu S (2021) Finger millet [Eleusine coracana (L.) Gaertn]: an orphan crop with a potential to alleviate the calcium deficiency in the semi-arid tropics of Asia and Africa. Front Sustain Food Syst 5:684447
Makuvara Z (2022) Millets (pearl and finger) as nutritional interventions for COVID 19 with focus on Zimbabwe: mini review. Cogent Food Agric 8:2111060
Malik DP, PAWAR N (2019) Production, returns & policy initiatives for pearlmillet in India. Green Farming 10:683–689
Mbinda W, Masaki H (2021) Breeding strategies and challenges in the improvement of blast disease resistance in finger millet. A current review. Front Plant Sci 11:602882
Meena RP, Joshi D, Bisht JK, Kant L (2021) Global scenario of millets cultivation. In: Millets and millet technology. Springer, Singapore, pp 33–50
Mekonnen A, Tessema A, Ganewo Z, Haile A (2021) Climate change impacts on household food security and farmers adaptation strategies. J Sci Agric Food 6:100197

Mgonja MA, Lenné JM, Manyasa E, Sreenivasaprasad S (eds) (2007) Finger millet blast management in East Africa. Creating opportunities for improving production and utilization of finger millet. In: Proceedings of the First International Finger Millet Stakeholder Workshop, Projects R8030 & R8445 UK Department for International Development – crop protection programme held 13–14 September 2005 at Nairobi. International Crops Research Institute for the Semi-Arid Tropics, Patancheru, p 196

Miki T, Eguchi M, Kurotani K, Kochi T, Kuwahara K, Ito R, Kimura Y, Tsuruoka H, Akter S, Kashino I, Mizoue T (2016) Dietary fiber intake and depressive symptoms in Japanese employees: the Furukawa Nutrition and Health Study. Nutrition 32:584–589

Millet Production by Country in 1000 MT. indexmundi.com (2022) https://www.indexmundi.com/agriculture/?country=ru&commodity=millet&graph=production

Millet Area Harvested by Country in 1000 HA. indexmundi.com (2022) https://www.indexmundi.com/agriculture/?country=td&commodity=millet&graph=area-harvested

Millet: acreage in Ukraine shrinks to 43,000 ha. Mundus Agri (2022) https://www.mundus-agri.eu/news/millet-acreage-ukraine-shrinks-43000-ha.n29916.html

Miyan MA (2015) Droughts in Asian least developed countries: vulnerability and sustainability. Weather Clim Extrem 7:8–23

Mubiru DN, Namakula J, Otim GA, Kashagama J, Nakafeero M (2019) Operational field guide for establishing and managing conservation agriculture demonstrations, Uganda 2nd edn. CIMMYT, pp 1–86

Muthamilarasan M, Prasad M (2021) Small millets for enduring food security amidst pandemics. Trends Plant Sci 26:33–40

Muthamilarasan M, Dhaka A, Yadav R, Prasad M (2016) Exploration of millet models for developing nutrient rich graminaceous crops. Plant Sci 242:89–97

N'Dri D, Mazzeo T, Zaupa M, Ferracane R, Fogliano V, Pellegrini N (2013) Effect of cooking on the total antioxidant capacity and phenolic profile of some whole-meal African cereals. J Sci Food Agric 93:29–36

Nakarani UM, Singh D, Suthar KP, Karmakar N, Faldu P, Patil HE (2021) Nutritional and phytochemical profiling of nutracereal finger millet (Eleusine coracana L.) genotypes. Food Chem 341:128271

Nani A, Belarbi M, Ksouri-Megdiche W, Abdoul-Azize S, Benammar C, Ghiringhelli F, Hichami A, Khan NA (2015) Effects of polyphenols and lipids from Pennisetum glaucum grains on T-cell activation: modulation of Ca2+ and ERK1/ERK2 signaling. BMC Complement Altern Med 15:1–11

NASS U (2022) Crop production 2021 summary (January 2022). USDA, Washington, DC. https://downloads.usda.library.cornell.edu/usda-esmis/files/k3569432s/sn00c1252/g158cj98r/cropan22.pdf

Nourou AIM, Saidou AK, Arifa W, Abdoulaye AO, Aune JB (2020) Intensification of pearl millet production in Niger through mechanized sowing and weeding, seed priming, seed treatment, and microdosing. Agronomy 10:629

Numan M, Serba DD, Ligaba-Osena A (2021) Alternative strategies for multi-stress tolerance and yield improvement in millets. Genes 12:739

Pakistan Millet Area Harvested (1000HA) indexmundi.com (2022). https://www.indexmundi.com/agriculture/?country=pk&commodity=millet&graph=area-harvested

Pavithra KM (2023) Review: in the 'international year of millets', here is everything you need to know about millets. https://factly.in/review-in-the-international-year-of-millets-here-is-everything-you-need-to-know-about-millets/

Pradhan A, Bhaskar A, Maske M (2017) Crop based demonstrations and trials under farming system for nutrition study in Wardha (2013–16): a report. M.S. Swaminathan Research Foundation, MSSRF Research Report, MSSRF/RR/17/42

Pucher A, Høgh-Jensen H, Gondah J, Hash CT, Haussmann BI (2014) Micronutrient density and stability in West African pearl millet—potential for biofortification. Crop Sci 54:1709–1720

Ramashia SE, Mashau ME, Onipe OO (2021) Millets cereal grains: nutritional composition and utilisation in Sub-Saharan Africa. IntechOpen, London, pp 1–14

Raza A, Razzaq A, Mehmood SS, Zou X, Zhang X, Lv Y, Xu J (2019) Impact of climate change on crops adaptation and strategies to tackle its outcome: a review. Plan Theory 8:34

Rodríguez JP, Rahman H, Thushar S, Singh RK (2020) Healthy and resilient cereals and pseudo-cereals for marginal agriculture: molecular advances for improving nutrient bioavailability. Front Genet 11:49

Sage RF, Zhu XG (2011) Exploiting the engine of C4 photosynthesis. J Exp Bot 62:2989–3000

Sage RF, Christi PA, Edwards EJ (2011) The C4 plant lineages of planet earth. J Exp Bot 62:155–169

Saini S, Saxena S, Samtiya M, Puniya M, Dhewa T (2021) Potential of underutilized millets as nutri-cereal: an overview. J Food Sci Technol 8:4465–4477

Saleem S, Mushtaq NU, Shah WH, Rasool A, Hakeem KR, Rehman RU (2021) Morpho-physiological, biochemical and molecular adaptation of millets to abiotic stresses: a review. Phyton 90:1363

Sarita ES, Singh E (2016) Potential of millets: nutrients composition and health benefits. J Sci Innov Res 5:46–50

Sarwar AKMG, Biswas JK (2021) Cereal grains of Bangladesh–present status, constraints and prospects. In: Cereal grains, vol 19. IntechOpen, London

Satyavathi CT, Ambawat S, Khandelwal V, Srivastava RK (2021) Pearl millet: a climate-resilient nutricereal for mitigating hidden hunger and provide nutritional security. Front Plant Sci 12:659938

Saxena R, Vanga SK, Wang J, Orsat V, Raghavan V (2018) Millets for food security in the context of climate change: a review. Sustainability 10:2228

Serba DD, Yadav RS, Varshney RK, Gupta SK, Mahalingam G, Srivastava RK, Gupta R, Perumal R, Tesso TT (2020) Genomic designing of pearl millet: a resilient crop for arid and semi-arid environments. In: Genomic designing of climate-smart cereal crops. SpringerLink, pp 221–286

Shah A, Nazari M, Antar M, Msimbira LA, Naamala J, Lyu D, Rabileh M, Zajonc J, Smith DL (2021) PGPR in agriculture: a sustainable approach to increasing climate change resilience. Front Sustain Food Syst 5:667546

Shah P, Dhir A, Joshi R, Tripathy N (2023) Opportunities and challenges in food entrepreneurship: in-depth qualitative investigation of millet entrepreneurs. J Bus Res 155:113372

Shanmugapriya A, Nazni P (2020) Effect of processing techniques on nutritional, viscosity and osmolarity of barnyard millet based diarrheal replacement fluids. Curr Res Nutr Food Sci 8:164

Sharma R, Sharma S, Dar BN, Singh B (2021) Millets as potential nutri-cereals: a review of nutrient composition, phytochemical profile and techno-functionality. Int J Food Sci Technol 56:3703–3718

Sher A, Nawaz A, Sarfraz M, Ijaz M, Ul-Allah S, Sattar A, Hussain S, Ahmad S (2019) Advanced production technologies of millets. Agronomic crops: production technologies. Semantic Scholar, pp 273–296

Shinggu C, Tunwari B, Gani M (2022) Finger millet (eleusine coracana (l.) gaertn) in sustainable food security in Nigeria: a review. FU W Trends Food Sci Technol J 1:332–336

Singh Y, Sidhu HS, Jat HS, Singh M, Chhokar RS, Setia R, Jat ML (2020) Conservation agriculture and scale of appropriate agricultural mechanization in smallholder systems. Borlaug Institute for South Asia (BISA), International Maize and Wheat Improvement Center (CIMMYT), p 180

Singh A, Bharath M, Kotiyal A, Rana L, Rajpal D (2022a) Barnyard millet: the underutilized nutraceutical minor millet crop. J Pharm Innov 11:115–128

Singh V, Lee G, Son H, Amani S, Baunthiyal M, Shin JH (2022b) Anti-diabetic prospects of dietary bio-actives of millets and the significance of the gut microbiota: a case of finger millet. Front Nutr. https://doi.org/10.3389/fnut.2022.1056445

Singh S, Singh AK, Singh K, Singh B, (2022c) Millets processing, nutritional quality & fermented product: A review. The Pharma Inn J 11:304–308

Sinha NM, Kumari N, Dikshit N, Kant S (2022) Ragi traditional but nutritional especially in the era of COVID-19. Int J Clin Pract 32:20–25

Sonko LB, Dibba L, Camara J, Trawally DNA, Faye MN, Jallow DB, Nyassi A, Jobe M (2022) Grain yield response of Fonio (Digitaria exilis) varieties to fertilizer and cost-effectiveness under two different production systems in the Gambia. African J Food Agric Nutr Dev 22:19062–19074

Sood S, Joshi DC, Pattanayak A (2020) Breeding advancements in barnyard millet. In: Gosal S, Wani S (eds) Accelerated plant breeding, vol 1. Springer, Cham, pp 391–409

Sujatha K (2018) Standardization of duration for accelerated ageing in barnyard millet cv. CO2 and MDU 1. Electron J Plant Breed 9:1234–1238

Taylor JR (2016) Millet pearl: overview. https://doi.org/10.1016/B978-0-12-394437-5.00011-5

Taylor JR, Duodu KG (2019) Traditional sorghum and millet food and beverage products and their technologies. In: Sorghum and millets. AACC International Press, pp 259–292

Ugare R, Chimmad B, Naik R, Bharati P, Itagi S (2014) Glycemic index and significance of barnyard millet (Echinochloa frumentacae) in type II diabetics. J Food Sci Technol 51:392–395

Ullah A, Ahmad A, Khaliq T, Akhtar J (2017) Recognizing production options for pearl millet in Pakistan under changing climate scenarios. J Integr Agric 16:762–773

Ullah A, Ahmad I, Ahmad A, Khaliq T, Saeed U, Habib-ur-Rahman M, Hussain J, Ullah S, Hoogenboom G (2019) Assessing climate change impacts on pearl millet under arid and semi-arid environments using CSM-CERES-Millet model. Environ Sci Pollut Res 26:6745–6757

Verma VC, Acharya S, Verma BC (2020) Millets for sustainable agriculture and nutritional security. Biot Res Today 10:1055–1057

Vetriventhan M, Azevedo VC, Upadhyaya HD, Naresh D (2019) Variability in the global proso millet (Panicum miliaceum L.) germplasm collection conserved at the ICRISAT genebank. Agriculture 9:112

Wakjira MT, Peleg N, Anghileri D, Molnar D, Alamirew T, Six J, Molnar P (2022) The connections between rainfall regimes and crop production in semi-arid and sub-humid climates: an example from Ethiopia (No. IAHS2022-244). Copernicus Meetings

Wiedemair V, Scholl-Bürgi S, Karall D, Huck CW (2020) Amino acid profiles and compositions of different cultivars of Panicum miliaceum L. Chromatographia 83:829–837

Wilson ML, VanBuren R (2022) Leveraging millets for developing climate resilient agriculture. Curr Opin Biotechnol 75:102683

World Health Organization (2020) The state of food security and nutrition in the world 2020: transforming food systems for affordable healthy diets, vol 2020. Food & Agriculture Organization, Rome

Xie H, Perez N, Anderson W, Ringler C, You L (2020) Can Sub-Saharan Africa feed itself? The role of irrigation development in the region's drylands for food security. In: Virtual water. Routledge, pp 80–98

Zhang Z, Zhang J, Lu P, Wu B, Liu M, Gao J, Wang C, Bai K, Guo G (2022) Six underutilized grain crops for food and nutrition in China. Plan Theory 11:2451

Chapter 11
Amaranth (*Amaranthus spp.*): Food Properties and Potential Health Benefits

Asad Abbas, Naveed Ahmad, Wisha Saeed, Muhammad Qamar, Muhammad Usama, Muhammad Zulqarnain Khan, and Tuba Esatbeyoglu

11.1 Introduction

Amaranthus species, grown as the principal grain crop by the Aztecs 5000–7000 years ago and as a vegetable by early civilizations over 2000 years ago (Kauffman and Weber 1990). The United States National Academy of Sciences reported an augmented interest in this plant in the 1980s due to its high nutritional value and agronomic potential (Ulbricht et al. 2009). It is utilized as food in various countries around the world i.e., USA (Southwest), China, India, Africa, Nepal, South Pacific Islands, the Caribbean, Greece, Italy, and Russia (Caselato Sousa and Amaya Farfán 2012; Das 2016). Amaranth grain is a highly nutritional "pseudo-cereal" having notable amount of protein, including lysine and methionine, minerals, dietary fiber, and bioactive compounds (Joshi et al. 2018; Venskutonis and Kraujalis 2013).

Slim amaranth contains pigments like β-cyanins, carotenoids, betalains, β-xanthins, and β-carotene which are vital prospects for the food industry regarding colorants and antioxidants. Phenolic acids and flavonoids are antiradical agents in amaranth (Repo-Carrasco-Valencia et al. 2010; Venskutonis and Kraujalis 2013; Sarker and Oba 2019). It decreases plasma cholesterol and glucose levels, stimulates the immune system, exerts an antitumor and anti-allergic activity, and improves hypertensive and anemic conditions (Caselato Sousa and Amaya Farfán 2012). The

A. Abbas · N. Ahmad
Multan College of Food & Nutrition Sciences, Multan Medical and Dental College, Multan, Pakistan

W. Saeed · M. Qamar (✉) · M. Usama · M. Z. Khan
Department of Food Science and Technology, Faculty of Food Science and Nutrition, Bahauddin Zakariya University, Multan, Pakistan

T. Esatbeyoglu
Institute of Food Science and Human Nutrition, Department of Food Development and Food Quality, Gottfried Wilhelm Leibniz University Hannover, Hannover, Germany

T. Ismail et al. (eds.), *Neglected Plant Foods of South Asia*,
https://doi.org/10.1007/978-3-031-37077-9_11

species have medicinal uses and reported to have antimicrobial, anticancer, antidiabetic, antimalarial properties, and also known as snake antidotes (Al-Mamun et al. 2016). Aforementioned bioactive compounds are likely to protect against cardiovascular diseases (CVD), atherosclerosis, emphysema, cataracts, arthritis, retinopathy, and neuron degeneration (Steffensen et al. 2011). It may also treat diarrhea, intestinal bleeding, and excessive menstruation (Sarker et al. 2018).

11.2 Common Edible Food-Grain Species

Among a rich assortment of crops globally, amaranth is indigenous to Africa and had remained an underutilized crop despite having excellent health-promoting commotions (Zhu 2020) with approximately 400 varieties and 70 species of *Amaranthus*, only a few (*A. caudatus*, *A. cruentus*, and *A. hypochondriacus*) have been adopted in different countries for essentially grain production (Fig. 11.1) while others (*A. blitum*, *A. tricolor*, *A. spinosus*, *A. dubius*, and *A. viridis*) are cultivated as leafy vegetables (Das 2012). The *A. cruentus* contains high protein contents with sulfur-containing amino acids complementing other cereals for snacks and food. The plant possesses a commercial status in South Africa, decorative in Nigeria, and a weed in urban centers (Ogwu 2020). It is also cultivated as a food crop in parts of Peru, Argentine, and Bolivia. *A. caudatus* also contains high protein, minerals, and fat compared to commonly used cereals, which can be milled for whole-meal Amaranth flour, gluten-free brown bread, cracker, cookies, and biscuits (Mekonnen et al. 2018).

Amaranth sprouts as novel functional food were assessed (Felicity's) at cultivation stage by supplementing selenium (as sodium selenite) at 10, 15, and 30 mg/L to identify its effect on Amaranth growth and lipid profile (Benzie and Strain 1996).

Fig. 11.1 Amaranthus whole plant, flowers, and seeds

The sprouts act as moderate selenium accumulators with improved essential fatty acids (α-linoleic and linolenic) levels which may serve as precursor for long-chain poly unsaturated fatty acids (PUFAs) thus suggesting amaranth sprouts as functional food (Carlson et al. 1989). As a dietary ingredient, the sprouts also serve as moderate antioxidant sources with no observed antimicrobial properties (Cartes et al. 2005). Vegetable and weedy *Amaranthus* species when tested for sterols revealed spinasterol as predominant component (46–54% by weight) (Fernando and Bean 1985).

11.3 Nutritional Perspectives of Amaranth

An obvious genetic diversity in amaranth in South America and Northern India has been observed. The leaves of amaranth (slightly sweet flavor) are used for cooking and in salads making, and are sometimes recommended to those having low red blood cell (RBC) count (Alemayehu et al. 2015). Amaranth with a balanced protein, lipids, and minerals profile comparing to corn, millet, rice, and wheat, makes this highly utilizable for human and animal feeding (Amare et al. 2015). The plant contains 13–19% protein, 5–13% lipids, 62–74% starch, 9% dietary fiber, and 2.14–2.91% ash contents (Chauhan et al. 2015). Amaranth is claimed to provide recommended levels of proteins as per FAO/WHO recommendations (Maurya and Arya 2018). Wheat, maize, and rice for an instance, contain less protein (13.3%, 9.2%, and 7.0%) than amaranth grain (Narwade and Pinto 2018). The average protein content also differs slightly among species (13.5% v/s 15.7% v/s 15.5 in *A. caudatus*, *A. cruentus*, and *A. hypochondriacus*) (Motta et al. 2019). Likewise, isoleucine (52 mg/g protein), lysine (76 mg/g protein), tryptophan (11 mg/g protein), threonine (51 mg/g protein), and sulfur-containing amino acids were higher in *A. caudatus* as compared to other species (Mota et al. 2016). The specie contained 12.5 mg/kg ascorbic acid, 28.0 mg/kg niacin, 4.5 mg/kg B_6, and 2.4 mg/kg B_2 in raw seeds with varying vitamin concentrations upon cooking, popping, drying, and germinating (Muriuki et al. 2014). The vitamin E isoforms (tocopherols and tocotrienols) ranged from 63.7 to 129.3 mg/kg in concentration (Tumwegamire et al. 2011).

Amaranthus species with vitamins, folate, carotene, and health-promoting bioactive compounds (protocatechuic, caffeic, hydroxybenzoic, ferulic acid, isoquercetin, routine, and nicotiflorin) seem promising niche to counter food insecurity (Mekonnen et al. 2018; Musa et al. 2011; Karamac et al. 2019). It's been evidenced that its consumption reduces cholesterol and anticipate anti-aging and inhibit memory loss due to age associated degenerative changes (Karamac et al. 2019; Alvarez et al. 2010; Alvarez et al. 2010). Amaranth consumption was claimed for improving health in children and improved milk production in lactating mothers (Tagwira et al. 2006). Globulins and albumins are predominant proteins with prolamins and gluten (contributing to an autoimmune disorder in celiac patients). Amaranth grains are thus free of prolamins and gluten proteins and make it a better dietar choice for celiac patients (Martínez Villaluenga et al. 2020). Anti-nutritional elements which include protease, oxalates, tannins, saponins, protease inhibitors, and nitrates making complexes with iron, zinc, and calcium, ultimately reducing their absorption have been reported to exist in relatively lower amounts than other non-conventional edible grains (Cuadrado et al. 2018). Amaranth plant's better adaptability to wider climatic conditions and its short production cycle make this crop as a "golden" vegetable with stand-out economic significance and kitchen gardening vegetables (Stetter et al. 2020).

Regarding vitamin C concentration, it ranged from 69 to 288 mg/100 g in *A. cruentus* (AG-122) and *A. hypochondriacus* (Rasna/grain type) and 62–209 mg/100 g in *A. tricolor* (Amar-peet) and *A. tricolor* (AV-101/vegetable type). Vitamin C in the leaf of amarnth also varied with its age. Oil contents in *A. hypochondriacus* seeds specimens ranged from 4.4% to 13.2% with a promising unsaturated fatty acid profile (Prakash et al. 1995). With 21% global deficiency, vitamin A deficiency accounts for the deaths of up to 800,000 children worlwide (Tumwegamire et al. 2011). According to Kenyan National Micronutrient Survey (KNMS) 70% of children <5 years were estimated to be iron deficient while 43.2% had the deficiency (Mwaniki et al. 1999). Among plant foods, cereals contribute ~62–68% of total dietary energy to young children and pre-schoolers, so introducing amaranth to these population segments particularly those from the resource-constrained economies may help evading micronutrient deficiencies (Black et al. 2008).

11.4 Strategies to Improve Nutrient Accessibility in Grain Amaranth

Nutrient bioavailability and digestibility in grain amaranth are generally affected by processing (Grundy et al. 2020; Aguilera Morales et al. 2018). Germinating (malting) the grain for 46 hours at 26 °C increased protein availability, total energy, reduced resistant starch, and increased linoleic acid by 8%, 11%, 70%, and 10% respectively (Najdi Hejazi and Orsat 2017). For reducing anti-nutrients in the grain, boiling, popping, roasting, milling, and soaking treatments were traditionally used (Njoki et al. 2014). Protein's digestibility and grains efficiency to be better utilized by the body are significantly enhanced by the heating or cooking operations (Bressani and Garcia Vela 1990).

Defatting amaranth flour with *n*-hexane also improves protein digestibility (Ramkisson et al. 2020). Popping treatment increases fat, ash, acid detergent fiber (ADF), and neutral detergent fiber (NDF) contents while fermentation increased protein (due to carbohydrate loss), fat, ash, iron, and magnesium content (Hedrén et al. 2002). Increased lipid content during fermentation are supposed to be associated with the type of fermenting medium like the yeast strain, which produces fat (Khetarpaul and Chauhan 1989). Fermentation and popping decrease mineral absorption inhibitor (IP6) to enhance mineral bioavailability (Amare et al. 2016). Phytates in amaranth are decreased by exogenous phytase from yeast and lactic acid bacteria (LAB) during the process of lactic acid fermentation (Baye et al. 2014).

Lactic acid fermentation generally improves the nutritive value of foods (Omamt et al. 2006). The specie (*Lactobacillus plantarum*) is effective in reducing anti-nutrients and enhancing nutrient accessibility (Sharma et al. 2021). Fermented flour of amaranth contains degraded phytate compared to grains, hence increasing the nutrient bioavailability from fermented flour. While fermentation in grain increases zinc and calcium content (Castro Alba et al. 2019). The malting process in Amaranth may also increase folate by 21% (Mota et al. 2016).

11.5 Global Market Potential of the Grain

Owing to potential health benefits of amaranth, the plant seems a promising facet for the food industry. In Zimbabwe and Europe, amaranth consumption benefited the communities with increased annual demand. Production volume and market linkage data are still scarce between producing and importing countries. The British Industry Confederation reported the import of 6000 tons of grains from developing countries most of which were amaranth seeds with Germany as a major market. The commodity is also being progressively identified in the UK, Sweden, France, Belgium, and the Netherlands. The major producers include Mexico, China, Russia, Peru, Kenya, and India (Ainebyona et al. 2012). Despite potential health aids, poorly characterized agronomic potential and high-input agriculture customary of staple crops still kept amaranth as an underutilized crop and hinders its widespread incorporation in food systems. Genetic limitations also exist for some underutilized (pseudo) cereal crops (Tsutsumi et al. 2017). In addition, these crops are often considered secondary with less than optimal growth season allocation comparing to the staple grain counterparts. Foremost promising food applications for amaranth include cooking, popping, extrusion, fermentation, bread baking and pasta production. The primitive forms of amaranth consumption include cooking in threefold of water or popped by dry and short heat with or without fat addition, which produces a nutty flavor with enhanced palatability.

Popped amaranth (soft and ready to be eaten) may complement cakes, cookies, granolas, and mueslis, and be incorporated into existing or new food formulations after milling. Popping induces partial gelatinization with an intact embryo and nutritional profile due to the short processing span. Gluten absence in Amaranth is a considerable aspect as it has no dough-forming properties and needs blending (up to 20%) with gluten-containing cereals for enriching pasta, biscuits, cookies, etc. Due to increased celiac disorders and gluten sensitivity, amaranth flour may be used in gluten-free bakery products. Amaranth and other pseudo-cereals are nutritionally superior to other gluten-free raw materials so amaranth based gluten-free products balance these products nutritionally. The rapidly growing gluten-free market seems one major driving factor toward enhanced production and utilization of amaranth.

Plant-derived (gluten-free) products with amaranth incorporation still need to be extensively researched in replacing major crops (De la Barca et al. 2010).

11.6 Amaranth's Adaptability to Climate Change

Amaranths' resilience to varied climatic conditions made the plant to be revived in modern agriculture. The crop is unique in tolerating droughts, salinity and frost both in hilly landscapes and plains (Parihar et al. 2021). The plant adopts seasonal uncertainty and uses a C_4 photosynthetic system with a higher rate enabling it to thrive even in nitrogen-poor soils (Kauffman 1992). Climate change is potentially tempting poor crop yields, desertification, deforestation, erosion, depletion of water quality and resources ultimately complicating food insecurity in climatic vulnerable areas (Delgado et al. 2011). Population explosion, increased energy and agricultural inputs cost together with previously mentioned factors require a more diverse cropping system with drought and stress-tolerant crops (Delgado et al. 2011). Amaranth as a suitable contender is probed to extend marginal agricultural land use in African and Asian regions to cope with increasing burden of food insecurity (Steffensen et al. 2011).

Minor crops with unique advantages are often ignored irrespective of the emerging challenges of climate change and food insecurity globally (Jacobsen et al. 2015). Global crop production supplies 75% of energy needs with three staple commodities (wheat, rice, and maize). Such a high dependency over conventional cereals is of serious concern (Schmidt et al. 2021). For an instance, in 2007 diminished food supplies led to price uncertainty and confusion in global food markets (Alemayehu et al. 2015). International initiatives are underway in Africa to expand agricultural food, feed, and biofuel production from conventional to non-conventional resources (Fanzo et al. 2012).

The current production of major staple food crops is not enough to fulfill the projected global food requirement (Gressel 2008). To relieve the pressure on major crops under a changing climate, alternate crops with the potential to meet global food requirements are needed. Amaranth and some other crops like quinoa are some underemployed species. Amaranth is well adapted to surplus land and holds significant potential for further adaptation and development due to genetic assortment and range of phenotypic plasticity (Ecker et al. 2010). Combining these attributes with its inherent tolerance to high temperatures, drought, poor soil conditions, and disease-resistance make it a highly acceptable crop under adverse growing conditions. Due to relatively high trading prices and the rising popularity of amaranth in the West, a stable income for bare-existence farmers may be generated through external markets (Emire and Arega 2012). Amaranth may easily be introduced as an unconventional and supplemental crop in Asia and Africa to the current cropping systems of these regions (Ecker et al. 2010).

11.7 Health-Promoting Properties

Since ancient times extracts of amaranth have been used to cure a variety of illnesses. However, interest in amaranth has increased recently. Amaranth extracts have been used as a diuretic and to cure a variety of ailments as urinary infections, gynaecological issues, diarrhoea, pain, respiratory illnesses and diabetes (Peter and Gandhi 2017). Amaranth can enhance consumer health because it contains bioactive substances. Bioactive compounds in amaranth grain include phenolic acids such as protocatechuic, hydroxybenzoic, caffeic acid, ferulic acid, nicotiflorin and isoquercetin. There is evidence that regular exercise slows down aging, quercetin protects against oxidation and nicotiflorin aids in memory function prevention (Aderibigbe et al. 2022). Due to amaranth capacity to scavenge free radicals and neutralize their degenerative effects, some species of it have been found to play significant roles in the regression of oxidative stress-induced illnesses (Jimoh et al. 2022). Aqueous extract from steamed red amaranth leaves may be utilized as a powerful nutritional supplement to fend off diabetic retinopathy according to recent research findings. Anti-glycative and anti-oxidative action of that extract against a high glucose-induced injury was examined in a human lens epithelial cell line HLE-B3 (Baraniak and Kania Dobrowolska 2022). Aqueous extracts prepared from red amaranth exhibited marked antiglycation effects, anti-α-amylase, anti-α-glucosidase, anti-lipase, and anti-AchE activities (Madadi et al. 2021). Sandoval-Sicairos et al. 2020 in Amaranth identified a total of 11 peptide sequences in the fractions evaluated exhibited potential biological activity against non-communicable diseases.

11.7.1 Anticancer

Anticancer effects of *A. mantegazzianus* isolated protein (MPI) and its mode of action against four tumour cells named MC3T3E1, UMR106, Caco-2, and TC7 were studied. MPI inhibited four cell lines with varied potentials wherein the most pronounced effect was against UMR 106 (Barrio and Anon 2010). In another study, *Amaranthus lividus* (AL) and *Amaranthus hybridus* (AH) in hemagglutination assay outlined notable antioxidant activity by inhibiting the Ehrlich's ascites carcinoma (EAC) cell growth by 45% and 43% at 100 μg/ml, respectively owing to their antioxidant potential (IC50 values of 28 ± 1.5 and 93 ± 3.23 μg/ml, separately) evaluated in the same study (Al-Mamun et al. 2016). The methanol extracts of aerial parts of *A. dubius* (AD), *A. spinosus* (AS), *A. tricolour* (AT), and *A. viridis* (AV) were explored for possible anticancer activities. AD methanol extract was recorded to have stronger anticancer activity against MCF7 cells with IC_{50} of 48.2 μg/mL but AS was more active against MDAMB231 cells with IC_{50} of 82.1 μg/mL (House et al. 2020). Proteins from amaranth seeds were extracted and hydrolyzed employing simulated gastrointestinal digestion with native and heat-denatured proteins to enhance digestibility and liberate amino acids and peptides carrying antioxidant features. Liberated hydrolysates (amino acids) showed a lot of antioxidant activity was considered for vitro anticancer activity in human triple-negative breast cancer cells. As a result protein hydrolysates inhibited cell growth with a GI_{50} of 48.3 0.2 μg/ml (Taniya et al. 2020). Another study compared the in vitro radical scavenging abilities and anti-cancer effects of amaranth protein hydrolysates (alcalase, trypsin, and pepsin). Trypsin hydrolysates outperformed controls and other two hydrolysates in antioxidant assays and anticancer activity against MCF-7, A549 and HEK 293 cell lines evaluated using MTT cytotoxicity experiment, suggesting it may be a cancer treatment (Ramkisson et al. 2020). Lately, methanol extract of three different medicinal plants named *Artocarpus lakoocha*, *Kigelia pinnata*, and *Amaranthus viridis* were evaluated for anticancer potential against three cancer cell lines i.e., Hela, A549 and MCF-7. *Amaranthus viridis* found much more effective against all three cancer cell lines when as compared to *Artocarpus lakoocha*, *Kigelia pinnata* with an IC_{50} of 48 μg/mL (MCF-7), 24 μg/mL (A549) and 13 μg/mL (Hela) (Ahmed et al. 2020).

11.7.2 Anti-hypertensive Effects

The Renin-angiotensin (a biochemical system) releases angiotensinogen to control blood pressure which depletes sodium from plasma by producing angiotensin-I (Ondetti and Cushman 1982) and cleaving dipeptide His-Leu by angiotensin-converting enzymes (ACE) and forming angiotensin-II (an effective vasoconstrictor) (Li et al. 2004). Scientists have agreed on the existence of several proteins which generate peptides consequently inhibiting ACE functioning (Silva Sánchez et al. 2008). In the line of this belief an in vitro research tried to explore the potential of amaranth protein isolate (API) and hydrolysates (APHs) against the activities of angiotensin-I converting enzyme (ACE), dipeptidyl peptidase-IV (DPP-IV) and α-glucosidase (AG). APHs mainly bromelain-4 h hydrolysate (B4) outlined distinguished potential in reducing the ACE, DPP-IV and AG when compared to API and other APHs. Phytochemicals investigation using mass spectrometry detected 116 peptides in B4 and predicted 17 to be biological active. Moreover 6 of them were projected to inhibit ACE whereas FPFPPTLGY and FPFPR found to the most active DPP-IV and AG hotspots respectively. It was suggested that APHs may be a source of antidiabetic and hypertension peptides for functional food composition (Kamal et al. 2021). Moreover, hypotensive effects of amaranth protein/peptides were evaluated using animal model i.e., in spontaneously hypertensive rats (SHR) and mode of action via in vivo and ex vivo. Study also examined how an O:W emulsion protected antihypertensive peptides against gastrointestinal digestion (GID). Results revealed that all the tested samples lowered systolic blood pressure. Rats treated with emulsion (G_E) and G_E + VIKP i.e., synthetic peptide reduced systolic blood pressure by 42 and 35 mmHg, respectively. The data implied that GID released or resisted peptides with biological activity. ACE, renin and the vascular system regulated systolic blood pressure using these peptides. These findings supported amaranth protein/peptides in hypertensive functional meals (Suárez et al. 2020). Antihypertensive peptides and exercise lower blood pressure. In a recent study, amaranth antihypertensive peptides did not alter the systolic blood pressure in normotensive rats but showed a notable decline in spontaneously hypertensive rats (SHR) after 8 week (42 mmHg) and results were comparable to group included low intensity physical exercise (42.7 mmHg) These results suggested safe and effective peptide-based functional meals (Cabrera Chávez et al. 2020).

11.7.3 Celiac Disease

The celiac disease is a clinical condition (gluten or food intolerance) and is triggered upon the ingestion of gluten-containing foods. The condition is only managed by complete adherence to gluten-free foods. As amaranth flour is low in prolamins (intolerant protein for celiac patient) a significant consideration has been given to this flour to formulate gluten-free products with substantial nutritional profile. It's been used by celiac patients causing no allergic responses in intestinal mucosa (Mariotti et al. 2009). Flour from amaranth grain has been suggested to develop non-allergenic foods to cope with food allergy. Components from amaranth grain as tested by in-vitro and in-vivo approach suppressed IgE production with increased cytokine Th1 synthesis resulting in reduced allergenicity for specific antigens. Laboratory studies have shown that water-soluble amaranth fraction promotes helper T cells (Th1) production, while amaranth's extract inhibited IgE production and allergic cascade (Hibi et al. 2003). Additionally, celiac patients have been reported with higher diabetes mellitus prevalence. Starch digestibility of an extruded product made from amaranth flour was investigated using glycemic and insulinemic responses in 11 women comparing it with white bread (the reference food). A mean glycemic index of 107 with enhanced insulin production was observed in women. Even after processing the digestibility profile indicated a suitable enzymatic action. Due to rapid digestibility amaranth-based snacks are also recommended for athletes and when consumed by celiac cum diabetic patient is supposed to provide adequate glycemic control (Alencar et al. 2017). A study characterized 40 varieties of amaranth for their possible digestibility and tolerance by immunoblotting/SDS-PAGE and ELISA in celiac patients. A non-significant difference in binding affinities to specific anti-gliadin antibodies was observed for all varieties with gluten-like protein contents <20 ppm. Amaranth when characterized on molecular basis seemed safe for celiac patients recommending varieties with the lowest protein contents having cross-reacting with antibodies (anti-gliadin) (Ballabio et al. 2011).

11.8 Value Added Products

The applied/industrial potential of amaranth is due to its high nutritional and functional properties. A lot of products i.e., bread, flatbread (gluten free), minerals rich snacks, muffins, nuggets, functional beverages has been made by adding amaranth flour and extracts (Fig. 11.2). In recent past, wheat flour was fortified with amaranth flour at 0–15% and bread was developed. The protein content of bread was significantly increased from 12.17% to 13.11%, fat from 2.16% to 2.77%, crude fiber from 1.11% to 1.72%, moisture content from 31.06% to 33.24%, and ash from 0.92% to 1.51%. The substitution of amaranth increased rheological aspects such as hardness, gumminess, chewiness, and cohesion (Nasir et al. 2020). Mukhtar et al. (2020) also developed amaranth flour-based fortified bread. The amaranth flour-based bread was nutritionally acceptable due to increased protein, lipid, and ash contents (Mukhtar et al. 2020). A novel gluten-free flatbread was developed using amaranth flour which was evaluated as possible healthy food. Flatbread with added amaranth flour has shown improved color and polyphenol proportions. Amaranth flour was added to rice and maize formulations to improve the technical, nutritional and

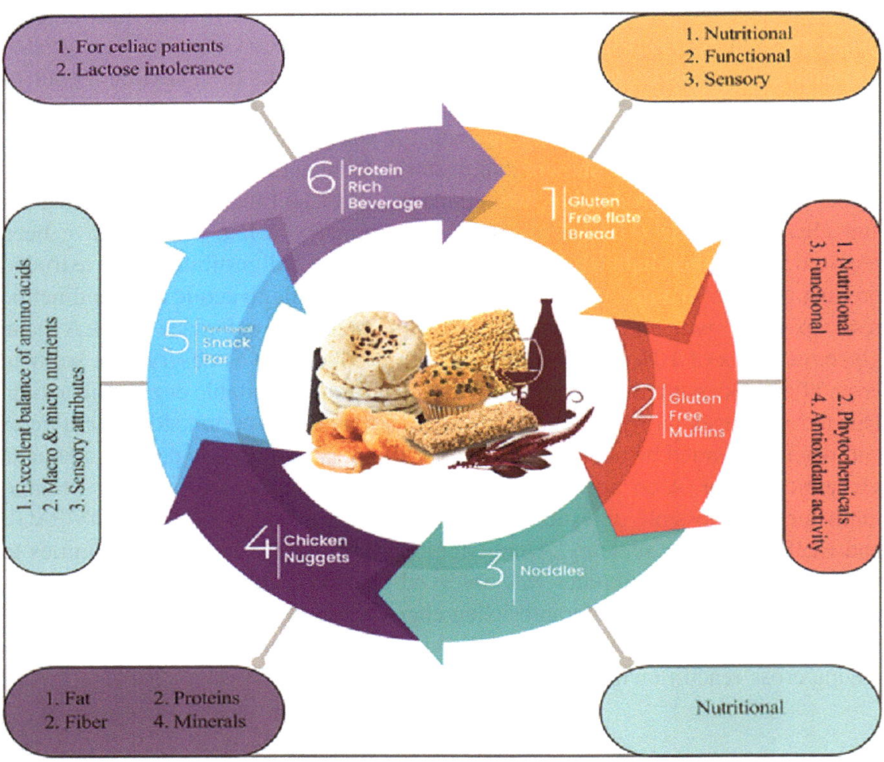

Fig. 11.2 Functional products made from Amaranthus

sensory qualities (Piga et al. 2021). Furthermore, Cotovanu and Mironeasa (2022) formulated composite flour using wheat and amaranth and considered technological parameters, dough rheological properties and nutritional composition of the bread. The novel product was considered an acceptable recipe with optimal particle size, technological and physical parameters. The protein extract from amaranth was used to formulate a nutritious and stable beverage using gums, gellan and xanthan with a similar composition to cow's milk (3.42% protein, 0.60% lipids, 1.9% fiber, 0.43% carbohydrates and 90.58% water). The tailored protein-rich beverage was suitable for vegans, celiac patients and lactose intolerants (Manassero et al. 2020). Amaranth flour was combined with black rice flour (0%, 10%, 20%, 30%, 40% and 50%) to improve the nutritional qualities of gluten-free muffins and their premix. For the 50% substituted formulation, the findings for total phenolics, flavonoids and antioxidant activities were found to be optimal. This combination resulted in a product with an acceptable nutritional profile (Bhatt et al. 2021). Yellow cassava is an inexpensive pseudo-cereal for formulating a healthy food with significant β-carotene content. Functional pasta from white and yellow cassava, fortified with 50 g/kg (w/w) amaranth dry leaf powder was evaluated for cultivar difference, processing and addition of Amaranth leaf enhanced antioxidant capability of the composite flour pasta (Lawal et al. 2021)

Noodles as a popular snack food among children and youngsters are not sufficient carriers of nutrients. The standard noodle recipe with 1%, 2%, and 3% amaranth leaf powder inclusion was developed and studied for nutritional content, physical quality and customer acceptance. The texture of the noodles enriched with amaranth leaf powder was noticeably softer (271.39 g) than the control noodles (609.08 g) (Qumbisa et al. 2022). The impact of replacing wheat flour with amaranth flour in the batter, chicken paste and chicken nuggets was studied. Amaranth flour dramatically enhanced the amount of protein, fat, minerals and fiber in the nuggets (Tamsen et al. 2018). A variety of ratios of pre-fermented amaranth, acha and pearl millet flours were blended to create the multigrain snack food. The snack bar had an excellent balance of amino acids, macro and micronutrients. It was potentially healthy, nutrient-dense, hypoglycemic and rich in antioxidants (Olagunju et al. 2022).

Snack bars using various amaranth grains, oats, and banana peel powder were developed for malnourished people. Oats (25%) and banana peel powder (15%) were shown to have greater sensory attributes and nutritive qualities than the bars developed with Amaranth grains. Biochemical composition of the snack bar's as evaluated by protein, mineral, dietary fiber, essential amino acid, phenolic, and antioxidant content was however boosted by the addition of amaranth grain, oat and banana peel powder. These functional snack bars were less expensive than market snack bars. It has been proposed that development of functional bars can serve as lead to industrial growth and promote the development and consumer acceptability of novel snack bars (Singh et al. 2022). The chikki made from 60% peanuts and 40% amaranth grains has been shown to have better nutritional and sensory properties. In terms of organoleptic qualities throughout the storage period, the chikki made from amaranth grain and peanut (40:60) in 90% jaggery and 10% liquid glucose was evaluated to hold better sensory and other physical properties (Pingale et al. 2019). Dietary fiber-rich amaranth and quinoa flour at varying proportions were added to refined wheat flour to make gluten-free goat meat nuggets. Despite the characteristics being highly acceptable, the product containing 1.5% quinoa flour had the highest overall acceptability score (Verma et al. 2019). The baked items enriched with antihypertensive components still need elucidation for their exact physiological impacts. Cookies with amaranth hydrolysates (alcalase-generated) were tested for their technological aspects like color, texture, expansion volume etc. and antihypertensive potential. Compounds' bioavailability was assessed in Balb/c mice while antihypertensive role in hypertensive rats. Resultantly, enriched cookies were more color-intense with significant ACE-1 activity reduction and antihypertensive potential. Amaranth hydrolysates kept their effect in enriched product making cookies a functional product (Ontiveros et al. 2020). Enzyme (alcalase) treated amaranth produces ACE-1 inhibitory peptides which may be added in foods for functionality. Amaranth hydrolysate was used to enrich pasta which was tested for properties like technological (cooking time, loss, texture), hedonic (texture, color) and antihypertensive. Amaranth-protein concentrate was added at 15% and 20%. The supplemented pasta though had reduced acceptability (cooking time, loss, color, firmness, taste) than the control sample however, the product was found to contribute significant antihypertensive potential (Valdez-Meza et al. 2019).

11.9 Conclusion

Amaranth species hold great potential to combat and alleviate hunger and associated illnesses. Promising nutritional profile, phytochemical contents and ability of amaranth to withstand harsh environmental conditions distinguish this crop from others and make it extremely useful for both domestic and commercial settings. Amaranth consumption is evidently associated with a myriad of health benefits such as antioxidant, anticancer, anti-inflammatory and antihypertensive properties and management of celiac disease as the most important among all. It can serve as wonderful ingredient for development of functional foods on industrial scale and can be exploited commercially in products such as bread, flatbread, snacks, pasta, muffins, nuggets and beverages. Contrary to it's miraculous food and health-promoting features, only a fraction of this crop has been explored so far for intended use of human. It is therefore necessary to make concerted and ongoing efforts in research as well as to coordinate the activities of all relevant stakeholders in making Amaranth a part of the existing food production system to alleviate the burden of food insecurity under the emerging challenges of climate change.

Conflicts of Interest All authors read and approved the final manuscript and declare that they have no conflict of interest.

References

Aderibigbe OR, Ezekiel OO, Owolade SO et al (2022) Exploring the potentials of underutilized grain amaranth (Amaranthus spp.) along the value chain for food and nutrition security: a review. Crit Rev Food Sci Nutr 62:656–669

Aguilera Morales ME, Canales Martínez MM, Ávila González E, Flores Ortíz CM (2018) Nutrients and bioactive compounds of the Lemna gibba and Ulva lactuca as possible ingredients to functional foods. Lat Am J Aquat Res 46:709–716

Ahmed M, Ramachandraiah K, Jiang GH, Eun JB (2020) Effects of ultra-sonication and agitation on bioactive compounds and structure of amaranth extract. Foods 9:1116

Ainebyona R, Mugisha J, Kwikiriza N et al (2012) Economic evaluation of grain amaranth production in Kamuli district, Uganda. J Agric Sci Technol 2:178

Alemayehu FR, Bendevis MA, Jacobsen SE (2015) The potential for utilizing the seed crop amaranth (Amaranthus spp.) in East Africa as an alternative crop to support food security and climate change mitigation. J Agron Crop Sci 201:321–329

Alencar NMM, de Morais EC, Steel CJ, Bolini HMA (2017) Sensory characterisation of gluten-free bread with addition of quinoa, amaranth flour and sweeteners as an alternative for coeliac patients. Int J Food Sci Technol 52:872–879

Al-Mamun MA, Husna J, Khatun M, Hasan R et al (2016) Assessment of antioxidant, anticancer and antimicrobial activity of two vegetable species of amaranthus in Bangladesh. BMC Complement Altern Med 16:157

Alvarez JL, Auty M, Arend EK, Gallagher E (2010) Baking properties and microstructure of pseudocereal flours in gluten free bread formulations. Eur Food Res Technol 230:437–445

Amare E, Mouquet Rivier C, Servent A et al (2015) Protein quality of amaranth grains cultivated in Ethiopia as affected by popping and fermentation. Food Sci Nutr 6:38

Amare E, Mouquet-Rivier C, Rochette I, Adish A, Haki GD (2016) Effect of popping and fermentation on proximate composition, minerals and absorption inhibitors, and mineral bioavailability of Amaranthus caudatus grain cultivated in Ethiopia. J Food Sci Technol 53:2987–2994

Ballabio C, Uberti F, Di Lorenzo C et al (2011) Biochemical and immunochemical characterization of different varieties of amaranth (Amaranthus L. ssp.) as a safe ingredient for gluten-free products. J Agric Food Chem 59:12969–12974

Baraniak J, Kania Dobrowolska M (2022) The dual nature of amaranth—Functional food and potential medicine. Foods 11:618

Barrio DA, Anon MC (2010) Potential antitumor properties of a protein isolate obtained from the seeds of Amaranthus mantegazzianus. Eur J Nutr 49:73–82

Baye K, Mouquet Rivier C, Icard Vernière C et al (2014) Changes in mineral absorption inhibitors consequent to fermentation of E thiopian injera: implications for predicted iron bioavailability and bioaccessibility. Int J Food Sci Technol 49:174–180

Benzie IF, Strain JJ (1996) The ferric reducing ability of plasma (FRAP) as a measure of "antioxidant power": the FRAP assay. Anal Biochem 239:70–76

Bhatt S, Kumari N, Abhishek V, Gupta M (2021) Elucidating the role of amaranth flour in formulation of gluten free black rice muffins and its premix: nutritional, physico-chemical and textural characteristics. J Food Meas Charact 15:675–685

Black RE, Allen LH, Bhutta ZA et al (2008) Maternal and Child Undernutrition Study Group. Maternal and child undernutrition: global and regional exposures and health consequences. Lancet 371:243–226

Bressani R, Garcia Vela LA (1990) Protein fractions in amaranth grain and their chemical characterization. J Agric Food Chem 38:1205–1209

Cabrera Chávez F, Lopez Teros V, Gutiérrez Arzapalo PY et al (2020) Antihypertensive effect of amaranth hydrolysate is comparable to the effect of low-intensity physical activity. Appl Sci 10:5706

Carlson CL, Kaplan DI, Adriano DC (1989) Effects of selenium on germination and radicle elongation of selected agronomic species. Environ Exp Bot 29:493–498

Cartes P, Gianfreda L, Mora ML (2005) Uptake of selenium and its antioxidant activity in ryegrass when applied as selenate and selenite forms. Plant Soil 276:359–367
Caselato Sousa VM, Amaya Farfán J (2012) State of knowledge on amaranth grain: a comprehensive review. J Food Sci 77:93–104
Castro Alba V, Lazarte CE, Perez Rea D et al (2019) Fermentation of pseudocereals quinoa, canihua, and amaranth to improve mineral accessibility through degradation of phytate. J Sci Food Agric 99:5239–5248
Chauhan A, Saxena DC, Singh S (2015) Total dietary fibre and antioxidant activity of gluten free cookies made from raw and germinated amaranth (Amaranthus spp.) flour. LWT – Food Sci Technol 63:939–945
Coţovanu I, Mironeasa S (2022) Features of bread made from different amaranth flour fractions partially substituting wheat flour. Appl Sci 12:897
Cuadrado C, Takacs K, Szabó EE, Pedrosa MM (2018) Non-nutritional factors: lectins, phytic acid, proteases inhibitors, allergens. In: Martin-Cabrejas MA (ed) Legumes: nutritional quality, processing and potential health benefits, 1st edn. The Royal Society of Chemistry, London, pp 152–176
Das S (2012) Systematics and taxonomic delimitation of vegetable, grain and weed amaranths: a morphological and biochemical approach. Genet Resour Crop Evol 59:289–303
Das S (2016) Amaranthus: a promising crop of future. Springer, Singapore
De la Barca AMC, Rojas Martínez ME, Islas Rubio AR, Cabrera Chávez F (2010) Gluten-free breads and cookies of raw and popped amaranth flours with attractive technological and nutritional qualities. Plant Foods Hum Nutr 65:241–246
Delgado MCO, Tironi VA, Añón MC (2011) Antioxidant activity of amaranth protein or their hydrolysates under simulated gastrointestinal digestion. LWT – Food Sci Technol 44:1752–1760
Ecker O, Weinberger K, Qaim M (2010) Patterns and determinants of dietary micronutrient deficiencies in rural areas of East Africa. Afr J Agric Resour Econ 4:175–194
Emire SA, Arega M (2012) Value added product development and quality characterization of amaranth (*Amaranthus caudatus L.*) grown in East Africa. Afr J Food Sci 3:129–141
Fanzo J, Cogill B, Mattei F (2012) Metrics of sustainable diets and food systems. Biodiversity International, Rome, pp 1–8
Fernando T, Bean G (1985) A comparison of the fatty acids and sterols of seeds of weedy and vegetable species of Amaranthus spp. J Am Oil Chem Soc 62:89–91
Gressel J (ed) (2008) Genetic glass ceilings: transgenics for crop biodiversity. JHU Press, Baltimore
Grundy MM, Momanyi DK, Holland C et al (2020) Effects of grain source and processing methods on the nutritional profile and digestibility of grain amaranth. J Funct Foods 72:104065
Hedrén E, Mulokozi G, Svanberg U (2002) In vitro accessibility of carotenes from green leafy vegetables cooked with sunflower oil or red palm oil. Int J Food Sci Nutr 53:445–453
Hibi M, Hachimura S, Hashizume S, Obata T, Kaminogawa S (2003) Amaranth grain inhibits antigen-specific IgE production through augmentation of the IFN-γ response in vivo and in vitro. Cytotechnology 43:33
House NC, Puthenparampil D, Malayil D, Narayanankutty A (2020) Variation in the polyphenol composition, antioxidant, and anticancer activity among different Amaranthus species. S Afr J Bot 135:408–412
Jacobsen SE, Sørensen M, Pedersen SM, Weiner J (2015) Using our agrobiodiversity: plant-based solutions to feed the world. Agron Sustain Dev 35:1217–1235
Jimoh MO, Okaiyeto K, Oguntibeju OO, Laubscher CP (2022) A systematic review on Amaranthus-related research. Horticulturae 8:239
Joshi DC, Sood S, Hosahatti R et al (2018) From zero to hero: the past, present and future of grain amaranth breeding. Theor Appl Genet 131:1807–1823
Kamal H, Mudgil P, Bhaskar B et al (2021) Amaranth proteins as potential source of bioactive peptides with enhanced inhibition of enzymatic markers linked with hypertension and diabetes. J Cereal Sci 101:103308
Karamac M, Gai F, Longato E et al (2019) Antioxidant activity and phenolic composition of amaranth (*Amaranthus caudatus*) during plant growth. Antioxidants 8:173
Kauffman CS (1992) Realizing the potential of grain amaranth. Food Rev Int 8:5–21

Kauffman CS, Weber LE (1990) Grain amaranth. Advances in new crops. Timber Press, Portland

Khetarpaul N, Chauhan BM (1989) Effect of fermentation on protein, fat, minerals and thiamine content of pearl millet. Plant Foods Hum Nutr 39:169–177

Lawal OM, van Stuijvenberg L, Boon N et al (2021) Technological and nutritional properties of amaranth-fortified yellow cassava pasta. J Food Sci 86:5213–5225

Li Y, Zhang W, Wang T et al (2004) Study on the polymorphism of angiotensin converting enzyme genes and serum angiotensin II level in patients with obstructive sleep apnea hypopnea syndrome accompanied hypertension. Clin Otolaryngol 18:456–459

Madadi M, Wang Y, Xu C et al (2021) Using Amaranthus green proteins as universal biosurfactant and biosorbent for effective enzymatic degradation of diverse lignocellulose residues and efficient multiple trace metals remediation of farming lands. J Hazard Mater 406:124727

Manassero CA, Añón MC, Speroni F (2020) Development of a high protein beverage based on amaranth. Plant Foods Hum Nutr 75:599–607

Mariotti M, Lucisano M, Pagani MA, Ng PK (2009) The role of corn starch, amaranth flour, pea isolate, and Psyllium flour on the rheological properties and the ultrastructure of gluten-free doughs. Food Res Int 42:963–975

Martínez Villaluenga C, Peñas E, Hernández Ledesma B (2020) Pseudocereal grains: nutritional value, health benefits and current applications for the development of gluten-free foods. Food Chem Toxicol 137:111178

Maurya NK, Arya P (2018) Amaranthus grain nutritional benefits: a review. J Pharmacogn Phytochem 7:2258–2262

Mekonnen G, Woldesenbet M, Teshale T, Biru T (2018) Amaranthus caudatus production and nutrition contents for food security and healthy living in Menit Shasha, Menit Goldya and Maji Districts of Bench Maji Zone, South Western Ethiopia. Nutr Food Sci Int J 7:10–19080

Mota C, Nascimento AC, Santos M et al (2016) The effect of cooking methods on the mineral content of quinoa (*Chenopodium quinoa*), amaranth (*Amaranthus sp.*) and buckwheat (*Fagopyrum esculentum*). J Food Compost Anal 49:57–64

Motta C, Castanheira I, Gonzales GB et al (2019) Impact of cooking methods and malting on amino acids content in amaranth, buckwheat and quinoa. J Food Compost Anal 76:58–65

Mukhtar H, Gull A, Ganaie TA et al (2020) Effect of supplementation of amaranth flour on various quality attributes and antioxidant activities of bread. Curr Nutr Food Sci 16:391–396

Muriuki EN, Sila DN, Onyango A (2014) Nutritional diversity of leafy amaranth species grown in Kenya. J Appl Biosci 79:6818–6825

Musa A, Oladiran JA, Ezenwa MI et al (2011) The effects of applied nitrogen fertilizer and leaf positions on levels of micronutrients, anti-nutrients and toxic substances in Amaranthus cruentus. Afr J Biotechnol 10:9857–9863

Mwaniki DL, Omwega AM, Muniu EM et al (1999) Anaemia and micronutrient status of iron vitamin A and zinc in Kenya. National micronutrient survey report

Najdi Hejazi S, Orsat V (2017) Optimization of the malting process for nutritional improvement of finger millet and amaranth flours in the infant weaning food industry. Int J Food Sci Nutr 68:429–441

Narwade S, Pinto S (2018) Amaranth—a functional food. CDVS 1(3):1–4

Nasir S, Allai FM, Gani M et al (2020) Physical, textural, rheological, and sensory characteristics of amaranth-based wheat flour bread. Int J Food Sci 2020:1–9

Njoki JW, Sila DN, Onyango AN (2014) Impact of processing techniques on nutrient and anti-nutrient content of grain amaranth (A. albus). Food Sci Qual Manag 25:10–17

Ogwu MC (ed) (2020) Value of Amaranthus [L.] species in Nigeria. In: Nutritional value of Amaranth. IntechOpen, London

Olagunju AI, Arigbede TI, Makanjuola SA, Oyebode ET (2022) Nutritional compositions, bioactive properties, and in-vivo glycemic indices of amaranth-based optimized multigrain snack bar products. FoodReview 7:100039

Omamt EN, Hammes PS, Robbertse PJ (2006) Differences in salinity tolerance for growth and water-use efficiency in some amaranth (*Amaranthus spp.*) genotypes. N Z J Crop Hortic Sci 34:11–22

Ondetti MA, Cushman DW (1982) Enzymes of the renin-angiotensin system and their inhibitors. Annu Rev Biochem 51:283–308

Ontiveros N, López-Teros V, de Jesús Vergara-Jiménez M et al (2020) Amaranth-hydrolyzate enriched cookies reduce the systolic blood pressure in spontaneously hypertensive rats. J Funct Foods 64:103613

Parihar M, Rakshit A, Chitara MK, Jatav HS, Rajput VD, Singh AK, Rana K, Jatav SS, Anjum M, Minkina T, Kumar U (2021) Role of plant-associated microbes in phytoremediation of heavy metal polluted soils. In: Aftab T, Hakeem KR (eds) Heavy Metal Toxicity in Plants: Physiological and Molecular Adaptations, 1st edn. CRC Press, Boca Raton, Florida, pp157–169

Peter K, Gandhi P (2017) Rediscovering the therapeutic potential of Amaranthus species: a review. Egypt J Basic Appl Sci 4:196–205

Piga A, Conte P, Fois S et al (2021) Technological, nutritional and sensory properties of an innovative gluten-free double-layered flat bread enriched with amaranth flour. Foods 10:920

Pingale AA, Kotecha PM, Chavan UD, Harsha HB (2019) Nutritional evaluation and sensory quality of amaranth grain and peanut blended chikki during storage. J Life Sci 16:27–30

Prakash D, Joshi BD, Pal M (1995) Vitamin C in leaves and seed oil composition of the amaranthus species. Int J Food Sci Nutr 46:47–51

Qumbisa ND, Ngobese NZ, Kolanisi U et al (2022) Effect of Amaranthus leaf powder addition on the nutritional composition, physical quality and consumer acceptability of instant noodles. S Afr J Bot 145:258–264

Ramkisson S, Dwarka D, Venter S, Mellem JJ (2020) In vitro anticancer and antioxidant potential of Amaranthus cruentus protein and its hydrolysates. Food Sci Technol 40:634–639

Repo-Carrasco-Valencia R, Hellstrom JK, Philava JM, Mattila PH (2010) Flavonoids and other phenolic compounds in Andean indigenous grains: quinoa (*Chenopodium quinoa*), kaniwa (*Chenopodium pallidicaule*) and kiwicha (*Amaranthus caudatus*). Food Chem 120:128–133

Sandoval Sicairos ES, Domínguez Rodríguez M, Montoya Rodríguez A et al (2020) Phytochemical compounds and antioxidant activity modified by germination and hydrolysis in Mexican amaranth. Plant Foods Hum Nutr 75:192–199

Sarker U, Oba S (2019) Salinity stress enhances color parameters, bioactive leaf pigments, vitamins, polyphenols, flavonoids and antioxidant activity in selected amaranthus leafy vegetables. J Sci Food Agric 99:2275–2284

Sarker U, Islam MT, Rabbani MG, Oba S (2018) Phenotypic divergence in vegetable amaranth for total antioxidant capacity, antioxidant profile, dietary fiber, nutritional and agronomic traits. Acta Agric Scand B Soil Plant Sci 68:67–76

Schmidt D, Verruma Bernardi MR, Forti VA, Borges MTMR (2021) Quinoa and amaranth as functional foods: a review. Food Rev Int 37:1–20

Sharma G, Barney JN, Westwood JH, Haak DC (2021) Into the weeds: new insights in plant stress. Trends Plant Sci 26:1050–1060

Silva Sánchez C, De La Rosa AB, León Galván MF et al (2008) Bioactive peptides in amaranth (Amaranthus hypochondriacus) seed. J Agric Food Chem 56(4):1233–1240

Singh A, Kumari A, Chauhan AK (2022) Formulation and evaluation of novel functional snack bar with amaranth, rolled oat, and unripened banana peel powder. J Food Sci Technol 59(9):3511–3521

Steffensen SK, Rinnan Å, Mortensen AG et al (2011) Variations in the polyphenol content of seeds of field grown Amaranthus genotypes. Food Chem 129:131–138

Stetter MG, Vidal-Villarejo M, Schmid KJ (2020) Parallel seed color adaptation during multiple domestication attempts of an ancient new world grain. Mol Biol Evol 37:1407–1419

Suárez S, Aphalo P, Rinaldi G et al (2020) Effect of amaranth proteins on the RAS system. In vitro, in vivo and ex vivo assays. Food Chem 308:125601

Tagwira M, Tagwira F, Dugger R, Okum B (2006) Using grain amaranth to fight malnutrition. RUFORUM Work Doc 1:201–206

Tamsen M, Shekarchizadeh H, Soltanizadeh N (2018) Evaluation of wheat flour substitution with amaranth flour on chicken nugget properties. LWT – Food Sci Technol 91:580–587

Taniya MS, Reshma MV, Shanimol PS et al (2020) Bioactive peptides from amaranth seed protein hydrolysates induced apoptosis and antimigratory effects in breast cancer cells. Food Biosci 35:100588

Tsutsumi N, Tohya M, Nakashima T, Ueno O (2017) Variations in structural, biochemical, and physiological traits of photosynthesis and resource use efficiency in Amaranthus species (NAD-ME-type C4). Plant Prod Sci 20:300–312

Tumwegamire S, Kapinga R, Rubaihayo PR et al (2011) Evaluation of dry matter, protein, starch, sucrose, β-carotene, iron, zinc, calcium, and magnesium in East African sweetpotato [*Ipomoea batatas (L.) Lam*] germplasm. HortScience 46:348–357

Ulbricht C, Abrams T, Conquer J et al (2009) An evidence-based systematic review of amaranth (*Amaranthus* spp.) by the natural standard research collaboration. J Diet Suppl 6:390–417

Valdez-Meza EE, Raymundo A, Figueroa-Salcido OG et al (2019) Pasta enrichment with an amaranth hydrolysate affects the overall acceptability while maintaining antihypertensive properties. Foods 8:28

Venskutonis PR, Kraujalis P (2013) Nutritional components of amaranth seeds and vegetables: a review on composition, properties, and uses. Compr Rev Food Sci Food Saf 12:381–412

Verma AK, Rajkumar V, Kumar S (2019) Effect of amaranth and quinoa seed flour on rheological and physicochemical properties of goat meat nuggets. J Food Sci Technol 56:5027–5035

Zhu F (2020) Dietary fiber polysaccharides of amaranth, buckwheat and quinoa grains: a review of chemical structure, biological functions and food uses. Carbohydr Polym 248:116819

Chapter 12
Moringa (*Moringa oleifera*): Multi-functional Role in Management of Malnutrition and Health Promotion

Saima Perveen, Muhammad Tauseef Sultan, Muhammad Amir, and Muhammad Usman Khalid

12.1 Introduction

Recent climatic calamities and pandemics shattered the conventional cropping system, pushed the nations to hidden hunger and increased the concerns of food security. Contrarily the existing chronic malnutrition due to political instabilities and hikes in prices in underdeveloped countries is responsible for half of the deaths in children under five (Hendriks et al. 2022). This panic situation spiked the fear of generations to think about not only for their survival but also to live a healthy life. Alternate plant food sources that can be available in unavoidable circumstances become the area of a quest by the researchers (Global Risks Report of World Economic Forum 2023). Many non-conventional plants i.e., moringa (*Moringa oleifera* L.) emerged to meet the need of the communities. Moringa is a rich source of high value protein and many health promoting non-nutritive compounds. It is mostly adapted to sub-Himalayan tracts and many countries of Africa, South East and Middle East Asia with great potential to alleviate malnutrition (Ma 2020).

Nutritional profiling revealed that moringa has carbohydrate contents of 28.50–32.079%, protein (25.02–29.36%), dietary fiber (6.00–11.83%), fat (4.03–10.42%), β-carotene (1.108 mg), vitamin B1 (326.4 μg/100 g), and vitamin C (187.56–279.54 mg/100 g). Moringa is also a rich source of minerals like iron, zinc and calcium @28 mg, 25.5–31.03 mg and 1.322–2.645 mg/kg respectively (González-Burgos et al. 2021; Sultana 2020; Chhikara et al. 2021; Olusanya et al. 2020). It contains anti-cancerous compounds like glycosides, isothiocyanates, glucosinolates, glycerol-1-9-octadecanoate and other phytochemicals like saponins,

S. Perveen · M. Amir (✉) · M. Usman Khalid
Department of Food Science & Technology, Faculty of Food Science & Nutrition, Bahauddin Zakariya University, Multan, Pakistan

M. T. Sultan
Department of Human Nutrition, Bahauddin Zakariya University, Multan, Pakistan

T. Ismail et al. (eds.), *Neglected Plant Foods of South Asia*,
https://doi.org/10.1007/978-3-031-37077-9_12

tannins, terpenoids, sterols, alkaloids, anthraquinones and flavonoids (Sreeja et al. 2021; Patil et al. 2022). Due to its low calorific value and high dietary fiber contents, moringa leaves are used to manage obesity, thwart colon cancer and digestive problems. The leaf extract also boosts immune system and acts as antiulcerative agent to reduce acidity in patients suffering from gastric ulcers (Meireles et al. 2020). The extract also proved to be neuroprotectant and is used to cure dementia by decreasing the acetylcholine esterase activity and improving cholinergic and memory functions (Arozal et al. 2022).

Moringa can be cultivated through seeds and stem cuttings. The stem cutting is mostly practiced to raise the plants for early fruiting at the age of 6–8 months. The plant is ever green and bears leaves around the year but flowers emerge in spring and summer seasons while seeds are produced in autumn and winter (Abdoun et al. 2022). Traditionally, almost all parts of moringa such as leaves, buds, flowers, immature pods and seeds have long been consumed as vegetables. Consumption patterns of *M. oleifera* as a food vary across cultures and geographical settings (Islam et al. 2021). Various fractions of moringa are therefore used to develop value added products such as cookies, rice crackers, brownie, bread and snacks compounded with protein, fiber, minerals, vitamins and other non-nutrient compounds of health significance (Oyeyinka and Oyeyinka 2018). More recently the moringa protein isolates are being extracted to supplement weaning foods which serve as an alternative to animal protein to reduce the occurrences of protein energy malnutrition (Javed et al. 2021). Moringa leave powder is available in the market for sale at high prices and is used as herbal remedy to manage the risk of many chronic diseases (Su and Chen 2020).

Apart from the food uses moringa leaf extract is used as an insecticide and fungicide on the standing crop for better productivity and on soils to improve porosity. Other promising applications of moringa include biosorption of toxic metal ions, water purification through particle sedimentation, plant growth enhancer through leaf extract sprays, biodiesel production from seeds and biogas production from leaves (Koul and Chase 2015).

12.2 Food and Nutritional Aspects of Different Edible Accessions of Moringa

Various parts of *M. oleifera* are being used as food and pharmaceutical product to manage the risk of food security and chronic metabolic disorders. These promising nutritional and health-promoting properties are because of remarkable macro and micronutrients and bioactive compounds of moringa which are present in different parts i.e., leaves, flowers, pods, roots and bark. Data presented in Tables 12.1, 12.2, 12.3, 12.4 and 12.5 highlights nutritional and phytochemical composition of various fractions of moringa.

Table 12.1 Nutritional composition of different edible fractions of moringa

	Leaves		Flowers		Seeds		Roots	
Nutritional composition (%)	Ali et al. (2019)	Mikore and Mulugeta (2017)	Igwilo et al. (2011)	Arise et al. (2014)	Melo et al. (2013)	Adegbe et al. (2016)	Verma and Rajini (2014)	Abdulkadir et al. (2016)
Protein	15.23	24.3–24.8	25.99	25.16	32.19	28.02	17	11.70–26.57
Crude Fiber	9.26	5.51–6.13	–	7.55	15.87	–	45	21.03–44.27
Fat	2.38	7.11–7.35	9.4	1.57	32.40	33.78	11	11.82–20.19
Ash	8.85	15.3–15.6	3.57	6.01	–	3.03	3	5.59–9.28
Moisture	4.42	6.6–6.8	10.43	5.8	–	6.40	9	8.1–8.4
Carbohydrates	59.86	40.2–40.4	50.57	53.67	15.96	28.7	15	8.16–31.12

Table 12.2 Minerals composition of *M. Oleifera*

	Leaves		Flower		Seeds		Roots
Minerals composition (mg/100 g)	Alba-Romero and Sierra-Campos (2015)	Mikore and Mulugeta (2017)	Kumssa et al. (2017)	Pasha et al. (2020)	Sodamade et al. (2017)	Ijarotimi et al. (2013)	Fakankun et al. (2013)
Ca	2016–2602.5	1880.3–1902.6	365	288	1.55	128.33	1283.4
Mg	322.5–340.6	427.3–437	283	240	5.36	26.33	–
Zn	1.0–1.6	2.13–3.1	3.27	2.0	–	–	1.58
Fe	7.07–19.37	8.13–8.16	25.3	288	0.97	7.33	49.2
K	1817–1845	957–1216	–	1710	–	52.33	1886.3

12.3 Moringa's Potential in Prevention and Management of Malnutrition

The industrial dawn and promotion of processed foods not only affected the dietary patterns but also surged a shift in prices in developing countries. This change increased the risk of multidimensional food security i.e., export of high value food products to the developed nations, economical inaccessibility to processed foods, nutritional imbalance, fewer storage facilities in the local market and most importantly the changes in metabolic pathways due to added preservatives and additives affected digestion, absorption and utilization (Udmale et al. 2020). Along these factors climatic variability and disasters worsen the food and nutrition situation of developing and under developed countries. All the causes contributed to public

Table 12.3 Amino acid profile of *M. Oleifera*

Amino acid (g/100 g)	Leaves		Flowers		Seeds		Roots	
	Olaofe et al. (2013)	Okereke and Akaninwor (2013)	Sánchez-Machado et al. (2010)	Igwilo et al. (2017)	Okereke and Akaninwor (2013)	Barakat and Ghazal (2016)	Okereke and Akaninwor (2013)	Alli et al. (2017)
Glutamate	15.1	15.14	17.0	7.20	14.76	1.52–7.04	13.53	13.18
Aspartate	6.27	6.86	12.3	4.00	6.14	1.15–1.63	6.01	6.02
Leucine	5.12	5.22	8.7	5.20	5.27	1.05–2.08	5.02	5.08
Arginine	8.22	1.88	20.1	6.50	8.06	0.7–5.33	1.74	1.81
Lysine	3.69	3.60	4.6	3.00	3.24	0.61–0.83	3.62	3.48
Valine	3.15	3.36	6.4	2.00	3.09	0.82–1.45	3.03	3.04
Phenylalanine	4.62	4.53	3.8	3.90	4.53	0.68–1.66	3.98	3.88
Alanine	3.29	3.43	8.1	4.0	3.23	0.7–1.42	3.36	3.34
Isoleucine	4.40	2.33	5.2	–	–	–	1.84	1.9

Table 12.4 Phytochemical constituents isolated from *Moringa oleifera Lam*

Edible fraction	Phytochemicals composition	References
Mature flowers	D-glucose, D-mannose, polysaccharide, ascorbic acid, quercetin, isoquercetin, kaemopherol, and kaempferitin	Paikra and Gidwani (2017)
Leaves	Niazirinin, glycoside niazirin, and three mustard oil glycosides, 4-[4′-O-acetyl- α -L-rhamnosyloxy) benzyl] isothiocyanate, niaziminin A and B	Taher et al. (2017)
Root	4-(α-L-rhamnopyranosyloxy)-benzylglucosinolate, benzylglucosinolate, aurantiamide acetate, Procyanidins, chlorogenic acid rhamnoglucoside quercetin, perygospermin	Liu et al. (2018)
Seed	Cysteine, 4-(α-L-rhamnopyranosyloxy)- benzylglucosinolate, moringyne, benzylglucosinolate, methionine, mono-palmitic and di-oleic triglyceride, niazirin, benzylglucosinolate, and niazimicin,	Paikra and Gidwani (2017)
Stem	Vanillin, sterols, β-sitosterone, β-sitosterol, 4-hydroxymellein, octacosanic acid, triterpenoids, and procyanidin, glycosides, tannins, alkaloids, β-sitosterol and octacosanoic acid	Liu et al. (2018)
Whole pods	Nitriles, thiocarbanates, isothiocyanate, methyl-p-hydroxybenzoate, 0-[2′-hydroxy-3′-(2″-heptenyloxy)]-propylundecanoate, 0-ethyl-4-[(α-1-rhamnosyloxy)-benzyl] carbamate, and β-sitostero	Taher et al. (2017)
Seed oil	Vitamin A, Beta carotene,	

Table 12.5 Concentration of other phenolic and flavonoids of *M.oleifera* (μg/g)

Compounds	Leaf	Flower	Seed	Root	Bark
Myricetin	1530	890	640	1100	1360
o-Coumaric acid	37	17	17	15	29
Quercetin	985	583	600	685	870
Naringin	22	12	21	8	30
Kaempferol	40	45	25	24	23
Catechin	25	12	20	20	13
Naringenin	14	14	14	23	29
Biochanin A	29	34	17	45	17
Resveratrol	19	25	20	19	14

health problems and increased disease burden. Approximately 155M children of 5 years and 45% of those below five are suffering from under nutrition while 41M are obese or overweight. This hidden enemy only can be managed by exploring the local flora and their utilization in the daily dietary routines (Pradhan et al. 2016).

Moringa olifera is one of the plants in tropical to sub-tropical regions that has prominent potential to manage malnutrition because of its valued nutritional profile (Sonewane et al. 2022). Moringa contains high value protein with essential amino acids like arginine and histidine that play a vital roles in cell division. Moringa leaves due to their high protein contents are being supplemented in various conventional recipes to improve the nutritional status of children, nursing mothers and pregnant women. James and Zikankuba (2017) observed that six tablespoons of Moringa leaves powder provide calcium and iron that can meet-up the daily breast-feeding requirements of pregnant women. Another study by Gopalakrishnan et al. (2016) reported that moringa leave powder supplementation in the daily diet produced higher concentrations of milk in lactating mothers. The lactogogue comprised of phytosterols works as a forerunner of various reproductive growth hormones and enhanced the milk formation rate in lactating mothers.

Moringa was reported to contain appreciable amounts of phytosterols i.e., kampesterol, sitosterol and stigma sterol which are good precursors of growth hormones. Moreover, these phytosterols enhance the concentrations of estrogen which stimulates the ducts of mammary glands to excrete the milk. Moringa leaves and pods are beneficial for the health of breast feeding and pregnant women, their feeding children and fetus accordingly. This commonly known miracle tree has the potential to combat malnutrition in both the mothers and child when used at optimum doses (Singh and Chaturvedi 2021).

Moringa leaves are rich in dietary fiber which modulates gut health and prevents epigastric regions from ulcerative changes. Apart from protein, carbohydrates and dietary fiber, moringa leaves and flowers are full of minerals i.e., it contains more potassium than bananas, iron than spinach, calcium than milk and more vitamin A than carrots. 100 g of moringa leaves powder can meet-up 1/3rd of daily requirements of calcium. The leaves are also good sources of zinc, copper, iron, and vitamin B that help to manage different metabolic diseases (Pokhrel et al. 2016). To eradicate the malnutrition among the populations, the identification of various locally available cheaper and more nutritious food sources, moringa is the demand of the current era. So from the above discussion we can conclude that Moringa a has high potential to reduce malnutrition among poor and lower middle income families.

12.4 Health Promoting Properties of Moringa

Moringa oleifera L. is cultivated plant in the developing countries of South Asia. In the region, various parts are used for the preparation of pickles i.e., buds and immature pods while leaves, buds and flowers are used for cooking purposes. *M. oleifera* has long been widely used to relieve various diseases, such as diabetes, obesity, hysteria, scurvy, and even tumors. The various parts of *M. oleifera* such as roots, roots bark, seeds, fruits, and leaves are used for the treatment of various skin infections, helminthic, abdominal tumors, prostates cancer, sores, and paralytic attacks (Islam et al. 2021). The plant also contains flavonoid, quercetin which was recently recommended for the prophylaxis and treatment of COVID-19. Moringa roots are used to reduce ulcerative colitis which is useful to prevent lining in the large intestine (Gholap et al. 2012). Research conducted on various health-promoting properties of moringa edible fraction is given in Table 12.6.

Table 12.6 Health benefits of edible fraction of *M.oleifera*

Reference study	Active compound	Part used	Dosage	Health benefits
Abou-zeid et al. (2021)	Eugenol, Isopropanol	Dry leaf	400–800 mg/kg	Anti-inflammatory Decreases nephrotoxicity
Anigboro et al. (2019)	Flavonoids	Dry leaves	100 mg/kg	Decreases risk of cancer
Mohammed et al. (2022)	Polyphenols, flavonoids, terpenoids	Fresh leaves powder	4 mg/L	Antibacterial Anticancer
Aziz et al. (2021)	Flavonoids, tannins	Leaf, bark, stem	100, 300, 1000 mg/kg	Decrease blood glucose level (anti diabetic level) Anti-diarrheal
Brar et al. (2022)	Phytochemicals, phytosterols (kaempferol, sitosterol, stigmetrol)	Leaves powder	14–30 g/day 0.5 g/day	Increase Hb in children Increase vitamin A in postmenopausal women and milk volume in women
Muzumbukilwa et al. (2019)	Flavonoids	Moringa leaf	500 mg/kg	Hepatoprotective Antidiabetic
Soltan et al. (2018)	Fiber part	Moringa roots	25 and 50 g/kg	Nutrient digestibility
Tshabalala et al. (2020)	Steroids, saponins, tannins, flavonoids	Root and leaf	100–400 mg/kg	Anti-microbial Anti-diabetic Antioxidant property

(continued)

12.5 Challenges and Opportunities for Moringa Growers in South Asia

Moringa is a perennial plant indigenous to South Asia and grows best in the arid to semiarid zones of East Africa, India, Afghanistan, Pakistan and Saudi Arabia where annual rainfall is below 300 mm. The plant is mostly found in the Himalayan foothills from northeastern Pakistan to northern West Bengal and India, especially in dry lands, semi-deserts and irrigated areas where the water table is high (Mahmood et al. 2010). Due to the region's rapid population increase, pervasive poverty, high reliance on natural resources, climate change and insufficient capacity for adaptation, South Asia is extremely vulnerable to food and nutritional security (Dhyani et al. 2021). This culturally rich area having a variety of races and ecosystems is totally dependent on their farm produce. Smallholder farmers, who dominate the agriculture sector in developing countries are being affected by abrupt onset of rainfalls, soil erosion, and intensive farming leading to loss of soil nutrient density and global warming. These circumstances are deadly to the conventional varieties resulting in lower productivity and less nutrient availability. The plants like moringa that are adapted to diverse climatic conditions have the potential to grow best with more nutrients availability. Moringa can yield in a variety of agroecological zones and was domesticated because of its adaptability to thrive in less fertile soils and is drought and frost-resistant. The specie is anticipated to be more stable and better adapted to future global climatic shifts as an alleviation intervention because of its wide range of adaptability (Sagona et al. 2020).

Despite of its adaptability to plains moringa may be grown at elevations up to 2000 m. It is propagated via stem cutting (1–2 m long) during the months of June and July in low mountains and from January to March in plains depending upon seasonal variability and flowers harvesting. In commercial farming, it is clipped up to three meters to make the harvest of the buds, flowers, pods and leaves easier (Pandey et al. 2011). It is an adaptive species, and may grow from 5 to 10 m tall in a short duration of time i.e., 4 m in just six months (Milla et al. 2021). Moringa has a lifespan of roughly 20 years and is a common plant in hot and low-lying semi-arid regions. The productivity of the Moringa species may be influenced by a number of critical factors, including soil quality and severe droughts. The specie needs loamy or sandy soils ranging from mildly acidic to alkaline (pH of between 5 and 9). However, it won't thrive on clay soils but amendment can boost productivity and growth (Devkota and Bhusal 2020).

Moringa is resistant to the majority of pests but cotton mealybug and hairy caterpillars can attack and cause root rot and defoliation and pose considerable risks, particularly to newly emerging leaves and planted stumps. The caterpillar (*Eupterote mollifera* Walker and *Noorda blitealis* Walker), can cause flower drop and the seasonal occurrence of fruit flies by Gitona spp. Perennial varieties face several production challenges, including a considerably longer pre-fruit bearing stage, the need for rainier days in areas with limited water supplies, and vulnerability to pests and diseases (Pandey et al. 2011). The worldwide programs for the establishment of moringa such as the Moringa Fund, Trees for Life, Global Moringa Network,

Table 12.6 (continued)

Reference study	Active compound	Part used	Dosage	Health benefits
Choudhary et al. (2013)	Alkaloids, carbohydrates, proteins, tannins, phenols, saponins, and steroids	Root bark	150, 350, and 500 mg/kg	Anti-ulcer
Bose (2007)	0.2% alkaloids	Root bark	1–10 mg/kg	Hypotensive and bradycardia effects Ovarian cancer
Cui et al. (2019)	Polysaccharides	Roots	20 g/kg	Increases in NO and TNF-α production that act as anti-inflammatory
Mohammed et al. (2022)	Flavonoids	Moringa whole plant	50 and 100 mg/day	Remedy to fight malnutrition. Decreased lipid peroxides and increased glutathione
Nwachukwu (2021)	Phytochemicals	Stem bark	200 mg/kg	Anti-diabetic
Niranjan et al. (2017)	Kaempferol, gallic acid, vanillic acid, coumaric acid and quercetin	Moringa leaves	150, 300 and 600 mg/kg	Free radical scavenging activity and reducing power
Famurewa et al. (2019)	Polyphenols, flavonoids	Moringa oliefera seed oil	5 ml/kg body weight/day	Antioxidant, anti-inflammatory and antiapoptotic mechanisms against anticancer drug 5-fluorouracil
Choudhary et al. (2013)	β-carotene, vitamin C, protein, iron, and potassium, tannins	Roots bark	150, 350, and 500 mg/kg	Decreased the ulcer Antisecretory, and cytoprotective activity
Satish et al. (2014)	Glutathione, a-Tocopherol polyphenols,	Root extract	15 g sample with 50 ml solvent	Antioxidant properties were established on free radicals; its ability to reduce iron and FeSO4-induced microsomal lipid peroxidation Protect DNA from oxidative damage
Abd-Rabou et al. (2017)	Carotenoids, phenolic compounds	Root core and root outer part	5, 10, or 15% w/w	Root core extract could kill the majority of cancer cells (approximately 70–80%)
Fatoba et al. (2013)	Alkaloid, Flavonoids, saponin and cyanogenic glycosides, tannins, and oxalates	Root	10, 15 mg/mL	Increases in sperm concentration in albino rat

Moringa Initiative, and Moringa News, have improved production and developed marketing strategies for value addition (Mridha 2015). Other than food uses various fractions of moringa can be used as a fertilizer and soil improver. Moringa leave extract is also used as bio-insecticide against aphids and jassids on wheat crop (Gandji et al. 2018).

12.6 Resilience to Climate Change

Climate change has significantly affected the floral diversity in a wide geographical setting and raised concerns of droughts and uneven rainfalls. Recent floods in Pakistan and coastal areas of India damaged many standing trees due to water stress. These constant threats are deadly to the regional flora (Pörtner et al. 2022). Plants like moringa have adopted to efficient utilization of water, oxygen, carbon dioxide and soil conditions. Even in a dry climate, the moringa plant and its bushes can thrive for months. This is due to its isohydric behavior i.e., when there is water stress moringa closes its stomata due to the accumulation of abscisic acid leading to inhibition of the photosynthetic pathways and negatively impacting the biomass production (Brunetti et al. 2018). But there is no impairment in the photochemistry due to which saturating photosynthesis and maximal fluorescence in the stressed leaves remain unaffected. This climate resilient behavior is due to the production of isoprene and their production increases three times during the stress conditions. The hypothesis is linked to the antioxidant and membrane stabilizing properties of isoprene as these are part of thylakoids responsible for photosynthesis (Siddique et al. 2016). Additionally, there observed an increased concentration of antioxidant flavonoid in the stressed leaves probably responsible for limiting the irreversible changes. These flavonoids and isoprenoids are considered as a key mechanism allowing moringa to lower lipid peroxidation and limit the photo-inhibitory process during severe water stress (Noreen et al. 2016). This mechanism also allows rapid recovery of photosynthesis when water is available to the root system and increases the growth rate. Phenotypic and metabolic plasticity also have been correlated with environmental adaptation of moringa in dry conditions, thus increasing its economic value. Moringa grows best in the temperature range of 25–35 °C but can also survive at elevated temperatures in the summer season i.e., 48 °C and is able to tolerate frost in winter for a short duration (Mashamaite et al. 2021). Under severe stress condition the young green leaves turn yellow due to the loss of chlorophyll and the synthesis of carotenoids. These carotenoids act as antioxidants and increase the survival of the plant. It was observed that calcium, iron, glucosinolates and antioxidant contents were higher in the leaves of the hot and humid season while zinc and carbohydrate contents were more in leaves of the cool-dry season. Some hydrolytic enzymes also play a vital role in the physiological maturity of the leaves and are responsible for the accumulation of more nutrients (Okumu et al. 2016). These factors are the base of the environmental adaptability of the moringa tree under climatic stress conditions.

12.7 Toxicological Aspects and Remedial Strategies

Anti-nutritional compounds are secondary metabolites mostly responsible for defense mechanisms in the plant system but can pose serious challenges to human health when consumed in abundance. These may have deleterious toxic responses in the absorption and utilization of nutrients and are linked with impaired metabolic pathways resulting in chronic disorders. In terms of toxicological burden, the moringa flower has phytic acid (0.436 mg/100 g), tannin (54.61–60 mg/100 g), saponins (15.20–46.51 mg/100 g), alkaloids (30.59 mg/100 g), oxalates (51.23 mg/100 g), and Cyanogenic glycosides of 4.31 mg/100 g respectively (Oyeyemi et al. 2017; Igwilo et al. 2011). These are generally and more specifically linked to poor absorption of iron and other divalent ions. Antinutrient contents in moringa leaves are limiting factors for utilization of minerals and are major safety challenges. Moringa leaves contain 9.0% phytate, 0.02% tannins, Alkaloids 0.33–2.80%, and saponins 0.45% which provides a bitter taste (Olumide et al. 2019; Arowosegbe 2019). High concentrations of these antinutrients adversely affect the absorption of trace elements in food sources and hinder protein digestion. Moreover, moringa leaves have a low proportion of trypsin inhibitors 1.45 TIU mg g^{-1}, oxalic acid 10.5 mg g^{-1} and 17 mg g^{-1} of nitrate (Teixeira et al. 2014). Moringa leaves have a low level of anti-nutritional factors compared to other leaf vegetables and produce no toxic effect on consumption. The low level of anti-nutritional factors compared to spinach and *Brassica spp* is a contributing factor towards the acceptability and wide uses of *M. oleifera* as a vegetable (James and Zikankuba 2017).

The potential toxicants in the edible fractions of moringa are a concern to the consumers because these are negatively correlated with metabolic disorders. There is a need to develop value added comer oriented products to enhance the consumption of moringa. Many approaches like fermentation and various physical treatments are being developed to reduce the toxicant burden in the final products (Saucedo-Pompa et al. 2018). Other treatments include water extraction, salt extraction (Stevens et al. 2015), differential solubility, enzyme treatment, microwave processing (Suhag et al. 2021), membrane processing (Fetzer et al. 2019), calcium chloride treatment (Tan et al. 2011), irradiation (Shawrang et al. 2011) and ion-exchange (Karlovsky et al. 2016). Processing methods, like blanching, sprouting, fermentation drying, and defatting resulted in reduced toxicological activity (Devisetti et al. 2016). Reduction in intrinsic toxicants or antinutrients' burden of moringa has proven to be promising in increasing nutritional quality of the plant.

12.8 Moringa Derived Value-Added Food Products

The high nutritional profile of various parts of moringa is the base of consumer approach. The leaves, flowers and seeds, in fresh and dry forms, have been explored as natural fortificant in cereals, dairy and confectionary products, and are the focus of recent era research (Raja et al. 2016). Moringa leaves contain high value protein

Table 12.7 Moringa derived value-added food products

Products	Parts used	Supplementation	Results	Author
Rice Crackers	ML	1, 2 and 5%	Improve shelf life, Increase the level of β carotene, calcium and Vitamin C	Milla et al. (2021)
Unleavened Flat Bread	ML	5, 10, 15, 20%	Acceptable <10% of leaf powder	Mushtaq et al. (2021)
Snack	LP	10–30%	Increased the nutritional values	Rweyemamu et al. (2015)
Noodles	LP	3, 4, 5, and 6%	Moringa leaves fiber noodles are known as low glycemic index food	Ganga et al. (2019)
Biscuits	MP ML MF	0.2–5% 5–15 g 11–15%	To increase the nutrient density and dietary fiber	Giuberti et al. (2021), Yadav et al. (2022)
Soup	LP Seeds flower	6% 0.5%	To improve the health and nutritional status	Mohajan et al. (2018) Radha et al. (2015)
Yogurt	Seeds extract Leaves extract	5, 10, 15 ml/L 1%	Yogurt has the potential to be exploited in commercial use	Quintanilha et al. (2021) El-Ziney et al. (2017)
Cheese	Seed extract	0.5, 1, 1.5, and 2%	Increased protein and minerals in cottage cheese	Mahami et al. (2012)

and are being used as herbal remedy in the form of powder. Moringa powder supplementation increased the letting of milk in breast feeding women. The protein isolates have been added to the weaning foods to reduce protein energy malnutrition (Javed et al. 2021). Various bakery products i.e., biscuits, bread, cereal porridge, cakes and leaf powder supplemented popcorn have been developed to mitigate chronic malnutrition and to test consumer acceptability. It provides a variety of minerals, essential amino acids, and vitamins to improve overall health. Moreover, consumer scale sensory revealed that moringa powder and protein isolate supplementation increased the organoleptic, physicochemical, quality characteristics and safety of food products (Adewumi et al. 2022). There is a need to bridge the gaps between the farmer and corporate sectors to improve the utilization of moringa by highlighting the market potential and health and economic perspectives of the moringa (Table 12.7).

12.9 Conclusion

Food security has become an inevitable challenge for the developing countries and is adversely affecting the health system in terms of increased disease burden and failing economies. Alternative plant food sources i.e., moringa a perennial plant cultivated in semiarid, tropical and subtropical areas have esteemed nutritional composition. There is a need to promote moringa cultivation to promote food security and to increase economic resources of the small land holder farmers. There should be policies at government level for value chain addition and export of moringa based products. The government must support the industries to take steps to develop moringa based products. The promotion of moringa at industrial and farmer level can reduce the risk of food security in changing climate.

Conflicts of Interest All authors read and approved the final manuscript and declare that they have no conflict of interest.

References

Abdoun K, Alsagan A, Altahir O, Suliman G, Al-Haidary A, Alsaiady M (2022) Cultivation and uses of Moringa oleifera as non-conventional feed stuff in livestock production: a review. Life 13:63

Abd-Rabou AA, Abdalla AM, Ali NA, Zoheir KM (2017) Moringa oleifera root induces cancer apoptosis more effectively than leave nanocomposites and its free counterpart. Asian Pac J Cancer Prev 18:2141

Abdulkadir AR, Zawawi DD, Jahan MS (2016) Proximate and phytochemical screening of different parts of Moringa oleifera. Russ Agric Sci 42:34–36

Abou-Zeid SM, Ahmed AI, Awad A, Mohammed WA, Metwally MM, Almeer R, Khalil SR (2021) Moringa oleifera ethanolic extract attenuates tilmicosin-induced renal damage in male rats via suppression of oxidative stress, inflammatory injury, and intermediate filament proteins mRNA expression. Biomed Pharmacother 133:110997

Adegbe AA, Larayetan RA, Omojuwa TJ (2016) Proximate analysis, physicochemical properties and chemical constituents characterization of Moringa oleifera (Moringaceae) seed oil using GC-MS analysis. Am J Chem 6:23–28

Adewumi OO, Felix-Minnaar JV, Jideani VA (2022) Functional properties and amino acid profile of bambara groundnut and Moringa oleifera leaf protein complex. PRO 10:205

Alba-Romero JJ, Sierra-Campos E (2015) Nutritional content and elemental and phytochemical analyses of Moringa oleifera grown in Mexico. J Chem 2015:1–9

Ali AM, Tajo ST, Zage AU, Ali M (2019) Phytochemical screening, proximate and mineral analysis of Moringa oleifera Leaf in Kano, Northern Nigeria. J Med Pharm Allied Sci 1:55–60

Alli IJ, Uzor BC, Korie MC (2017) Microbiological and nutritional analysis of roots and seeds of Moringa oleifera. Int J Res Pharm Biosci 4:19–24

Anigboro AA, Avwioroko OJ, Ohwokevwo OA, Pessu B (2019) Bioactive components of Ficus exasperata, Moringa oleifera and Jatropha tanjorensis leaf extracts and evaluation of their antioxidant properties. EurAsian J Biosci 13:1763–1769

Arise AK, Arise RO, Sanusi MO, Esan OT, Oyeyinka SA (2014) Effect of Moringa oleifera flower fortification on the nutritional quality and sensory properties of weaning food. Croat J Food Sci Technol 6:65–71

Arowosegbe S (2019) Comparative studies on the phytochemical content and proximate composition of different parts of Moringa oleifera plant grown in Ado-Ekiti, Nigeria. Bull Pure Appl Sci Bot 38:1–10

Arozal W, Purwoningsih E, Lee HJ, Barinda AJ, Munim A (2022) Effects of Moringa oleifera in two independents formulation and as neuroprotective agent against scopolamine-induced memory impairment in mice. Front Nutr 9:799127

Aziz NA, Jayasuriya N, Fan L, Al-Gheethi A (2021) A low-cost treatment system for underground water using Moringa oleifera seeds and Musa cavendish peels for remote communities. J Chem Technol Biotechnol 96:680–696
Barakat H, Ghazal GA (2016) Physicochemical properties of Moringa oleifera seeds and their edible oil cultivated at different regions in Egypt. Food Sci Nutr 7:472
Bose CK (2007) Possible role of Moringa oleifera Lam. root in epithelial ovarian cancer. Medscape J Med 9:26
Brar S, Haugh C, Robertson N, Owuor PM, Waterman C, Fuchs GJ III, Attia SL (2022) The impact of Moringa oleifera leaf supplementation on human and animal nutrition, growth, and milk production: a systematic review. Phytother Res 36:1600–1615
Brunetti C, Loreto F, Ferrini F, Gori A, Guidi L, Remorini D, Centritto M, Fini A, Tattini M (2018) Metabolic plasticity in the hygrophyte Moringa oleifera exposed to water stress. Tree Physiol 38:1640–1654
Chhikara N, Kaur A, Mann S, Garg MK, Sofi SA, Panghal A (2021) Bioactive compounds, associated health benefits and safety considerations of Moringa oleifera L.: an updated review. Nutr Food Sci 51:255–277
Choudhary MK, Bodakhe SH, Gupta SK (2013) Assessment of the antiulcer potential of Moringa oleifera root-bark extract in rats. J Acupunct Meridian Stud 6:214–220
Cui C, Chen S, Wang X, Yuan G, Jiang F, Chen X, Wang L (2019) Characterization of Moringa oleifera roots polysaccharide MRP-1 with anti-inflammatory effect. Int J Biol Macromol 132:844–851
Devisetti R, Sreerama YN, Bhattacharya S (2016) Processing effects on bioactive components and functional properties of moringa leaves: development of a snack and quality evaluation. J Food Sci Technol 53:649–657
Devkota S, Bhusal KK (2020) Moringa oleifera: a miracle multipurpose tree for agroforestry and climate change mitigation from the Himalayas–a review. Cogent Food Agric 6:1805951
Dhyani S, Murthy IK, Kadaverugu R, Dasgupta R, Kumar M, Adesh Gadpayle K (2021) Agroforestry to achieve global climate adaptation and mitigation targets: are South Asian countries sufficiently prepared? Forests 12:303
El-Ziney MG, Shokery ES, Youssef AH, Mashaly RE (2017) Protective effects of green tea and moringa leave extracts and their bio-yogurts against oxidative effects of lead acetate in albino rats. JNHFS 5:1–11
Fakankun OA, Babayemi JO, Utiaruk JJ (2013) Variations in the mineral composition and heavy metals content of Moringa oleifera. Afr J Environ Sci Technol 7:371–379
Famurewa AC, Asogwa NT, Aja PM, Akunna GG, Awoke JN, Ekeleme-Egedigwe CA, Nwoha PA (2019) Moringa oleifera seed oil modulates redox imbalance and iNOS/NF-κB/caspase-3 signaling pathway to exert antioxidant, anti-inflammatory and antiapoptotic mechanisms against anticancer drug 5-fluorouracil-induced nephrotoxicity in rats. S Afr J Bot 127:96–103
Fatoba TA, Faleyimu OI, Adebayo AJ (2013) Effects of increasing aqueous root extract of Moringa oleifera on sperm production of Albino Rats. Agrosearch 13:29–36
Fetzer A, Herfellner T, Eisner P (2019) Rapeseed protein concentrates for non-food applications prepared from pre-pressed and cold-pressed press cake via acidic precipitation and ultrafiltration. Ind Crop Prod 132:396–406
Gandji K, Chadare FJ, Idohou R, Salako VK, Assogbadjo AE, Kakaï RG (2018) Status and utilisation of Moringa oleifera Lam: a review. Afr Crop Sci J 26:137–156
Ganga MU, Karthiayani A, Vasanthi G, Baskaran D (2019) Study on development of fiber-enriched noodles using Moringa leaves (Moringa olifera). Asian J Dairy Food Res 38:145–149
Gholap PA, Nirmal SA, Pattan SR, Pal SC, Mandal SC (2012) Potential of Moringa oleifera root and Citrus sinensis fruit rind extracts in the treatment of ulcerative colitis in mice. Pharm Biol 50:1297–1302
Giuberti G, Bresciani A, Cervini M, Frustace A, Marti A (2021) Moringa oleifera L. leaf powder as ingredient in gluten-free biscuits: nutritional and physicochemical characteristics. Eur Food Res Technol 247:687–694

Global Risks Report of World Economic Forum (2023). https://www3.weforum.org/docs/WEF_Global_Risks_Report_2023.pdf
González-Burgos E, Ureña-Vacas I, Sánchez M, Gómez-Serranillos MP (2021) Nutritional value of Moringa Oleifera Lam. leaf powder extracts and their neuroprotective effects via antioxidative and mitochondrial regulation. Nutrients 13:2203
Gopalakrishnan L, Doriya K, Kumar DS (2016) Moringa oleifera: a review on nutritive importance and its medicinal application. Food Sci Hum Wellness 5:49–56
Hendriks SL, Montgomery H, Benton T, Badiane O, de la Mata GC, Fanzo J, Guinto RR, Soussana JF (2022) Global environmental climate change, covid-19, and conflict threaten food security and nutrition. BMJ 378:e071534
Igwilo IO, Ezeonu FC, Udedi SC, Ubaoji KI, Ozumba NA, Ogbunugafor HA (2011) Nutrient composition and anti-nutritional factors in a local cultivar of Moringa oleifera (Lam) flowers. Nat Prod: Indian J NPAIJ 7:86–90
Igwilo IO, Ugochukwu GC, Ezekwesili CN, Nwenyi V (2017) Comparative studies on the nutrient composition and antinutritional factors in different parts of Moringa oleifera plant found in awka, Nigeria. Biosci J 5:1–12
Ijarotimi OS, Adeoti OA, Ariyo O (2013) Comparative study on nutrient composition, phytochemical, and functional characteristics of raw, germinated, and fermented Moringa oleifera seed flour. Food Sci Nutr 1:452–463
Islam Z, Islam SM, Hossen F, Mahtab-ul-Islam K, Hasan M, Karim R (2021) Moringa oleifera is a prominent source of nutrients with potential health benefits. Int J Food Sci. https://doi.org/10.1155/2021/6627265
James A, Zikankuba V (2017) Moringa oleifera a potential tree for nutrition security in sub-Sahara Africa. Am J Res Commun 5:1–12
Javed MS, Amjad A, Shah M, Shah FUH, Sardar H, Tariq MR, Khan AA, Sajid MW, Ali U, Amir M, Nasir F (2021) Isolation and characterization of moringa oleifera l. Flower protein and utilization in functional food bars. Food Sci Technol 41:643–652
Karlovsky P, Suman M, Berthiller F, De Meester J, Eisenbrand G, Perrin I, Dussort P (2016) Impact of food processing and detoxification treatments on mycotoxin contamination. Mycotoxin Res 32:179–205
Koul B, Chase N (2015) Moringa oleifera Lam.: Panacea to several maladies. J Chem Pharm Res 7:687–707
Kumssa DB, Joy EJ, Young SD, Odee DW, Ander EL, Broadley MR (2017) Variation in the mineral element concentration of Moringa oleifera Lam. and M. stenopetala (Bak. f.) Cuf.: role in human nutrition. PLoS One 12:e0175503
Liu Y, Wang XY, Wei XM, Gao ZT, Han JP (2018) Values, properties and utility of different parts of Moringa oleifera: an overview. Chin Herb Med 10:371–378
Ma ZF, Ahmad J, Zhang H, Khan et al (2020) Evaluation of phytochemical and medicinal properties of Moringa (Moringa oleifera) as a potential functional food. S. Afr J Bot 129:40–46
Mahami T, Ocloo FCK, Odonkor ST, Owulah C, Coffie SA (2012) Preliminary study on the influence of moringa seed extracts supplementation on the yield and quality of cottage cheese. Int J Recent Trends Sci Technol 2:4–8
Mahmood KT, Mugal T, Haq IU (2010) Moringa oleifera: a natural gift-A review. J Pharm Sci Res 2:775
Mashamaite CV, Pieterse PJ, Mothapo PN, Phiri EE (2021) Moringa oleifera in South Africa: a review on its production, growing conditions and consumption as a food source. S Afr J Sci 117:1–7
Meireles D, Gomes J, Lopes L, Hinzmann M, Machado J (2020) A review of properties, nutritional and pharmaceutical applications of Moringa oleifera: integrative approach on conventional and traditional Asian medicine. Adv Trad Med 20:495–515
Melo V, Vargas N, Quirino T, Calvo CMC (2013) Moringa oleifera L. An underutilized tree with macronutrients for human health. Emir J Food Agric 25:785–789

Mikore D, Mulugeta E (2017) Determination of proximate and mineral compositions of Moringa oleifera and Moringa stenopetala leaves cultivated in Arbaminch Zuria and Konso, Ethiopia. Afr J Biotechnol 16:808–818
Milla PG, Peñalver R, Nieto G (2021) Health benefits of uses and applications of Moringa oleifera in bakery products. Plan Theory 10:318
Mohajan S, Orchy TN, Farzana T (2018) Effect of incorporation of soy flour on functional, nutritional, and sensory properties of mushroom–moringa-supplemented healthy soup. Food Sci Nutr 6:549–556
Mohammed AB, Mohamed A, El-Naggar NEA, Mahrous H, Nasr GM, Abdella A, Ali AS (2022) Antioxidant and antibacterial activities of silver nanoparticles biosynthesized by Moringa oleifera through response surface methodology. J Nanomater 2022:1–15
Mridha MAU (2015) Prospects of moringa cultivation in Saudi Arabia. J Appl Environ Biol Sci 5:39–46
Mushtaq BS, Pasha I, Omer R, Hussain MB, Tufail T, Shariati MA, Kharitonov DV (2021) Characterization of Moringa oleifera leaves and its utilization as value added ingredient in unleavened flat bread (chapatti). J Microbiol Biotechnol Food Sci 2021:750–755
Muzumbukilwa WT, Nlooto M, Owira PMO (2019) Hepatoprotective effects of Moringa oleifera Lam (Moringaceae) leaf extracts in streptozotocin-induced diabetes in rats. J Funct Foods 57:75–82
Niranjan A, Ngpoore NK, Anis N, Kumar A, Lehri A, Shirke PA, Tewari SK (2017) Simultaneous quantification of six phenolic compounds in various parts of Moringa oleifera Lam. using high-performance thin-layer chromatography. J Planar Chromatogr – Mod TLC 30:502–509
Noreen S, Mahmood S, Athar HR, Zafar Z, Ashraf M (2016) Potential usage of antioxidants, hormones and plant extracts. In: Water stress and crop plants: a sustainable approach. Wiley, pp 124–141. https://doi.org/10.1002/9781119054450.ch9
Nwachukwu CF (2021) Partial elucidation on the remedial effect of stem-bark extract of Moringa oleifera on diabetic rats. GSC Biol Pharm Sci 15:111–116
Okereke CJ, Akaninwor JO (2013) The protein quality of raw leaf, seed and root of Moringa oleifera grown in Rivers State, Nigeria. Ann Biol Res 4:34–38
Okumu MO, Mbaria JM, Kanja LW, Gakuya DW, Kiama SG, Ochola FO (2016) Phytochemical profile and antioxidant capacity of leaves of Moringa oleifera (Lam) extracted using different solvent systems. J Pharmacogn Phytochem 5:302
Olaofe O, Adeyeye EI, Ojugbo S (2013) Comparative study of proximate, amino acids and fatty acids of Moringa oleifera tree. Elixir Appl Chem 54:12543–12554
Olumide MD, Ajayi OA, Akinboye OE (2019) Comparative study of proximate, mineral and phytochemical analysis of the leaves of Ocimum gratissimum, Vernonia amygdalina and Moringa oleifera. J Med Plant Res 13:351–356
Olusanya RN, Kolanisi U, Van Onselen A, Ngobese NZ, Siwela M (2020) Nutritional composition and consumer acceptability of Moringa oleifera leaf powder (MOLP)-supplemented mahewu. S Afr J Bot 129:175–180
Oyeyemi SD, Arowosegbe S, Famosa MA (2017) Phytochemical constituents and nutritional evaluation of three selected edible flowers in Ado-Ekiti, Nigeria. ChemSearch J 8:41–48
Oyeyinka AT, Oyeyinka SA (2018) Moringa oleifera as a food fortificant: recent trends and prospects. J Saudi Soc Agric Sci 17:127–136
Paikra BK, Gidwani B (2017) Phytochemistry and pharmacology of Moringa oleifera Lam. J Pharmacopuncture 20:194
Pandey A, Pradheep K, Gupta R, Nayar ER, Bhandari DC (2011) 'Drumstick tree' (Moringa oleifera Lam.): a multipurpose potential species in India. Genet Resour Crop Evol 58:453–460
Pasha SN, Shafi KM, Joshi AG, Meenakshi I, Harini K, Mahita J, Sowdhamini R (2020) The transcriptome enables the identification of candidate genes behind medicinal value of Drumstick tree (Moringa oleifera). Genomics 112:621–628
Patil SV, Mohite BV, Marathe KR, Salunkhe NS, Marathe V, Patil VS (2022) Moringa tree, gift of nature: a review on nutritional and industrial potential. Curr Pharma Rep 8:262–280

Pokhrel CP, Timilsina A, Yadav RKP, Khanal R (2016) Moringa Oleifere: a potential cash crop in Nepal. International symposium on Healthy Society & Healthy World. Kathmandu, pp 33–41
Pörtner HO, Roberts DC, Adams H, Adler C, Aldunce P, Ali E, Begum RA, Betts R, Kerr RB, Biesbroek R, Ibrahim ZZ (2022) Climate change 2022: impacts, adaptation and vulnerability. IPCC, Geneva, p 3056
Pradhan PM, Dhital R, Subhani H (2016) Nutrition interventions for children aged less than 5 years following natural disasters: a systematic review. BMJ Open 6:e011238
Quintanilha GEO, Baptista ATA, Gomes RG, Vieira AMS (2021) Yogurt production added ultrafiltered seed extract of Moringa oleifera Lam. Biocatal Agric Biotechnol 37:102159
Radha C, Ogunsina BS, Hebina Babu KT (2015) Some quality and micro-structural characteristics of soup enriched with debittered Moringa oleifera seeds flour. Am J Food Technol 3:145–149
Raja W, Bera K, Ray B (2016) Polysaccharides from Moringa oleifera gum: structural elements, interaction with β-lactoglobulin and antioxidative activity. RSC Adv 6:75699–75706
Rweyemamu LM, Yusuph A, Mrema GD (2015) Physical properties of extruded snacks enriched with soybean and moringa leaf powder. Afr J Food Sci 6:28–34
Sagona WCJ, Chirwa PW, Sajidu SM (2020) The miracle mix of Moringa: status of Moringa research and development in Malawi. S Afr J Bot 129:138–145
Sánchez-Machado DI, Núñez-Gastélum JA, Reyes-Moreno C, Ramírez-Wong B, López-Cervantes J (2010) Nutritional quality of edible parts of Moringa oleifera. Food Anal Methods 3:175–180
Satish A, Reddy PV, Sairam S, Ahmed F, Urooj A (2014) Antioxidative effect and DNA protecting property of Moringa oleifera root extracts. Int J Geogr Inf Syst 20:209–220
Saucedo-Pompa S, Torres-Castillo JA, Castro-López C, Rojas R, Sánchez-Alejo EJ, Ngangyo-Heya M, Martínez-Ávila GCG (2018) Moringa plants: bioactive compounds and promising applications in food products. Food Res Int 111:438–450
Shawrang P, Mansouri MH, Sadeghi AA, Ziaie F (2011) Evaluation and comparison of gamma- and electron beam irradiation effects on total and free gossypol of cottonseed meal. Radiat Phys Chem 80:761–762
Siddique Z, Jan S, Imadi SR, Gul A, Ahmad P (2016) Drought stress and photosynthesis in plants. In: Water stress and crop plants: a sustainable approach. Wiley, pp 1–11. https://doi.org/10.1002/9781119054450.ch1
Singh A, Chaturvedi N (2021) Medicinal value, herbal gold to combat malnutrition and health benefit of moringa leaf and multipurpose uses of all parts–a review. Plant Arch 21:1327–1332
Sodamade A, Owonikoko AD, Owoyemi DS (2017) Nutrient contents and mineral composition of Moringa oleifera Seed. Int J Commun Syst 5:205–207
Soltan YA, Hashem NM, Morsy AS, El-Azrak KM, El-Din AN, Sallam SM (2018) Comparative effects of Moringa oleifera root bark and monensin supplementations on ruminal fermentation, nutrient digestibility, and growth performance of growing lambs. Anim Feed Sci Technol 235:189–201
Sonewane K, Chouhan SS, Rajan M, Chauhan NS, Rout OP, Kumar A, Baghel GS, Gupta PK (2022) Pharmacological, ethnomedicinal, and evidence-based comparative review of Moringa oleifera Lam.(Shigru) and its potential role in the management of malnutrition in Tribal Regions of India, especially Chhattisgarh. World J Tradit Chin Med 8:314
Sreeja M, Jayasri P, Keerthi N, Yeshashwini J, Praveen J (2021) Moringa oleifera: a review on nutritive importance and its potential use as nutraceutical plant. J Med Plants 9:15–17
Stevens CG, Ugese FD, Otitoju GT et al (2015) Proximate and anti-nutritional composition of leaves and seeds of Moringa oleifera in Nigeria: a comparative study. Agric Sci1 4(2):9–17
Su B, Chen X (2020) Current status and potential of Moringa oleifera leaf as an alternative protein source for animal feeds. Front Vet Sci 7:53
Suhag R, Dhiman A, Deswal G, Thakur D, Sharanagat VS, Kumar K, Kumar V (2021) Microwave processing: a way to reduce the anti-nutritional factors (ANFs) in food grains. LWT – Food Sci Technol 150:111960
Sultana S (2020) Nutritional and functional properties of Moringa oleifera. Metabol Open 8:100061

Taher MA, Nyeem MAB, Ahammed MM, Hossain MM, Islam MN (2017) Moringa oleifera (Shajna): the wonderful indigenous medicinal plant. Asian J Med Biol Res 3:20–30
Tan SH, Mailer RJ, Blanchard CL, Agboola SO (2011) Canola proteins for human consumption: extraction, profile, and functional properties. J Food Sci 76:16–28
Teixeira EMB, Carvalho MRB, Neves VA, Silva MA, Arantes-Pereira L (2014) Chemical characteristics and fractionation of proteins from Moringa oleifera Lam. leaves. Food Chem 147:51–54
Tshabalala T, Ndhlala AR, Ncube B, Abdelgadir HA, Van Staden J (2020) Potential substitution of the root with the leaf in the use of Moringa oleifera for antimicrobial, antidiabetic and antioxidant properties. S Afr J Bot 129:106–112
Udmale P, Pal I, Szabo S, Pramanik M, Large A (2020) Global food security in the context of COVID-19: a scenario-based exploratory analysis. Prog Disaster Sci 7:100120
Verma KS, Rajni N (2014) Nutritional assessment of different parts of Moringa oleifera Lam collected from Central India. J Nat Prod Plant Res 4:81–86
Yadav KC, Bhattarai S, Shiwakoti LD, Paudel S, Subedi M, Pant BR, Upadhyaya J (2022) Sensorial and chemical analysis of biscuits prepared by incorporating Moringa flower powder and leaf powder. Int J Food Prop 25:894–906

Chapter 13
Date Palm (*Phoenix dactylifera*): A Review of Economic Potential, Industrial Valorization, Nutritional and Health Significance

Aijaz Hussain Soomro, Asadullah Marri, and Nida Shaikh

13.1 Introduction

Early human civilization reveals pattern of acquiring food that were transited from hunting animals or gathering food to the emergence of agricultural practices. This advent of agriculture prospect was proved to be as a key advancement and a potent source of a better survival of mankind on the earth. From the beginning of agriculture in primitive times till now, the innovations in plant cultivation and animal husbandry are continuing to happen. According to Gros-Balthazard and Flowers (2021) the earliest domesticated crops (cereal and pulses) dates back to around 12,000 years ago, while the major Mediterranean fruit crops (date palms, grapes, and olives etc.) were domesticated on later periods. Since that time and till now, these fruit crops are an indispensable portion of the human diet and are critically imperative for food security (Food and Agriculture Organization 2013) and for good health.

Among different perennial fruit crops, the date palm (*Phoenix dactylifera* L.) is a major and the oldest crop of many hot and arid regions around the world. The *Phoenix dactylifera* L. is one among the 14 known species of genus *Phoenix* (Al-Alawi et al. 2017). The date palm Phoenix dactylifera, belongs to the family Palmae (Arecaceae), is one of oldest cultivated plant which has played a significant role in day-to-day life of people of Arabian Peninsula from last 7000 years (Ahmed et al. 1995). The *Phoenix dactylifera* L. is popular due to its sweet, nutritious, and palatable fruit. The *Arecaceae* family comprises of some 200 different genera and around 2500 species. The *Phoenix* species are mainly native to the tropical and subtropical Asian and African regions (Krueger 2021). This species name "*dactylifera*" means "finger-bearing" since the term indicates fruit clusters which are produced by date palm tree. The term *dactylifera* is emerged from a Greek ("dactylus" meaning

A. H. Soomro (✉) · A. Marri · N. Shaikh
Institute of Food Sciences and Technology, Sindh Agriculture University, Tandojam, Pakistan
e-mail: ahsoomro@sau.edu.pk

T. Ismail et al. (eds.), *Neglected Plant Foods of South Asia*,
https://doi.org/10.1007/978-3-031-37077-9_13

"finger") and a Latin word ("ferrous" meaning "bearing") (Ashraf and Hamidi-Esfahani 2011).

Date palm mainly grows in arid to semiarid hot areas of Middle East, Arabian Peninsula and North Africa etc. It is likely to presume that the date palm cultivation was originated around the outskirt regions of Iraq (Mesopotamia), however since ancient times its cultivation was gradually propagated to many different countries. Ancient scripts recovered from Iraq indicate establishment of date culture to date back around 3000 BC (Chao and Krueger 2007). The most probable date palm homeland is Persian Gulf (Świąder et al. 2020). However, date palm tree is considered as world's first cultivated tree that was primarily cultivated for its fruit. At beginning, the date palm domestication was started in today's Iraq. The wild dates species may hybridize and are morphologically related to the domesticated form of date while the wild dates have prominent characteristic of their smaller fruit size (Zohary and Hopf 2000). From the ancient days of its cultivation till today, fruit from date tree is a great source of sweetness, deliciousness, and trade.

13.2 Cultivation of Date Palm

Phoenix dactylifera, a prominent member of family *Arecaceae* due to its edible fruit. The species other than *Phoenix dactylifera* also produce edible date fruits (The information pertaining to some *Phoenix* species with edible fruits are presented in Table 13.1). Date palm tree is mainly cultivated for producing sweet, delicious, and nutritious date fruit (Iqbal et al. 2011). Other than the main cultivating areas, the date palm can also be cultivated in many parts of world under variable climatic situations and soil conditions (Farooq et al. 2021) such as in Australia, Arizona, Mexico, Southern Africa, South America, United States, southern California, and Texas etc. (Hazzouri et al. 2015). However, it may be cultivated in groves, home gardens, and in oases as a mono or inter-crop (Fatima et al. 2016). The cultural practices needed to cultivate date palm are often considered as labor-intensive and these practices includes pollination, pruning, and bunch management.

Date palm can be sexually propagated through its seeds and vegetatively by the offshoots but both methods exhibit limitations (Bekheet and El-Sharabasy 2015). The seed of date palm remain dormant for years and may start to grow as it achieves favorable conditions (Tengberg 2012). The pollination in date palm may be natural (mainly through wind or insects), artificial or manual (Ortiz-Uribe et al. 2019) while a single adult and healthy male date palm tree is enough to pollinate around 50 female date palm trees (Bekheet and El-Sharabasy 2015). From pollination to the date fruit maturation, date palm takes around 6–7 months (Saleem and Baloch 2004). Date palm tree may grow on many soil types while best yield may be acquired on sandy loams (Świąder et al. 2020). The tree may be cultivated in the areas with suitable climatic conditions for example, in hot, arid, or semi-arid regions and under appropriate soil conditions (Farooq et al. 2021) along with slight raining and humidity during pollination (Chao and Krueger 2007). The crop needs sufficient water

Table 13.1 Edible date fruit bearing species of *Phoenix* their geographical regions, fruit description, nutritional and medicinal properties

Phoenix species	Geographical region	Fruit description	Nutritional/medicinal properties	References
Phoenix canariensis	Canary Islands	Oval, yellow to orange drupe 2 cm long, 1 cm in diameter, with single large seed	Antioxidant activity	Spennemann (2019)
Phoenix atlantica	Cape Verde Islands	Oval, pink drupe 2 cm long, 1 cm in diameter, with single large seed, scarce fruit pulp	Not-known	Henderson et al. (2006)
Phoenix loureiroi	Southern Asia: Bhutan, Burma, Cambodia, India, Philippines, Taiwan, southern Thailand, Vietnam, Pakistan, China	Single-seeded drupe, bluish black in color	Antioxidant activity	Deborah and Anand (2017)
Phoenix pusilla	Southern India, Sri Lanka	Fruits in clusters, orange-red turn black drupes, fleshy, sweet, mealy pulp	Fruit may be used to cure hyperdipsia, burning sensation, consumption, cardiac disorder, fevers, seminal impotence, gastropathy etc.	Sankar and Shoba (2014)
Phoenix reclinate	Tropical Africa, Arabian Peninsula, Madagascar, Comoro Islands	Fruit are appeared in large clusters, oblong fruit, orange in color, 2.5 cm in diameter with single seed	Not-known	Roodt and Veronica (1992)
Phoenix roebelenii	South-eastern Asia, South-western China Laos, Northern Vietnam	Edible drupe 1 cm, small with thin meal	Antioxidant activity	Wanderley et al. (2012)

either as irrigation water or groundwater (Krueger 2021). Approximately, an adult palm tree requires 150–200 liters of water per day (Świąder et al. 2020). The flower and fruit development of date palm requires higher temperature (Allbed et al. 2017).

The tree grows in hot, arid, and sandy environments and requires sufficient water for its proper growth and for appropriate fruit production. Date palm trees grow on an average temperature of 12.7–27.5 °C, enduring up to 50 °C and sustaining minimal frost conditions (at temperatures as low as −5 °C). On average, temperature required for growing date palm may ranges from 21 to 27 °C during pollination till ripening of fruit (Chao and Krueger 2007). Date palm is a monocotyledonous and angiosperm plant with having some unique anatomical characteristics and environmental adaptation qualities. The date palm tree yields fruit at an average 5 years of age. An average estimated fruit yield from a tree is about 400–600 kg/year. A date palm tree continues to produce fruits till 50 years of its life. Some important date fruit quality attributes for grading are flavor, color, moisture, sugar percent, and an absence of any imperfections like insect invasion, physical damages, or cracks (Al-Alawi et al. 2017). The regions with different climatic conditions yields date fruits with different characteristics such as size, color, shape, taste and weight etc. (Barghini et al. 2007). In date fruit, the stage of fruit maturity, its yield, physical characteristics, organoleptic properties, and nutritional composition are all affected by agricultural practices, and the climate variability (Alharbi et al. 2021).

13.3 Date Palm Tree

Date palm trees are dioecious with having separate male and female trees (Chao and Krueger 2007). Date palm tree merely bear one point in apical meristem for vegetative growth. The flowers of the tree are unisexual (i.e., staminate in male and pistillate in female). The male tree of date palm produces pollen whereas female produces flowers. The *Phoenix* species range in height from stemless or dwarf (*Phoenix acaulis*) to the tall (*Phoenix dactylifera*) reaching over the height of 30 m (Johnson 2011). The trunk of *Phoenix dactylifera* tree is flexible with diameter up to 50 cm (Świąder et al. 2020). An average economic life of a date palm garden is around 40–50 years (Chao and Krueger 2007). The tree provides maximum fruits after 30–40 years, while a tree may survive for more than a hundred years. On average, a date palm tree may produce over 40 kg date fruit/year. The roots of its tree are deep enough, strong and may survive in oases or in the deserts (Świąder et al. 2020). The tree bears feather-like, gray to green leaves. These leaves are beautiful and not usually fall off until the tree dies (Sharif et al. 2010). On a palm tree, the number of leaves may be 80–120 which live for up to 7 years and dries. The inflorescences of both male and female trees differed in morphology and are covered in a hard structure known as spathe. This spathe shields flower from sunlight or heat during early stages of flower development (Chao and Krueger 2007).

Date palm tree usually flowers when shade temperature increases from 18 °C while fruit is formed at 25 °C in female date palm tree (Zaid and de Wet 2002). In female trees, the flowers are borne on peduncle or fruit stalk. The flowers are small, yellow colored and are directly connected to a spikelet that develops as a date palm fruit (Biglari et al. 2007). An immature date fruit is green in color while during ripening process the fruit changes its color from yellow to red or brown. The date fruit is an ovate-cylindrical berry comprising of fruit skin, pulp or flesh, an inner layer, and a seed with an average length of 7 cm (Świąder et al. 2020). The fruit of date palm exhibits five different ripening or maturity phases from fruit setting to fruit maturity (Iqbal et al. 2011), and these phases are usually referred to as (i)

Hababouk, (ii) Kimri, (iii) Khalal or Bisr, (iv) Rutab, and (v) Tamar. Date fruit drops 25–35 days after the spathe cracks.

The fruit is edible, single seeded, and drupe-like berry. The seed (pit or kernel) is enclosed into a fibrous parchment (or endocarp) around fleshy pulp (or mesocarp), and fruit skin or peel (or epicarp). The bulk of the fruit is mesocarp consisting of 85–90% while seed represents merely 6–15% of the fruit weight however, Afiq et al. (2013) stated that on an average nearly 11–18% of whole date fruit is comprised of date seed. Date seed weight is cheifly dependent on the size of the date fruit. Date seed is bony, odorless, light brown and slightly bitter in taste (Ahmed et al. 2014). The fruit has usually an oblong shape although several varieties of date bears fruit with having spherical shape (Al-Alawi et al. 2017). The color of date fruit may be yellow to black. The color of date fruit is considered as one of the crucial external characters of fruit for examining its maturity and for sorting or grading purpose (Sarraf et al. 2021).

13.4 Date Fruit Maturation Phases

After pollination and fertilization of female date palm flowers, generally one carpel (or rarely two carpels of three) of pistillate flower (female) develops into date fruit (Othmani et al. 2020). Growth and development of this fruit involve numerous internal and external alterations which influence different characteristics the fruit such as aroma, color, flavor, size, texture, shape, appearance, and chemical composition. Based on the changes that occur during the entire maturation period, date fruit maturation had been categorized into four phases. These phases are popular with their internationally adapted Arabic names such as Kimri, Khalal or Bisr, Rutab, and Tamar (Alharbi et al. 2021). Whereas date fruit maturation process may also be distinguished into five distinct phases viz., Hababouk, Kimri, Khalal or Bisr, Rutab, and Tamar (Al-Mssallem et al. 2013; Al-Alawi et al. 2017; Othmani et al. 2020). The characteristics of different maturation phases of date fruit are described as under

13.4.1 Hababouk

At this early stage of date fruit maturation, the fruit undergoes cell division (Othmani et al. 2020). This phase begins from fruit set and has slower growth rate. The size of fruit is round, very small (pea sized) with light green color while it may have 85–90% moisture content and lower sugar concentration. The duration of this phase lasts for 4–5 weeks (Świąder et al. 2020).

13.4.2 Kimri

The Kimri is a green, firm, hard, astringent, and an immature stage of date fruit (Haider et al. 2018). Kimri is usually formed 2–5 months after the pollination (Othmani et al. 2020), while duration of phase lasts from 9 to 14 weeks (Świąder et al. 2020). Date fruit at this stage has some beneficial impact on the human health in comparison to Tamar stage. The Kimri dropped from the date tree prior to thorough maturation is valued as an animal feed by the farming communities. Date fruit at Kimri stage has proven to be the carrier of protein (5.5–6.4%), total polyphenols and tannins with promising antioxidant activity (Othmani et al. 2020).

13.4.3 Khalal

In Khalal stage, the date fruit is physiologically mature, edible, have an attractive yellow, or red color, hard and crispy texture, sweet taste with more than 50% moisture while its duration lasts for 3–5 weeks (Świąder et al. 2020). Date fruit is very perishable at this stage. At Khalal stage, some date fruit cultivars with sweet taste, low tannin and astringency are generally utilized for commercial trade. Therefore, for expanding market potential of Khalal date, it is mandatory to acquire new post-harvest strategies for improving keeping quality of the fruit (Mortazavi et al. 2015).

13.4.4 Rutab

The date fruit at Rutab stage involves physiological processes which turns fruit to completely ripe (Lo'ay et al. 2021). Hence Rutab date may be termed as a semi-ripe fruit. Date fruit at this phase is edible and is considered as the fresh form of date fruit which is mainly loaded with sugars. It may have amber, brown, or black color with semi soft texture (Świąder et al. 2020). The moisture content in Rutab stage is comparatively lower than Khalal stage but higher than Tamar stage (Khodabakhshian and Khojastehpour 2021). The maturity phase duration of Rutab lasts for 2–4 weeks.

13.4.5 Tamar

The Tamar is the final maturation stage of date fruit when fruit is considered as completely ripe. The fruit is soft and brown with lower moisture content at this stage (Świąder et al. 2020). The Tamar may be traditionally stored for later consumption and processing. Tamar fruit has highest total soluble sugars (particularly sucrose), titratable acidity and the pectin content. Date fruit industry usually stores Tamar date fruit at −3 °C for up to a year (Al-Yahyai and Al-Kharusi 2012).

13.5 Date Palm Climate Friendly Cultivation

Owing to have unique cultural characteristics and environmental adaptation traits, date palm tree is no doubt a blessing from the God. The crop has many beneficial attributes such as its climate friendly nature that can help the tree to grow even in high temperature water stress soils. The date palm tree contributes for achieving an environmental balance since it may grow under severe or strict climatic conditions. On an average, date palm tree absorbs and stores carbon dioxide (CO_2) to a significant degree when compared to the other fruiting trees. This ability of date palm tree makes it a climate friendly flora to fight against global warming and its consequential impacts on the face of earth (Sharif et al. 2010). Besides having many important features, the one amongst many unique properties of the date palm tree is its resistance against soil alkalinity (Chao and Krueger 2007) salinity, drought conditions and water conservation ability to prevent decline in soil productivity (Ibrahim 2010). Due to its adoptability to saline water, the tree is also considered as a halophytic tree (Krueger 2021). The date palm tree may reduce atmospheric temperature and load of pollutants from the atmosphere. A distinctive symmetry of date palm tree adds another dimension to its impact on the future improvement in environment (Ibrahim 2010). The crop of date palm has tremendous capacity to grow in diverse climatic conditions. To fight against the heatwaves (or global warming), the tree of date palm may be grown along the roadsides, residential areas, remote regions, and into the industrial sites (Salama et al. 2019).

13.6 Cultivar and Crop Production Risks

13.6.1 Date Palm Cultivars

There are about 6000 varieties of date palm around the globe (Dayang et al. 2014) indicating a significant intra-species diversity of the fruit (Świąder et al. 2020). Among the listed number of cultivars, merely a few cultivars have been examined for assessing fruit quality (Mrabet et al. 2019). A substantial number of date cultivars and varieties have emerged through the history of cultivation of date palm. The date fruit varieties have different shapes, size, and organoleptic properties (Świąder et al. 2020). Generally, the date cultivars are characterized by morphological characteristics of its trees and its fruits (such as leaves, spines, fruit shape). Moreover, date fruit cultivars may vary in their color, flavor, texture, and particularly owing to their genotypic characteristics (El Hadrami and Al-Khayri 2013). Despite lack of commercialization of many date palm varieties, the date fruit is still a significant marketable commodity (Mrabet et al. 2019). Some prominent economically significant date palm cultivars from main date producing countries are listed in Table 13.2.

Table 13.2 Some prominent and economically important cultivars of date palm from main date producing countries

Name of cultivars	Country/region	Description for fruit characteristics	References
Almehtari	Iran	Yellow-light brown colored soft date	Karizaki (2017)
Amhat	Egypt	Soft date fruit	Dorria et al. (2012)
Amry	Egypt	Semi-dry date fruit	Rabie et al. (2021)
Anbara	Saudi Arabia	Maroon-red, large, long wrinkled date fruit	Aleid et al. (2014), Zhang et al. (2015)
Ajwa	Saudi Arabia	Medium-oval shaped, medium-sized date with black, wrinkled skin	Zhang et al. (2015), Khalid et al. (2017)
Aseel	Pakistan	Semi-dry date fruit with sweet taste	Markhand et al. (2010)
Barhi	Bahrain, India, Iran, Iraq, Saudi Arabia, Syria, Sudan	Red-brown, ovate to round soft date fruit	Świąder et al. (2020)
Barni Madinah	Saudi Arabia	Cylindrical shaped, brown colored date fruit	Zhang et al. (2015)
Bartamoda	Egypt	Dry date fruit	El–Merghany and Zaen El-Daen (2013)
Beid	Saudi Arabia	Date fruit with average weight of 7.14 g	Gasim (1994)
Begum Jangi	Pakistan	Date fruit with average weight of 5.51 g	Qadri et al. (2016)
Bint-Eisha	Egypt	Soft date, red colored at *Khalal* stage	Rabei et al. (2012)
Dayri	Iran	Dark brown colored, large, thin dry date fruit	Karizaki (2017)
Dhakki	Pakistan	Large sized, fine texture date fruit with average weight of 25.11 g	Suhail et al. (2020)
Deglet Noor	Algeria, Saudi Arabia, Syria	Oblong-ovate or elliptical shaped, light brown, semi-dry date fruit with unique taste	Zhang et al. (2015), Świąder et al. (2020)
Gantar	Iran	Brown-red colored, elliptical shaped soft date fruit	Karizaki (2017)
Gargoda	Egypt, Sudan	Dry date with average weight of 7.35 g	Elshibli and Korpelainen (2009)
Gondaila	Egypt, Sudan	Dry date with 2.72% sucrose	Elshibli and Korpelainen (2009)
Hayany	Egypt	Black, shiny, and oblong soft date	Omar and El-Ashry (2015), Świąder et al. (2020)
Hulwa	Saudi Arabia	Medium, long soft date fruit	Zhang et al. (2015)
Khalas	Kuwait, Oman, Saudi Arabia, Syria	Red-brown colored, elliptical to ovate shaped soft date	Świąder et al. (2020)

(continued)

Table 13.2 (continued)

Name of cultivars	Country/region	Description for fruit characteristics	References
Khodry	Saudi Arabia	Sweet date fruit with average weight 9.94 g	Habib and Ibrahim (2009)
Lulu	United Arab Emirates	Dark amber colored, oblong-oval shaped soft date fruit	Świąder et al. (2020)
Mabroom	Saudi Arabia	Brown colored, medium to large, long date fruit	Zhang et al. (2015)
Malkaby	Egypt	Dry date	Rabie et al. (2021)
Medjool	Morocco, India, Israel, Kuwait, Palestine, Saudi Arabia, Syria	Light to dark brown colored, soft date fruit with attractive appearance	Świąder et al. (2020)
Meneifi	Saudi Arabia	Light brown colored, medium sized, oval shaped date fruit	Zhang et al. (2015)
Mozafati	Iran, Pakistan	Dark brown to black colored, cylindrical shaped soft date fruit	Świąder et al. (2020)
Nabtat Ali	Saudi Arabia	Maroon colored, medium to large, oval shaped date fruit	Zhang et al. (2015)
Oraiby	Egypt	Soft date fruit	Rabie et al. (2021)
Piarom	Iran	Dark brown -black colored, large, thin, semi-dry date fruit	Karizaki (2017)
Rabbi	Iran	Dark brown or red colored, long, thin, fleshy semi-dry date fruit	Karizaki (2017)
Sakkoty	Egypt	Dry date fruit	Rabie et al. (2021)
Samany	Egypt	Soft date fruit	Rabie et al. (2021)
Sayer	Iran, Iraq	Brown-red colored, rectangular ellipse shaped, semi-dry date fruit	Karizaki (2017)
Shahani	Iran	Brown-golden colored, long, thin headed semi-dry date fruit	Karizaki (2017)
Zaghloul	Egypt, India, Syria	Soft date fruit	Rabie et al. (2021)
Zehidi	Iran	Yellow-brown colored, pseudo-elliptical shaped dry or semi dry date fruit	Karizaki (2017)

13.6.2 Date Palm Crop Production Risks

The risks associated to date palm crop are due to obstacles in the domain of crop production, protection, and processing. The sector of date palm cultivation is badly affected owing to its poor pest and disease control, climate change, unimproved cultivation practices, pre- and post-harvest losses. The insects, pests and fungal diseases are responsible for causing great economic losses. Other factors such as low-quality fruit, improper conventional propagation methods, ground water quality, poor marketing services, inappropriate post-harvest management and lack of modern styled cold storage facilities etc. are among some common obstacles in date fruit production and trade (Bekheet and El-Sharabasy 2015). An insufficient integrated pest management is responsible for insects, pests, and disease infestation in date crop. In past few years, date palm production has declined in date palm tree growing regions including Arab region, Egypt, Iraq, and Pakistan owing to invasion of insects, pests, and diseases. Around 30% the global crop production has been potentially affected by pests and diseases. The date crop protection sector in Egypt and Gulf countries is facing severe risks associated with Red Palm. In North Africa, the Bayoud disease (developed by a parasitic fungus) is a major threat to date palm crop (El-Juhany 2010).

In Saudi Arabia, the Al Wijam disease (a phytoplasma disease) leads to severe failure in date fruit yield (Alhudaib et al. 2007). It has also been reported in some Asian countries including Pakistan and India where early monsoon raining prior to date fruit maturation adds to existing threats to the fruit crop production. Irregularity in rainfall destroys the entire crop while the measures for controlling fruit loss such as covering of fruit bunch increase the cost of the fruit production. On the other side, in Arab region the conditions like severe drought, low rainfall, drying of water wells are imposing serious threats to the crop (Alhudaib et al. 2007). An abnormal flowering, incomplete pollination and fertilization, low fruit set and the high rate of fruit drop prior maturation (Al-Qurashi et al. 2012) are amongst some serious problems in date crop production.

Date palm cultivars improvement is mandated to mitigate pre and postharvest losses of the fruit, and to prevent substantial economic losses among farming communities. The conventional breeding programs have limitation in this context owing to date palm's long generation period and its heterozygosity. A rapid progress in developing improved date palm cultivars may be achieved by modern plant biotechnological or genetical techniques (Bekheet and El-Sharabasy 2015) such as tissue culturing, molecular characterization, and genetic engineering. Advanced biotechnological techniques along with conventional breeding programs may contribute to develop new date palm cultivars with higher yield, good quality fruit and resistance against biotic and abiotic stresses.

13.7 Climate Change and Date Palm Crop

The global climate changes are badly affecting all forms of life including different plant species at a massive and rapid pace, creating complex ecosystems and disturbance in the existing species (Kaky et al. 2020). Agriculture is a main sector facing the problem of climate change. Some recent studies are predicting that patterns of crop production (Schierhorn et al. 2020), areas of crop production (Chemura et al. 2020), and the type of fruit crops (Shabani et al. 2017) will get modified owing to the changing states of climate (Kogo et al. 2019). A considerable number of research studies have been performed on understanding climate change influence on the productivity of agricultural produces. The reported findings suggest that climate change is responsible in lowering agricultural crop productivity (Allbed et al. 2017). Like other food crops, date palm cultivation are also facing various productivity constrains.

Alongside the negative impact of climate change, a persistently increasing figure of global population and food demand are declining the natural resources of arable land, soil and water. Another risk to date palm cultivation is an elevated heavy metal content in date palm cultivating soils. The heavy metals are imposing serious impacts on the crop production due to plant stress, and the quality of food that may also pose a detrimental effect on consumers' health (Salama et al. 2019). Heavy metals such as aluminum (Al), arsenic (Ar), boron (B), cadmium (Cd), chromium (Cr), copper (Cu), lead (Pb), and zinc (Zn) are toxic and can exert deleterious effects on the health. This necessitates monitoring of heavy metals in date palm cultivation as a critical quality aspect for date fruit industry. Some other man-made ecological factors such as intensive use of fertilizers, sewage irrigation, mining, industrial emissions or effluents and metallurgical operations are critically deteriorating quality of the date palm cultivation lands (Abass et al. 2016).

The date palm production is continuously being affected by changes in the state of climate (Farooq et al. 2021). Therefore, an identification of suitable date palm production practices under the changing climate states are needed. Looking into the severity of climate change issues, many researchers have proposed various distribution models for studying the impact of climate change on species distribution e.g., Boosted Regression Tree (Radinger et al. 2015), CLIMEX (Allbed et al. 2017), Ecological Niche Models (Silva et al. 2014), Forest (Vincenzi et al. 2011), Global Climate Models (Porfirio et al. 2014), SPECIES (Hampe 2004), and Random MaxEnt (Nazeri et al. 2012).

13.8 Role of Date Palm Cultivation on Rural Household Food Security

It is evident from the history of agriculture that primarily date palm was domesticated with the advent of agriculture, later implementation of new agricultural technologies has shown improved cultivation and this crop was widely broadcasted and became world's major tree crop (Johnson et al. 2013). The fruit from date palm may offer strong opportunity for improving household food security more particularly in rural agriculture context (Ghnimi et al. 2017). The rural household food security through date palm depends largely on its evaluation, utilization, conservation of the genetic resources, assessing potential date cultivars, endorsing good cultivation practices, processing/value addition, marketing, and expanding the range of date palm tree and fruit derived value-added products. Date fruit is an important source of subsistence for its nutritional value and as a livelihood mean for the people of rural areas. Younas et al. (2020) reported that consuming a few dates daily are sufficient to significantly improve nutritional security. This considered as a major cash crop in date fruit producing regions. Food and Agriculture Organization of the United Nations recognizes socio-economic and ecological importance of the date palm crop and is working along with governments and local organizations for promoting production of the date crop (Arias et al. 2016). The role of date palm industry in achieving rural household food security is discussed here:

- The sector of date palm industry may contribute its crucial role in food security in rural areas of many under-developed countries. The date fruit is energy dense and nutritionally rich functional food, and has tremendous capacity to combat malnutrition, micro-nutrient deficiencies, and anemia among marginal populace of rural areas.
- Date palm industry is a labor-intensive industry and engages a large size of rural populace, women in particular, as a mean of rural livelihood subsistence.

13.9 International Trade Potential of Date Palm Fruit

The market economy value of date palm represents much more than its importance as a nutritionally rich and a high-energy fruit (Johnson 2011). Date palm crop is considered as a cash crop in many countries (Allbed et al. 2017) mainly due to its fruit and other valuable by-products. In international trade of agricultural produces, the sector of date palm is oldest one contributing vital role in the agriculture economy of date palm producing countries. Date palm production sector is also providing employment opportunities, and significantly contributing to income generation, trade, and industrialization (Frija et al. 2017). The date palm crop is globally marketed as fruits (fresh consumption, powder development and for high value confectionery production), and the by-products. In prominent date palm producing countries, many parts of date palm tree are commonly utilized for human consumption and as animal feed, medicines, cosmetics, woodwork, fuel etc. (Hamad et al. 2015).

Production of date palm has therefore expanded on global scale in the recent decades and is likely to increase in future (Krueger 2021). According to Food and Agriculture Organization Corporate Statistical Database (2020), Egypt remained the largest date fruit producing country (1,690,959 tonnes) followed by Saudi Arabia (1,541,769 tonnes), Iran (1,283,499 tonnes), Algeria (1,151,909 tonnes), Iraq (735,353 tonnes) and Pakistan (543,269 tonnes) while main date fruit exporting countries on international market are Tunisia, Algeria, Egypt, and UAE. In Gulf Cooperation Council (GCC) countries such as Bahrain, Kuwait, Oman, Qatar, Saudi Arabia and UAE, the date palm sector is the oldest economic sector anticipating strategic significance the nation's economic development (Dhehibi et al. 2018).

13.10 Nutritional Significance of Date Fruit

Date fruit is a staple food (Tang et al. 2013) with high calorific value packed with constituents like sugar, dietary fiber, minerals, vitamins, fat, and protein (Ahmed et al. 2014). The nutritional value of dates is due to their high sugar (50–60%), potassium (2.5 times more than bananas), calcium, magnesium, iron, thiamine, riboflavin and niacin content.

The dates are rich in sugar on dry weight basis ranging from 65% to 80% mostly of inverted form (glucose and fructose). Inverted sugars are high in fresh varieties, and in equal amount in the semi dried varieties, while higher sucrose content is present in dried varieties. The moisture content of the fresh and dried date varies in range from 7% to 79%. Due to higher sugar content dried dates can easily be stored and preserved. Therefore, date is believed to be nutritious fruit that promotes human health when consumed in combination of other foods (Lambiote 1982) or as fresh and/or dried. A more detailed information on nutritional significance of date fruit is given in the following sections.

13.10.1 Date Fruit Flesh

Carbohydrates

The soluble sugars and dietary fibers are predominant carbohydrates found in date fruit flesh followed by moisture, minerals, vitamins, lipids, and proteins (Mrabet et al. 2019). The date fruit owing to its rich high bioavailable carbohydrate contents, delivers rapid energy to the consumer (Habib and Ibrahim 2011). The most prominent carbohydrate that account for up to 70–80% of the date fruit flesh are simple sugars such as glucose, fructose, xylose, arabinose, and sucrose (Nehdi et al. 2010). Among neutral sugars, the date flesh contains about 8.8–9.4% glucose and 8.7–9.2% xylose (Elleuch et al. 2008). An average 100 g of dried and fresh date flesh can supply 314 and 213 kcal of energy (Al-Farsi and Lee 2008). As per USDA National Nutrient Database, a 100 g serving of date fruit gives about 75 g of carbohydrates. Change in sugar profile of date fruit at its various stages of maturation is a natural phenomenon. During fruit maturation phases from Kimri to Tamar, the concentration of simple sugars (glucose and fructose) gradually increases in the fruit (Ahmed et al. 2014).

The rate of sugar conversion in date fruit is dependent on temperature, relative humidity, physiological activities, and other climatic conditions. Furthermore, the date cultivars also distinctly contribute variability in proportion of different sugars (Ali et al. 2009). Date fruit flesh is an excellent source of dietary fiber such as insoluble cellulose, partly soluble hemicelluloses, pectin, hydrocolloids, and lignin (Mrabet et al. 2019). Elleuch et al. (2008) reported 14.4–18.4% dietary fiber in date flesh of some Tunisian date varieties. The type and amount of dietary fibers in date fruit is dependent on fruit maturation stage and the cultivar. The total fiber content decreases in date fruit when it becomes soft since enzymatic hydrolysis of polysaccharides during ripening results in the formation of more soluble compounds, decrease in fiber content, and fruit softening (Mrabet et al. 2012). High popularity of some date fruit varieties is also attributed to their dietary fiber content (Mrabet et al. 2019).

Proteins and Amino Acids
Although date flesh is low in protein but has a considerable proportion of good protein (Al-Farsi and Lee 2008). In comparison to many other fruits, date fruit contains higher levels of protein. Among different date fruit maturation phases, the highest protein content i.e., 5.5–6.4% had been reported at Kimri phase (Ahmed et al. 2014). Net protein contents decline in later stages of maturation and reach up to 2.5% at peak maturity (Habib and Ibrahim 2011). This decrease in protein content during final stages of fruit maturation have been attributed to tannin precipitation and non-enzymatic browning or Maillard's reactions. The proteins in date fruit have molecular weight ranging from 12,000 to 72,000 Da. In date flesh, the main proteins are water soluble albumins. Al-Farsi and Lee (2008) reported amino acid profile of date fruit which revealed the presence of all essential amino acids in the fruit irrespective of different cultivar. Date fruit flesh is rich in acidic amino acids with having lower sulfur containing amino acids (Ishurd et al. 2004). The amino acid profile of date fruit also varies with variable maturation phases. The Kimri stage represents the higher concentration of alanine, aspartic acid, glutamic acid, lysine, leucine, and serine whereas aspartic acid, glutamic acid, lysine, leucine, glycine, and proline are most abundantly occurring amino acids during ripening of date fruit (Tang et al. 2013).

Fats and Fatty Acids
The flesh of date contains very negligible amounts i.e., 0.2–0.5% of the edible lipids (Ahmed et al. 2014). Moreover, the fat proportion may be different among different date cultivars. The fat content in 100 g of fresh and dried date fruit flesh have been cited as 0.14 and 0.38 g, respectively (Al-Farsi and Lee 2008). Most of the fat in date fruit is present in fruit skin. Among the fatty acids reported in date fruit flesh, a few promising fatty acids include lauric acid, myristic acid, oleic acid, and palmitic acid (Habib et al. 2013).

Vitamins, Minerals and Antioxidants
Date flesh is also a plentiful source of water and fat-soluble vitamins. Aslam et al. (2013) reported a significant variation in water-soluble vitamins among different date palm cultivars whereas mature date fruits were noticed to have vitamin B-complexes (B1, B3, B5, and B6) and vitamin C (Dahlan and Ardhi 2021). The flesh of date fruit is considerably rich in minerals and may contains 15 different minerals mainly selenium, potassium, magnesium, and copper (Ahmed and Ramaswamy 2006). It is estimated that 100 g of date fruit can meet over 15% of daily RDA of some important minerals (Al-Farsi and Lee 2008). Other minerals such as iron, calcium, manganese, and phosphorus are moderately present in date fruits and may fulfill about 7% of their RDAs (Nasir et al. 2015). Date fruit is a substantial source of

natural antioxidants (Saleh et al. 2011) and therefore serves as an excellent functional food. The antioxidants occurring in different date cultivars are mainly phenolic acids, benzoic acids and derivatives, flavonoids glycosides and esters, anthocyanins and proanthocyanidins (Ghnimi et al. 2017).

13.10.2 Date Seed

The seeds recovered from date fruits are considered as byproduct in date fruit industry (Afiq et al. 2013). Date seeds are mainly used as a feed for poultry and livestock in many countries whereas in some date producing countries, date seeds are considered as waste material despite having an appreciable content of dietary fiber, minerals, protein, fat (Besbes et al. 2004), phenolic compounds, and antioxidants (Mrabet et al. 2015). Factors like the cultivar of date palm, origin, harvesting time and type of fertilizer used for cultivating date crop may affect the nutrient content of date seed. Date seeds are plentiful source of dietary fibers. According to Aldhaheri et al. (2004), total dietary fiber found in date seed are around 58% of which 53% are water insoluble dietary fiber. Hamada et al. (2002) reported 65–69/100 g of total dietary fiber in different varieties of date seed. Date seeds are therefore suggested as a fiber enrichment source to develop value added products of human nutrition and health significance. Date seed may be also used to extract date seed oil (Afiq et al. 2013).

Fat content of date seed may range between 5.7% and 12.7% (Habib and Ibrahim 2009). Oleic acid constitutes more than 50% of the total fatty acid contents in date seed followed by linoleic acid, lauric acid, and palmitic acid (Nehdi et al. 2010). Fatty acid composition of date seed oil may be influenced either with the date palm variety or by the date seed roasting process. Date seeds also contain appreciable quantity of dietary proteins including albumin, globulin, glutelin, and prolamin (Hamada et al. 2002). Date seeds have also been suggested as promising carrier of minerals such as calcium, cadmium, copper, chromium, iron, lead, magnesium, manganese, phosphorus, potassium, sodium and zinc (Habib and Ibrahim 2009; Nehdi et al. 2010).

13.11 Functional and Health Promoting Properties of Date Fruit and the Seed

Date palm fruit is ranked among superfoods in some countries on account of its ability to preserve good health, and as indicated in Quran and by the prophet Muhammad (SAW). In Surah An-Nahl it is stated that, “And from the fruits of the palm trees and grapevines, you take intoxicant and good provision. Indeed, in that is a sign for a people who reason.” (AlQuran 16:66) (Nadiah et al. 2018). Date fruit contain plentiful therapeutic bioactive agents and the functional compounds demonstrating the fruit to exhibit anticancer, antioxidant, anti-inflammatory, antimutagenic, antidiabetic, gastroprotective, nephroprotective, hepatoprotective, and immunostimulant activities (Tang et al. 2013). The dietary fibers, minerals, vitamins, and antioxidants are considered as main functional constituents of the date fruit demonstrating many health promoting attributes in addition to the few listed above (Al-Farsi and Lee 2008; Maqsood et al. 2020). Traditional Egyptian medicine practice however report the date seed as a folk remedy for managing many disorders such as those of endocrinal, liver and gastrointestinal disorders. This section gives the brief insight of the functional and health promoting properties of date fruit.

13.11.1 Antioxidant Activity

The antioxidant activities of date fruit are chiefly attributed due to its valuable phytochemicals including anthocyanins, carotenoids, phenolic acids, flavonoids, and selenium etc. (Yeh et al. 2009). According to Tang et al. (2013), the fruit extract may act as an inhibitor against protein oxidation and may neutralizes free radicals. The fruit has great antioxidant capacity to exert protective properties for maintaining the functions of major body organs such as liver and kidney (Ali and Abdu 2011). Allaith (2008) reported that date fruit at its unripe stage have higher antioxidant activity than ripe fruit. In another study, Al-Farsi et al. (2005) evaluated three date fruit cultivars and recorded that antioxidant activity of fresh date fruit is higher than dried date fruit. Furthermore, Ardekani et al. (2010) studied seeds from 14 Iranian date palm varieties and reported relatively high antioxidant activity due to the immense concentration of phenolic compounds in seeds. Cancer is a fatal disease and is mainly caused by viral infections, chemicals (pesticides), radiation, environmental factors (pollutants), and dietary factors etc. The leading conventional treatment of cancer includes surgery, chemotherapy, and radiotherapy etc. Date fruit is rich in effective nutraceutical polyphenols and β-glucan. The date fruit extract showed potent anticancer activity with its valuable bioactive compounds (El-Far et al. 2016). Date seeds are also rich in phenolic compounds (flavonoids etc.) bearing medicinal properties against different ailments (Khattak et al. 2020).

13.11.2 Anti-inflammatory Activity

Inflammation is an important defensive mechanism in human body which functions against various infection, toxicities, allergens (Sharma et al. 2011). Date fruit (cultivar Ajwa) is a good source of natural anti-inflammatory agents. The ethyl acetate in date fruit, its methanolic and water extracts are effective against lipid peroxidation (Ferguson et al. 2004) and other inflammatory mechanisms. It has been reported that date fruit pollens (Vinson et al. 2005), leaves (Rock et al. 2009), fruit (Ranilla et al. 2008) and seeds (Bouhlali et al. 2020) exhibit potent anti-inflammatory activities.

13.11.3 Anti-mutagenic Activity

The term mutagenicity referred to as an initiation of permanent modification in DNA sequence in an organism. The antimutagenic compounds either natural or synthetic have potential to neutralize the deleterious effects of mutagens. Date seed is also a substantial source of important bioactive compounds such as antimutagenic agents and has gained little attention for exploring its potential as a functional and health promoting date fruit by-product. Selim et al. (2021) discovered the date seed of cultivar Ajwa for its antimutagenic activities due to gallic acid and reportedly determined the significant antimutagenic activities of gallic acid against *Salmonella* species. Similarly aqueous extract of date fruit also showed antimutagenic properties (Vayalil 2002).

13.11.4 Antidiabetic Activity

Diabetes is generally accompanied by an increased oxidative stress or hyperglycaemia (Tang et al. 2013). The use of minerals and antioxidants may be an effective treatment strategy against diabetes (Dembinska-Kiec et al. 2008). Date fruit may exert antidiabetic effect in preventing and controlling diabetes mellitus since it contains antioxidants and minerals. In date fruit, the phenolic compounds inhibit the activities of alpha-glucosidase and alpha-amylase (Ranilla et al. 2008) since these enzymes are responsible for carbohydrate digestion.

13.11.5 Gastroprotective/Antiulcer Activity

The gastric or peptic ulcers are common but preventable disorders of the gastrointestinal tract which happened owing to over secretion of gastric acids in the stomach (Kuna et al. 2019; Al-Gabri et al. 2022). The fruit of date palm is conventionally consumed to defend gastric mucosa from the detrimental impacts of stomach acids (Al-Qarawi et al. 2005). It has proven that aqueous and ethanolic extracts from date fruit are effective in ameliorating ethanol-induced gastric ulcer (Al-Qarawi et al. 2003). The antiulcer activity of date fruit is linked with antioxidant compounds namely flavonoids, beta-carotene, beta-sitosterol, pro-anthocyanidins, and selenium (Iwasaki et al. 2004). The antiulcer activity of date seed extract was studied by Al-Gabri et al. (2022) which indicates the positive effect of date seeds against the gastric ulcer.

13.11.6 Nephroprotective Activity

The nephrotoxicity is a common side effect generated by many pharmacological compounds (analgesics, antibiotics, and cytostatic). Date fruit and its seed may serve as a good source of bioactive agents against nephrotoxicity. An *in-vivo* study conducted on rat model (with gentamicin treated nephrotoxicity) exhibited significant reduction in the plasma creatinine and urea levels when administered with date flesh and seed extracts (Al-Qarawi et al. 2008).

13.11.7 Hepatoprotective Activity

Liver is a main metabolic organ of the body, and its damage may cause serious health concerns in human body. The date fruit and its seeds exhibit hepatoprotective activities (Tang et al. 2013). Abdelaziz and Ali (2014) stated that the hepatoprotective effect of date seeds might be attributed to free radical scavenging activities. It was further reported by Saafi et al. (2011) that extracts from date fruit could reduce the levels of hepatic marker enzymes such as alkaline phosphatase gamma-glutamyl transferase, lactate dehydrogenase and transaminases. Al-Qarawi et al. (2004) further reported that feeding rats on date fruit extract significantly decreased CCl_4-induced plasma enzyme and bilirubin level.

13.11.8 Cardioprotective Activity

Date fruit has proved to be as a good cardioprotective food however studies on cardioprotective activities of date fruit are not sufficiently demonstrated. In an in-vivo study Ahmed et al. (2016) examined the anti-hyperlipidemic effect of extracts from Aseel date cultivar on albino rats and noticed a significant decrease in serum lipid profile, triglyceride, low-density lipoprotein, and in alkaline phosphate level.

13.12 Date Palm Fruit Consumption and Risk of Nutritional Anemia

Anemia is a common problem with lower concentration of hemoglobin and numbers of red blood cells in an individual. This condition affects one third population of world and contributing an increased rate of morbidity, mortality, birth defects, reduced cognitive and neurological development (Chaparro and Suchdev 2019). Among different forms of anemia, the iron deficiency anemia (IDA) is usually occurring owing to dietary deficiency of iron or poor iron absorption in blood stream through small intestine etc. The proportion of iron deficiency anemia IDA differs in population groups since have different geographical setting, burden of infectious diseases, and prevalence of other types of anemia. Particularly, preschool children and women of reproductive age are among the vulnerable group of anemia. IDA is highly predominant in women (i) with heavy menstrual bleeding, (ii) having pregnancy and (iii) at their postpartum state (Mirza et al. 2018). Although IDA is preventable in individuals consuming their diet as per recommended daily intake of iron however it is hardly to manage among the people with marginal groups in most developing countries. Anemia is most prevalent condition in underdeveloped countries mainly owing to nutritional deficiencies, infections, hemoglobinopathy, and micronutrient deficiencies (Sari et al. 2018).

Hemoglobin is a conjugated protein which carries oxygen and is mainly formed in red blood cells and spinal cord. The deficiency of iron directly affects the hemoglobin synthesis. Anemia can be treated with three different ways i.e., pharmacologically (through an oral administration of iron supplements) and non-pharmacologically (through blood transfusion and consuming iron containing diets). An oral administration of iron supplements is an efficient but a cost-effective method to treat IDA. Furthermore, this method displays side effects (i.e., nausea, vomiting, stomachache, gastrointestinal irritation, and constipation) which restrict its prolonged use. Also, oral iron supplements are not properly assimilated in blood stream through small intestine owing to an overexpression of a peptide hormone which plays chief role in maintaining iron homeostasis (Cotoraci et al. 2021).

Date palm fruit is a major source of iron among different fruits (Sahari et al. 2007). About 100 grams of date fruit may have 3.6 mg of Iron. Date fruit is also a reasonable source of iron, vitamin B12, B9, and C that are responsible for synthesizing hemoglobin. The consumption of date fruit may significantly contribute to fulfilling the daily dietary requirement of minerals such as Fe, Se, Ca, P, Zn, Cu, and I and providing reasonable proportion of vitamins such as B-complexes (Vayalil 2012). Therefore, date fruit consumption on daily basis may be a good alternative for treating deficiency-related diseases such as IDA, goiter, osteomalacia, rickets, and beriberi etc. Since the fruit is rich in iron (Eldaim and Elnadi 2014) and is therefore reportedly involved in increase the hemoglobin concentration (Zen et al. 2013). Treating experimental animals with date fruit extract increased red blood cells count, platelets count, and packed cell volume. According to Abdelsalam et al. (2014) date fruit extract significantly raised hemoglobin concentration by 13% in comparison to control. Dahlan and Ardhi (2021) experimented that iron supplement and date fruit can improve hemoglobin concentration among pregnant women in their third trimester and thus recommended the consumption of iron supplement and date fruit for maintaining proper nutrition during pregnancy. In another study, Rahayu (2021) found that hemoglobin concentration significantly raised in adolescent females who were intervened on 10 ml of date fruit juice three times/day for 14 days.

13.13 Date Palm Fruit-Based Value-Added Food Products

The date palm tree including its major parts such as leaves, trunks, buds, fruit and seed are associated in one way or the other for human welfare (Chao and Krueger 2007). The dried leaves of the tree are useful in making baskets, hand fans, mats, ropes and packaging materials for agricultural produces (PHDEB 2008). The trunks of the tree can be used in construction of cottage home, bridges, and for making packing materials (Anwar 2006). The terminal buds and tender young leaves of the date palm can be cooked as a vegetable whereas rachises are used in paper manufacturing (El Hadrami and Al Khayri 2013). The date palm tree is mainly grown for its valuable fruit that is preferred to be consumed in raw as a delicious and sweet fruit. Date palm is an abundantly cultivated fruiting tree with hundreds of cultivars available for fruit production and its valorization. The utilization of date fruit for developing value-added food products should depends on its nutritional and functional characteristics. Date fruit has been extensively used for the development of many value-added food products such as date juice concentrates (syrup, spread, and liquid sugar), fermented-date products (alcohol, wine, vinegar, and organic acids) and date pastes (mostly used by bakery and confectionary industries). Moreover, food processing industries produce different date products such as date-dip, date-jam, and date-bars. Pectin and dietary fiber are the main date by-products used as thickening or gelling agents in different confectionery products, jams, jellies, soft cheeses, and yoghurt etc. For producing date paste, the date fruits are destoned, steamed, macerated to a semi-solid form with having approximately 20–23% moisture and less than 0.6 water activity (Ahmed et al. 2005). Date paste has a great economical value and may be utilized as a filling agent, sugar substitute in different food formulations and in confectioneries (Alhamdan and Hassan 1999). Date syrup (dibis) may also be utilized for producing jams, marmalades, chocolates bars, ice cream, concentrated beverages, confectionery items, sweet meats, snack foods, bakery products and other functional foods (Riedel 1986). Date fruits in their mature stage are also processed for making date bars and syrup (Abd El-Mohsen and Nezam El-Din 1995).

Date seed is also a valuable part of the fruit. It has been reported that some Arabian date palm producers have commercially introduced roasted date seeds for making caffeine free coffee substitute and as mix with the real coffee in some Arabian markets. Moreover, date seed oil has been suggested as edible oil and for producing margarine due to its high thermal stability and prolonged shelf life or storability (Afiq et al. 2013). In many date producing countries, a significant proportion of date fruit is wasted due to the inferior quality of the fruit, fruit damage and improper storage (Besbes et al. 2009). Such an underutilization of the fruit can contribute to major economic loss to the date palm growers. It has been proposed that intelligent utilization of relatively low-quality date fruit, not intended to be consumed as raw fruit, may be processed for the extraction of valuable components such as dietary fibers, phenolic compounds, and other pharmaceutical agents. Intensive research attempts are thus desired to be placed in practice to promote date fruit valorization and ensure sustainability of the date fruit processing industry (Sirisena et al. 2015). A brief set of information on the types of date fruit value-added products, processing methods, nutritional value and modes of use is available in Table 13.3.

Table 13.3 Information regarding types of dates fruit-based value-added food products, their processing methods, nutritional value and means of utilization

Value-added food products	Processing methods	Nutritional value	Utilization	Reference
Date Pickle-in-oil (DPO)	The deseeded date fruit at Kimri or Khalal stage should be added with various spices, salt and mustard oil for few days	–	DPO can be used to consume in mealtime as a condiment food	Hamad and Yousif (1986), Tang et al. (2014), Ghnimi et al. (2017)
Salt-stock date pickle (SSDP)	Date fruit at Kimri or Khalal stage is treated with vinegar and fermented	–	SSDP can be used to consume in mealtime as a condiment food	Ghnimi et al. (2017)
Date chutney (DC)	Date fruit at Kimri or Khalal stage is grounded with onion, tomato, mint leaves, garlic cloves, green chili and salt etc.	–	DC can be consumed as a condiment in mealtime	Tang et al. (2014)
Dried date fruit (DDF)	Date fruit (whole or sliced) should be dried using a laboratory cabinet oven with air velocity 1.5 m/s and temperatures of 60–80 °C until date fruit dries thoroughly	The phenolic compounds in dried date are pyrogallol, catechein, benzoic acid, chlorogenic, caffeine, catechhol, coumarin, vanillic, gallic acid etc.	DDF can be consumed as such and may be transformed to powder and may be used as sucrose replacer and functional ingredient	İzli (2016)
Date paste (DPA)	Date fruits should be washed, scalded at 100 °C (1:1 water/fruit) for 3 min, deseeded, peeled. The fruit flesh should be grounded until a smooth and homogenous date paste obtained	DPA is composed of 3.62% protein, 1.54% fat, 1.73% ash, 4.48% crude fiber, 77.55% total sugar	In baking industry, DPA is used in filling pastries, cakes, breads, cookies, biscuits etc. In dairy industry, DPA can be used in making ice cream, puddings, milk shakes etc. In confectionary industry uses DPA as a source of sucrose for making confections, date bars, date candy etc.	Ghnimi et al. (2017)

(continued)

Table 13.3 (continued)

Value-added food products	Processing methods	Nutritional value	Utilization	Reference
			In meat industry, DPA is used to develop meat-based products (such as bologna-type meat products) for enhancing nutritional and technological properties of meat products	
Date juice and syrup (DJ and DS)	Deseeded date fruit should be washed, soaked in hot water at 70 °C for 30–45 min, hand blended, cloth strained, squeezed to completely strain DJ. This DJ should be concentrated by boiling and continuous stirred to obtain 53 °Brix value in refractometer to be transformed as DS. It should be cooled and stored	The DS was composed of fat 0.005%, protein 1.43%, ash 2.18%, glucose 27.75%.	As a sweetener, DS can be used as sucrose replacer In dairy industry, DS can be used in making flavored yoghurt and ice cream etc. In beverage industry, DJ can be used to prepare carbonated and non-carbonated beverages, date wine. DJ may can be used for preparing jam and jelly	Kaushik et al. (2016), Farahnaky et al. (2018)
Date flesh powder (DFPO)	Date fruit should be washed, cut into slices, dried at 90 °C for 2 h, milled and sieved to obtain DFPO	The DFPO of H'loua cultivar is consisting of fat 2.3%, protein 2.8%, ash 2.26%, carbohydrate 85.80%.	DFPO can be used as sucrose replacer, functional food ingredient for making bread, cake, cookies, biscuits, yoghurt etc.	Hasan et al. (2022)
Date seed powder (DSP)	For preparing DSP, the date seeds should be separated from flesh, soaked in water for 6 h, dried in an oven (50 °C for 4–6 h), and milled to obtain DSP	The date seed of Deglet Nour cultivar is consisting of fat 10.19%, protein 5.56%, ash 1.15%, carbohydrate 83.1%.	DSP can be used as a functional food ingredient for food fortification such as in making, cake, biscuits, bread etc.	Besbes et al. (2004)

(continued)

Table 13.3 (continued)

Value-added food products	Processing methods	Nutritional value	Utilization	Reference
Roasted date seed coffee powder (RDSCP)	For preparing RDSCP, date seeds should be separated from flesh, soaked in water (100 °C for 1 h), dried at (50 °C for 24 h) roasted using natural pre-heated convection oven (160–200 °C for 10–30 min), cooled at room temperature, grounded using mill, defatted, and finally brewed	–	RDSCP can be used to prepare coffee as a substitute of real coffee	Fikry et al. (2019)
Date seed oil margarine (DMAR)	The date seed oil is hydrogenated to obtain margarine	–	DMAR may be used instead of conventional margarine as a bread spread etc.	Nehdi et al. (2010)
Date seed oil mayonnaise (DMAY)	The ingredients (such as eggs, salt, sugar, mustard, white pepper, vinegar) should be mixed using an electric mixer and date seed oil should be gradually added in the mixture till desirable consistency of mayonnaise is obtained	–	DMAY may be used instead of conventional mayonnaise as a salad dresser, sandwich spread etc.	Basuny and Al-Marzooq (2011)

13.14 Barrier in Establishing Sustainable Date Palm Fruit Industry

A sustainable date palm fruit industry needs attention from the government, growers, exporters, and other stakeholders to overcome the constraints that are involved in its destabilization in date palm fruit producing countries. The date fruit processing industry has great economic growth potential; however, the constraints in the growth of this sectors are needed to be genuinely countered. Emphasis is desired to be given to improving characteristics of date cultivars, expanding date palm plantation programs, increasing date palm fruit yield, improving pre and postharvest technologies, date fruit-based value-added food product development, and marketing sector for the development of date palm industry. Almost all major date fruit producing countries has some limitations in establishing a sustainable date palm fruit industry. In Egypt, despite having an invaluable role of date palm fruit to the country's agriculture sector, date palm industry is countering issues like absence of proper information system on the status of date palm industry, genetic diversity, post-harvest fruit loss, high disease and pest burden (Bekheet and El-Sharabasy 2015).

The date palm industry in Iraq has been facing biotic and abiotic stress such as drought, salinity, genetic diversity, inadequate replacement of old trees with the new plantation, high burden of diseases and pests, and post war associated economic crises. Such a situation is immensely contributing to the sustainability of date fruit industry in Iraq. The devastation faced by Iraq during wars seriously destructed date palm groves (Al-Jboory 2007). The major problems of date palm production in Saudi Arabia are diseases, pests, insufficient adaptation of new technologies and inadequate marketing strategies. Saudi date fruits are also facing competition in international markets. The water scarcity has also vigorously contributed to decline in date palm productivity and even fruiting plants death. The date palms allegedly consume plenty of irrigation water (El-Juhany 2010).

The situation is also not different in United Arab Emirates where problems like limitation of trained and qualified staff, inadequate research studies in the field of date palm pests and disease control, lack of good quality date cultivars, poor irrigation, and poor adaptation to innovative postharvest and processing technologies for the date fruits are devastating the date palm fruit industry. In Tunisia, a major threat to date palm crop is fungal diseases, insects, pests, and nematode species (Hammadi et al. 2009). The old age palm tree groves are also responsible for reduced fruit production in Tunisia. A similar situation alongside irregular raining system is the major constraints to date palm crop in Pakistan. To establish sustainable date palm fruit industry particularly in the main date fruit producing countries there is dire need to adapt newly emerging technologies, sowing of improved date fruit cultivars and promising infrastructure are mandatory steps.

13.15 Conclusions

Date palm is a perennial fruiting tree of arid and semi-arid regions with enormous potential of genetic diversity, drought and salinity resistance, climate resilience. The date fruit has immense nutritional significance alongside therapeutic properties against nutritional and other health disorders. The fruit being widely acceptable in different cultures of the world has potential to be utilized for the development of value-added food products. On the other side, the date crop is facing many preventable production challenges linked to climate change, date palm and fruit disease, insects, pest infestation, inappropriate water management, and pre-and post-harvest fruit losses. In the light of the given facts, intensive engagement of the policy making institutions, scientists, technologists, growers, food processors, exporters, and other stakeholders are solicited to protect and promote date palm production, and introduce innovative technologies aimed at fruit valorization and strengthening date fruit processing industry.

Conflicts of Interest All authors read and approved the final manuscript and declare that they have no conflict of interest.

References

Abass MH, Neama JD, Al-Jabary K (2016) Biochemical responses to cadmium and lead stresses in date palm (Phoenix dactylifera L.) plants. Advan Agricul Botan 8(3):92–110

Abd El-Mohsen M, Nezam El-Din M (1995) Technological study on Dibis production from the Siwi date. Egypt J Food Sci 23(3):229–239

Abdelaziz DH, Ali SA (2014) The protective effect of Phoenix dactylifera L. seeds against CCl4-induced hepatotoxicity in rats. J Ethnopharmacol 155(1):736–743

Abdelsalam MM, Zeitoun MM, Ateah MA et al (2014) Impact of probiotic fermented milk, palm date extract and their mixture supplementation on neonatal traits and hematological parameters of late pregnant Nadji ewes. Int J Biol Chem 8:37–47

Afiq AMJ, Abdul Rahman R, Che Man YB et al (2013) Date seed and date seed oil. Int Food Res J 20(5):2035–2043. https://doi.org/10.3390/foods9060787

Ahmed J, Ramaswamy HS (2006) Physico-chemical properties of commercial date pastes (*Phoenix dactylifera*). J Food Eng 76(3):348–352. https://doi.org/10.1016/j.jfoodeng.2005.05.033

Ahmed IA, Ahmed AW, Robinson RK (1995) Chemical composition of date varieties as influenced by the stage of ripening. Food Chem 54(3):305–309. https://doi.org/10.1016/0308-8146(95)00051-J

Ahmed J, Ramaswamy HS, Khan AR (2005) Effect of water activity on glass transitions of date paste. J Food Eng 66(2):253–258. https://doi.org/10.1016/j.jfoodeng.2004.03.015

Ahmed J, Al-Jasass FM, Siddiq M (2014) Date fruit composition and nutrition. In: Siddiq M, Aleid SM, Kader AA (eds) Dates: postharvest science, processing technology and health benefits. Wiley, New York, pp 261–228. https://doi.org/10.1002/9781118292419.ch11

Ahmed S, Khan AR, Jamil S (2016) Anti-hyperlipidemic and hepatoprotective effects of native date fruit variety Aseel(*Phoenix dactylifera*). Pak J Pharma Sci 29(6):1945–1950

Al-Alawi RA, Al-Mashiqri JH, Al-Nadabi JS et al (2017) Date palm tree (Phoenix dactylifera L.): natural products and therapeutic options. Front Plant Sci 8:845. https://doi.org/10.3389/fpls.2017.00845

Aldhaheri A, Alhadrami G, Aboalnaga N et al (2004) Chemical composition of date pits and reproductive hormonal status of rats fed date pits. Food Chem 86(1):93–97. https://doi.org/10.1016/j.foodchem.2003.08.022

Aleid SM, Hassan BH, Almaiman SA, Al-Kahtani SH, Ismail SM (2014) Microbial loads and physicochemical characteristics of fruits from four Saudi date palm tree cultivars: conformity with applicable date standards. Food Nutr Sci 5:42653

Al-Farsi MA, Lee CY (2008) Nutritional and functional properties of dates: a review. Crit Rev Food Sci Nutr 48(10):877–887. https://doi.org/10.1080/10408390701724264

Al-Farsi M, Alasalvar C, Morris A et al (2005) Comparison of antioxidant activity, anthocyanins, carotenoids, and phenolics of three native fresh and sun-dried date (*Phoenix dactylifera L.*) varieties grown in Oman. J Agric Food Chem 53(19):7592–7599

Al-Gabri N, Elnagar GM, Saghir SA et al (2022) Preliminary study of gastroprotective effect of aloe perryi and date palm extracts on pyloric ligation-induced gastric ulcer in experimental rats. BioMed Res Intern. https://doi.org/10.1155/2022/9246785

Alhamdan AM, Hassan BH (1999) Water sorption isotherms of date pastes as influenced by date cultivar and storage temperature. J Food Eng 39:301–306. https://doi.org/10.1016/S0260-8774(98)00170-8

Alharbi KL, Raman J, Shin HJ (2021) Date fruit and seed in nutricosmetics. Cosmetics 8(3):59. https://doi.org/10.3390/cosmetics8030059

Alhudaib K, Arocha Y, Wilson M et al (2007) "Al-Wijam", a new phytoplasma disease of date palm in Saudi Arabia. Bull Insectol 60(2):285–286

Ali A, Abdu S (2011) Antioxidant protection against pathological mycotoxins alterations on proximal tubules in rat kidney. Funct Foods Health Dis 1(4):118–134

Ali A, Al-Kindi YS, Al-Said F (2009) Chemical composition and glycemic index of three varieties of Omani dates. Int J Food Sci Nutr 60(4):51–62. https://doi.org/10.1080/09637480802389094

Al-Jboory IJ (2007) Survey and identification of the biotic factors in the date palm environment and its application for designing IPM-Program of date palm pests in Iraq. Univ Aden J Nat Appl Sci 11(3):423–457

Allaith AAA (2008) Antioxidant activity of Bahraini date palm (*Phoenix dactylifera L.*) fruit of various cultivars. Int J Food Sci Technol 43(6):1033–1040. https://doi.org/10.1111/j.1365-2621.2007.01558.x

Allbed A, Kumar L, Shabani F (2017) Climate change impacts on date palm cultivation in Saudi Arabia. J Agric Sci 155:1203–1218

Al-Mssallem IS, Hu S, Zhang X et al (2013) Genome sequence of the date palm Phoenix dactylifera L. Nat Commun 4:2274. https://doi.org/10.1038/ncomms3274

Al-Qarawi AA, Ali BH, Al-Mougy SA et al (2003) Gastrointestinal transit in mice treated with various extracts of date (*Phoenix dactylifera L.*). Food Chem Toxicol 41(1):37–39. https://doi.org/10.1016/S0278-6915(02)00203-X

Al-Qarawi AA, Mousa HM, Ali BH et al (2004) Protective effect of extracts from dates (*Phoenix dactylifera L.*) on carbon tetrachloride-induced hepatotoxicity in rats. Int J Appl Res Vet Med 2(3):176–180

Al-Qarawi AA, Abdel-Rahman H, Ali BH et al (2005) The ameliorative effect of dates (*Phoenix dactylifera L.*) on ethanol-induced gastric ulcer in rats. J Ethnopharmacol 98(3):313–317. https://doi.org/10.1016/j.jep.2005.01.023

Al-Qarawi AA, Abdel-Rahman H, Ali BH et al (2008) Nephroprotective action of Phoenix dactylifera in gentamicin-induced nephrotoxicity. Pharm Biol 46(4):227–230. https://doi.org/10.1080/13880200701739322

Al-Qurashi AD, Awad MA, Elsayed MI (2012) Pre-harvest fruit drop, bunch weight and fruit quality of 'Rothana'and 'Ghur' date palm cultivars as affected by some plant growth regulators. Afr J Biotechnol 11(81):14644–14651

Al-Yahyai R, Al-Kharusi L (2012) Physical and chemical quality attributes of freeze-stored dates. Int J Agric Biol 14:97–100

Anwar MA (2006) Phoenix dactylifera L: a bibliometric study of the literature on date palm. Malays J Libr Inf Sci 11(2):41–60

Ardekani MRS, Khanavi M, Hajimahmoodi M et al (2010) Comparison of antioxidant activity and total phenol contents of some date seed varieties from Iran. Iran J Pharm Res 9(2):141–146

Arias E, Hodder AJ, Oihabi A (2016) FAO support to date palm development around the world: 70 years of activity. Emir J Food Agric 28(1):1–11. https://doi.org/10.9755/ejfa.2015-10-840

Ashraf Z, Hamidi-Esfahani Z (2011) Date and date processing: a review. Food Rev Int 27(2):101–133. https://doi.org/10.1080/87559129.2010.535231

Aslam J, Khan SH, Khan SA (2013) Quantification of water-soluble vitamins in six date palm (*Phoenix dactylifera L.*) cultivar's fruits growing in Dubai, United Arab Emirates, through high performance liquid chromatography. J Saudi Chem Soc 17(1):9–16. https://doi.org/10.1016/j.jscs.2011.02.015

Barghini P, Di Gioia D, Fava F et al (2007) Vanillin production using metabolically engineered Escherichia coli under non-growing conditions. Microb Cell Factories. https://doi.org/10.1186/1475-2859-6-13

Basuny AMM, Al-Marzooq MA (2011) Production of mayonnaise from date pit oil. Food Nutr Sci 2:938–943

Bekheet SA, El-Sharabasy SF (2015) Date palm status and perspective in Egypt. In: Al-Khayri J, Jain S, Johnson D (eds) Date palm genetic resources and utilization. Springer, Dordrecht, pp 75–123. https://doi.org/10.1007/978-94-017-9694-1_3

Besbes S, Blecker C, Deroanne C et al (2004) Date seeds: chemical composition and characteristic profiles of the lipid fraction. Food Chem 84(4):577–584. https://doi.org/10.1016/S0308-8146(03)00281-4

Besbes S, Drira L, Blecker C et al (2009) Adding value to hard date (*Phoenix dactylifera L.*): compositional, functional, and sensory characteristics of date jam. Food Chem 112(2):406–411. https://doi.org/10.1016/j.foodchem.2008.05.093

Biglari F, Al Karkhi AFM, Easa AM (2007) Antioxidant activity and Phenolic content of various date palm (*Phoenix dactylifera*) fruits from Iran. Food Chem 107(4):1636–1164. https://doi.org/10.1016/j.foodchem.2007.10.033

Bouhlali ET, Hmidani A, Bourkhis B et al (2020) Phenolic profile and anti-inflammatory activity of four Moroccan date (*Phoenix dactylifera L.*) seed varieties. Heliyon 6(2):e03436. https://doi.org/10.1016/j.heliyon.2020.e03436

Chao CT, Krueger RR (2007) The date palm (*Phoenix dactylifera L.*): overview of biology, uses, and cultivation. HortSci 42(5):1077–1082. https://doi.org/10.21273/HORTSCI.42.5.1077

Chaparro CM, Suchdev PS (2019) Anemia epidemiology, pathophysiology, and etiology in low- and middle-income countries. Ann N Y Acad Sci 1450(1):15–31. https://doi.org/10.1111/nyas.14092

Chemura A, Schauberger B, Gornott C (2020) Impacts of climate change on agro-climatic suitability of major food crops in Ghana. PLoS One 15(6):e0229881. https://doi.org/10.1371/journal.pone.0229881

Cotoraci C, Ciceu A, Sasu A et al (2021) Natural antioxidants in anemia treatment. Int J Mol Sci 22(4):1883. https://doi.org/10.3390/ijms22041883

Dahlan FM, Ardhi Q (2021) The effect of Fe tablet and date palm on improving hemoglobin level among pregnant women in the third semester. J Midwifery 5(2):32–38. https://doi.org/10.25077/jom.5.2.32-38.2020

Dayang JF, Reuben CR, Raji F (2014) Nutritional, socioeconomic and health benefits of dates. Int J Food Nut Sci 3:63–73

Deborah S, Anand SP (2017) In vitro antibacterial activity of Phoenix loureiroi Kunth against selected Gram negative and Gram positive pathogenic bacteria. Int J Sci Eng Res 8:1412–1421

Dembinska-Kiec A, Mykkänen O, Kiec-Wilk B et al (2008) Antioxidant phytochemicals against type 2 diabetes. Br J Nutr 99:ES109–ES117. https://doi.org/10.1017/S000711450896579X

Dhehibi B, Salah MB, Frija A (2018) Date palm value chain analysis and marketing opportunities for the Gulf Cooperation Council (GCC) Countries. In: Kulshrestha SN (ed) Agricultural economics - current issues. IntechOpen. https://doi.org/10.5772/intechopen.82450

Dorria AM, Ashour NE, Mostafa EAM, Saleh MMS, Hassan HSA (2012) Yield and fruit quality of Amhat date palms as affected by spraying some vitamins. J Appl Sci Res 8:4922–4926

El Hadrami A, Al-Khayri JM (2013) Socioeconomic and traditional importance of date palm. Emir J Food Agric 25(4):371–385

Eldaim NAEN, Elnadi AH (2014) Determination of iron content in portulaca, dates and grewia. Int J Res Rev Appl Sci 19(1):29–35

El-Far AH, Shaheen HM, Abdel-Daim MM et al (2016) Date palm (*Phoenix dactylifera*): protection and remedy food. Curr Trends Nutraceutical 1(2):1–10

El-Juhany LI (2010) Degradation of date palm trees and date production in Arab countries: causes and potential rehabilitation. Aust J Basic Appl Sci 4(8):3998–4010

Elleuch M, Besbes S, Roiseux O et al (2008) Date flesh: chemical composition and characteristics of the dietary fibre. Food Chem 111(3):676–682. https://doi.org/10.1016/j.foodchem.2008.04.036

El–Merghany S, Zaen El-Daen EMA (2013) Evaluation of some date palm cultivars grown under toshky conditions. J Plant Prod 4(8):1207–1218

Elshibli S, Korpelainen H (2009) Biodiversity of date palms (Phoenixdactylifera L.) in Sudan: chemical, morphological and DNA polymorphisms of selected cultivars. Plant Genet Resour 7:194–203

Farahnaky A, Mardani M, Mesbahi G, Majzoobi M, Golmakani MT (2018) Some physicochemical properties of date syrup, concentrate, and liquid sugar in comparison with sucrose solutions. J Agric Sci Technol 18:657–668

Farooq S, Maqbool MM, Bashir MA et al (2021) Production suitability of date palm under changing climate in a semi-arid region predicted by CLIMEX model. J King Saud Univ Sci 33(3):101394. https://doi.org/10.1016/j.jksus.2021.101394

Fatima G, Khan IA, Buerkert A (2016) Socio-economic characterisation of date palm (Phoenix dactylifera L.) growers and date value chains in Pakistan. SpringerPlus 5(1):1222. https://doi.org/10.1186/s40064-016-2855-4
Ferguson LR, Philpott M, Karunasinghe N (2004) Dietary cancer and prevention using antimutagens. Toxicology 198(1–3):147–159. https://doi.org/10.1016/j.tox.2004.01.035
Fikry M, Yusof YA, Al-Awaadh AM, Rahman RA, Chin NL, Mousa E, Chang LS (2019) Kinetics modelling of the colour, hardness, grinding energy consumption and oil yield changes during the conventional roasting of palm date seeds. Food Sci Technol Res 25:351–362
Food and Agriculture Organization (2013) Perennial crops for food security: proceedings of the FAO expert workshop. Rome, Italy. Accessed 28–30 Aug 2013
Frija A, Dhehibi B, Salah MB et al (2017) Competitive advantage of GCC date palm sector in the international market: market shares, revealed comparative advantages, and trade balance indexes. Int J Market Stud 9(6):1–13. https://doi.org/10.5539/ijms.v9n6p1
Gasim AA (1994) Changes in sugar quality and mineral elements during fruit development in five date palm cultivars in AI-Madinah AI-Munawwarah. Science 6:29–36
Ghnimi S, Umer S, Karim A et al (2017) Date fruit (Phoenix dactylifera L.): an underutilized food seeking industrial valorization. NFS J 6:1–10. https://doi.org/10.1016/j.nfs.2016.12.001
Gros-Balthazard M, Flowers JM (2021) A brief history of the origin of domesticated date palms. In: Al-Khayri JM, Jain SM, Johnson DV (eds) The date palm genome, Compendium of plant genomes, vol 1. Springer, Cham, pp 55–74. https://doi.org/10.1007/978-3-030-73746-7_3
Habib HM, Ibrahim WH (2009) Nutritional quality evaluation of eighteen date pit varieties. Int J Food Sci Nutr 60(1):99–111. https://doi.org/10.1080/09637480802314639
Habib HM, Ibrahim WH (2011) Nutritional quality of 18 date fruit varieties. Int J Food Sci Nutr 62(5):544–555. https://doi.org/10.3109/09637486.2011.558073
Habib HM, Kamal H, Ibrahim WH et al (2013) Carotenoids, fat soluble vitamins and fatty acid profiles of 18 varieties of date seed oil. Ind Crops Prod 42:567–572. https://doi.org/10.1016/j.indcrop.2012.06.039
Haider MS, Khan IA, Jaskani MJ et al (2018) Pomological and biochemical profiling of date fruits (*Phoenix dactylifera L.*) during different fruit maturation phases. Pak J Bot 50(3):1069–1076
Hamad AM, Yousif AK (1986) Evaluation of brine and salt-stock pickling of two date varities in the Kimri stage. In: Proceeding of the second symposium on date palm. Second symposium on date palm
Hamad I, AbdElgawad H, Al Jaouni S et al (2015) Metabolic analysis of various date palm fruit (*Phoenix dactylifera L.*) cultivars from Saudi Arabia to assess their nutritional quality. Molecules 20(8):13620–13641. https://doi.org/10.3390/molecules200813620
Hamada JS, Hashim IB, Sharif FA (2002) Preliminary analysis and potential uses of date pits in foods. Food Chem 76(2):135–137. https://doi.org/10.1016/S0308-8146(01)00253-9
Hammadi H, Mokhtar R, Mokhtar E et al (2009) New approach for the morphological identification of date palm (*Phoenix dactylifera L.*) cultivars from Tunisia. Pak J Bot 41(6):2671–2681
Hampe A (2004) Bioclimate envelope models: what they detect and what they hide. Glob Ecol Biogeogr 13(5):469–476. https://www.jstor.org/stable/3697578
Hasan F, Nazir A, Sobti B, Tariq H, Karim R, Al-Marzouqi AH, Kamal-Eldin A (2022) Dehydration of date fruit (*Pheonix dactylifera* L.) for the production of natural sweet powder. NFS J 27:13–20
Hazzouri KM, Flowers JM, Visser HJ et al (2015) Whole genome re-sequencing of date palms yields insights into diversification of a fruit tree crop. Nat Commun 6:8824. https://doi.org/10.1038/ncomms9824
Henderson SA, Billotte N, Pintaud JC (2006) Genetic isolation of Cape Verde Island Phoenix atlantica (Arecaceae) revealed by microsatellite markers. Conserv Genet 7:213–223
Ibrahim KM (2010) The role of date palm tree in improvement of the environment. Acta Hortic 882:777–778

Iqbal M, Munir M, Niamatullah M (2011) Physio-chemical characteristics of date palm (*Phoenix dactylifera L.*) cultivars at various maturity stages under environmental conditions of Dera Ismail Khan. J Agric Res 49(2):249–260

Ishurd O, Zahid M, Xiao P et al (2004) Protein and amino acids contents of Libyan dates at three stages of development. J Sci Food Agric 84(5):481–484. https://doi.org/10.1002/jsfa.1600

Iwasaki Y, Matsui T, Arakawa Y (2004) The protective and hormonal effects of proanthocyanidin against gastric mucosal injury in Wistar rats. J Gastroenterol 39(9):831–837. https://doi.org/10.1007/s00535-004-1399-5

İZLİ G (2016) Total phenolics, antioxidant capacity, colour and drying characteristics of date fruit dried with different methods. Food Sci Technol 37:139–147

Johnson DV (2011) Introduction: date palm biotechnology from theory to practice. In: Jain S, Al-Khayri J, Johnson D (eds) Date palm biotechnology. Springer, Dordrecht, pp 1–11

Johnson DV, Al-Khayri JM, Jain SM (2013) Seedling date palms (*Phoenix dactylifera L.*) as genetic resources. Emir J Food Agric 25(11):809–830. https://doi.org/10.9755/ejfa.v25i11.16497

Kaky E, Nolan V, Alatawi A et al (2020) A comparison between Ensemble and MaxEnt species distribution modelling approaches for conservation: a case study with Egyptian medicinal plants. Ecol Inform 60. https://doi.org/10.1016/j.ecoinf.2020.101150

Karizaki VM (2017) Iranian dates and ethnic date-based products. J Ethn Foods 4:204–209

Kaushik A, Kostaki E, Kyriakoppoulos M (2016) The stigma of mental illness in children and adolescents: a systematic review. Psychiatry Res 243:469–494

Khalid S, Ahmad A, Kaleem M (2017) Antioxidant activity and phenolic contents of Ajwa date and their effect on lipo-protein profile. Funct Food Health Dis 7:396–410

Khattak MNK, Shanableh A, Hussain MI et al (2020) Anticancer activities of selected Emirati date (*Phoenix dactylifera L.*) varieties pits in human triple negative breast cancer MDA-MB-231 cells. Saudi J Biol Sci 27(12):3390–3396. https://doi.org/10.1016/j.sjbs.2020.09.001

Khodabakhshian R, Khojastehpour M (2021) Characteristics changes of date fruits during ripening period on palm. Agric Eng Int 23(4):243–255

Kogo BK, Kumar L, Koech R et al (2019) Modelling impacts of climate change on maize (*Zea mays L.*) Growth and productivity: a review of models, outputs and limitations. J Geosci Environ Prot 7:76–95

Krueger RR (2021) Date palm (Phoenix dactylifera L.) biology and utilization. In: Al-Khayri JM, Jain SM, Johnson DV (eds) The date palm genome, Compendium of plant genomes, vol 1. Springer, Cham, pp 3–28. https://doi.org/10.1007/978-3-030-73746-7_1

Kuna L, Jakab J, Smolic R, Raguz-Lucic N, Vcev A, Smolic M (2019) Peptic ulcer disease: a brief review of conventional therapy and herbal treatment options. J Clin Med 8(2):179. https://doi.org/10.3390/jcm8020179

Lambiote B (1982) Some aspects of the role of dates in human nutrition. Proceedings of the first international symposium on Date palm. King Faisal University, Saudi Arabia, 23-25 March 1982

Lo'ay AA, Elgammal RE, Alhaithloul HA et al (2021) Enhance fruit ripening uniformity and accelerate the rutab stage by using ATP in 'Zaghloul' dates during the shelf life. Foods 10(11):2641. https://doi.org/10.3390/foods10112641

Maqsood S, Adiamo O, Ahmad M et al (2020) Bioactive compounds from date fruit and seed as potential nutraceutical and functional food ingredients. Food Chem 308:125522. https://doi.org/10.1016/j.foodchem.2019.125522

Markhand GS, Abul-Soad AA, Mirbahar AA, Kanhar NA (2010) Fruit characterization of Pakistani dates. Pak J Bot 42(6):3715–3722

Mirza FG, Abdul-Kadir R, Breymann C et al (2018) Impact and management of iron deficiency and iron deficiency anemia in women's health. Expert Rev Hematol 11(9):727–736. https://doi.org/10.1080/17474086.2018.1502081

Mortazavi SMH, Karami Z, Mostaan A (2015) Use of ethylene scavenger sachet in modified atmosphere packaging to maintain storage stability of khalal date fruit. Int J Postharvest Technol Innov 5(1):52–63. https://doi.org/10.1504/IJPTI.2015.072447

Mrabet A, Rodríguez-Arcos R, Guillén-Bejarano R et al (2012) Dietary fiber from Tunisian common date cultivars (*Phoenix dactylifera* L.): chemical composition, functional properties, and antioxidant capacity. J Agric Food Chem 60(14):3658–3664. https://doi.org/10.1021/jf2035934

Mrabet A, Rodríguez-Gutiérrez G, Guillén-Bejarano R et al (2015) Valorization of Tunisian secondary date varieties (*Phoenix dactylifera L.*) by hydrothermal treatments: new fiber concentrates with antioxidant properties. LWT-Food Sci Technol 60(1):518–524. https://doi.org/10.1016/j.lwt.2014.09.055

Mrabet A, Hammadi H, Rodríguez-Gutiérrez G et al (2019) Date palm fruits as a potential source of functional dietary fiber: a review. Food Sci Technol Res 25(1):1–10. https://doi.org/10.3136/fstr.25.1

Nadiah MZN, Nazefah AH, Asralwirda AA et al (2018) Beneficial effects of date palm (*Phoenix Dactylifera*) in iron deficiency anaemia: a systematic review. Curr Topics Nutraceutical Res 16(4):245–252. https://www.nchpjournals.com/Manuscript?id=1509

Nasir MU, Hussain S, Jabbar S et al (2015) A review on the nutritional content, functional properties and medicinal potential of dates. Sci Lett 3(1):17–22

Nazeri M, Jusoff K, Madani N et al (2012) Predictive modeling and mapping of malayan sun bear (*Helarctos malayanus*) distribution using maximum entropy. PLoS One 7:e48104. https://doi.org/10.1371/journal.pone.0048104

Nehdi I, Omri S, Khalil M et al (2010) Characteristics and chemical composition of date palm (*Phoenix canariensis*) seeds and seed oil. Ind Crops Prod 32(3):360–365. https://doi.org/10.1016/j.indcrop.2010.05.016

Omar AK, El-Ashry HA (2015) Impact of pollen source on yield and fruit quality of Hayany date palm (Phoenix dactylifera L.). Egypt J Hort 42:357–366

Ortiz-Uribe N, Salomón-Torres R, Krueger R (2019) Date palm status and perspective in Mexico. Agric 9(3):46. https://doi.org/10.3390/agriculture9030046

Othmani A, Jemni M, Kadri K et al (2020) Preharvest fruit drop of date palm (*Phoenix dactylifera L.*) Cv. Deglet Nour at Kimri Stage: development, physico-chemical characterization, and functional properties. Int J Fruit Sci 20(3):414–432. https://doi.org/10.1080/15538362.2019.1651241

PHDEB, Pakistan Horticulture Development and Export Board (2008) Dates marketing strategy. www.phdec.org.pk/MktStrategies/Dates.pdf. Accessed 5 Aug 2014

Porfirio LL, Harris RM, Lefroy EC et al (2014) Improving the use of species distribution models in conservation planning and management under climate change. PLoS One 9(11):e113749. https://doi.org/10.1371/journal.pone.0113749

Qadri RWK, Waheed S, Haider MS, Khan I, Naqvi SA, Bashir M, Khan MM (2016) Physicochemical characterization of fruit of different date palm (*phoenix dactylifera* l.) varieties grown in Pakistan. JAPS: J Anim Plant Sci 26:1268–1277

Rabei S, Said WM, Rizk RM, El SF (2012) Morphometric taxonomy of date palm diversity growing. Egypt J Bot 189:175–189

Rabie S, Hassanine RMES, Hassan AA, Abo Elhussien O, EL-Mahdi BM (2021) Morphological and molecular characterization of Sarcocystis cameli and Sarcocystis ippeni from the muscles of One-Humped Camel (*Camelus dromedarius*) in New valley Governorate, Egypt. SVU- Int J Vet Sci 4:103–118

Radinger J, Wolter C, Kail J (2015) Spatial scaling of environmental variables improves species-habitat models of fishes in a small, sand-bed lowland river. PLoS One 10(11):e0142813. https://doi.org/10.1371/journal.pone.0142813

Rahayu TB (2021) Increasing of hemoglobin level in adolescent female through giving of date palm fruit (*Phoenix dactylifera*). Matern Neonatal Health J 5(2):49–56. https://doi.org/10.36696/mikia.v5i2.76

Ranilla LG, Kwon Y, Genovese MI et al (2008) Antidiabetes and antihypertension potential of commonly consumed carbohydrate sweeteners using in vitro models. J Med Food 11(2):337–348. https://doi.org/10.1089/jmf.2007.689

Riedel HR (1986) Dates, where they come from and how they can be used in the confectionery industry. Kakao-und-Zucker 38(12):16–19

Rock W, Rosenblat M, Borochov-Neori H et al (2009) Effects of date (Phoenix dactylifera L., Medjool or Hallawi Variety) consumption by healthy subjects on serum glucose and lipid levels and on serum oxidative status: a pilot study. J Agric Food Chem 57(17):8010–8017. https://doi.org/10.1021/jf901559a

Roodt, Veronica (1992) Phoenix reclinata. In: The shell field guide to the common trees of the Okavango Delta and Moremi Game reserve. Shell Oil Botswana, Gaborone, p 112

Saafi EB, Louedi M, Elfeki A et al (2011) Protective effect of date palm fruit extract (*Phoenix dactylifera L.*) on dimethoate induced-oxidative stress in rat liver. Exp Toxicol Pathol 63(5):433–441. https://doi.org/10.1016/j.etp.2010.03.002

Sahari MA, Barzegar M, Radfar R (2007) Effect of varieties on the composition of dates (*Phoenix dactylifera L.*)-note. Food Sci Technol Int 13(4):269–275. https://doi.org/10.1177/1082013207082244

Salama KF, Randhawa MA, Al Mulla AA et al (2019) Heavy metals in some date palm fruit cultivars in Saudi Arabia and their health risk assessment. Int J Food Pro 22(1):1684–1692. https://doi.org/10.1080/10942912.2019.1671453

Saleem SA, Baloch AK (2004) Influence of hot water on ripening/curing of Dhakki dates. Pak J Biol Sci 7(12):2034–2038

Saleh EA, Tawfik MS, Abu-Tarboush HM (2011) Phenolic contents and antioxidant activity of various date palm (*Phoenix dactylifera L.)* fruits from Saudi Arabia. Food Nutr Sci 2(10):1134–1141

Sankar V, Shoba FG (2014) Quantitative and qualitative analysis of various phyto-constituents in Phoenix pusilla unripe fruit ethanolic extract. Asian J Pharma Clin Res 7:5–8

Sari A, Pamungkasari EP, Dewi YLR (2018) The addition of dates palm (*Phoenix dactylifera*) on iron supplementation (Fe) increases the hemoglobin level of adolescent girls with anemia. Bali Med J 7(2):356–360

Sarraf M, Jemni M, Kahramanoğlu I et al (2021) Commercial techniques for preserving date palm (*Phoenix dactylifera*) fruit quality and safety: a review. Saudi J Biol Sci 28(8):4408–4420. https://doi.org/10.1016/j.sjbs.2021.04.035

Schierhorn F, Hofmann M, Adrian I et al (2020) Spatially varying impacts of climate change on wheat and barley yields in Kazakhstan. J Arid Environ 178:104164. https://doi.org/10.1016/j.jaridenv.2020.104164

Selim S, Abdel-Mawgoud M, Al-sharary T et al (2021) Pits of date palm: bioactive composition, antibacterial activity and antimutagenicity potentials. Agron 12(1):54. https://doi.org/10.3390/agronomy12010054

Shabani F, Kumar L, Solhjouy-Fard S (2017) Variances in the projections, resulting from CLIMEX, Boosted Regression Trees and Random Forests techniques. Theor Appl Climatol 129(3):801–814

Sharif AO, Sanduk M, Taleb HM (2010) The date palm and its role in reducing soil salinity and global warming. Acta Hortic 882:59–64

Sharma GN, Dubey SK, Sati N (2011) Anti-inflammatory activity and total flavonoid content of Aegle marmelos seeds. Int J Pharma Sci Drug Res 3(3):214–218

Silva DP, Vilela B, De Marco JP et al (2014) Using ecological niche models and niche analyses to understand speciation patterns: the case of sister neotropical orchid bees. PLoS One 9(11):e113246. https://doi.org/10.1371/journal.pone.0113246

Sirisena S, Ng K, Ajlouni S (2015) The emerging Australian date palm industry: date fruit nutritional and bioactive compounds and valuable processing by-products. Comp Rev Food Sci Food Safety 14(6):813–823. https://doi.org/10.1111/1541-4337.12162

Spennemann DH (2019) The connective potential of vertebrate vectors responsible for the dispersal of the Canary Island date palm (*Phoenix canariensis*). Flora 259:151468

Suhail M, Durrani Y, Hashmi MS, Muhammad A, Ali SA, Uzair M, Shahid M (2020) Postharvest losses of dhakki dates during supply chain in Pakistan. Fresenius Environ Bull 29:299–309

Świąder K, Białek K, Hosoglu I (2020) Varieties of date palm fruits (*Phoenix dactylifera L.*), their characteristics and cultivation. Technol Prog Food Process 1:173–179

Tang ZX, Shi L, Aleid SM (2013) Date fruit: chemical composition, nutritional and medicinal values, products. J Sci Food Agric 93(10):2351–2361. https://doi.org/10.1002/jsfa.6154
Tang ZX, Shi LE, Aleid SM (2014) Date and their processing byproducts as substrates for bioactive compounds production. Braz Arch Biol Technol 57:706–713
Tengberg M (2012) Beginnings and early history of date palm garden cultivation in the Middle East. J Arid Environ 86:139–147. https://doi.org/10.1016/j.jaridenv.2011.11.022
Vayalil PK (2002) Antioxidant and antimutagenic properties of aqueous extract of date fruit (*Phoenix dactylifera L. Arecaceae*). J Agric Food Chem 50(3):610–617
Vayalil PK (2012) Date fruits (*Phoenix dactylifera Linn*): an emerging medicinal food. Crit Rev Food Sci Nutr 52(3):249–271. https://doi.org/10.1080/10408398.2010.499824
Vincenzi S, Zucchetta M, Franzoi P et al (2011) Application of a Random Forest algorithm to predict spatial distribution of the potential yield of Ruditapes philippinarum in the Venice lagoon, Italy. Ecol Model 222(8):1471–1478. https://doi.org/10.1016/j.ecolmodel.2011.02.007
Vinson JA, Zubik L, Bose P et al (2005) Dried fruits: excellent in vitro and in vivo antioxidants. J Am Coll Nutr 24(1):44–50. https://doi.org/10.1080/07315724.2005.10719442
Wanderley CS, Faria RT, Ventura MU (2012) Adubação química, adubação orgânica e extrato pirolenhoso no desenvolvimento de mudas de palmeiras fênix (Phoenix roebelenii). Semina: Ciências Agrárias 33:2233–2240
Yeh C, Ching L, Yen G (2009) Inducing gene expression of cardiac antioxidant enzymes by dietary phenolic acids in rats. J Nutr Biochem 20(3):163–171. https://doi.org/10.1016/j.jnutbio.2008.01.005
Younas A, Naqvi SA, Khan MR et al (2020) Functional food and nutra-pharmaceutical perspectives of date (*Phoenix dactylifera L.*) fruit. J Food Biochemist 44(9):e13332. https://doi.org/10.1111/jfbc.13332
Zaid A, De Wet PF (2002) Climatic requirements of date palm. Date palm cultivation, Food and Agriculture Organization Plant Production and Protection Paper No 156, pp 57–72
Zen ATH, Pertiwi D, Chodidjah C (2013) The effect of date (Phoenix dactylifera) juice on haemoglobin level an experimental study in iron supplemented rats. Sains Medika 5(1):17–19
Zhang L, Gu L, Ringler P, Smith S, Rushton PJ, Shen QJ (2015) Three WRKY transcription factors additively repress abscisic acid and gibberellin signaling in aleurone cells. Plant Sci 236:214–222
Zohary D, Hopf M (2000) Domestication of plants in the old world, 3rd edn. Oxford University Press, Oxford

Chapter 14
Quinoa (*Chenopodium quinoa*): Potential of the "Golden Grain" for Food and Nutritional Security in South Asia

Muhammad Younis, Dur-e-Shahwar Sattar, and Raheel Suleman

14.1 Introduction

Quinoa (*Chenopodium quinoa*), a member of the Amaranthaceous family and native to the Andes, is grown for its edible seeds which are called quinoa and pronounced as KEEN-wah. Humans are consuming quinoa from the last 5000–7000 years while natural distribution of quinoa has been extended from Northern Colombia to the Southern regions of Chile (Alandia et al. 2020; Martínez-Villaluenga et al. 2020). Crop species, referred to as "Major Crops," are the key contributors to the world's food supply. Approximately 30 such crops provide around 95% of the food needs of people at planet earth. In the recent past, needs for improving biodiversity of food crops and the genetic improvement have been pounded due to reduction in the size of food basket and high reliance on major food crops. As a result, scientists from around the world have emphasized the importance of preserving and improving the usage of crops that have been overlooked by science, technology, agro food market. The terms "Neglected and Underutilized Crops" (NUCS), also referred to as "Orphan Crops," are used to describe these plants. One such underutilized grain is quinoa, a pseudocereal that is native to South America's Andean highlands. It was listed as one of the most promising crops for humans by the Food and Agricultural Organization (FAO) in 1996. According to the recent statistics, quinoa production has been estimated as 146,735 tons from 173,242 hectares with an average crop yield of 847 kg/ha (FAO 2017). Historically, quinoa was an important crop for the Inca Empire and was referred as the "mother of all grains". Although it has been a staple diet in South America for thousands of years, it just recently gained popularity and attained "super-food status". Now a days, quinoa and goods produced with

M. Younis · D. Sattar · R. Suleman (✉)
Department of Food Science and Technology, Faculty of Food Science & Nutrition, Bahauddin Zakariya University, Multan, Pakistan
e-mail: dr.raheel@bzu.edu.pk

T. Ismail et al. (eds.), *Neglected Plant Foods of South Asia*,
https://doi.org/10.1007/978-3-031-37077-9_14

it are available all over the world, especially in cuisines developed with natural food resources.

Three cultivars of quinoa namely black, red and white quinoa are cultivated worldwide (Fig. 14.1). Quinoa protein has a high biological value (73%) than conventional staple cereals like white rice (56%), wheat (49%) and corn (36%), and is equivalent to that of beef (74%). Quinoa has 12.9–16.5% protein content and is compounded with good concentration of all essential amino acids (Fuentes and Paredes-Gónzalez 2013; Bastidas et al. 2016). Recent evidences suggest quinoa as an advantageous food crop with a range of nutritive and health promoting bioactive substances which can anticipate preventive and therapeutic properties against a range of nutritional and health ailments (Razzeto et al. 2019; Mohyuddin et al. 2019).

Given that there is evidence linking certain bioactive substances, including phenolic compounds, polysaccharides, and saponins, to a variety of biological activities, including antioxidant activities, anti-inflammatory and anticancer activities presence of these substances suggests that quinoa may have beneficial effects (Ong et al. 2020). Triterpenoids (phytosterols, phytoecdysteroids and saponins), glycine betaine, phenolics and beta-lains are the main categories of secondary metabolites associated with quinoa. UHPLC analysis with photo-diode range can be used to evaluate biologically active quinoa constituents including 7-isopentenyloxycoumarin, auraptene, umbelli-prenin, and 4′-geranyloxyferulic acid (Fig. 14.2).

Fig. 14.1 Worldwide cultivated quinoa varieties

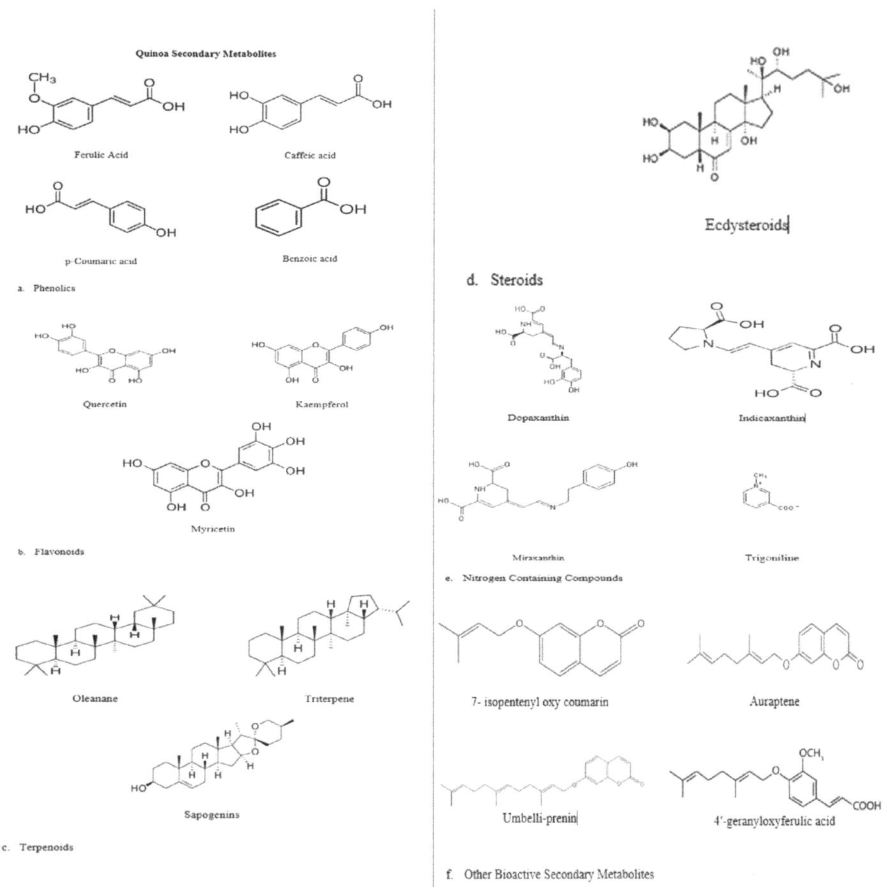

Fig. 14.2 Biologically active substances found in quinoa

14.2 Quinoa Adaptation, Cultivation, and Production Trends in South Asia

Population expansion necessitates higher food production, which raises the demand for agricultural products in terms of both quantity and quality. There is a growing push to include quinoa into various marginal agricultural production systems around the world since it has been identified as a very valuable crop that is climate adaptable. Quinoa has a flexible ability to adapt to various environmental conditions during its vegetative cycle, which ranges from 150 to 240 days. Due to their relative indifference to photoperiod and altitude, quinoa cultivars can be grown anywhere from the sea level to 3900 meters above sea. They are also highly resilient to a number of abiotic stimuli, such as cold, salt, and water stress, and can withstand a wide range of soil temperatures (−4 to ~38 °C) and pH values from 6 to 8.5. As a result, quinoa farming has been adopted in countries all the way from Chile (47°S) to Colombia (2°N). The plant can withstand and flourish in environments with high salinity (40 mS/cm) and little precipitation (50 mm/year). As a result, quinoa may be a source of nutrients for the majority of the world's population. Given that growth conditions vary depending on location and necessitated plant adaptation, the diversity of cultivation locations, altitudes, and rainfall patterns has resulted in a significant richness of quinoa species (Tanwar et al. 2019).

More than 60% of the quinoa is grown on the Southern Altiplano, primarily as Royal Quinoa, which is mainly exported. Quinoa is gaining interest from other countries that may help growers to boost their agro food market, improve exports, bolstering the processing industry, and creating more jobs. World widely in some countries such as Bolivialess than 2% of total cultivated land and less than 5% of the land used for cereal cultivation are specifically set aside for the cultivation of quinoa. Furthermore, it contributes very little to the farming industry. Quinoa cultivation has not expanded as much as it should have for a number of reasons more importantly absence of state initiatives to support quinoa cultivation. On top of that, almost 80% of the rural farms that grow quinoa are run by resource-constrained farmers, many of whom practice subsistence farming. The Altiplano's rural residents depend on the quinoa crop for food sustenance. Quinoa production is lower than other crops including soy, maize, wheat, rice, potatoes, sugar cane, and barley since it is grown on a smaller crop cultivation area. However, quinoa's better adaptability and tolerance to severe soil and climatic conditions are reasoned for its higher popularity. Excellent grain quality and the robust production potential in difficult agroecological settings are the added features of quinoa attracting attention of the farming communities of the many countries. Quinoa is relatively new in South Asia while its cultivation was introduced for the first time in the year 2009 in Pakistan. Mass scale cultivation of quinoa was initiated later than 2019 when first quinoa variety was registered (Afzal et al. 2022; Thiam et al. 2021; Angeli et al. 2020).

Royal Quinoa due to its larger grain size i.e., 2.5 mm is considered the most promising cultivar of quinoa. Moreover, it is relatively more resistance to frosts and droughts, and can tolerate challenging environmental conditions. The cost of processing Royal Quinoa is high due to the high saponins content of the grain that anticipates bitterness to the grains when consumed raw or unprocessed. Nonetheless, high saponins concentration gives the grain some defense against insect or disease attacks, particularly against birds. Consequent to multiple genetic improvement attempts, the first sweet cultivar named as "Sajama" devoid of saponin was obtained in 1970. This cultivar is now grown in nearly all of Bolivia's Central and Northern Altiplano as well as high-altitude regions of Peru. The Sajama cultivar is advantageous for growers since it has a commercial yield of 1500 kilos per hectare and is saponin-free.

According to the Additive Main Effects and Multiplicative Interaction analysis (AMMI), the genotype explains 23% of the variability in grain yield, whereas the location by variety interaction explains 32%. In order to develop new lines with superior crop attributes and higher production for the South Asian markets, it is necessary to expand the variety of quinoa germplasm and establish an active breeding program. Scaling up quinoa production in South Asia requires mechanized production, processing methods, and a controlled market. In order to develop an environment that is beneficial for quinoa production and marketing, integration between the key production and marketing activities, and collaboration among the decision-makers, research institutes, farmers associations, and the commercial sector are necessary (Afzal et al. 2022; Morales and Higuchi 2022).

14.3 Nutritional Perspectives of Quinoa and Comparison with Traditional Cereals

Quinoa grains are superior to traditional cereals in terms of their higher and better quality protein levels, which provide them exceptional nutritional worth. Being carrier of all essential amino acids, quinoa protein is considered substitute to milk proteins (Felipe et al. 2003). Quinoa is a possible plant protein source that has a reported protein content that is over two times higher than that of wheat and that is qualitatively and quantitatively superior to other plant proteins (Ceccato et al. 2011). Granular starch of quinoa makes up 54% of the grains' carbohydrates and due to high water retention properties, high expansion rate and better viscosity, quinoa starch performs better as a thickening agent than wheat and barley starch. These characteristics make quinoa starch perfect candidate for making ready-to-eat frozen infant foods since it has good freeze-thawing stability (Ruales et al. 2002). Approximately 58–64% of the quinoa dry matter is constituted with starch, of which 11% is amylose. As mentioned in Table 14.1 quinoa also has significant levels of D-xylose and maltose and low levels of glucose and fructose (Bastidas et al. 2016).

Among the essential fatty acids, linoleic acid is one of the most common polyunsaturated fatty acids present in quinoa, and has been associated with an increased risk of insulin sensitivity and a decreased risk of heart disease (Bastidas et al. 2016). Quinoa contains 14.5% of its total weight as lipids, with roughly 70% of those lipids being unsaturated. Linoleic and oleic acids roughly constitute 39% and 28% of the total seed lipids, respectively. Various genotypes of quinoa in addition to oleic and linoleic acid also contain a variety of fatty acids such as palmitic acid (11.39–13.25%) and α–linoleic acid (4.4–7.71%) (Nasir et al. 2015). Dietary fibers are categorized as total, soluble, and insoluble fibers. Quinoa is good carrier of dietary fibers which makes up 2.6–10% of the total grain's weight, and contains 78% insoluble and 22% soluble fiber (Bastidas et al. 2016). Quinoa seeds processing for debittering and for the seeds extrusion significantly reduces its total fiber contents (Morales and Higuchi 2022). Comparing with different conventional cereals, quinoa seeds contain higher amount of thiamine, niacin, folates, ascorbic acid, tocopherol and many other water soluble vitamins (Table 14.1). Quinoa contains 0.20 mg of pyridoxine, 0.61 mg of pantothenic acid, 78.1 mg of folic acid, 0.4 mg of thiamine, and 1.4 mg of ascorbic acid per 100 g of the edible grains. Quinoa also contains 37.5–59.8 mg/g of vitamin E, and this seed also contains the tocopherol isoforms -tocopherol (17–26 mg/g), -tocopherol (47–53 mg/g) and -tocopherol (5 mg/g). Beta carotene

Table 14.1 Quinoa flour nutritional contents

Mineral (mg per 100 g sample) in quinoa flour		Vitamin (mg per 100 g sample) in quinoa flour		Proximate composition (g per 100 g sample) in quinoa flour		Sugar contents (mg per 100 g sample) in quinoa flour	
Iron	2.60	Vit. A	0.90–1.00	Proteins	13.50–15.70	Glucose	19
Copper	7.50	Vit. B1	0.36–0.40	Carbohydrates	58.30–69.80	Fructose	19.6
Calcium	86.00	Vit. B2	0.30–0.35	Fat	6.30–7.60	D-Ribose	72
Zinc	3.80	Vit. B3	1.40–1.52	Dietary Fiber	9.50–14.10	D-Galactose	61
Sodium	93.00	Vit. B5	0.48–0.61	Total Ash	1.20–7.70	Maltose	101
Manganese	Nd	Vit. B6	0.20–0.49	Moisture	11.20–14.50	D-Xylose	120
Led	Nd	Vit. B9	78.1–184.0	Energy	368.00 Kcal		
Magnesium	232.00	Vit. C	1.20–1.40				
Potassium	714.00	Vit. E	37.49–59.82 μg/1 g				
Phosphorus	22.00						

Ogungbenle (2003), Vidueiros et al. (2015), Bastidas et al. (2016)

contents of quinoa have been reported up to a level of 0.39 mg/100 g dry weight that is far higher than observed in wheat and barley grains (Morales and Higuchi 2022).

Quinoa grains also hold appreciable amount of calcium, magnesium, potassium, iron, manganese, copper, and molybdenum which are virtually low in gluten-free cereals (Table 14.1). Quinoa contains 874 mg/kg of calcium, 948.5 mg/kg of iron, 2735 mg/kg of phosphorus, 9562 mg/kg of potassium and 1901.5 mg/kg magnesium contents (Bastidas et al. 2016). In comparison to most grain crops, quinoa has higher amounts of minerals in the outer bran layers. The quinoa embryo contains minerals like magnesium, potassium, and phosphorus, whereas the pericarp alongside pectic substances contains calcium and phosphorus. The quinoa embryo contains a homogeneous distribution of sulfur. Since iron in quinoa has been found to be highly soluble, people with iron deficiency anemia may have biologically high access to it (Morales and Higuchi 2022).

14.4 Application of Quinoa in Management of Various Forms of Nutritional Disorders

With every passing day, quinoa being a concentrated nutrient energy source is gaining more popularity as potential alternate to rice and many other conventional cereals. On account of high protein contents, quinoa based food formulations or food preparations are considered as ideal option for the vegetarians or health conscious persons planning weight loss or maintaining weight as quinoa improves satiety function, increases metabolism and promotes weight loss. At the moment the pseudocereal is potential source of food due to its gluten devoid high quality proteins. However, combinations of wheat flour and quinoa are employed to make baked goods because the quinoa proteins can't support dough making properties as like gluten. That's why addition of quinoa into leavened and pasta products are limited. The amount of high-protein possessing an attractive amino acid balance of lysine and methionine represents befits in improving nutritional value of the consumer goods, and achieving sensory acceptability of the finished products. Conflict of opinion exists among the scientists on considering quinoa protein as complete protein rather nearly complete since it holds relatively lower amount of some essential amino acids like lysine (Craine and Murphy 2020). Quinoa protein quality also adheres quite closely to the Food and Agricultural Organization's recommended amino acid profile and amino acid balance (FAO) (Filho et al. 2017). The World Health Organization (WHO) and the Food and Agricultural Organization (FAO) of the United Nations claim that quinoa meets the daily requirements of adults for the amino acids such as cysteine and methionine (212%), lysine (338%), histidine (180%), isoleucine (274%) and phenylalanine (320%) and tyrosine. Poor nutritional quality of conventional gluten-free food products which largely comply with the food demands of celiac patients and those with Irritable bowel syndrome (IBS) are advised to replace their gluten free flours with quinoa and quinoa flour based products (Peñas et al. 2014; Tanwar et al. 2019). Comparing a quinoa-based newborn meal (100 g twice day for 15 days) against no therapy could aid in lowering childhood malnutrition. It is known that IGF-1, a marker of rising plasma levels and an indication of malnutrition, speeds up the rise of body weight (Srujana et al. 2019; Brito et al. 2022). With above referred recommendations, quinoa may be considered as a nutrient dense food that may be exploited in developing nourishing food formulations for the infant, young children (Bastidas et al. 2016). The National Aeronautics and Space Administration (NASA) has used quinoa in developing space food due to its versatility in meeting human nutritional needs during space missions. Quinoa entire grains have been used to make a wide range of dishes, including porridges, drinks, savory soups, sauces, desserts and soufflés; the grains can also be boiled and eaten like rice. The grains are germinated for sprouts and malted for beer production. The fermented beverage made from quinoa seeds are called *chicha* and the coarse bread prepared is called *kispina*. It is also served as starchy material, and as protein concentrate while quinoa flours for various food uses are also marketed.

14.5 Health Benefits of Quinoa as a Functional Food

Quinoa as a wonderful carrier of complete proteins, polyunsaturated fatty acids, carotenoids, dietary fibers, minerals and other plant bioactive compounds, and is thought to provide significant nutritional health advantages to the consumers.

Quinoa seeds are also considered a "functional food" due to their high level of polyphenols, including phenolic acids or flavonoids, and vitamins (A, B, and E), which collectively contribute a strong antioxidant capacity to quinoa based products. These antioxidants may guard consumers against chronic illnesses like cancer, heart disease, and help to shield healthy cells from damage brought on by the free radicals. Higher magnesium contents of quinoa support cardiovascular health and aids in managing blood pressure (Farinazzi-Machado et al. 2012; Pasko et al. 2009; Tang and Tsao 2017). Quinoa is rich in phytochemicals including nicotiflorin and rutin, which support lowering high blood pressure.

Quinoa has a variety of phytochemicals with neuroprotective effects that reduce cytokines and support the healing of injured brain tissues. Quinoa's greater fiber and phytosterol content can lower cholesterol and triglyceride levels, as well as help to promote satiety and control digestion. Studies also demonstrate that quinoa consumption has a positive effect on metabolic health as it decreases LDL or bad cholesterol, improves insulin and triglycerides and aids in the management of diabetes, especially during the summer (Arts and Hollman 2005; Gawlik-Dziki et al. 2013; Singh 2019). Quinoa extracts have significant amounts of ferulic, gallic, and sinapinic acids as well as isorhamnetin, rutin, and kaempferol, which have an effect on prostate cancer cell motility and proliferation (Gawlik-Dziki et al. 2013).

Quinoa grain also contains phytoestrogens, which shield against chronic illnesses including osteoporosis, breast cancer, heart disease, and women health associated issues brought on by a loss of estrogen during menopause. Phytoestrogens are present in the majority of cereals and soya bean. A study of this kind has not yet been conducted on quinoa as it may have promosing levels of phytoestrogens, though statement needs confirmation by scientific exploration of the golden grains. The Royal Quinoa Huallata, Pandela, Mok'o Rosado, Achachino, Chillpi Amapola, K'ellu, Rose Blanca, and Chillpi Rosado kinds are used to cure bone fractures by using unprocessed milled quinoa grain (Royal Quinoa Catalogue 2003). Since quinoa has a low glycemic index and is absorbed gradually from the gut to blood, it may help controlling instant rise in postprandial blood sugar levels. Fiber and protein in moderation are essential for regulating blood sugar levels (FAO, CIRAD 2015).

14.6 Processing Technologies to Improve Safety and Nutritional Significance of Quinoa

Despite its nutritional importance, quinoa contains a variety of anti-nutritional components such as saponins, phytic acid and oxalates, which can affect its digestibility and nutritional value (El Hazzam et al. 2020; Saypol Anwar and Zulkifli 2021), and make people hesitant to consume it. Attempts to include quinoa in food items around the world have been impeded by the presence of saponins, which are responsible for reducing product acceptability due to their bitter flavor and/or aftertaste (Suárez-Estrella et al. 2018). Saponins are somewhat dangerous chemical substance that may toxicate consumer once entered the bloodstream at levels more than the normal. Saponins have been proven to affect protein digestibility by precipitating with the protein component and forming complexes (Samtiya et al. 2020). In the grain pericarp, there are numerous saponins and other bitter and astringent compounds that can hamper the digestion and absorption of nutrients (Wu et al. 2021).

To get rid of these anti-nutrients in the quinoa, several processing techniques such as soaking, washing, cooking, fermenting, sprouting, pearling and roasting are deployed to effectively reduce saponins contents (Samtiya et al. 2020). Due to their great solubility in water, washing is the most popular method for removing saponins and other impurities from seeds at the household level. In rural settings, hand washing has historically been done in water, either in water tanks that may be alkalined to improve saponin extraction. Commercial washing also makes use of tanks with moving blades for turbulence washing. Water being used for washing quinoa or removing saponins can create risks for the cold blooded animals and hence raises questions on environmental effects of quinoa washing in open water streams. Damp seeds are dried as soon as possible to prevent the growth of mold and reduce their potent germination ability. Roasting, scarifying, washing, or chemically treating the grain are some other viable techniques to remove maximum saponins load from the quinoa seeds. In "processing plants" that are frequently owned by the wholesalers themselves, the majority of conventional quinoa is prepared for marketing purposes using conventional methods.

The bitterness of quinoa can be lessened or modified to increase its palatability and hence its consumption. Many different procedures have been suggested in addition to the development of sweet quinoa variants. Techniques such as washing, pearling, and a combination of these two can anticipate reduction of saponins to an acceptable level. Others, like fermentation and germination, can mask bitterness by producing aroma compounds and/or sugar. Drying method is also a beneficial to remove saponins effectively with utilizing heat at high temperature i.e., 70 °C but the quinoa perisperm swells at this temperature and may cause the starch gelatinization at 50 °C. Moreover, many essential nutrients are also lost during this treatment. Recent research has shown that extracting phenolic contents from the outer layers of grains through pearling is an effective technique. Quinoa seeds have been pearled successfully to lessen the amount of saponins present in the seed's outer layers. Pearling is more environmentally friendly than washing because it uses no water, doesn't dry the seeds with heat, and produces no waste. Another benefit of the abrasion method is that it requires less time and effort. Around 8–12% of the grain weight is made up of pearled by-products, which have a saponin concentration of 200–300 g/kg and can be used to create medicines, detergents, and insecticides. On the other hand when abrasion increases, the amount of fiber and phenolic compounds decreases. Quinoa loses fewer phenolic compounds after pearling than other cereals do (Suárez-Estrella et al. 2018). Bivalent minerals like calcium, iron, zinc, and copper, as well as proteins and enzymes, can all be bound by phytic acid. Rural diets in underdeveloped nations and vegetarian diets are frequently linked to high phytate contents. The bioavailability of vital divalent elements including iron, calcium, and zinc can be increased by reducing the phytate level of quinoa through quinoa fermentation. Fermented quinoa to be widely accepted, it must not only be nutritionally complete, secure, and nutritious, but also have favorable sensory qualities (Lazarte et al. 2021). The favorable impacts of processing quinoa include increased digestibility, nutrient bioavailability, taste, flavor, texture, and palatability of these seeds (Motta et al. 2019).

14.7 Value – Addition and Industrial Uses of Quinoa

Quinoa is increasingly in demand worldwide, particularly in the food and health sector, but its supply cannot keep up with the demand, in particular for some consumers like the elderly, children, high-performance athletes, diabetics, celiacs, and those who are lactose or gluten intolerant (Valcárcel-Yamani and Lannes 2012). Quinoa flour can be produced and combined with wheat flours to produce delicious savory or sweet dishes in both solid and liquid forms. It is used to prepare high-nutrient cookies, "barritas," pies, whips, cakes, spaghetti, pasta, flakes, etc. Moreover, it is used to make batters, improving them, preserving their wetness and giving a very pleasant flavor as well as a fine and distinctive texture. Using conventional pasta-making techniques, quinoa flour is combined with other flours, such as maize or rice flour, and then made into pasta. Similar to puffed rice, quinoa is cleaned, dried, then heated under high pressure and temperature to puff it into a light, crispy, and puffed product.

Quinoa is a source of compounds with potential pharmaceutical properties such as flavonoids and saponins. These compounds are being studied for their potential anti-inflammatory, antioxidant, anti-cancer, and anti-diabetic properties. The herbal and pharmaceutical industries have developed quinoa-based products to address the celiac disease, sports performance and fitness enhancement, weight loss and/or metabolic health associated disorders such as diabetes, obesity, hypertension, and hyperlipidemia, postmenopausal, skin and hair care, and drug absorption. Several quinoa-based traditional food products (Table 14.2) such as quinoa noodles, quinoa flakes, pies, biscuits etc. have been developed so far. Quinoa flakes, which are manufactured by drying (~15% moisture) the grains and pressing them between two convergent rollers, are one of the latest quinoa products. Although quinoa varieties have different plasticity characteristics, the final characteristics of the flakes are determined by the variety. Quinoa flakes can be incorporated to many other dishes because they cook more quickly than other grains (such as juices, soups, pies, and cakes) (Quiroga et al. 2013).

In a study by De Carvalho et al. (2014), consuming 25 g/d quinoa flakes (QF) for 4 weeks has shown a positive effect on metabolic health markers. QF consumption improved protein and fiber intake but not total calorie intake. GSH were found increased whereas triglycerides, TBARS, and cholesterol are decreased (. Recent developments in extruded quinoa products might be recognized as novel quinoa food items. High pressure, high temperatures, and quick shearing are all factors in the extrusion of quinoa. Quinoa's starch and protein composition are restructured

Table 14.2 Traditional quinoa food products

Type of product	Description of the Product
Quinoa Soup	Quinoa prepared with meat, vegetables, and tubers
Lawa	a dish comprised of raw wheat, limewater, and animal fat that resembles porridge
P'esque	Grain of quinoa prepared without salt and served with milk or cheese, grated
Kispiña	Buns that have been steam-cooked
Tacti o tactacho	Doughnut-like fried buns made with llama fat and flour
Mucuna	Seasoning-filled quinoa flour balls that have been steam-cooked
Phiri	Quinoa flour that has been roasted and mildly moistened
Phisara	Grain of cooked and lightly toasted quinoa
Q'usa	A macerated cold beverage called quinoa chicha
El Ullphu, Ullphi	A cold drink prepared from quinoa grain that has been toasted
Kaswira de quinua	White quinoa and flattened bread are cooked in oil with lime.
Kaswira de ajara	ajara or black quinoa, along with flattened bread, fried in oil.
K'api kispiña	Steamed bun with quinoa flour
Turucha quispiña o Polonca	Large, steaming quinoa and katahui breads with gentle milling (chama)
Mululsito quispiña	Steamed bread prepared from quinoa and katahui flour
Quichi quispiña	Quinoa and katahui flour are used to make steamed and fried bread.
Juchcha	An Andean soup made with quinoa flour and katahui
Chiwa	In salads and soups, young quinoa leaves are utilized as a vegetable.

Bojanic (2011), Angeli et al. (2020)

throughout this process, creating various textures. For those who have celiac disease or gluten intolerance, quinoa noodles are an alternative to conventional noodles with minor organoleptic variations (Wang and Zhu 2016; Tanwar et al. 2019). In addition to the above stated quinoa products, there are several other potential quinoa by-products available in global food market. These items are developed by extracting chemicals from quinoa and developing value added products. Starches, bioactive substances, protein concentrates and isolates, and speciality oil are a few examples of these value added products. Many processes, such as starch gelatinization and dextrinization, protein texturing, partial denaturation of any vitamins present, melting and plasticization of the food, and expansion by flash evaporation of moisture, take place during the processing of quinoa.

Quinoa that has been pearled is subjected to milling and sifting using specialized milling equipments in order to obtain high-quality flour, even though there is a slight drop in the protein content, notably because of the separation of the germ. To make the end product, the dough is extruded through the machine. A rotary cutter is then used to shear the dough at the exit (Quiroga et al. 2013). Without affecting the final product's texture or flavor, the flour can be used up to 20–25% in place of wheat flour when creating bread, biscuits and pastries. Quinoa consumption in processed variants as given above is advised to the children, adults and expecting mother on account of their better digestibility. To expand the volume, the pearled grains are heated to a high pressure within a chamber; this process is also referred to as insufflation. This product can be flavor-infused with various flavors and made available for quick consumption. Typically, quinoa serves as a raw ingredient in the creation of wholesome feeds for cattle, birds and other animals. The entire plant can be utilized as green fodder, and harvest leftovers can be given to the animals (Srujana et al. 2019; Adetunji et al. 2021). Quinoa can be used as a protein source in animal feed, particularly for fish and poultry. In the poultry industry, this is typically utilized as a direct food source, while it is occasionally used to provide balanced meals for cattle and other animals.

Quinoa is a versatile crop that is used by agro-industrialists all over the world, therefore classifying it and learning more about its practical applications can help farmers earn more money. The agro-industry-focused organizations participate in a small number of the agro-industrial trials and use their resources to support the work in the agronomic field. Quinoa is one amongst the foods which are recognized as most nutrient-dense, and that may be more advantageous against the crises of global hunger. Quinoa plant on account of having 30–60 times less carbon and water footprints than other foods like meat, is also an environmental friendly and sustainable food crop. Several quinoa cultivars have been studied for nutritional value, water use efficiency and salt tolerance. Positive findings of these trials have boosted interest of the farmers to opt quinoa cultivation on a global scale. Due to its diversity in adaptability to survive in different climatic condition and suitability for friendly processing and consumption has made the grain a super food for many nations across the globe to eradicate the issues of food insecurity (Bastidas et al. 2016). In the Middle East and North Africa and Central Asia regions, where soil and water salinity are rapidly becoming a barrier to agricultural development, quinoa has emerged as an alternative crop with enormous potential to play a critical role in maintaining agriculture sustainability and food security (Choukr-Allah et al. 2016).

14.8 Challenges and Opportunities to Improve Quinoa Production in South Asia

Global warming has reached a critical level in recent decades, placing some people at risk, particularly in terms of the food supply. Other dangers to food security include population growth and a drop in precipitation. According to production statistics of the grain crops, yield of many food crops is expected to drop between 3.1% and 7.4% for every 1 °C increase in temperature unless breeders create new varieties for warmer temperatures (Tanwar et al. 2019). Ecosystems are predicted to face higher climate variability from year to year, and a greater frequency of extreme events in the coming years (Perez et al. 2010), which will put pressure on dependable food supply. Thus, there is a greater challenge to be addressed by researching and cultivating novel crops which are better adapted and resilient to the harsh climatic conditions. As the world's population rises from 6 billion to over 9 billion people by the year 2050, the demand for food is anticipated to rise by 70–100%

(Tilman et al. 2002; Godfray et al. 2010). According to the FAO, IFAD, and WFP (2014), the prevalence of diabetes, obesity, and other metabolic diseases has reached epidemic levels worldwide, and almost 1 in 8 people today suffer from chronic undernourishment (Nguyen and Lau 2012; Zimmet et al. 2014). Likewise, according to projections made by Lutz et al. (2008), the median age of the global population will rise from 26.6 years in 2000 to 31.1 years in 2050. Age-related illnesses like frailty, cardiovascular conditions, and osteoporosis are predicted to increase in frequency (Lunenfeld and Stratton 2013; Graf et al. 2015).

Under the given circumstances, quinoa on account of it's superior production, nutrition and health promoting characteristics is recognized as a superfood with the potential to bridge up future global food and nutritional security gaps. According to the FAO, quinoa has the ability to significantly contribute to maintaining global food security due to its superior nutritional value and resilience to a variety of climatic situations. Given its capacity to endure unfavorable climatic conditions, quinoa, a model plant with a halophytic habit, has the potential to address the problems of climate change and salinization (Granado-Rodríguez et al. 2021; Afzal et al. 2022). Quinoa is still produced in an agricultural setting under inadequate technological circumstances. There are no seed development programs to enable uniform harvesting, extend the crop cycle to avoid being negatively impacted by frosts, allow for the growth of black grains known as "ajaras," and thus optimize the selection and extraction process, among other things. Increased deterioration (erosion, desertification, salinization) and low yield are directly related to improper soil management and inadequate soil fertility. Quinoa does not require a lot of water, but certain studies have shown the viability of irrigation systems, which can significantly boost the yield and guarantee maximum levels of production, especially during droughts.

To implement the technological developments aimed at increasing production and competitiveness of quinoa crop, training and technical help of the quinoa growers is mandated. Although the institutions offering the services are now working extremely hard, there have been very few results due to the restricted coverage and the fact that the approaches used haven't always matched the beneficiaries' expectations, therefore these procedures haven't been long-lasting. Many of the production and quality related challenges with quinoa crop can be lessened with advancements in production technology and production that is sustainable in terms of time and space. This will improve resource conservation and management and lead to the identification of quinoa varieties which are resistant to adverse climatic factors and insects, pests and plant diseases invasion. The yield of the plant and the nutrient makeup of quinoa seeds are both influenced by a number of variables. The productivity and nutritional value of quinoa is largely influenced by genetic and environmental factors. There are needs for better quinoa breeding efforts to create high-yielding varieties with beneficial nutritional traits which can equally work in different agroecological zones (Reguera et al. 2018; Tanwar et al. 2019). When consumers think quinoa is expensive, they spend more money on common meals like rice, pasta, and potatoes. Improving production to an extent of other conventional staple food crops can address issue of market supply and it's affordability (Morales and Higuchi 2022).

The duration of the growing season and crop yield of quinoa must be improved through improved breeding programs and expansion of its worldwide cultivation. One of the most significant breeding objectives for the future is the creation of saponin-free sweet cultivars, which will increase acceptability of the quinoa as competitive crop to the other staple foods of South Asia and other continents of the world. Breeding this trait into quinoa cultivars is still difficult since it is difficult to choose the proper alleles and check saponin levels before anthesis (Suárez-Estrella et al. 2018). The primary impediments to the cultivation and commercialization of the crop include weed pressure, terminal heat stress, stem lodging, bold grain size, and an unorganized market. It is crucial to include community resilience and socio-economic constraints in the quinoa market when considering environmental issues in the quinoa value chain. Moreover, rural communities lack in knowledge for traditional culinary preparations of this grain and are unaware of its sensory features and nutritional advantages Some other barriers to the widespread adoption of quinoa for cultivation include lack of knowledge best crop production technology, lack of trainings on modern crop production methods, and poor access to plant protection procedures.

14.9 Conclusion

Quinoa has the potential to significantly increase food and nutritional security in South Asia because of its nutritional benefits and ability to adapt to diverse climatic conditions. Quinoa may be ranked among the viable food crops for small-scale farmers of South Asia. To fully utilize the potential of quinoa, a number of challenges to crop production and consumption must be resolved, such as increasing access to high-quality seeds, creating suitable production methods, and setting up effective supply chains, fulfilling requirements for proper processing methods and the educational campaigns on awareness of quinoa among farmers and consumers. In order to encourage the production quinoa production in South Asia, governments through their agriculture ministries may start programs and initiatives to facilitate farmers to opt for quinoa cultivation by offering incentives such as subsidized seeds quality provisions, access to agriculture services, establishing quinoa value chain and linking farmers with national and international industry.

Conflicts of Interest All authors read and approved the final manuscript and declare that they have no conflict of interest.

References

Adetunji CO, Michael OS, Varma A, Oloke JK, Kadiri O, Akram M, Olisaka FN et al (2021) Recent advances in the application of biotechnology for improving the production of secondary metabolites from quinoa. In: Varma A (ed) Biology and biotechnology of quinoa. Springer, Singapore, pp 373–396

Afzal I, Basra SMA, Rehman HU, Iqbal S, Bazile D et al (2022) Trends and limits for quinoa production and promotion in Pakistan. Plan Theory 11:1–20
Alandia G, Rodriguez JP, Jacobsen SE, Bazile D, Condori B et al (2020) Global expansion of quinoa and challenges for the Andean region. Glob Food Sec 26:1–10
Angeli V, Miguel Silva P, Crispim Massuela D, Khan MW, Hamar A, Khajehei F, Piatti C et al (2020) Quinoa (*Chenopodium quinoa* Willd.): an overview of the potentials of the "Golden Grain" and socio-economic and environmental aspects of its cultivation and marketization. Foods 9:1–13
Arts IC, Hollman PC (2005) Polyphenols and disease risk in epidemiologic studies. Am J Clin Nutr 81:317S–325S
Bastidas EG, Roura R, Rizzolo DAD, Massanés T, Gomis R et al (2016) Quinoa (*Chenopodium quinoa* Willd), from nutritional value to potential health benefits: an integrative review. J Nutr Food Sci 6:1–10
Bojanic A (2011) Quinoa: An ancient crop to contribute to world food security. FAO Fiat Panis, Regional Office for Latin America and the Caribbean
Brito IDL, Chantelle L, Magnani M, Cordeiro AMTDM et al (2022) Nutritional, therapeutic, and technological perspectives of quinoa (*Chenopodium quinoa* Willd.): a review. J Food Process Preserv 46:1–20
Ceccato DV, Bertero HD, Batlla D et al (2011) Environmental control of dormancy in quinoa (*Chenopodium quinoa*) seeds: two potential genetic resources for pre-harvest sprouting tolerance. Seed Sci Res 21:133–141
Choukr-Allah R, Rao NK, Hirich A, Shahid M, Alshankiti A, Toderich K, Butt KUR et al (2016) Quinoa for marginal environments: toward future food and nutritional security in MENA and Central Asia regions. Front Plant Sci 7:1–11
Craine EB, Murphy KM (2020) Seed composition and amino acid profiles for quinoa grown in Washington State. Front Nutr 7:1–16
De Carvalho FG, Ovídio PP, Padovan GJ, Jordão Junior AA, Marchini JS, Navarro AM et al (2014) Metabolic parameters of postmenopausal women after quinoa or corn flakes intake–a prospective and double-blind study. Int J Food Sci Nutr 65:380–385
El Hazzam K, Hafsa J, Sobeh M, Mhada M, Taourirte M, El Kacimi K, Yasri A et al (2020) An insight into saponins from quinoa (*Chenopodium quinoa* Willd): a review. Molecules 25:1–22
FAO (2017) Future of food and agriculture: trends and challenges; Annual Report, Rome
FAO, CIRAD (2015) State of the art report of quinoa in the world in 2013, by Bazile D, Bertero C. Nieto, eds, Rome
FAO, IFAD, WFP (2014) The state of food insecurity in the world 2014. Strengthening the enabling environment for food security and nutrition, Rome
Farinazzi-Machado FMV, Barbalho SM, Oshiiwa M, Goulart R, Pessan Junior O (2012) Use of cereal bars with quinoa (*Chenopodium quinoa* W) to reduce risk factors related to cardiovascular diseases. Food Sci Technol 32:239–244
Felipe CX, Guamis LB, Quevedo TJM, Trujillo MAJ, Inventors (2003) Liquid product of vegetable origin as milk substitute. European Patent Application, Published in Accordance with Art. 158(3) EPC
Filho AMM, Pirozi MR, Borges JTDS, Pinheiro Sant'Ana HM, Chaves JBP, Coimbra JSDR (2017) Quinoa: nutritional, functional, and antinutritional aspects. Crit Rev Food Sci Nutr 57:1618–1630
Fuentes F, Paredes-Gónzalez XIMENA (2013) Nutraceutical perspectives of quinoa: biological properties and functional applications. State of the art report on quinoa around the world pp 286–299
Gawlik-Dziki U, Świeca M, Sułkowski M, Dziki D, Baraniak B, Czyż J (2013) Antioxidant and anticancer activities of Chenopodium quinoa leaves extracts–in vitro study. Food Chem Toxicol 57:154–160
Godfray HCJ, Beddington JR, Crute IR, Haddad L, Lawrence D, Muir JF, Toulmin C (2010) Food security: the challenge of feeding 9 billion people. Science 327:812–818

Graf BL, Rojas-Silva P, Rojo LE, Delatorre-Herrera J, Baldeón ME, Raskin I (2015) Innovations in health value and functional food development of quinoa (*Chenopodium quinoa* Willd.). Compr Rev Food Sci Food Saf 14:431–445

Granado-Rodríguez S, Aparicio N, Matías J, Pérez-Romero LF, Maestro I, Gracés I, Reguera M et al (2021) Studying the impact of different field environmental conditions on seed quality of quinoa: the case of three different years changing seed nutritional traits in Southern Europe. Front Plant Sci 12:1–21

Lazarte CE, Castro-Alba V, Granfeldt Y et al (2021) Quinoa fermentation and dry roasting to improve nutritional quality and sensory properties. Biology and biotechnology of quinoa. Springer, Singapore, pp 325–343

Lunenfeld B, Stratton P (2013) The clinical consequences of an ageing world and preventive strategies. Best Pract Res Clin Obstet Gynaecol 27(5):643–659

Lutz W, Sanderson W, Scherbov S et al (2008) The coming acceleration of global population ageing. Nature 451:716–719

Martínez-Villaluenga C, Peñas E, Hernández-Ledesma B (2020) Pseudocereal grains: nutritional value, health benefits and current applications for the development of gluten-free foods. Food Chem Toxicol 137:1–26

Mohyuddin SG, Riaz A, Qamar A, Ali SH, Hu C, Wu L, Ju XH (2019) Quinoa is beneficial to the comprehensive nutritional value of potential health. Pak J Sci 71:69–74

Morales LE, Higuchi A (2022) Who is eating quinoa? How consumer characteristics and beliefs affect the expenditure on this functional food versus traditional staple items. J Sens Stud 37:1–11

Motta C, Castanheira I, Gonzales GB, Delgado I, Torres D, Santos M, Matos AS (2019) Impact of cooking methods and malting on amino acids content in amaranth, buckwheat and quinoa. J Food Compost Anal 76:58–65

Nasir MA, Pasha I, Butt MS, Nawaz H (2015) Biochemical characterization of quinoa with special reference to its protein quality. Pak J Agric Sci 52:731–737

Nguyen T, Lau DC (2012) The obesity epidemic and its impact on hypertension. Can J Cardiol 28:326–333

Ogungbenle HN (2003) Nutritional evaluation and functional properties of quinoa (*Chenopodium quinoa*) flour. Int J F Sci Nutr 54:153–158

Ong ES, Pek CJN, Tan JCW, Leo CH (2020) Antioxidant and cytoprotective effect of quinoa (*Chenopodium quinoa* Willd.) with pressurized hot water extraction (PHWE). Antioxidants 9:1–16

Pasko P, Bartoń H, Zagrodzki P, Gorinstein S, Fołta M, Zachwieja Z (2009) Anthocyanins, total polyphenols and antioxidant activity in amaranth and quinoa seeds and sprouts during their growth. Food Chem 115:994–998

Peñas E, Uberti F, di Lorenzo C, Ballabio C, Brandolini A, Restani P et al (2014) Biochemical and immunochemical evidence supporting the inclusion of quinoa (*Chenopodium quinoa* Willd.) as a gluten-free ingredient. Plant Foods Human Nutr 69:297–303

Perez C, Nicklin C, Dangles O, Vanek S, Sherwood SG, Halloy S, Forbes GA (2010) Climate change in the high Andes: implications and adaptation strategies for small-scale farmers. Int J Environ Cult Econ Soc Sustain 6:71–88

Quiroga C, Escalera R, Aroni G, Bonifacio A, González JA, Villca M, Ruiz A et al (2013) Traditional processes and technological innovations in quinoa harvesting, processing and industrialization. State of the art report on quinoa around the world, Section 3 Nutritional and Technical Aspects pp 216–249

Razzeto G, Uñates M, Moreno J, Lucero López R, Aguilar E, Sturniolo H, Escudero N et al (2019) Evaluation and comparative study of the nutritional profile and antioxidant potential of new quinoa varieties. Asian J Agric Hortic Res 3:1–11

Reguera M, Conesa CM, Gil-Gómez A, Haros CM, Pérez-Casas MÁ, Briones-Labarca V, Bascuñán-Godoy L (2018) The impact of different agroecological conditions on the nutritional composition of quinoa seeds. PeerJ 6:1–20

Royal Quinoa Catalogue La Paz (2003) Mcknight Foundation. MACIA, SINARGEAA, PROINPA Foundation, Bolivia

Ruales J, Grijalva YD, Lopez-Jaramillo P, Nair BM (2002) The nutritional quality of an infant food from quinoa and its effect on the plasma level of insulin-like growth factor-1 (IGF-1) in undernourished children. Int J F Sci Nutr 53:143–154

Samtiya M, Aluko RE, Dhewa T (2020) Plant food anti-nutritional factors and their reduction strategies: an overview. Food Prod Proc Nutr 2:1–14

Saypol Anwar FN, Zulkifli M (2021) A review on the effects of processing methods on the nutritional composition and antinutritional components of quinoa (*Chenopodium quinoa*) and buckwheat (*Fagopyrum esculentum*). SoRIF 2021:33–36

Singh D (2019) Quinoa (*Chenopodium quinoa* willd): Apotential crop for future food, health security, livelihood generation and poverty eradication. Scientific Publishers (India), pp 11–19

Srujana MNS, Kumari BA, Suneetha WJ, Prathyusha P et al (2019) Processing technologies and health benefits of quinoa. Pharm Innov J 8:155–160

Suárez-Estrella D, Torri L, Pagani MA, Marti A (2018) Quinoa bitterness: causes and solutions for improving product acceptability. J Sci Food Agric 98:4033–4041

Tang Y, Tsao R (2017) Phytochemicals in quinoa and amaranth grains and their antioxidant, anti-inflammatory, and potential health beneficial effects: a review. Mol Nutr Food Res 61:1–16

Tanwar B, Goyal A, Irshaan S, Kumar V, Sihag MK, Patel A, Kaur I et al (2019) Quinoa. Whole grains and their bioactives: composition and health. Wiley Online Library, pp 269–305

Thiam E, Allaoui A, Benlhabib O (2021) Quinoa productivity and stability evaluation through varietal and environmental interaction. Plan Theory 10:1–14

Tilman D, Cassman KG, Matson PA, Naylor R, Polasky S et al (2002) Agricultural sustainability and intensive production practices. Nature 418:671–677

Valcárcel-Yamani B, Lannes SDS (2012) Applications of quinoa (*Chenopodium quinoa* Willd.) and amaranth (Amaranthus spp.) and their influence in the nutritional value of cereal based foods. Food Public Health 2:265–275

Vidueiros SM, Curti RN, Dyner LM, Binaghi MJ, Peterson G, Bertero HD, Pallaro AN et al (2015) Diversity and interrelationships in nutritional traits in cultivated quinoa (*Chenopodium quinoa* Willd.) from Northwest Argentina. J Cereal Sci 62:87–93

Wang S, Zhu F (2016) Formulation and quality attributes of quinoa food products. Food Bioprocess Technol 9:49–68

Wu LG, Wang A, Shen R, Qu L (2021) Effect of heating under pressure treatment on the antioxidant of quinoa. Int J Food Eng 17:795–804

Zimmet PZ, Magliano DJ, Herman WH, Shaw JE (2014) Diabetes: a 21st century challenge. Lancet Diab Endrocrinol 2:56–64

Chapter 15
Sweet Potato (*Ipomoea batatas*): An Intervention Food in Management of Food and Nutritional Security in South Asia

Muhammad Waseem, Wisha Saeed, and Muhammad Ammar Khan

15.1 Introduction

Sweet potato (*Ipomoea batatas* Lam) is botanically a dicotyledonous tuber crop and inherited to Convolvulaceae family, genus Ipomoea and sub-genus eriospermum (Niu et al. 2019). Global statistics rank the sweet potato third and fifth in terms of food significance and and calories contributor for the human diet, respectively (FAOSTAT 2013; Muñoz-Rodríguez et al. 2018). Sweet potato is well-known as a crop against the crisis as tubers have been in use during natural disasters and wars to overcome the food insecurity and malnutrition challenges. Sweet potato tuber is ranked among the top tuber crops for human consumption due to its significant nutrition, under-exploitation and health promising photochemical profile. The global food crops statistics reveal annual sweet potato production to exceed 90 million tonnes with the China as its leading producer followed by India and Bangladesh (FAOSTAT 2020). Since ages, the tuber crop is native to Latin America, Central America, South America and is also under cultivation in Asia (China, Pakistan, India, Srilanka, Bangladesh and Nepal), South Africa, Japan, the Caribbean, and Papua New Guinea (Zhang et al. 2018).

The morphological classification of sweet potato identifies its four major varieties on the basis of flesh color i.e., white, purple and yellow. In addition to changes in flesh and/or skin color, these varieties also exhibit difference in nutritional and antioxidant profile (Wang et al. 2017). Earlier, studies have reported sweet potato to

M. Waseem (✉) · M. A. Khan
Department of Food Science & Technology, Faculty of Agriculture and Environment, The Islamia University Bahawalpur, Bahawalpur, Punjab, Pakistan
e-mail: mwaseem@iub.edu.pk; ammar@iub.edu.pk

W. Saeed
Department of Food Science & Technology, FFS&N, Bahauddin Zakariya University, Multan, Punjab, Pakistan

T. Ismail et al. (eds.), *Neglected Plant Foods of South Asia*,
https://doi.org/10.1007/978-3-031-37077-9_15

exhibit a wide variability in its nutrient composition as it embodies higher magnitudes of dietary proteins, dietary fibers, vitamins (vitamin A, B_1, B_6, C and E), coloring pigments (i.e., β-carotenes), mineral elements (i.e., iron, zinc, calcium, and magnesium) and numerous dietary bioactive molecules including polyphenols (Alam et al. 2020; Sun et al. 2019). Sweet potato inedible fractions such as seeds, leaves, peel and roots also embody a wide array of health promoting bioactive molecules including a variety of essential amino acids (i.e. lysine and isoleucine), carotenoids, tocopherols and tocotrienols, anthocyanins (i.e., anthocyanidins), phenolics (i.e., flavonoids, flavonols and flavones) and tannins (Dong et al. 2017; Suarez et al. 2020; Tang et al. 2021). The roots also contain fermentable simple sugars (i.e., fructose, glucose and sucrose), soluble fibers, lipids, essential amino acids (Marques et al. 2014), and twice the vitamin C, calorie contents than the ordinary potatoes (Ellong et al. 2014). The attractive nutritional and biomolecular profile of sweet potatoes consisting of higher polysaccharides, simple sugars, phenolics, pro-vitamin A and peptides has gained immense popularity, and has attracted scientists and agro-food sector towards exploiting this resource as a healthier and safer food and fuel of the future generations (Tanaka et al. 2017).

Phytochemical profile of sweet potato skin reveals a wide array of distinguishing pigments i.e., cream, orange, purple, white, red, yellow and pink. These coloring compounds are at large responsible for better product quality, higher consumer preferences and enhanced antioxidant potential of the product (Alam 2021). Orange fleshed sweet potato is considered to hold significantly higher concentrations of b-carotene i.e., 277 μg/g (Alam et al. 2020). Earlier, available literature report sweet potato to embody higher levels of b-carotene as well as lutein as compared to the spinach, kale, broccoli, peas and lettuce (Ishiguro 2019). Kurata et al. (2019) has reported the data on polyphenols reports the presence of higher phenolic contents in sweet potato than other common vegetables like eggplant, cabbage, spinach, cauliflower, amaranth, kale, peas and lettuce.

Functional bioactive compounds of sweet potato have been listed as polyphenols, saponins, steroids, terpenoids, alkaloids and glycosides (Alam et al. 2020; Luo et al. 2021). The most predominant phenolics includes phenolic acids (e.g., chlorogenic acids and caffeic acid), p-coumaric acid, derivatives of caffeoylquinic acids, p-anisic acids, sinapic acid, hydroxybenzoic acids, flavonoid derivatives (such as luteolin, myricetin, quercetin and apigenin), and anthocyanins derivatives (such as cyanidins, pelargonidins and peonidins) (Kurata et al. 2019; Luo et al. 2021). unique diversified biochemical profile of this vegetable crop and its fractions anticipates anti-inflammatory, antioxidative, anticancer, immunomodulatory, antitumour, anti-obesity, antiulcer, antimicrobial, hepatoprotective and antidiabetic properties to prevent and alleviate numerous health malignancies (Alam 2021).

For last few decades, abundant research has been performed on the health promising potential of sweet potato and its anatomical segments. This chapter comprehensively provides the latest updates on ability of sweet potato roots and its fractions to be utilized as nutrition and health ameliorative food resources, and their role as natural food ingredients for value addition aimed at addressing the triple burden of malnutrition and food insecurity. Furthermore, the study also spotlights pragmatic use of sweet potato as therapeutic agent in ameliorating several acute to chronic health maladies.

15.2 Sustainable Cultivation of Sweet Potato as Food, Fodder and Fuel Crop

Globally, the sweet potato is ranked on second place among the tropical food crops exhibiting the annual global total production of 140 million tonnes on an area of about 9 million hectares. The tuber crop is being cultivated in more than 110 countries of the world. Sweet potato producing countries includes Asia, Indonesia, China, Brazil, Africa, Philippines, Nigeria, Uganda, Tanzania, Argentina and Cuba (Thatoi et al. 2016). Statistical data on sweet potato also ranks its production at number seven after the wheat, potato, rice, maize, cassava and barley. Asia is reported as world's largest sweet potato producing continent producing 129 million tonnes of tuber each year (Thatoi et al. 2016). The cultivation span or the vegetative growth cycle of sweet potatoes ranges between 3 and 5 months. Morphological composition of the crop indicates it as a hard tuber with multiple uses (Gruneberg et al. 2015).

Sweet potato is produced in all tropical and subtropical parts of the world, and grows well all the year with minimum crop supervision requirements. Sweet potato crop has the ability to grow even in poor growth conditions i.e., poor soil nitrogen and harsh environmental conditions. The crop maturity period is around 4 months and it varies with the cultivar and external environmental factors (Ellong et al. 2014). When compared for various growth requirements, the sweet potato cultivation has several benefits over other crops as it requires lesser agriculture inputs including water, fertilizer and the pesticides. The crop can withstand drought conditions and tends to grow on marginal lands (Zhu and Xie 2018).

Globally, alternative sources of sustainable food, feed and fuel are receiving higher attention. Sweet potato holds higher significance in terms of productivity, nutrition and biological value of these nutrients. The tubers of sweet potatoes and the dry grasses have above 90% digestibility and are fed to the ruminants as fresh, dehydrated chips, chopped, silage and/or in ground form to fulfill the energy requirements of the ruminants. The tips and skins of sweet potato are claimed to hold higher magnitudes of dietary proteins i.e., 18–21% and 50–90%, respectively. The leaves and tips also contain higher concentrations of the essential amino acids like cysteine, cysteine and methionine. Leaves and vines of the sweet potato are commonly fed to the pigs in numerous countries of the world including Taiwan, China, Papua New Guinea, Philipines, South Korea, and India. Roots of this viable crop are used as remarkable source of energy for the poultry supplying about 3500 kcal/kg. High quality fodder is characterized with the higher contents of inorganic residues, dietary fibers and proteins (Murugan et al. 2012).

Biochemical profile of the leaves and roots of sweet potato have been reported edible and are considered as plentiful contributors to the nutrients and energy needs of human consumers. Varietal stratus orients distinguished quality features among the cultivars, as orange-fleshed sweet potato being rated for its higher beta- carotene (pro-vitamin A) contents (Gruneberg et al. 2015). Globally, different parts of sweet potatoes after conversion in various forms are used as feed for cattles, pig, fish, rabbits and goats. Vines, damaged roots, surplus roots, leaves, culls, and root leftovers are used as fresh green fodder or ground fodder in Africa (Kenya, Rwanda, Uganda), Egypt, South America, (Argentina, Brazil, Ecuador, Haiti, Peru and Venezuela) and Jamaica (FAO 2011). Sweet potato is fed to animals as feed due to its higher contents of dry matter (d.w.b) (i.e., 24–34%), proteins (4–6%) and inorganic residues (2–4%) (Tripura and Sankaran 2008). Among different cultivars of sweet potato, the white flesh sweet potato has higher starch contents (i.e., 25–40%) however due to its lower palatable quality, it is not considered as main food crop (Thatoi et al. 2016).

Agricultural Research Service at USDA has reported the sweet potatoes as auspicious sources of carbohydrates linked with two to three-times higher carbohydrates contents than maize and is used as biofuel feedstock for ethanol production (Ibeto et al. 2011). Developments in crop breeding program have been made to improve starch contents of the sweet potato aimed at enhancing the fermentable sugars contents for industrial production of ethanol. A study has reported sweet potato crop to yield 4500–6500 L/ha ethanol than 2800–3800 L/ha from the maize crop (Ziska et al. 2009). Ethanol production from the conventional sweet potato cultivar has been reported in range between 4800 and 4900 L/ha (Lareo et al. 2013). Sweet potato biomass has emerged as an environment friendly bioethanol source wherein the natural starches and sugars are fermented using microorganisms to produce the renewable natural fuel to meet energy needs of the future (Thatoi et al. 2016).

Earlier literature have reported the generation of economical and environment friendly biofuel using the natural food crops as plant biomass such as soybean, sugar beet, corn, wheat and rapeseed. Bioethanol is conventionally derived from maize and sugarcane crops while the biodiesel is being obtained on a large scale form oilseed crops which thereby are used as natural source of energy, food and feed. When compared with the other approaches, renewable energy production using this particular approach is recommended as safe, health and environment friendly (Koçar and Civaş 2013). Data on sweet potato production indicate 95% of the total crop production to belong to the developing countries of which 50% is produced in Asia, and is used as food and animal feed (Consultative Group for International Agricultural Research 2005).

Use of plant biomass for the production of the renewable energy for the future is being rated as best and sustainable approach and has a positive environment conservation impact. The natural reserves of fossil fuels around the world are depleting rigorously and resulting in global warming; in this connection exploration of some environment friendly biofuel is direly needed to replace the depleting fossil fuels as viable source of energy. However, plant biomass is now being considered as a promising replacement of fossil fuels for generating renewable energy resources due to their cost effective, economic and less CO_2 emitting ability. Plant biomass meets

14% global energy requirements and is significantly contributing to stabilization of economy, social and ecological sustainability (Rutz and Janssen 2007). South East Asian countries are highly dependent on conventional energy resources that create a huge impact on economy and the environment. Since Asia is the major producer of sweet potato, the crop may be used as promising source for biofuel production and can significantly lessen the fuel import burden.

15.3 Conventional Food Applications of Sweet Potato in South Asia

The crop is used as food and feed to mitigate nutritional security and malnutrition, and also exploited as bio-fuel for generating income, and used in the industrial production of flours, syrups, starches, colorants and juices (USDA 2017). Tubers and roots are considered as the third most valued food crops for humans following the cereals and legumes. These crops are preferred for consumption as staple or subsidiary food in various global parts such as Africa, South America and South Asia, and East Asia. Sweet potato, yams, colocasia or taro, cassava are categorized in the tropical roots and tuber crops. Asia is ranked as the largest producer of tuber crops. Considering the remarkable nutritional significance, sweet potatoes are used as natural food supplement in development of several value added fabricated foods. Dietary fibers and essential peptides are being used as supplementing ingredients or the additives in development of functional beverages, poultry and animal feed (Ray and Sivakumar 2009). Starches, dietary fibers and essential peptides are integral constituents of sweet potatoes (Ray and Ravi 2005).

In previous few decades, a significant drift have been observed for the industrial production, characterization, and utilization of sweet potatoes based starches in novel product development (Zhu and Wang 2014). Sweet potatoes are exploited as food ingredients for value addition in development of starches, flours, syrups, juices, and colorants, and as feed for animals to ameliorate malnutrition and nutritional insecurity and as bio-fuel for generating income (USDA 2017). Among functional foods, Shochu is a fermented Japanese alcoholic distilled food product made with barley, sweet potatoes or rice. Sochu contains ~25% higher alcohol content than wine and sake while lower than observed in whisky (Yamakawa 2000). Similarly, Koji is a hydrolyzed alcoholic product prepared by *Aspergillus niger* or *Aspergillus kawachii* led fermentation of sweet potatoes or steamed rice (Bindumole et al. 2000).

Studies from India have reported lactic acid fermentation led pickles production from sweet potatoes naturally enriched with anthocyanins and β-carotenes in the presence of salt @ 5–20% (w/v) (Panda et al. 2009). Lactic acid fermentation-based pickles have reported with the better flavor and sensorial appeals along with the preservation of the useful compounds like ascorbic acid, phenols, β-carotenes and anthocyanins exhibiting anti-oxidants potential (Shivashankara et al. 2004). Physico-chemcial analysis of anthocyanin–rich sweet potato based lacto-pickles holds pH (i.e., 2.5–2.8), titratable acidity (i.e., 1.5–1.7 g/kg), anthocyanins (i.e., 780 mg/kg), lactic acid (i.e., 1.0–1.3 g/kg), starch (i.e., 56–58 g/kg) and (Panda et al. 2009). A study revealed the use of boiled sweet potato roots pulp extract in development of lacto-fermented juice and the results elucidated significant improvement in β-carotene and anthocyanin contents of the functional juice (Panda et al. 2009).

Fermented food products like yogurt and curd are widely consumed by lactose intolerant population due to better digestibility. Earlier, studies on curd manufacturing have reported the use of pulp of boiled sweet potatoes @ 8–16% and are naturally enriched with anthocyanins and β-carotene (Panda et al. 2006). Anthocyanins enriched sweet potatoes are used in the production of functional beverages like yellow, red, and black beer, wine and sparkling liquors in Japan (Yamakawa 2000). Sweet potato and its flour are also used to produce lactic acid using *Lactobacillus sp* (Panda and Ray 2008). Sweet potato starches are used for the citric acid production in the industry using *Aspergillus niger* (Bindumole et al. 2000). In China, most significant food product prepared with sweet potatoes includes starch, alcohol. Noodles and the citric acid (Jiang et al. 1993). Sweet potato derived citric acid and ascorbic acid has been used as flavor enhancer in drink in Japan (Woolfe 1992).

A study by Terahara et al. (2003) has reported production of high antioxidant activity bearing red vinegar from the purple fleshed sweet potatoes via fermentation. Soy sauce is a fermented salty dark brown liquid product being used as flavoring agent. The soy sauce is fermented by *Aspergillus oryzae* and/or *A. sojae* and made with the combination of soybeans and wheat. Recent studies have shown the use of sweet potato flour as a substitute to the wheat flour in the production of soy sauce (Data et al. 1986). Monosodium glutamate is a flavor enhancer used in meat and meat based foods. Sweet potato starches have also been used in development of monosodium glutamate (MSG). The mechanism of MSG production includes degradation of sweet potato starches into glutamic acid and simple carbohydrates using *Brevibacterium glutamicum* (Jiang et al. 1993).

15.4 Nutritional Composition of Sweet Potato and Its Genetically Modified Varieties

Sweet potatoes roots and its fractions are dense source of nutrients and phytonutrients. The nutritional composition of sweet potatoes manifests presence of carbohydrates (20.1 g), sugars (4.2 g), starch (12.7 g), fibers (3.0 g), fat (0.1 g), proteins (1.6 g), vitamin A equivalent (709 μg), β-carotene (8509 μg), thiamine (0.1 mg), riboflavin (0.1 mg), niacin (0.61 mg), pantothenic acid (0.8 mg), B6 (0.2 mg), folates (11 μg), ascorbic acid (2.4 mg), vitamin E (0.26 mg), calcium (30 mg), iron (0.6 mg), magnesium (25 mg), phosphorous (47 mg), potassium (337 mg), sodium (55 mg), zinc (0.3 mg) and 360 kcal energy per 100 g (Mohanraj and Sivasankar 2014).

15.4.1 Carbohydrates

Polysaccharides in General

Polysaccharides are the large biomolecules consisting of sweet and lower carbohydrates like glucose units. Sweet potatoes, potatoes and beetroots are considered as enriched sources of polysaccharides. Earlier studies have reported the presence of glucose, D-xylose, rhamnose and galactose residues in purple sweet potatoes, however; data is scant to validate occurrence of these carbohydrates in other cultivars of sweet potatoes (Wu et al. 2015). The available literature validate the presence of water soluble (i.e., 5–6%) and water insoluble (1–2%) polysaccharides in purple sweet potatoes cultivars (Qu et al. 2014). Water soluble polysaccharides dominated by rhamnose, glucose and galactose were isolated from sweet potato tubers and modified using selenylation to develop modified polysaccharides i.e., selenyle based sweet potato polysaccharide (Se-SWP). Se-SWP feeding to tumor induced mice @ 100 mg/kg body weight/day anticipated significant improvement in the antioxidant potential, and was found effective against the cancer cells i.e., IL-2, TNF-α and VEGF (Yuan et al. 2017).

A study by Sun et al. (2018) reported significant effect of different doses of purple sweet potato complex polysaccharides fed to mice with CCl_4 induced hepatic injury @ 100, 200 and 400 mg/kg/day for 30 days. The study reported significant decline in the serum enzymes (i.e., alkaline phosphatase, aspartate transaminase and alanine transaminase) and liver lipid peroxidation. Likewise, the results suggest increase in hepatic antioxidant enzymes (i.e., catalase, glutathione peroxidase & superoxide dismutase). Hepato-protective effect of alkaline polysaccharides was also reported quite higher tha water soluble polysaccharides.

A study indicated polysaccharides of purple sweet potato tuber to act as antioxidative agent against free radicals. The study also reported the inhibition properties of polysaccharides against albumin non-enzymatic glycosylation (Guo et al. 2017). Another study by Tang et al. (2018) on the immune-modulatory effect of water-soluble and alkaline-based polysaccharides in mice on different dose exposures @ 100, 200 and 400 μg/mL via lipopolysaccharide treated RAW264.7 macrophage cell models. The outcomes reported the significant decline in abnormal phagocytosis and apoptosis caused by lipopolysaccharides and retarded the production of tumor cells i.e., TNF-α and IL-6. Likewise, the study also showed significant effect against inflammation and reduced the expression of lipopolysaccharide stimulated IL-1b @ 400 μg/mL.

Likewise, another study Tang et al. (2018) also showed the feeding of water-soluble and alkaline-based polysaccharides to cyclo-phosphamide (CTX) treated and normal mice @ 400 mg/kg b.w. daily resulted in improvement of immunity cytokines (IL-2 and IL-6), spleen health in the CTX-treated mice. The findings also revealed that these polysaccharides also modulated gut microbiota in targeted mice by enhancing the level of Lachnospiraceae, Ruminococcaceae, Oscillospira and Ruminococcus whereas reducing the concentration of Proteobacteria, Firmicutes, Sutterella and Alcaligenaceae in CTX-treated mice. Enhanced levels of intestinal acids such as acetic, propionic and butyric acid were also reported in the healthy mice showing the significance of feeding polysaccharides isolated from purple sweet potato.

Dietary Fibers

Dietary fibers are plant polysaccharides consisted of cellulose, hemicellulose and pectin which are generally resistant to digestion by human gastrointestinal enzymes. These fiber components are linked with formation of cellular wall in plants (Theuwissen and Mensink 2008). The dietary intake of fibers from the sweet potato may aid in balancing the water status in the body. Fibers linked with the sweet potatoes may get affected by the external environmental factors, growing conditions and genetic makeup of the plant. The dry matter after removal of starch and other residues shows the presence of about 75% dietary fiber in that dry matter. The most prominent polysaccharide residues used in the agro-industry are celluloses, lignins, pectins and hemicelluloses (Mei et al. 2010). Freeze dried sweet potatoes fiber contents may vary between 1.9% and 2.4% (Ji et al. 2015). Additionally, dietary intake of fibers present in the sweet potato is known to relieve the constipation and related health challenges (Ruhee et al. 2020).

Pectin – a hetero-polysacharide exists in cell walls of plants such as sweet potato and is a carrier of L-arabinose and galactose (Takamine et al. 2007). Sweet potato pectin generally resembles apple pectin and holds substantial antioxidant properties. The antioxidant potential has been reported to be elevated by depolymerization of pectin using numerous techniques such as sonication and ultrasonic (Ogutu and Mu 2017). Various prebiotic components such as raffinose, fructo-oligosaccharides and inulin have also been revealed in several cultivars of sweet potatoes. Dietary fibers anticipate better prebiotic activities than FOS and inulin towards the *Bifidobacterium longum* JCM 1217 and *Lactobacillus plantarum* Mut7 (Lestari et al. 2013).

Dietary Starches

Starches as polysaccharides are found in the highest magnitudes in sweet potatoes. Nutritional value of sweet potatoes is directly linked with digestion rate, polymers composition and metabolism during absorption of monomers (Abegunde et al. 2013). Sweet potato-based starches are considered as readily digestible starches as these can be completely hydrolyze into glucose in 20 minutes as compared to starches of other tuber crops which digests slowly and may take up about 120 minutes for complete breakdown into glucose. Beneficial gut microbiota ferments the resistant starches inside the intestinal mucosa within 120 minutes (Keenan et al. 2015). Starch accounts for 50–80% availability in the sweet potato tuber and is responsible for distinctive physicochemical and functional features in sweet potato (Zhang et al. 2018; Zhu and Wang 2014).

Sweet potato starches exhibits higher amylose to amylopectin ratios when compared with the other tuber crops such as potatoes. High amylose carrier sweet potatoes cultivars are considered appropriate for diabetes patients due to their ability to increase blood glucose levels slowly than the normal sugars (Mohanraj and Sivasankar 2014). Resistance starches are helpful in mediating the insulin levels and also mediate the metabolic activities, and prevent hypercholesterolemia (Bindels et al. 2015). Resistant starch is fermented in the large intestine by intestinal gut microbiota functioning as prebiotic; producing beneficial short chain fatty acids (Keenan et al. 2015). Resistant starches recovered from purple sweet potato elucidated improvement in the *Bifidobacterium spp.* and acidity in the gut thus linking prebiotic properties to sweet potato resistant starches (Zheng et al. 2016). Modification in particle size of the sweet potato starches by milling operations has been reported to improve digestibility, functional, nutritional and other quality attributes of the starch (Hung et al. 2014). Presence of health promoting phenolics, anthocyanins, oligosaccharides, polysaccharides and resistant starches in sweet potato tubers thus suggests sweet potatoes to hold prebiotic potential (Haydersah et al. 2012).

15.4.2 Proteins

When compared with the other vegetables like spinach, sweet potato contains significantly higher magnitudes of protein. Foliar composition of sweet potato depicts a wide variety in its amino acid profile and is considered to hold good quality proteins (Suarez et al. 2020; Tang et al. 2021). Earlier researches have reported 4–27% and 1–9% protein contents in roots and leaves of sweet potatoes, respectively (Dong et al. 2017). The tryptophan, valine and threonine are main building block of proteins in sweet potatoes (Sun et al. 2012). Several studies are available on health beneficial aspects of peptides and protein of sweet potatoes and its cultivars. Sporamin protein is reported as the main nitrogenous protein found in sweet potato roots which comprise the 80% of total proteins of the root crop (Senthilkumar and Yeh 2012).

The role of sweet potato protein in health promotion is undeniable as the earlier researchers have validated the anti-hypertensive properties of sweet potato proteins. A study by Ishiguro et al. (2012) reported decline in blood pressures of albino rats on feeding the rats with peptides @ 100 and 500 mg/kg at after 4–8 hours of peptide administration. In another study the lipid metabolism attributes of the sweet potato-based peptides were also noticed indicating improved lipid metabolism in mice on administration of the high fat diet accompanied with the sweet potato protein peptides @ 5% for 28 days. The findings revealed appreciable decline in the TNF-α, very low-density lipoproteins (VLDL) and low density lipoproteins (LDL) contents (Ishiguro et al. 2016). Another study by Zhang et al. (2012) elucidated significant role of protein hydrolysates enzymatically extracted from sweet potato roots on the antioxidant potential (i.e., free radical-scavenging activities). The comparative assessment of sweet potato root proteins with the enzymatically extracted protein hydrolysates of sweet potato suggested significantly higher antioxidant activities of the sweet potato root protein hydrolysates. Retrospective studies by Zhang and Mu (2018) and Li et al. (2013) have validated the food and nutritional exploitation of protein hydrolysates, recombinant proteins (i.e., MT-I and MT-II) and simulation derived defensin proteins (i.e., GFR, GPCSR, CFCTKPC and MCESASSK; Tainong 57) of sweet potato against colon, colorectal and lung carcinoma cells.

15.4.3 Minerals and Vitamins

Earlier studies have reported sweet potatoes as viable sources of mineral elements. On the basis of dietary intake of a specific food nutrient per serving as percent daily value (%DV), Food and Drug Administration (FDA) has classified the food nutrients as low nutrient (i.e., <5% DV), good nutrient (i.e., 10–19% DV), and rich nutrient (i.e., >20% DV) source foods. Sweet potato is considered as viable source of the minerals i.e., iron, copper, magnesium manganese, potassium, and calcium, and vitamins i.e., B_1, B_6, C and E. Available data on nutritional composition of sweet potato portrays it to be used as a nutrient rich food which could be helpful in prevention and alleviation of malnutrition and related health challenges (Alam et al. 2020). The recommended daily allowances for macro minerals in sweet potatoes are i.e., calcium (1000 mg), magnesium (130 mg), phosphorous (500 mg), potassium (3000 mg) while the minerals exhibits recommended daily allowances for microminerals in sweet potatoes are; iron (10 mg), zinc (5 mg), and sodium (1500 mg). However, a study by Sanoussi et al. (2016) reported the presence of macro-minerals as; Ca (24–30 mg), Mg (24–26 mg), P (42–44 mg), K (311–316 mg), and microminerals as; Fe (0.6–0.7 mg), Zn (0.2–0.7 mg), and Na (31–33 mg).

Since last few decades, the micronutrient deficiencies are the leading cause of mortalities around the globe. In this nexus, the researchers are in quest of the cheap, nutrient rich and neglected food sources with better nutrition. Vitamin A deficiency has affected millions of children and adults around the world as its deficiency is directly associated the impaired vision and growth (Neela and Fanta 2019). Sweet potatoes are also enriched in pro-vitamin A i.e., beta-carotene which is known to contribute its significant role against night blindness. Earlier reports reveal that the dietary intake of about 125 g of orange-fleshed sweet potato per day may improve vitamin A status in humans and thereby can decrease night blindness in children (Alam et al. 2020). Higher dietary consumption of sweet potatoes may alleviate the micronutrient malnutrition. Magnesium from sweet potato helps in intestinal and digestive problems and provides soothing effect. The higher calcium, potassium, vitamin C, B-complex vitamins, and β-carotene of sweet potatoes are helpful in ameliorating stomach ulcers (Ruhee et al. 2020; Mohanraj 2018). Earlier studies have reported the role of sweet potatoes against anaemia and hematinous (Osime et al. 2009). Montejo et al. (2015) has revealed dietary administration of the sweet potato leaves powder to mice resulted in increase of the hemoglobin contents, packed cells, and red blood cells.

15.4.4 Bioactive Compounds

Sweet potato is a well k known vital food commodity containing several health-promoting functional bioactive compounds such as β-carotene, polyphenols, steroids, saponins, lutein, terpenoids and glycosides (i.e., alkaloids) (Alam et al. 2020; Luo et al. 2021), phenolics (i.e., caffeic acid, monocafeoyl quinic, p-anisic acids p-coumaric acid, hydroxybenzoic acids, sinapic acid and CQA derivatives), flavonoids (apigenin, myricetin, quercetin and luteolin) and anthocyanins derivatives (such as peonidins, pelargonidins and cyanidins) and β-carotene (i.e., 277 μg/g) (Ishiguro 2019), secondary plant metabolites consisting of chalcones, anthocyanins, flavonols, flavanes, flavanones and flavones (Gould et al. 2008). Phenolic compounds such as flavonoids, phenolic acids, stilbenes and lignans are commonly present in all varieties of sweet potatoes however the anthocyanins are only present in purple-fleshed sweet potatoes (Kurata et al. 2019). Sweet potatoes leaves are known to hold higher concentrations of phenolics than the spinach, lettuce, cabbage, cauliflower, green peas, amaranth, purple corn, kale and eggplant, elderberry, grape skin, and red cabbage (Kurata et al. 2019; Makori et al. 2020; Kano et al. 2005).

A study by Wang et al. (2018) screened the phytochemical metabolites in different varieties of sweet potatoes using the liquid chromatography electrospray ionization mass spectrometry. The data reported wide array of health promoting phytochemical bioactive compounds in all cultivars of the sweet potatoes. The findings depicted a total tally of about 213 metabolites consisting of 27 phenolic acids and 29 flavonoids (Wang et al. 2018). Also, Park et al. (2016) elucidated the sweet potato to contain health featuring phytochemicals and the researchers detected 41 polar primary metabolites consisting of 1 amine group, 7 sugar alcohols, 13 organic acids and 17 amino acids. Presence of these bioactive compounds in sweet potatoes exert useful biological activities against obesity, cancer, tumor, diabetes, pathogens, inflammation, cardio vascular disorders and hepatic problems (Prakash and Gupta 2009).

15.5 Sweet Potato as an Intervention Food for Management of Nutritional Disorders

Sweet potato is a nutritive food crop which is cultivated and consumed globally due to a wide array of essential nutrients which play its significant role in ameliorating the food insecurity, malnutrition and associated health maladies. It has a rich nutritional profile characterized by the presence of several bioactive compounds (Mohanraj 2018). The significant role of sweet potato and its key fractions in ameliorating the health malignancies are summarized here under;

15.5.1 Anti-cancer

Cancer is the most prevalent health disorder connected with higher magnitudes of mortalities per annum i.e., ten million. Recent data on cancer reveals that its spread will rise by about 70% by the end of the upcoming decade. Consumption of fresh fruits and vegetables may help in mitigating the cancer insurgencies and associated mortalities. Notable change is being observed among researchers on utilization of cheaper and new food sources to ameliorate cancer (Khan et al. 2020). However, sweet potato has been found a tremendous source of health preventive food constituents in protecting human health against cancer by inducing apoptosis and inhibiting the cellular proliferation. Scant data is in hand on the in-vitro exploration of anticancer features and pathways of sweet potato compounds and its fractions. Caffeoyl compounds in sweet potatoes are in close nexus to ameliorate cancer (Sun et al. 2018).

Asadi et al. (2017) reported in-vivo anti-carcinogenic activities in anthocyanins of sweet potato. Sweet potato protease-treated hydrolysates demonstrated anti-proliferation activities with the IC_{50} values varying between 120 and 422 μg/mL in HT-29 human colon cancer cells (Zhang and Mu 2018). Li et al. (2019) spotlighted appreciable decline in cancer cell lines i.e., MCF-7 in mice on dietary feeding of sweet potato glycoprotein SPG-56 (i.e., @ 240 mg/kg/day). Likewise, SPG-8700, a glycoprotein extracted from sweet potato also exhibited the anti-colorectal cancer activity against the human colon cancer cells i.e., HCT-116 cells via modulation of Bcl-2 and Bax expressions and resulted in subsequent decrease in the CA242, CA125, and CA199 tumor markers by 24%, 35%, and 16%, respectively (Tian et al. 2019).

Sweet potato phytochemicals have been reported to play ameliorative role against several cancer cells such as liver HepG2 cells (Sun et al. 2019; Wang et al. 2019), cervical cancer HeLa cells (Vishnu et al. 2019), colon cancer HCT-116 cells (Vishnu et al. 2019), and breast cancer MDA-MB-231 cells (Wang et al. 2019). The sweet potato-methanol extracts significantly ($p < 0.05$) inhibited the proliferation of human prostate cancer cell lines (i.e. DU145, C4-2B, LNCaP, PC-3, C4-2) with the IC_{50} varied between 145 and 315 μg/mL (Karna et al. 2011). The anti-carcinogenic effects of these extracts were associated with their interference in cell cycle progression, modulation in the cell cycles and molecular regulation for apoptosis induction (Karna et al. 2011). Anthocyanin-rich extracts reported the inhibition of cancer cells i.e., CQA derivatives such as 3,4,5-triCQA (Kurata et al. 2007).

15.5.2 Anti-diabetic

Diabetes Mellitus (DM) is a metabolic dysfunction categorized by elevated blood glucose associated with impaired insulin secretion. Dietary habits, lifestyle and other factors like stress, sedentary lifestyle, obesity, or genetics are the main causes of diabetes (Alam et al. 2020). It has been reported that currently diabetes is prevailing in more than 450 million people around the world and these cases are expected to increase to 700 million by the end of 2045 (International Diabetes Federation 2019). Globally, 50% of diabetes cases remain undiagnosed and the health disorder prevails more in low income developing countries. Earlier studies have reported the anti-diabetic properties of sweet potatoes due to their health improving compounds anthocyanins, flavonoids, phenols, saponins, terpenes and alkaloids (Luo et al. 2021; Zhang and Mu 2016). These compounds exhibit anti-diabetic effect through their action on pancreatic beta cells, hence facilitating the production of insulin along with successful modulation of biochemical reactions involved in glucose metabolism (Alam et al. 2020). In-vivo studies are indicative of the anti-diabetic properties of sweet potato and its connection with the genes like acetyl-CoA carboxylase 2, carnitine palmitoyltransferase 1, myocyte enhanced factor 2A, nuclear respiratory factor 1, and glucose transporter 4 (GLUT4). In this nexus, a study on orange fleshed sweet potato exhibited a significant effect on T2DM genes in insulin resistant C2C12 cells (Ayeleso et al. 2016; Esatbeyoglu et al. 2017).

Shih et al. (2020) explored hypoglycemic and insulin modulating properties of anthocyanins and cyaniding-3-caffeoyl-phydroxybenzoylsophoroside-5-glucosides of purple fleshed sweet potatoes in mice. Sweet potato leaves in different forms act as anti-diabetic agent which could be attributed to the phytochemicals including phenolic acids, anthocyanins and flavonoids (Luo et al. 2021; Zengin et al. 2017). Clinical trials have documented the sweet potato leaves based hot water extracts in management of diabetes, hence decreasing the insulin resistance in diabetes induced mice (Olowu et al. 2011). A study by Nagamine et al. (2014) elucidated positive correlation of leaf extracts of sweet potato on decreasing glucose levels in KK-Ay mice and results also showed increment in glucagon like peptide-1 (GLP-1) secretion. Shih et al. (2020) reported the blood sugar regulation and an enhanced insulin signaling in diabetic rats using the leaves obtained from white-fleshed sweet potatoes. Likewise, Zhao et al. (2013) reported glucose lowering effects of anthocyanins obtained from purple fleshed sweet potato against streptozotocin induced diabetic mice. The results elucidated significant decline in blood sugar levels in mice on completion of study trials i.e., 60 days.

15.5.3 Cardio-protective

Cardiovascular diseases (CVDs) are chronic health disorders associated with high blood pressure, excessive alcohol intake, diabetes, hyperlipidemia, obesity and other metabolic health calamities. Globally, CVD cases are rising owing to several factors such as poor lifestyle and bad dietary habits (Tang et al. 2017). On account of global statistics, cardiovascular disorders accounts for about 31% mortalities (Tang et al. 2017). The factors associated with the cardiovascular health disorders may include low-density lipoproteins (LDL) oxidation, plaque and atherosclerosis (Wang et al. 2018). Clinical studies by Shafe et al. (2016) reported the dietary intake of sweet potato resulted in significant cardio-protective effects. Also, Jawi et al. (2020) the dietary feeding of the purple sweet potato-based tablets fed to the Wistar rats @ 200 mg/kg body weight/day showed significant ($p < 0.05$) decline in low density lipoprotein (LDL) cholesterol and triglycerides due to presence of trypsin inhibitor and sporamin compounds in purple sweet potatoes. Anthocyanins are abundantly found in sweet potato leaves and play a significant role in suppressing the risk of CVDs (Wang et al. 2018).

Dietary intake of anthocyanin-rich purple sweet potato resulted in reduced expression of vascular cell adhesion molecule, thereby implying its role as an anti-atherogenic agent (Satriyasa 2017). Sweet potato is an enriched source of pyridoxine and vitamin B_6 which aids in breakdown of homocysteine. Moreover, sweet potatoes are also an abundant source of potassium which helps in regulation of body fluid balance and blood pressure (Mohanraj 2018). In a study by Cui et al. (2011), the freeze-dried sweet potato samples inhibited the lactate dehydrogenase, creatine and ACE activities (Cui et al. 2011). Furthermore, Chen et al. (2011) reported the clinical trials on impacts of purple sweet potato leaves powder in humans resulted in the reduced cellular proliferation and MMP-22 activities against the umbilical vascular endothelial cells.

15.5.4 Anti-inflammatory and Immuno-modulatory

Inflammation is an underlying cause of morbid health disorders which results in deregulation of several biochemical and physiological phenomena such as functioning of tumor necrosis factor (TNF), thromboxanes, cytokines, prostaglandins and interleukins (IL) associated with cancer, asthma, atherosclerosis and rheumatoid arthritis (Majid et al. 2018). Sweet potato is considered as nutrient rich food which could play its ameliorative role against inflammatory and associated health disorders (Majid et al. 2018; Wang et al. 2013; Wang et al. 2018). Handayani and Elya (2019) revealed significant effect of white, purple and orange sweet potato varieties as anti-inflammatory agents against 5-Lipoxygenase (5-LOX) inhibition which thereby resulting in the release of Leukotriene and helps in prevention of inflammation. Extracts of sweet potato cultivars were recorded to inhibit the inflammatory properties against inflammatory cells i.e., 5-LOX with the IC_{50} values of 46, 52 and 64 µg/mL in mice. Moreover, the results also elucidated significant ($p < 0.05$) decrease in pro-inflammatory cytokines i.e., (TNF- α, IL-1, and IL-6), while increase in anti-inflammatory cytokine (i.e., IL-10) (Sun et al. 2020). Sweet potato water-soluble polysaccharides were observed to control the adaptive immunity in healthy and immunosuppressed rats by stimulating release of IL-10 and immunoglobulins (IgM, IgA, sIgA and IgG), and reducing the production of IL-6, TNF-α and IL-1β in blood (Sun et al. 2020; Tang et al. 2019).

Earlier studies have validated the utilization of sweet potato leaf extracts in modulating the immune health of mice. Phenolic extracts of sweet potato leaves @ 50–500 µg/mL notably decreased the NO levels between 11% and 44% in

LPS-treated RAW264.7 macrophage cell models (Cho et al. 2020). In human aortic models, purple sweet potatoes significantly inhibited the TNF-α-induced adhesion of monocyte-endothelial cells (Chao et al. 2013). Zengin et al. (2017) investigated the anti-inflammatory properties of leaves extracts of sweet potatoes and their role in decreasing the PGE2, nitrite levels and iNOS genetic expressions in the male C57BL6 mice which referred the polyphenols, flavonoids and anthocyanins in sweet potatoes to contribute to the anti-inflammatory properties in mice (Zengin et al. 2017). Moreover, a study by Mohanraj (2018) reported sweet potatoes as an abundant source of vitamin A, vitamin E and minerals (i.e., phosphorus, manganese and magnesium) elucidating their role in immunity boosting and lowering the disease burden.

15.5.5 Anti-obesity

Globally, the obesity has emerged as an alarming concern with prevalence of around 1.9 billion overweight individuals (Singh et al. 2020). The metabolic disorder is characterized by over-accumulation of body fats leading to health disorders such as cardiovascular diseases, diabetes mellitus, and metabolic syndrome. Bioactive compounds naturally occurring in plant foods are claimed prevent the health risks of obesity (Mir et al. 2019). Dietary intake of sweet potato and its cultivars in different forms could act as a viable food to ameliorate obesity and associated health challenges. A study by Kim et al. (2020) reported sweet potato extracts to exhibit anti-obesity properties both in-vivo with obese mice induced by a high-fat diet and in-vitro with 3T3-L1 cells. Findings of the study elucidated that the extracts doses of 100 μg/mL remarkably reduced the fats and prevented adipogenesis in C57BL/6J mice fed a high-fat diet and resulted in significant decrease of body weight and improved the lipid profile (Ju et al. 2017). Likewise, Ju et al. (2017) also elucidated significant impact of purple fleshed sweet potato feeding to mice @ 15% and 30% with the high-fat diet for 12 weeks resulted in decrease of body fat, weight and improved the fatty acid profile with the modulation of energy by exhibiting the anti-obesity properties against the obesity cells i.e., C57BL/6J.

Kurata et al. (2017) and Lee et al. (2015) reported significant role of sweet potato feeding to rats in declining the cholesterol, body weight, triglycerides and obesity of rats. Earlier studies by Kim et al. (2020) have shown positive impact of oral feeding of carotenoids and anthocyanins extracts of sweet potatoes to the mice on decreasing the triglycerides cells and protein expressions of peroxisome proliferator-activated receptor gamma and ultimately improved the lipid profile including triglycerides to HDL-C ratio. Flavonoids, phenolic acids, carotenoids, anthocyanin, organosulfurs, terpenoids, and phytosterols are natural phytochemicals which holds the promising abilities as anti-obesity agents (Singh et al. 2020). The sweet potato bioactive nutrients have the capacity to modulate oxidative, cell proliferative mechanisms, inflammatory responses and obesity (Singh et al. 2020).

15.6 Sweet Potato Based – Value Added Food Products

Food value addition has emerged as a viable strategy to help in management of health challenges associated with malnutrition, and food insecurity. However, researchers are exploring natural foods with better health significance to develop value added functional foods. The transition in innovative product development has opened new avenues for food producers and consumers to develop fabricated food products using the natural ingredients such as sweet potato in development of nutrient enriched value-added juices, ice-cream, snacks, and pastas. Earlier studies have validated the use of sweet potato as a viable ingredient of choice in the development of value added food products and preferred by the consumers in raw and or processed forms. Globally, the sweet potatoes are being consumed as staple foods, value added in snacks and bakery products (Onwude et al. 2019). Commercially, the valued added products of sweet potatoes include starch, composite ingredient-based foods, indigenous products and industrial and food products. Although, sweet potato-based value added finished products are available in markets, however further exploration is required to explore novel marketing opportunities for its application in dairy products (i.e., fermented milk and yogurt), noodles, beverages (i.e., alcoholic and non-alcoholic) and baking products (i.e., bread) (Bovell-Benjamin 2010).

Earlier studies have spotlighted the economic and nutritional significance of sweet potatoes. The roots have been considered as a suitable substitute of wheat. Sweet potato has been preferred as an energy and nutrient dense food exhibiting higher concentrations of pro-vitamin i.e., β-carotene. These are used in development of composite flours and purees in Sub Saharan African countries (Wanjuu et al. 2018). Sweet potato varieties hold higher magnitudes of dry matter (35–41%), extractable starch (20–23%) and total starch (25–27%). Due to these sweet potatoes are of immense significance for utilization in starch processing industries. On account of total global production of starch, only Asia reveals 26% production of that starch. Several fabricated finished foods are manufactured with sweet potato based extracted starches such as vermicelli, conventional noodles, sugar syrups and thickening agents, monosodium glutamate (MSG), citric acid, microbial enzymes, lactic acid, and alcohol (Truong et al. 2018).

The tubers are excellent carrier of starches which are used to produce ethanol by enzymatic hydrolysis which could be used as an appropriate substitute of gasoline. One ton of sweet potato starch can produce approx., 400–450 L ethanol which shows 40% higher capacity of producing ethanol than sweet sorghum, potato, Jerusalem artichoke, sugar beet and fodder beet. This humungous ethanol production may revolutionize the agro-industrial sector to reduce the burden over petroleum (El-Sheikha and Ray 2017). Consumer preferences towards food utilization of synthetic coloring pigments are decreasing due to availability of natural coloring pigments in markets. Anthocyanins and β-Carotene from orange and purple fleshed sweet potatoes, respectively, are endorsed for food utilization in value added foods such as dairy drinks, plant-based beverages, candies, ice creams, and cereal products owing to their vibrant color and economical production. The color profile of sweet potatoes encourages its utilization in preparation of the supplemented food products like confectioneries and tea (Alam 2021).

15.7 Challenges to Sweet Potato Production and Marketing in South Asia

Globally, production data of sweet potato accounts; Asia (71.1%), Africa (24.3%), America (3.7%), Europe (0.1%), and Ocaiana (0.9%) (Alam 2021). Tuber and root crops have significant importance in agriculture sector which play their pivotal role in ameliorating food security and malnutrition. In 2017, approximately 494.6 million tons of tubers and roots were produced worldwide. Globally, the sweet potatoes have been grown in more than 115 countries and reported the total production i.e., 112.8 million tons in 2017, China being the largest producer accompanied with the Tanzania, Nigeria, Uganda and Indonesia (FAOSTAT 2019). Since previous few years the consumption of sweet potatoes is increased in Asia, Africa, South America and Caribbean Islands. Sweet potato is among the most cultivated food crops of Africa. It is the third most essential food crop cultivated in seven central and eastern African countries. Sweet potato is ranked on fourth in terms of food significance in African countries (International Potato Center 2017). Sweet potato is considered as "poor man's crop" because it is typically grown and utilized by lower income generating population groups especially by women-headed families (Alam 2021).

Despite of higher adaptability against drought and harsh environmental conditions, the root crop has short cultivation period of 3–5 months and yields 50–60 tons/ha. Drought, soil in-fertility, improper planting materials, inferior cultivars, low seed viability and poor insect, pest management, and post-harvest problems affect its yield (Etana et al. 2019). The optimum temperature required for growth of sweet potatoes may vary between 20 and 35 °C. Exposure of maximum flux of light during a day is favorable for the development of sweet potato. The preferred pH for sweet potato production range between 5.5 and 6.5 (Etana et al. 2019). Poor nutrients availability to the plant from soil is responsible for lower yields. Application of fertilizers, manures and low input cultivation practices may also influence the growth of the sweet potatoes. A study by Agbede and Adekiya (2011) suggested the optimum storage conditions for sweet potatoes growth and development: temperature (i.e., 15 °C), relative humidity (i.e., 85–90%) and pH (5.5–6.5). Little fluctuation in storage conditions may result in internal breakdown, decay and inedibility.

15.8 Conclusion

Sweet potato is a universal staple food known for its good nutritional and medicinal quality. Compared with the other traditional vegetables sweet potato contains a wide array of essential vitamins, dietary minerals, prebiotic fibers and high quality proteins. Retrospective studies have reported the health promoting bioactive compounds of sweet potatoes to exhibit anti-diabetic, anti-tumor, anti-cancer, anti-cardiovascular disorders and hepato-protective properties. Consumption of sweet potato may be helpful in ameliorating the health challenges associated with malnutrition and food insecurity. It is also noteworthy ability of sweet potato to maintain gut microbiota proven its prebiotic potential and amino acid profile worked exponentially well against cancerous cells, LDL and VLDL levels in human body which are linked with management of immunity imbalance, obesity, cholesterol, cancer, and diarrhea. However, the tuber crop could be valued for its possible utilization in food and pharma sector. Despite its cheap cultivation cost, short ripening time, good yield and informal availability, it is under-rated/neglected food around the globe. Conclusively, further research could assist in exploration of the beneficial aspects of this miraculous root crop in development of value added nutrient enriched finished food, feed and fuel.

Conflicts of Interest All authors read and approved the final manuscript and declare that they have no conflict of interest.

References

Abegunde OK, Mu TH, Chen JW et al (2013) Physicochemical characterization of sweet potato starches popularly used in Chinese starch industry. Food Hydrocoll 33(2):169–177. https://doi.org/10.1016/j.foodhyd.2013.03.005

Agbede TM, Adekiya AO (2011) Evaluation of sweet potato (*Ipomoea batatas* L.) performance and soilproperties under tillage methods and poultry manure levels. Emir J Food Agric 23(2):164–177. https://doi.org/10.9755/ejfa.v23i2.6454

Alam MK (2021) A comprehensive review of sweet potato (*Ipomoea batatas* [L.] Lam): revisiting the associated health benefits. Trends Food Sci Technol 115:512–529. https://doi.org/10.1016/j.tifs.2021.07.001

Alam MK, Sams S, Rana ZH et al (2020) Minerals, vitamin C, and effect of thermal processing on carotenoids composition in nine varieties orange-fleshed sweet potato (*Ipomoea batatas* L.). J Food Compos Anal 92:103582. https://doi.org/10.1016/j.jfca.2020.103582

Asadi K, Ferguson LR, Philpott M et al (2017) Cancer-preventive properties of an anthocyanin-enriched sweet potato in the APCMIN Mouse Model. J Cancer Prev 22(3):135. https://doi.org/10.15430/JCP.2017.22.3.135

Ayeleso TB, Ramachela K, Mukwevho E (2016) A review of therapeutic potentials of sweet potato: pharmacological activities and influence of the cultivar. Trop J Pharm Res 15(12):2751–2761. https://doi.org/10.4314/tjpr.v15i12.31

Bindels LB, Walter J, Ramer-Tait AE (2015) Resistant starches for the management of metabolic diseases. Curr Opin Clin Nutr Metab Care 18(6):559–565. https://doi.org/10.1097/MCO.0000000000000223

Bindumole VR, Kiran KS, Balagopalan C (2000) Production of citric acid by the fermentation of sweet potato using *Aspergillus niger*. J Root Crops 26(1):38–42

Bovell-Benjamin AC (2010) Sweet potato utilization in human health, industry and animal feed systems. In: Sweet potato: post-harvest aspects in food, feed and industry. Nova Science Publishers Inc, New York, pp 193–224

Chao PY, Huang YP, Hsieh WB (2013) Inhibitive effect of purple sweet potato leaf extract and its components on cell adhesion and inflammatory response in human aortic endothelial cells. Cell Adhes Migr 7(2):237–245. https://doi.org/10.4161/cam.23649

Chen CM, Li SC, Chen CYO et al (2011) Constituents in purple sweet potato leaves inhibit in vitro angiogenesis with opposite effects ex vivo. Nutrition 27(11–12):1177–1182. https://doi.org/10.1016/j.nut.2011.01.005

Cho HD, Rabbani MB, Islam S et al (2020) Comparison of total phenolic contents, antioxidant and anti-inflammation activity in sweet potato leaves obtained from 24 different cultivars. Curr Dev Nutr 4(2):380–380

Consultative Group for International Agricultural Research (2005) Research and impact: areas of research: sweet potato. http://www.cgiar.org/impact/research/sweetpotato.html. Accessed 20 May 2008

Cui L, Liu CQ, Li DJ et al (2011) Effect of processing on taste quality and health-relevant functionality of sweet potato tips. Agric Sci China 10(3):456–462. https://doi.org/10.1016/S1671-2927(11)60025-4

Data ES, Diamante JC, Forio EE (1986) Soy sauce production utilizing root crop flour as substitute for wheat flour (100% substitution) [Philippines]. Ann Trop Pathol (Philippines). https://agris.fao.org/agris-search/search.do?recordID=PH871055688

Dong JU, Mu TH, Sun HN (2017) Sweet potato and potato residual flours as potential nutritional and healthy food material. J Integr Agric 16(11):2632–2645. https://doi.org/10.1016/S2095-3119(16)61601-5

Ellong EN, Billard C, Adenet S (2014) Comparison of physicochemical, organoleptic and nutritional abilities of eight sweet potato (*Ipomoea batatas*) varieties. Food Nutr Sci 5(2):1–6. https://doi.org/10.4236/fns.2014.52025

El-Sheikha AF, Ray RC (2017) Potential impacts of bioprocessing of sweet potato. Crit Rev Food Sci Nutr 57(3):455–471. https://doi.org/10.1080/10408398.2014.960909

Esatbeyoglu T, Rodríguez-Werner M, Schlösser A et al (2017) Fractionation, enzyme inhibitory and cellular antioxidant activity of bioactives from purple sweet potato (*Ipomoea batatas*). Food Chem 221:447–456. https://doi.org/10.1016/j.foodchem.2016.10.077

Etana MB, Gonfa MT, Duresa CO (2019) Production constraints of sweet potato (*Ipomoea batatas* L.): a review. Int. J. Agric 5(2):37–46

FAO (2011) Food and agriculture organization, FAOSTAT database. Availabale online www.fao.org.com

FAOSTAT (2013) Food balance sheets; Food systems for better nutrition Rome. Available at http://faostat.fao.org/site/10021/DesktopDefault.aspx?PageID. Accessed in March 2016

FAOSTAT (2019) Food Agriculture and Organization (FAOSTAT). Retrieved from http://www.fao.org/faostat/en/#data/QC

FAOSTAT (2020) Statistics division of Food and Agriculture Organization of the United Nations. http://www.fao.org/faostat/en/#data

Gould K, Davies KM, Winefield C (eds) (2008) Anthocyanins: biosynthesis, functions, and applications. Springer Science & Business Media

Gruneberg WJ, Ma D, Mwanga ROM et al (2015) Advances in sweetpotato breeding from 1992 to 2012. In: Potato and sweetpotato in Africa: transforming the value chains for food and nutrition security. CABI International, Oxfordshire, pp 3–68. https://doi.org/10.1079/9781780644202.0003

Guo H, Kong F, Yan C (2017) Optimization of polysaccharide ultrasonic extraction conditions using purple sweet potato tubers based on free radical scavenging and glycosylation inhibitory bioactivities. Pharmacogn Mag 13(51):504. https://doi.org/10.4103/0973-1296.211044

Handayani R, Elya B (2019) Inhibitory effects of different varieties of sweet potato (*Ipomoea batatas* L.) tubers extracts on lipoxygenase activity. Pharm J 11(6):1195–1198. https://doi.org/10.5530/pj.2019.11.185

Haydersah J, Chevallier I, Rochette I et al (2012) Fermentation by amylolytic lactic acid bacteria and consequences for starch digestibility of plantain, breadfruit, and sweet potato flours. J Food Sci 77(8):466–472. https://doi.org/10.1111/j.1750-3841.2012.02811.x
Hung PV, My NTH, Phi NTL (2014) Impact of acid and heat–moisture treatment combination on physicochemical characteristics and resistant starch contents of sweet potato and yam starches. Starch Stärke 66(11–12):1013–1021. https://doi.org/10.1002/star.201400104
Ibeto CN, Ofoefule AU, Agbo KE (2011) A global overview of biomass potentials for bioethanol production: a renewable alternative fuel. Trends Appl Sci Res 6(5):410–425
International Diabetes Federation (2019) IDF diabetes atlas, 9th edn. International Diabetes Federation, Brussels. Retrieved from https://www.diabetesatlas
International Potato Center (2017) Sweet potato in Africa. Retrieved from https://cipotato.org/research/sweetpotato-in-africa/
Ishiguro K (2019) Sweet potato carotenoids. In: Sweet potato. Academic Press, pp 223–241
Ishiguro K, Sameshima Y, Kume T et al (2012) Hypotensive effect of a sweetpotato protein digest in spontaneously hypertensive rats and purification of angiotensin I-converting enzyme inhibitory peptides. Food Chem 131(3):774–779. https://doi.org/10.1016/j.foodchem.2011.09.038
Ishiguro K, Kurata R, Shimada Y et al (2016) Effects of a sweetpotato protein digest on lipid metabolism in mice administered a high-fat diet. Heliyon 2(12):00201. https://doi.org/10.1016/j.heliyon.2016.e00201
Jawi IM, Yasa IWPS, Mahendra AN et al (2020) Effective dose and safety profile of purple sweet potato tablet preparation in rats with high cholesterol diet. Biomed Pharmacol J 13(2):665–672. https://doi.org/10.13005/bpj/1930
Ji H, Zhang H, Li H et al (2015) Analysis on the nutrition composition and an- tioxidant activity of different types of sweet potato cultivars. Food Nutr Sci 6:161–167. http://creativecommons.org/licenses/by/4.0/
Jiang L, Rozelle S, Huang J (1993) Production, technology and post-harvest processing of sweet potatoes in Sichuan Province. Unpublished Report of CIP Sub-project 6211, 63 pp
Ju R, Zheng S, Luo H et al (2017) Purple sweet potato attenuate weight gain in high fat diet induced obese mice. J Food Sci 82(3):787–793. https://doi.org/10.1111/1750-3841.13617
Kano M, Takayanagi T, Harada K et al (2005) Antioxidant activity of anthocyanins from purple sweet potato, *Ipomoea batatas* cultivar Ayamurasaki. Biosci Biotechnol Biochem 69:979–988. https://doi.org/10.1271/bbb.69.979
Karna P, Gundala SR, Gupta MV et al (2011) Polyphenol-rich sweet potato greens extract inhibits proliferation and induces apoptosis in prostate cancer cells in vitro and in vivo. Carcinogenesis 32(12):1872–1880. https://doi.org/10.1093/carcin/bgr215
Keenan MJ, Zhou J, Hegsted M et al (2015) Role of resistant starch in improving gut health, adiposity, and insulin resistance. Adv Nutr 6:198–205. https://doi.org/10.3945/an.114.007419
Khan T, Ali M, Khan A et al (2020) Anticancer plants: a review of the active phytochemicals, applications in animal models, and regulatory aspects. Biomol Ther 10(1):47. https://doi.org/10.3390/biom10010047
Kim HJ, Koo KA, Park WS et al (2020) Anti-obesity activity of anthocyanin and carotenoid extracts from color-fleshed sweet potatoes. J Food Biochem 44:e13438. https://doi.org/10.1111/jfbc.13438
Koçar G, Civaş N (2013) An overview of biofuels from energy crops: current status and future prospects. Renew Sust Energ Rev 28:900–916. https://doi.org/10.1016/j.rser.2013.08.022
Kurata R, Adachi M, Yamakawa O et al (2007) Growth suppression of human cancer cells by polyphenolics from sweetpotato (*Ipomoea batatas* L.) leaves. J Agric Food Chem 55(1):185–190. https://doi.org/10.1021/jf0620259
Kurata R, Kobayashi T, Ishii T et al (2017) Influence of sweet potato (*Ipomoea batatas* L.) leaf consumption on rat lipid metabolism. Food Sci Technol Res 23(1):57–62. https://doi.org/10.3136/fstr.23.57

Kurata R, Sun HN, Oki T et al (2019) Sweet potato polyphenols. In: Mu TH, Singh J (eds) Sweet potato: chemistry, processing and nutrition. Academic Press, pp 177–222. https://doi.org/10.1016/B978-0-12-813637-9.00007-7

Lareo CMD, Ferrari M, Guigou L et al (2013) Evaluation of sweet potato for fuel bioethanol production: hydrolysis and fermentation. Springer Plus 2:493

Lee SL, Chin TY, Tu SC et al (2015) Purple sweet potato leaf extract induces apoptosis and reduces inflammatory adipo-kine expression in 3T3-L1 differentiated adipocytes. J Evid Based Complement Alternat Med 2015:1–9. https://doi.org/10.1155/2015/126302

Lestari LA, Soesatyo MHNE, Iravati S et al (2013) Characterization of Bestak sweet potato (*Ipomoea batatas*) variety from Indonesian origin as prebiotic. Int Food Res J 20:2241–2245

Li P, Mu T, Deng L (2013) Anticancer effects of sweet potato protein on human colorectal cancer cells. World J Gastroenterol 19:3300–3308. https://doi.org/10.3748/wjg.v19.i21.3300

Li Z, Yu Y, Wang M, Xu H et al (2019) Anti-breast cancer activity of SPG-56 from sweet potato in MCF-7 bearing mice in situ through promoting apoptosis and inhibiting metastasis. Sci Rep 9(1):146. https://doi.org/10.1038/s41598-018-29099-x

Luo D, Mu T, Sun H (2021) Profiling of phenolic acids and flavonoids in sweet potato (*Ipomoea batatas* L.) leaves and evaluation of their anti-oxidant and hypoglycemic activities. Food Biosci 39:100801. https://doi.org/10.1016/j.fbio.2020.100801

Majid M, Nasir B, Zahra SS et al (2018) *Ipomoea batatas* L. Lam. ameliorates acute and chronic inflammations by suppressing inflammatory mediators, a comprehensive exploration using in vitro and in vivo models. BMC Complement Med Ther 18(1):216. https://doi.org/10.1186/s12906-018-2279-5

Makori SI, Mu TH, Sun HN (2020) Total polyphenol content, antioxidant activity, and individual phenolic composition of different edible parts of 4 sweet potato cultivars. Nat Prod Commun 15(7):1–12. https://doi.org/10.1177/1934578X20936931

Marques JM, Silva TF, Vollu RE et al (2014) Plant age and genotype affect the bacterial community composition in the tuber rhizosphere of field-grown sweet potato plants. FEMS Microbiol Ecol 88:424–435. https://doi.org/10.1111/1574-6941.12313

Mei X, Mu TH, Han JJ (2010) Composition and physicochemical properties of dietary fiber extracted from residues of 10 varieties of sweet potato by a sieving method. J Agric Food Chem 58:7305–7310. https://doi.org/10.1021/jf101021s

Mir SA, Shah MA, Ganai SA et al (2019) Understanding the role of active components from plant sources in obesity management. J Saudi Soc Agric Sci 18(2):168–176. https://doi.org/10.1016/j.jssas.2017.04.003

Mohanraj R (2018) Sweet potato: bioactive compounds and health benefits. In: Mérillon JM, Ramawat K (eds) Bioactive molecules in food, Reference series in phytochemistry. Springer, Cham, pp 919–934. https://doi.org/10.1007/978-3-319-54528-862-1

Mohanraj R, Sivasankar S (2014) Sweet potato (*Ipomoea batatas* [L.] Lam) – a valuable medicinal food: a review. J Med Food 17:733–741. https://doi.org/10.1089/jmf.2013.2818

Montejo JF, Mondonedo JAB, Lee MGA et al (2015) Hematological effects of *Ipomoea batatas* (camote) and Phyllanthus niruri (*Sampa sampalukan*) from Philippines in the ICR mice (Mus musculus). Asian Pac J Trop Biomed 5(1):29–33. https://doi.org/10.1016/S2221-1691(15)30166-0

Muñoz-Rodríguez P, Carruthers T, Wood JR et al (2018) Reconciling conflicting phylogenies in the origin of sweet potato and dispersal to Polynesia. Curr Biol 28:1246–1256. https://doi.org/10.1016/j.cub.2018.03.020

Murugan S, Paramasivam SK, Nedunchezhiyan M et al (2012) Sweet potato as animal feed and fodder. Fruit Veg Cereal Sci Biotechnol 6:106–114

Nagamine R, Ueno S, Tsubata M et al (2014) Dietary sweet potato (*Ipomoea batatas* L.) leaf extract attenuates hyperglycaemia by enhancing the secretion of glucagon-like peptide-1 (GLP-1). Food Funct 5(9):2309–2316. https://doi.org/10.1039/c4fo00032c

Neela S, Fanta SW (2019) Review on nutritional composition of orange-fleshed sweet potato and its role in management of vitamin A deficiency. Food Sci Nutr 7(6):1920–1945. https://doi.org/10.1002/fsn3.1063

Niu S, Li XQ, Tang R et al (2019) Starch granule sizes and degradation in sweet potatoes during storage. Postharvest Biol Technol 150:137–147. https://doi.org/10.1016/j.postharvbio.2019.01.004

Ogutu FO, Mu TH (2017) Ultrasonic degradation of sweet potato pectin and its antioxidant activity. Ultrason Sonochem 38:726–734. https://doi.org/10.1016/j.ultsonch.2016.08.014

Olowu A, Adeneye A, Adeyemi O (2011) Hypoglycaemic effect of *Ipomoea batatas* aqueous leaf and stem extract in normal and streptozotocin-induced hyperglycaemic rats. J Nat Pharm 2(2):56. https://doi.org/10.4103/2229-5119.83951

Onwude DI, Hashima N, Abdana K et al (2019) The effectiveness of combined infrared and hot-air drying strategies for sweet potato. J Food Eng 241:75–87. https://doi.org/10.1016/j.jfoodeng.2018.08.008

Osime E, Ediale G, Omoti C et al (2009) Effect of sweet potato leaf (*Ipomoea batatas*) extract on some haematological parameters using rabbits. J Med Biomed Res 7(1–2):12–15. https://doi.org/10.4314/jmbr.v7i1-2.44540

Panda SH, Ray RC (2008) Direct conversion of raw starch to lactic acid by Lactobacillus plantarum MTCC 1407 in semi-solid fermentation using sweet potato (*Ipomoea batatas* L.) flour. J Sci Ind Res 67:531–537

Panda SH, Naskar SK, Ray RC (2006) Production, proximate and nutritional evaluation of sweet potato curd. J Food Agric Environ 4:124–127

Panda SH, Naskar SK, Sivakumar PS et al (2009) Lactic acid fermentation of anthocyanin-rich sweet potato (*Ipomoea batatus* L.) into lacto-juice. Int J Food Sci 44:288–296. https://doi.org/10.1111/j.1365-2621.2007.01692.x

Park SY, Lee SY, Yang JW et al (2016) Comparative analysis of phytochemicals and polar metabolites from colored sweet potato (*Ipomoea batatas* L.) tubers. Food Sci Biotechnol 25(1):283–291

Prakash D, Gupta KR (2009) The antioxidant phytochemicals of nutraceuticals of importance. Open Nutraceuticals J 2:20–35. https://doi.org/10.2174/1876396000902010020

Qu HS, Wu QY, Jia JQ (2014) Ultrasonic-assisted enzymatic extraction polysaccharide of purple-fleshed sweet potato residue and its bioactivities research. Food Ind 35:47–52

Ray RC, Ravi S (2005) Post harvest spoilage of sweet potato and its control measures. Crit Rev Food Sci Nutr 35:623–644. https://doi.org/10.1080/10408390500455516

Ray RC, Sivakumar PS (2009) Traditional and novel fermented foods and beverages from tropical root and tuber crops. Int J Food Sci 44(6):1073–1087. https://doi.org/10.1111/j.1365-2621.2009.01933.x

Ruhee RT, Sams S, Begum A et al (2020) Dietary fibre content in ethnic and unconventional vegetables and fruits growing in Bangladesh. Potravinarstvo Slovak J Food Sci 14:33–41. https://doi.org/10.5219/1202

Rutz D, Janssen R (2007) Biofuel technology handbook. WIP Renewable Energies, Munchen

Sanoussi AF, Adjatin A, Dansi A et al (2016) Mineral composition of ten elites sweet potato (*Ipomoea batatas* [L.] Lam.) Landraces of Benin. Int J Curr Microbiol Appl Sci 5(1):103–115

Satriyasa BK (2017) Purple sweet potato ethanolic extract reduces aortic VCAM expression in rabbit with high-cholesterol diet. Bali Med J 6(3):S26–S28. https://doi.org/10.15562/bmj.v6i3.713

Senthilkumar R, Yeh K (2012) Multiple biological functions of sporamin related to stress tolerance in sweet potato (*Ipomoea batatas* Lam). Biotechnol Adv 30:1309–1317. https://doi.org/10.1016/j.biotechadv.2012.01.022

Shafe MO, Eze ED, Ubhenin AE et al (2016) Effects of aqueous tuber extract of *Ipomea batatas* on cardiac enzymes, lipid profile and organ weights in Wistar rats. Int J Sci Basic Appl Res 2(4):414–417

Shih CK, Chen CM, Varga V et al (2020) White sweet potato ameliorates hyperglycemia and regenerates pancreatic islets in diabetic mice. Food Nutr Res 64:3609. https://doi.org/10.29219/fnr.v64.3609

Shivashankara KS, Isobe S, Al-Haq MJ et al (2004) Fruit antioxidant activity, ascorbic acid, total phenol, quercetin, and carotene of Irwin mango fruits stored at low temperature after high electric field pretreatment. J Agric Food Chem 52:1281–1286

Singh M, Thrimawithana T, Shukla R et al (2020) Managing obesity through natural polyphenols: a review. Future Foods 1–2:100002. https://doi.org/10.1016/j.fufo.2020.100002

Suarez S, Mu T, Sun H et al (2020) Antioxidant activity, nutritional, and phenolic composition of sweet potato leaves as affected by harvesting period. Int J Food Prop 23(1):178–188. https://doi.org/10.1080/10942912.2020.1716796

Sun M, Mu T, Zhang M et al (2012) Nutritional assessment and effects of heat processing on digestibility of Chinese sweet potato protein. J Food Compos Anal 26:104–110. https://doi.org/10.1016/j.jfca.2012.03.008

Sun H, Zhang P, Zhu Y et al (2018) Antioxidant and prebiotic activity of five peonidin-based anthocyanins extracted from purple sweet potato (*Ipomoea batatas* (L.) Lam.). Sci Rep 8(1):5018. https://doi.org/10.1038/s41598-018-23397-0

Sun Y, Pan Z, Yang C et al (2019) Comparative assessment of phenolic profiles, cellular antioxidant and antiproliferative activities in ten varieties of sweet potato (*Ipomoea batatas*) storage roots. Molecules 24(24):4476. https://doi.org/10.3390/molecules24244476

Sun J, Gou Y, Liu J et al (2020) Anti-inflammatory activity of a water-soluble polysaccharide from the roots of purple sweet potato. RSC Adv 10(65):39673–39686. https://doi.org/10.1039/D0RA07551E

Takamine K, Abe J, Shimono K et al (2007) Physicochemical and gelling characterizations of pectin extracted from sweet potato pulp. J Appl Glycosci 54:211–216. https://doi.org/10.5458/jag.54.211

Tanaka M, Ishiguro K, Oki T et al (2017) Functional components in sweetpotato and their genetic improvement. Breed Sci 67:52–61. https://doi.org/10.1270/jsbbs.16125

Tang GY, Meng X, Li Y et al (2017) Effects of vegetables on cardiovascular diseases and related mechanisms. Nutrients 9(8):857. https://doi.org/10.3390/nu9080857

Tang C, Sun J, Zhou B et al (2018) Immunomodulatory effects of polysaccharides from purple sweet potato on lipopolysaccharide treated RAW 264.7 macrophages. J Food Biochem 42:e12535. https://doi.org/10.1111/jfbc.12535

Tang C, Sun J, Liu J et al (2019) Immune-enhancing effects of polysaccharides from purple sweet potato. Int J Biol Macromol 123:923–930. https://doi.org/10.1016/j.ijbiomac.2018.11.187

Tang CC, Ameen A, Fang BP et al (2021) Nutritional composition and health benefits of leaf-vegetable sweet potato in South China. J Food Compos Anal 96:103714. https://doi.org/10.1016/j.jfca.2020.103714

Terahara N, Matsui T, Fukui K et al (2003) Caffeoylsophorose in a red vinegar produced through fermentation with purple sweet potato. J Agric Food Chem 51:2539–2543. https://doi.org/10.1021/jf026146u

Thatoi H, Dash PK, Mohapatra S et al (2016) Bioethanol production from tuber crops using fermentation technology: a review. Int J Sustain Energy 35(5):443–468. https://doi.org/10.1080/14786451.2014.918616

Theuwissen E, Mensink RP (2008) Water-soluble dietary fibers and cardiovascular disease. Physiol Behav 94:285–292. https://doi.org/10.1016/j.physbeh.2008.01.001

Tian C, Wang M, Liu S et al (2019) A new glycoprotein SPG-8700 isolated from sweet potato with potential anti-cancer activity against colon cancer. Nat Prod Res 33(16):2322–2328. https://doi.org/10.1080/14786419.2018.1446007

Tripura A, Sankaran M (2008) ICAR research complex for NEH region, production techniques and utilization of tuber crops in Tripura. ICAR Research Complex for NEH Region

Truong VD, Avula RY, Pecota KV et al (2018) Sweetpotato production, processing, and nutritional quality. In: Siddiq M, Uebersax MA (eds) Handbook of vegetables and vegetable processing. Wiley, pp 811–838. https://doi.org/10.1002/9781119098935.ch35

USDA (2017) Sweet potato. United States Department of Agriculture online database: 02-22. Retrieved on March, 2017. www.usda.gov

Vishnu VR, Renjith RS, Mukherjee A et al (2019) Comparative study on the chemical structure and in vitro antiproliferative activity of anthocyanins in purple roots and leaves of sweet potato (*Ipomoea batatas*). J Agric Food Chem 67(9):2467–2475. https://doi.org/10.1021/acs.jafc.8b05473

Wang H, Fan W, Li H et al (2013) Functional characterization of dihydroflavonol-4-reductase in anthocyanin biosynthesis of purple sweet potato underlies the direct evidence of anthocyanins function against abiotic stresses. PLoS One 8(11):78484. https://doi.org/10.1371/journal.pone.0078484

Wang L, Zhao Y, Zhou Q et al (2017) Characterization and hepatoprotective activity of anthocyanins from purple sweet potato (*Ipomoea batatas* L. cultivar Eshu No. 8). J Food Drug Anal 25:607–618

Wang A, Li R, Ren L et al (2018) A comparative metabolomics study of flavonoids in sweet potato with different flesh colors (*Ipomoea batatas* (L.) Lam). Food Chem 260:124–134. https://doi.org/10.1016/j.foodchem.2018.03.125

Wang YT, Yang CH, Huang TY et al (2019) Cytotoxic effects of chlorophyllides in ethanol crude extracts from plant leaves. J Evid Based Complement Alternat Med 2019:9494328. https://doi.org/10.1155/2019/9494328

Wanjuu C, Abong G, Mbogo D et al (2018) The physiochemical properties and shelf-life of orange-fleshed sweet potato puree composite bread. Food Sci Nutr 6(6):1555–1563. https://doi.org/10.1002/fsn3.710

Woolfe JA (1992) Sweet potato. An untapped food resource. Cambridge University Press, New York

Wu Q, Qu H, Jia J, Kuang C et al (2015) Characterization, antioxidant and antitumor activities of polysaccharides from purple sweet potato. Carbohydr Polym 132:31–40. https://doi.org/10.1016/j.carbpol.2015.06.045

Yamakawa O (2000) New cultivation and utilization system for sweet potato toward the 21st century. In: Nakatani M, Komaki K (eds) Potential of root crops for food and industrial resources. ISTRC, pp 8–13

Yuan B, Yang XQ, Kou M et al (2017) Selenylation of polysaccharide from the sweet potato and evaluation of antioxidant, antitumor, and antidiabetic activities. J Agric Food Chem 65:605–617. https://doi.org/10.1021/acs.jafc.6b04788

Zengin G, Locatelli M, Stefanucci A et al (2017) Chemical characterization, antioxidant properties, anti-inflammatory activity, and enzyme inhibition of *Ipomoea batatas* L. leaf extracts. Int J Food Prop 20:1907–1919. https://doi.org/10.1080/10942912.2017.1357127

Zhang M, Mu T (2016) Optimisation of antioxidant hydrolysate production from sweet potato protein and effect of in vitro gastrointestinal digestion. Int J Food Sci Technol 51:1844–1850. https://doi.org/10.1111/ijfs.13156

Zhang M, Mu TH (2018) Contribution of different molecular weight fractions to anticancer effect of sweet potato protein hydrolysates by six proteases on HT-29 colon cancer cells. Int J Food Sci Technol 53(2):525–532. https://doi.org/10.1111/ijfs.13625

Zhang M, Mu T, Wang Y et al (2012) Evaluation of free radical-scavenging activities of sweet potato protein and its hydrolysates as affected by single and combination of enzyme systems. Int J Food Sci Technol 47:696–702. https://doi.org/10.1111/j.1365-2621.2011.02895.x

Zhang L, Zhao L, Bian X, Guo K, Zhou L, Wei C (2018) Characterization and comparative study of starches from seven purple sweet potatoes. Food Hydrocoll 80:168–176

Zhao JG, Yan QQ, Lu LZ et al (2013) In vivo antioxidant, hypoglycemic, and anti-tumor activities of anthocyanin extracts from purple sweet potato. Nutr Res Pract 7(5):359–365. https://doi.org/10.4162/nrp.2013.7.5.359

Zheng Y, Wang Q, Li B et al (2016) Characterization and prebiotic effect of the resistant starch from purple sweet potato. Molecules 21:932–943. https://doi.org/10.3390/molecules21070932
Zhu F, Wang S (2014) Physicochemical properties, molecular structure, and uses of sweet potato starch. Trends Food Sci Technol 36(2):68–78. https://doi.org/10.1016/j.tifs.2014.01.008
Zhu F, Xie Q (2018) Rheological and thermal properties in relation to molecular structure of New Zealand sweet potato starch. Food Hydrocoll 83:165–172
Ziska LH, Runion GB, Tomeceka M et al (2009) An evaluation of cassava, sweet potato and field corn as potential carbohydrate sources for bioethanol production in Alabama and Maryland. Biomass Bioenergy 33(11):1503–1508

Chapter 16
Taro (*Colocasia spp.*): Applications in Food Production and Improving Nutrition in South Asia

Md. Mokter Hossain, Md. Asaduzzaman, Muhammad Ammar Khan, and Limu Akter

16.1 Introduction

Taro (*Colocasia esculenta,* L. Schott) belongs to Araceae family, is one of the edible aroids dispersed throughout the worldwide. It was originated in the Bay of Bengal region of South-east Asia and found in subtropical and tropical areas where it was substantially a major dietary source on the Pacific islands, particularly New Zealand, west to Indonesia and Hawaii. Although grown as an annual, taro is a perennial herb with a tuberous thick underground stem whose leaves are long-petioled, simple, and broad. Petioles and unopened leaves are generally prepared just like spinach leaves and consumed as a vegetable, on the other hands the tubers can be boiled and steamed. In contrast to the other tuber and root crops, taro has a greater nutritional value. Bothe leaves and corms are a good source of protein, potassium, calcium, iron and phosphorus. The corms of taro also contain a fine-grained digestible starch, ash and lipids content. This chapter will review the use of taro in the food industry and its role in improving human nutrition.

Md. M. Hossain (✉)
Department of Horticulture, Faculty of Agriculture, Bangladesh Agricultural University, Mymensingh, Bangladesh
e-mail: mokter.agr@bau.edu.bd

Md. Asaduzzaman · L. Akter
Horticulture Research Center, Bangladesh Agricultural Research Institute, Joydebpur, Gazipur, Bangladesh

M. A. Khan
Department of Food Science & Technology, Faculty of Agriculture and Environment, The Islamia University Bahawalpur, Bahawalpur, Pakistan

T. Ismail et al. (eds.), *Neglected Plant Foods of South Asia*,
https://doi.org/10.1007/978-3-031-37077-9_16

16.2 Climate Resilient Cultivation of Taro

The intensification and yield potential of different methods of wet taro growing are clearly distinct. Low-density, nomadic groups benefit most from simple irrigation. Other techniques, such as pond field cultivation, may offer nearly limitless intensification potential. Planting in minimally-prepared stream bed gardens is also prevalent in pond fields, which alternate between wet and dry states throughout the planting cycle. Research further demonstrates that low island and atoll pit agriculture is more variable than previously thought. The research into the ecological, demographic, and sociocultural causes of this variation is still in its early stages. There is also a great deal of variation in the frequency of irrigation on some of the bigger islands, as well as within specific archipelagoes. This is a more plausible scenario.

Heat stress accompanied with excessive humidity and high temperatures is responsible for seed components of the plant to dry out. Such a situation results in drying out of the taro plants and yields small, poor-quality tubers with little market value (Chakraborty and Newton 2011). At the tuber development stage, heat stress on taro reduces crop yield because taro is very sensitive to temperature increases. Furthermore, excessive rains improve the occurrence of late blight and bacterial wilt, lowering yields. Floods make planting and land preparation more difficult and cause seeds to decay when sown (Mwongera et al. 2019). To mitigate these consequences, some farmers discarded the variety or improve the amount of herbicides sprayed, both of which are costly for smallholder farmers. Climate change particularly increment of temperature during growth of taro plant enhance the harvesting time resulting lower yield of crop. Crimp et al. (2017) noticed that the influence of temperature on the harvest time of taro herb and observed that higher temperature decrease a maturity duration. They found that each 1 °C increase of average temperature reduced 2-weeks in a maturity duration.

Climate Smart Agriculture (CSA) refers to various functionaries in the agriculture sector that are pushing farming communities to opt novel methods and technologies alongside conventional subsisting strategies to counteract the impacts of climate rotation (Chandra et al. 2018). Crop rotation, agroforestry, intercropping, soil cover maintenance, climate-resilient crop varieties, minimum tillage, residue retention, water conservation, enhanced livestock management are among the practices and technologies that can help farmers adapt to future harsh conditions. In comparison to traditional agricultural production methods, CSA has been shown to produce consistent and greater yields, as well as substantial revenue from farming, resulting in high durability in some areas (Wekesa et al. 2018).To adopt climate change and variability, farmers must use a variety of climate smart innovations. Climate wise agricultural innovations which have the ability to remit climate-associated apprehension such as sustentation of rainwater and producing climate-resilient seed genotypes (Senyolo et al. 2018). On the contrary, seed genotypes suited to global warming with regard to maturation period and pest and disease defiance, are found to be less expensive and has relatively less management needs when assisted by appropriate agronomic practices (Atlin et al. 2017). Farmers favor enhanced crop types as the dominant way to adapt with climate change according to Fadina and Barjolle (2018).

16.3 Status of Taro Production in South Asia as an "Orphan Crop"

Taro, also known as an "orphan crop" is a valuable source of minerals, carbohydrates, dietary fibers, proteins, resistant starch, and vitamins (Li et al. 2020). Taro is considered as a smart future food under the classification of underutilized tuber and root crops to reduce hidden hunger and malnutrition and in the Asia (Siddique et al. 2021). Taro (Colocasia spp.) is native to Southeast Asia (Kolchar 2006) and one of the traditional, herbaceous perennial plant also known as "Eddoe", tropical plant predominantly cultivated as a root crop for its starchy corms, and as well as a leafy vegetable (Van Wyk 2005). This is an ancient crop species which grown in the Bangladesh and Eastern India (Yen and Wheeler 1968). It is considered that the wild type Colocasia esculenta var. aquatilis, either in South East Asia or North East India are the origins of domesticated taro (Matthews 1991). Colocasia, Alocasia and Cyrtosperma taro is believed to grow in Asia (Ramanatha et al. 2010). Taro is known as "Potato of the Tropics" and cultivated throughout the sub-tropics and tropics. Taro is a nutritive vegetable grown throughout India and is also sometimes known as potato of the humid tropics.

In India, taro is grown in localized areas of various states of West Bengal, Madhya Pradesh, Uttar Pradesh, Andhra Pradesh, Orissa, and Kerala. Generally, taro is grown throughout India by the small and marginal farmers. It is usually cultivated as a mono crop in Madhya Pradesh, Orissa and Andhra Pradesh while in some places taro is cultivated as an intercrop. According to the First Advance Estimates in 2021–22 (All India), the area under cultivation of taro and its production are estimated as 35 thousand hectares and 819 thousand metric tons, respectively. In Bangladesh, the Araceae family comprise on 109 species and 27 genera of which twenty-nine are cultivated and eighty-one species are wild (Ara 2016). The cultivated aroids species are used as food. The edible part of taro is enlarged root that acts as a storage organ. The edible aroids are belonging to five genera such as Cyrtosperma, Amorphophallus, Colocasia, Alocasia and Xanthosoma. The Bengali name of taro is Kachu. It has many types based on the typical variations in rhizome and corm sizes. Among the species, *Colocasia esculenta* L. (Taro: Mukhikachu (Fig. 16.1), Panikachu (Figs. 16.2 and 16.3), Panchamukhi kachu (Fig. 16.4), Ol kachu) are commercially growing in different parts of Bangladesh while *Alocasia macrrorrhiza* (Giant taro) is grown in the homestead and at the road sides (Fig. 16.5).

The corms and cormels of Mukhi kachu are rich source of starch and can be eaten in a way similar to potatoes. Moreover, stems, stolons, leaves, leaf petiole of taro are also consumed as good source of vitamins and minerals particularly iron

Fig. 16.1 Mukhikachu in the field (**a**) and corm & cormels of Mukhikachu (**b**)

Fig. 16.2 Panikachu in the field (**a**), Harvested rhizome (**b**), Slicing of rhizome (**c**) and ready to cook rhizome (**d**). (Photo credit Dr. Md. Mokter Hossain, BAU, Mymensingh)

Fig. 16.3 Panikachu in the field (**a**, **b**), and stolon of Panikachu (Loti Raj) (**c**)

Fig. 16.4 Plant (**a**) and corm of Panchamukhi kachu (**b**, **c**)

Fig. 16.5 Plant of Giant taro (**a**) and Rhizome of Giant Taro (**b**) usually grown in homestead and road sides. (Photo credit Dr. Md. Mokter Hossain, BAU, Mymensingh, Bangladesh)

Table 16.1 Area and production of different types of aroids

Types of aroids/Taro	Species	Area (ha.)	Production (MT)
Mukhi kachu	*Colocasia esculenta* L.	9550.61	109803.00
Panikachu		5703.59	62210.30
Panchamukhi kachu[a]		–	–
Elephant foot yam/*Ol kachu*	*Amorphophallus campanulatus* (Roxb.)	898.64	17037.58
Giant taro /*Maankachu*/*Fankachu*	*Alocasia macrrorrhiza*	1940.09	16348.75

[a]Data not yet recorded in BBS (2021)

(Fe). Bangladesh Agricultural University (BAU) has developed five varieties of kachu namely, BAU kachu-1 (Panchamukhi), BAU kachu-2 (Poidnal), BAU Ol kachu-1, BAU Maankachu-1 and BAU Salad kachu. Bangladesh Agricultural Research Institute (BARI) has addressed several genotypes of kachu such as BARI panchamukhi kachu-1 to BARI panchamukhi kachu-6, BARI mukhikachu (Bilashi), BARI Panikachu (Latiraj).

According to BBS (Bangladesh Bureau of Statistics) (2021), the total area and production of different types of Taro/ariods (kachu) in Bangladesh are given in Table 16.1.

Despite the common uses of Taro as a vegetable it is considered as a neglected crop in Pakistan and many other countries of the region. It is commonly known as arbi and arvi are grow as a summer crop. Taro corms are generally consumed as a vegetable and its young leaves are occasionally consumed after cooking, both in the urban and rural settings. The area under taro cultivation and yield of taro gradually declined during the last 5 years. In 2014–15, area and production were 2757 ha and 30,575 metric tons which were declined to 2388 ha. and 24,281 metric tons in 2018–19 (GoP 2020). Taro is widely grown in upland regions of Nepal while some wild varieties are cultivated in muddy soil whose leaves are generally used as green vegetables. Taro has been known an underutilized herb in Nepal because it failed to attract attentiveness of farmers for its cultivation and also government agricultural programs emphasizing least with no significant investment on promotion of taro production.

In Nepal, it is called as Karkalo or Pidalu. The aerial leafy part of taro is called Karkalo and the underground part is known as pidalu. Taro is a cardinal traditional vegetable crop throughout the hills and plain of Nepal (Pandey 1999). It is well adapted in marginal and harsh environments, therefore, taro is widely adopted and cultivated as a vegetable crop. The crop is grown for its nutritious leaves, petioles, flowers and tubers on marginal lands of almost all households as a preferred food choice, and is consumed either in fresh or cooked form. Taro cultivars are selected and produced by the indigenous farming communities considered as the curator of both the genetic and cultural diversity. Though the crop is growing in both homestead and open field condition, but it is uniquely adapted to homestead gardens enabling peoples to satisfy their dietary energy needs and ensure household food security. Various types of taro are cultivated as food and other reasons in different parts of Nepal. Farmers, particularly women, prepare various types of products from taro for income generation.

16.4 Nutritional Composition of Taro

Taro has a rich nutritive value as compared to other tuber and root crops. The stems, leaves, leaf petiole, corms, cormels contain good protein quality and are good sources of phosphorus, potassium, phosphorus, iron and phosphorus. The corms and cormels also possess fine grained digestible starch. Starch extracted from corm of taro is distinctive because of its very tiny particle size ranges from 1–5 μ (Jane et al. 1992). Taro is considered as a low glycemic index food, good energy supplier (Huang et al. 2000) and a good reservoir of carbohydrates for diabetics (Onwueme 1978). Taro gives higher carbohydrates, energy and vitamin A as compared to potatoes and sweet potato. Leaves of taro have maximum vitamin A contents in contrast to shoots, corms, and other tuber crops. The corms are also well-known to supply good amounts of vitamin C, riboflavin, thiamine, protein, niacin and dietary fiber (Niba 2003). Taro leaves taro contains iron, folic acid β-carotene which protects in case of anemia and is vital source of vitamins and proteins (Sukamoto 2003). Taro leaves contain high vitamin A, in contrast to shoots corms and other tuber crops. Corms and leaves of taro possess significantly higher magnesium, potassium and calcium than other tuber crops (USDA 2021).

The major coloring pigments present in taro herbs are carotenoids and anthocyanins which contribute varying colors to the corm including yellow, white, green, red, pink, dark purple, light purple etc. Taros possess different carotenoids pigments incorporating zeaxanthin, phytoene, carotene, and lutein (Champagne et al. 2013). These phyto-pigments devote to various biological functions including disease prevention and health promotion. The edible aroids have maximum amounts of amino acids and protein as compared to other tropical root crops (Bradbury 1983). The nutritive composition of taro varies significantly depending upon climatic condition, and other agronomical factors of the cultivation and variety. However, per 100 g fresh taro contains 68–70 g water, 3.7–3.9 g crude protein, 1.16–1.53 g crude fibre, 21.7–24.4 g carbohydrates.

Taro also contains different valuable minerals such as potassium (382–1049 mg/100 g), phosphorus (114–171 mg/100 g), copper (0.19–0.51 mg/100 g), iron (3.24–11.72 mg/100 g), zinc (3.01–5.63 mg/100 g), manganese (0.54–6.82 mg/100 g), calcium (58–266 mg/100 g), magnesium (78–96 mg/100 g), and sodium (8.4–39.6 mg/100 g). Taro leaves contain total antioxidant capacity (130 mg/100 g) and total polyphenol (120 mg/100 g). However, taro has some anti-nutritional elements such as oxalates and trypsin inhibitors. The oxalates compounds are responsible for mordacity (Soudy et al. 2010), which reduce the acceptability as well as palatability of taro.

16.5 Industrial and Food Applications of Taro Starch

Taro starch has numerous applications in food industries. Taro is a unique ingredient being used as a stabilizer, emulsifier, fat substitute, filler agent and a good source prebiotic for the production of different products. Taro is potential source for the utilization mucilage in making different kinds of food products. Taro mucilage is widely utilized as edible coating or films in edible food packaging production. It has application in bakery industry as a fat replacer. Taro mucilage is being using as a dye remover in dye and food industries. Mijinyawa et al. (2018) noticed that under an ultrasonication mucilage of taro can produce methylene blue at basic pH. Different studies indicated that taro has major bioactive components and those components are being used for the production of a various nutraceuticals and functional foods. Taro starch is also used for the production of baby foods and is traditionally used as a complimentary food for the young child in various cultures. Bakery and confectionery industries have a great use of taro starch. Certain genotypes of taro are utilized as a source of red kapa dye. Taro starch is used in manufacturing biodegradable plastics.

Taro starch is also used to produce modified starches namely phosphorylated, acetylated, and succinylated. Gelatinization properties of modified starch can be applied for improving functional aspects of easily frozen and heated foods, ready-to-eat, and for preparing food products with longer shelf life. Taro starch content has been studied as a new reservoir for the value-addition of food products because of its functional and physicochemical properties. Taro starch can form more consistent and stable hydrogel gel because of its possessed good viscosity characteristic. In industrial perspectives, taro starches are mainly used as fillers or modifiers for plastics in plastic modifier industries. In cosmetic industry, taro starch is used as a several vanishing cream and preparation of different cosmetic products.

16.6 Traditional Food Uses of Taro in South Asia

Taro corms and leaves have many traditional applications in South Asian cultures. According to a different location, taro can be used in various ways. Taro is a very famous vegetable in Bangladesh. It is generally prepared with Hilsha fish and small prawns into a curry and some other dishes are prepared with dried fish. Boiled leaves and stems of taro are generally ground to form a paste and finely minced to develop shak. Taro stolons or its stems are prepared with dried fish, shrimp, and the head of the Hilish fish. Kachu shak vorta (paste) is another food preparation made with taro and is a popular village food in Bangladesh. Corm of elephant foot yam is utilized mostly in curries and mashed forms, and rarely as pickles. Leaves of taro with elephant foot yam petiole are also used as a green leafy vegetable.

In India regions, taro herb is a common dish served in variety of ways in different provinces of the country. During a season of lean, the collection of stems and leaves from wild genotypes is a famous tradition in eastern India and the efflorescence is also utilized for cooking (Teron 2019). It is very common among the Naga community in Northeast India and is consumed all year in processed, semi processed and raw forms. The stalks, corms and leaves are used as dawl bai. Moreover, the stalks and leaves are often generally preserved to be consumed in dry season. Taro curry is also a popular spicy food which commonly use in Southern India. In some locations, taro leaves are boiled in water and rolled with gram and corn flour and served as delicious food item. In another province, the young leaves of taro are sun-dried, steamed, and stored for later utilization. Taro leaves and stems are pickled. Crushed stems and leaves are variegated with black lentils and then dried in form of small balls for future use. The stems and leaves of certain genotypes of taro are also utilized as a vegetable in Kerala. In Karnataka province, taro is utilized as a breakfast dish. In West Bengal, corms of taro are slightly sliced and fried to prepare chips.

In Nepal, taro has also traditional food applications. Arikanchan is a famous traditional food of Eastern Terai of Nepal and is developed from black lentils paste, spices and taro leaves. Locally, the taro corms are considered as pindalu however the petioles with leaves are considered as gava, kaichu and also karkalo in Maithili. Almost all fraction of taro are eaten in varieties of dishes in Nepal. Boiled taro corm is usually served with spices, chilies and salts. Taro is a most familiar dish in the hilly Nepal regions. Chopped petioles and leaves are mixed with Urad bean flour to prepare dried balls called maseura. Corms are mixed with turmeric and dried fish, and then dried in cakes called sidhara which are curried with chili, garlic, and radish and with other seasonings, and served with rice.

In Pakistan, taro is a very well-known dish served without and with gravy. A popular dish is *arvi gosht* wherein corm of taro is cooked in mutton and lamb meat. The leaves of taro are folded along with batter of gram flour and then steamed and fried to prepare a dish called *Pakora*, which is finished by tempering with red chilies and carrom (*ajwain*) seeds. *Arvi* or taro is also prepared with minced spinach. The dish called *Arvi Palak* is another most renowned dish made of taro. In Sri Lanka, corms of taro are eaten after boiling them and preparing them into a curry with coconut milk. The leaves of some selected taro genotypes are also consumed by Sri Lankan.

16.7 Application of Taro for Producing Infant Flours

Taro flour is one of the major processed forms of taro. Since, taro starch is considered as an easily digestible source of energy and hence considered suitable to be utilized as infant food or for preparing infant food products. Taro has some essential amino acids although it lacks histidine, lysine, isoleucine, tryptophan and methionine. Traditional infant foods made of grains and cereals are bulky and may be low in nutrients including, vitamin A, protein iron or zinc, and may also possess high amount of antinutrients decreasing mineral bioavailability. Moreover, taro starches are easily gelatinized and produce soft and clear food products similar to potato starch. Unique functional properties of taro starch therefore attract food industry of many Asian countries in high value food production (Wang 1983). Taro also has soluble dietary fiber that has many desirable functional properties and fairly high amount of minerals such as potassium, phosphorus and zinc. In general, taro based infant young child preparations may contain appreciable amount of many nutrients of health significance, and also serve as a cost effective high quality food for the impoverished and food insecured populations (Patrick et al. 1999). Studies suggest taro utilization in infant formulation to not merely reduce burden on conventional food crops but also promote dietary diversity, sustainability and diversified nutrients delivery. Taro is a promising and preferred food crop of South East Asia, and hence serves as an alternate, healthier and cost effective food choice to address household food insecurity of the rural populations especially children.

16.8 Taro Based Value Added Food Products

Taro starch has wider application in manufacturing value added products including taro flour, cereals, taro paste, canned taro, cakes and bread, taro flakes, dry instant beverage powders, taro slices, taro meal, frozen taro, and taro starch based extruded foods. Thinly slice taro root are baked in oven or fried in air fryer to produce crunchy taro chips, a potential alternate to potato chips. Besides, taro French fries, taro has also been used in developing frozen yogurt or ice cream, milk tea and pancake. In some parts of the world, taro has also been used for producing ice cream and drinks (Femina 2013). The taro flour has also been proposed as fundamental ingredient in food formulations like instant and flavored poi, taro bread, taro cookies, baby foods, pasta.

Unlike some other tropical root crops like sweet potato and cassava, taro is still a neglected food crop. Conventionally, taro is used for producing edible goods at household and industrial level, and has also been exploited in animal feed production and for the extraction of non-starch polysaccharides. Beside its application for the production of value - added food commodities at industrial scale, taro starch is used as fillers or modifiers for plastics. In animal feeds, attention must be given to the problem of acridity. Taro has been least explored as a source of non-starch polysaccharides necessitating concerted efforts to be paid for its production and industrial uses. Intercropping of taro has been suggested to fill seasonal food gaps. Taro has good potential to yield reasonably in intercropping under conditions where other crops may be unable to give enough yield (Cartherwood et al. 2007). Taro leaves and the tubers are not suggested to be eaten raw on account of calcium oxalates based acridity.

Taro itself is a significant source of oxalates which limits eating quality of this food crop while acridity linked with raw taro consumption may cause mouth swelling and throat. Other nutrition and health concerns with oxalates include reduced biological availability of calcium and increased risks of kidney stone formation. Studies suggest cooking including boiling of high oxalates carrying food crops may help improve their palatability, safety and digestibility. Other potential strategies to mitigate oxalates and other antinutrients in raw foods include but not limited to chemical treatment and genetic improvements of the food crops (Paull et al. 1999; Fufa et al. 2021). Researchers have been conducted studies on the application of taro as food ingredient, in extruded pasta and canned products. These incorporate taro's application in manufacturing taro flour, canned taro, baby foods, taro meal and grits, extruded products (noodles, macaroni and rice), gum replacer and alcoholic fermented beverage. Hence, food processors have not yet considered listed products for commercial production.

16.9 Future Prospects of Taro Value Addition

Taro roots, including its edible and non - edible fractions can gain substantial application at industrial scale, and can potentially contribute in improving food security situation of the rural economies. There are several products which can substantially increase utilization of taro.

16.9.1 Ready-to-Eat (RTE) Taro Patties and Chunks

Higher acceptability of the consumers has been recorded for boiled and baked taro served with or without coconut cream. Expansion of RTE and ready to cook food market has also been considered as potential business area for taro based products such as chunks and patties. Promoting establishment of cottage snack industries in taro producing regions can further help addressing problems of processing a large volume of corm at the household scale. Subject to the cost – effective availability of taro and taro based food products; food service organizations can also increase the use of this underutilized plant food in conventional menus.

16.9.2 Taro Flour as a Food Ingredient

Low grade corms of taro which are not suitable for the chip industry of taro and for supply to the fresh vegetables market, can be processed to taro flour – a value – added product. Flour of taro can then be utilized in other food products preparations such as taro cookies, pasta, baby food, taro bread, flavored or instant poi, and other products.

16.9.3 Taro Starch

Due to the small granulated size, taro starch may impart itself to specialty markets like as the food, cosmetic and plastic industries. There should be some opportunities available for industrial exploitation of taro starch if its physiochemical properties and suitability to be incorporated in various foods or non – food formulations are well known.

16.9.4 Taro Mucilage

Taro possesses about 10 percent mucilage on a dry weight basis. This mucilage may have some potentiality in the dietary fiber and gum market, and can further be explored for their use in food, nutraceutical and other allied industry.

16.9.5 Taro Peels

The taro peels produced from the preparation of food products of taro must be completely exploited in order to increase the utilization of the corm of taro. The peels of taro may be utilized as feed of animal, to produce bio-energy, and dried to mulch for the agriculture industry.

16.10 Challenges in Taro Production and Commercialization

Although taro is not resistant to saline, but it favors slightly acidic (pH 5.5–6.5) condition. Production of taro is greater when it is exposed to straightly sunlight in its later growth stages. In flood conditions, the rate of leaf senescence is reduced. Under optimum conditions and with good cultivars, taro crop can generate a yield up to 38 t/ha. During the first 3 months after planting, taro has a hard time competing with weeds. Another disadvantage is that it is susceptible to pests and diseases such as plant caterpillars, cocoyam root rot. mites, leaf hoppers, and taro leaf blight disease (Chukwu et al. 2008). Taro is traditionally vegetatively reproduced, however it can also reproduce from the seeds (Ubalua et al. 2016). Producing taro propagating materials for next planting season has been documented with many challenges which includes but not limited to leaf blight diseases infestation during cultivation, low market price during harvesting, lack of storage & poor marketing facilities.

The commercialization of taro crop production necessitates rigorous strategies such as engagement of small scale producers, capacity building of the stakeholders to better understand market behavior, exploring existing channels and establishing new one for taro marketing. Commercialization of taro production further demands appropriate policy, organizational, institutional, and technological measures to ensure that small farmers may yield higher benefits from commercialization process. Processed tuber and roots crop commodities have been produced in response to rising requirements from urban populations and present great potential for tuber and root crops value-addition and enhanced earnings to the taro producers. Promotional activities such as pilot scale commercial operations which may attract larger growers, producers, processers and consumer market are also promising avenues that demand serious inputs such as financial assistance, technological and logistic support and mass scale promotion Once commercialization of taro begins, it is critical to monitor and evaluate these operations. Taros like many other root crops have the reputation of being low-protein and inferior food crops that can't compete with cereal crops like wheat, rice, and maize. Commercialization of several tropical tuber and root crops is therefore quite weak, despite the fact that they are substantial carbohydrate sources, and hence needs serious consideration for better valorization.

16.11 Conclusion

Taro is a vital tuber crop which is widely grown as a good starch source and promising carrier of vitamins and minerals. Irrespective of the fact that taro is of tremendous nutritional value, it is still ranked as a neglected crop in south Asia. In a backdrop of prevailing situation of hunger, food insecurity and triple burden of malnutrition, promoting dietary diversity is of immense significance to secure nutritional and health status of the people of the developing countries. In given circumstances, taro could play an important role to provide food and nutritional security under the changing climate. Taro is widely adapted for cultivation and it can be cultivated with minimum care under various agro-climatic conditions.

Conflicts of Interest All authors read and approved the final manuscript and declare that they have no conflict of interest.

References

Ara H (2016) Taxonomic studies in the Family Araceae from Bangladesh. Ph. D Dissertation, Department of Botany, University of Dhaka, Bangladesh

Atlin GN, Cairns JE, Das B (2017) Rapid breeding and varietal replacement are critical to adaptation of cropping systems in the developing world to climate change. Glob Food Sec 12:31–37. https://doi.org/10.1016/j.gfs.2017.01.008

BBS (2021) Yearbook of agricultural statistics of Bangladesh. Bangladesh Bureau of Statistics, Ministry of Planning, Dhaka

Bradbury JF (1983) Identification of cultivable bacteria from plants and plant tissue cultures by use of simple classical methods. Acta Hortic 225:27–37. https://doi.org/10.17660/ActaHortic.1988.225.2

Cartherwood DJ, Savage GP, Mason SM et al (2007) Oxalate content of cormels of Japanese taro (*Colocasia esculenta*) and the effect of cooking. J Food Compost Anal 20(3):147–151. https://doi.org/10.1016/j.jfca.2005.12.012

Chakraborty S, Newton AC (2011) Climate change, plant diseases and food security: an overview. Plant Pathol 60(1):2–14. https://doi.org/10.1111/j.1365-3059.2010.02411.x

Champagne A, Legendre L, Lebot V (2013) Biofortification of taro (*Colocasia esculenta*) through breeding for increased contents in carotenoids and anthocyanins. Euphytica 194:125–136. https://doi.org/10.1007/s10681-013-0980-5

Chandra A, McNamara KE, Dargusch P (2018) Climate-smart agriculture: perspectives and framings. Clim Policy 18(4):526–541. https://doi.org/10.1080/14693062.2017.1316968

Chukwu GO, Nwosu KI, Madu TU et al (2008) Development of gocing storage method for Cocoyam. https://mpra.ub.uni-muenchen.de/17444/ MPRA Paper No. 17444. Accessed 25 Sep 2009 12:27 UTC

Crimp S, Lisson S, Gleadow R et al (2017) Understanding the response of taro and cassava to climate change. Final report no. FR2019–72, ACIAR, Canberra ACT 2601, Australia

Fadina AMR, Barjolle D (2018) Farmers' adaptation strategies to climate change and their implications in the Zou Department of South Benin. Environments 5(1):15. https://doi.org/10.3390/environments5010015

Femina (2013) Advantage of taro and sweet potato (in Bahasa Indonesia). www.femina.co.id/diet/nutrisi/keunggulan.talas.dan.ubi.jalar/003/001/12

Fufa TW, Oselebe HO, Abetw WG et al (2021) Review of acridity in taro [*Colocasia esculenta* (L) Schott]: its health and economic impact, and possible reduction mechanisms. J Agric Sci 6(5):156–164
GoP (2020) Fruit, vegetables and condiments statistics of Pakistan 2018–19. Ministry of National Food Security and Research, Government of Pakistan, Islamabad. http://www.amis.pk/files/Fruit%20&%20Vegetable%20Condiments%20of%20Pakistan%202018-19.pdf
Huang AS, Titchenal CA, Meilleur BA (2000) Nutrient composition of Hawaiian taro corms and breadfruit. J Food Compost Anal 13(5):859–864. https://doi.org/10.1006/jfca.2000.0936
Jane J, Shen L, Chen J et al (1992) Physical and chemical studies of taro starches and flours. Cereal Chem 69(5):528–535
Kolchar K (2006) Economic botany in the tropics. Macmillan, Delhi
Li X, Yadav R, Siddique KHM (2020) Neglected and underutilized crop species: the key to improving dietary diversity and fighting hunger and malnutrition in Asia and the Pacific. Front Nutr 7:593711. https://doi.org/10.3389/fnut.2020.593711
Matthews PJ (1991) A possible tropical wild type taro: *Colocasia esculenta* var. *aquatilis*. IPPA Bullet 11:69–81
Mijinyawa AH, Durga G, Mishra A (2018) Isolation, characterization, and microwave assisted surface modification of *Colocasia esculenta* (L.) Schott mucilage by grafting polylactide. Int J Biol Macromol 119:1090–1097. https://doi.org/10.1016/j.ijbiomac.2018.08.045
Mwongera C, Nowak A, Notenbaert AMO et al (2019) Climate-smart agricultural value chains: risks and perspectives. In: Rosenstock TS, Nowak A, Girvetz E (eds) The climate-smart agriculture papers: investigating the business of a productive, resilient and low emission future. Springer, pp 235–245. https://doi.org/10.1007/978-3-319-92798-5_20
Niba LL (2003) Processing effects on susceptibility of starch to digestion in some dietary starch sources. Int J Food Sci Nutr 54:97–109. https://doi.org/10.1080/0963748031000042038
Onwueme IC (1978) In the tropical tuber crops, yam, cassava, sweet potato and cocoyam. Wiley, New York and Brisbane Toronto, pp 199–227
Pandey YR (1999) Ethnobotany and diversity of Taro in Nepal. Advances in Agricultural Research in Nepal. Proceedings of the first Society of Agricultural Scientists (SAS) Convention, Kathmandu, 29–31 March 1999
Patrick MM, Grace WN, Christine HSK (1999) Traditional food plants of Kenya. National Museum of Kenya, Nairobi, p 288
Paull RE, Tang CS, Gross K et al (1999) The nature of the taro acridity factor. Postharvest Biol Technol 16(1):71–78. https://doi.org/10.1016/S0925-5214(98)00099-4
Ramanatha RV, Matthews PJ, Eyzaguirre PB et al (2010) The global diversity of taro: Ethnobotany and conservation. Rome, Italy: BIJ 2010
Senyolo MP, Long TB, Blok V et al (2018) How the characteristics of innovations impact their adoption: an exploration of climate-smart agricultural innovations in South Africa. J Clean Prod 172:3825–3840. https://doi.org/10.1016/j.jclepro.2017.06.019
Siddique KHM, Li X, Gruber K (2021) Rediscovering Asia's forgotten crops to fight chronic and hidden hunger. Nat Plants 7:116–122
Soudy ID, Delatour P, Grancher D (2010) Effect of traditional soaking on the nutritional profile of taro flour (*Colocasia esculenta* L. Schott) produced in Chad. Rev Med Vet 1:37–42
Sukamoto LA (2003) Development of early matursing and leaf blight resistant cocoyam (*Colocasia esculenta* (L.) Schott) with improved taste. Paper presented at the final research coordination meeting organized by the joint FAO/IAEA division of nuclear techniques in food and agriculture, Pretoria, South Africa, p 19–23
Teron R (2019) Ethnobotanical study of dietary use and culinary knowledge of Aroids (family Araceae) in in Karbi Anglong district, Assam. NeBIO 10(2):80–84
Ubalua AO, Ewa F, Okeagu OD (2016) Potentials and challenges of sustainable taro (*Colocasia esculenta*) production in Nigeria. J Appl Biol Biotechnol 4(01):53–59. https://doi.org/10.7324/JABB.2016.40110
USDA (2021) USDA Food data central. https://fdc.nal.usda.gov/

Van Wyk BE (2005) Food plants of the world: identification, culinary uses and nutritional value. Briza Publications, Pretoria, p 2005
Wang JK (1983) Taro: a review of *Colocasia esculenta* and its potential. University of Hawaii, Honolulu
Wekesa BM, Ayuya OI, Lagat JK (2018) Effect of climate-smart agricultural practices on household food security in smallholder production system: micro-level evidence from Kenya. Agric Food Secur 7(1):80. https://doi.org/10.1186/s40066-018-0230-0
Yen DE, Wheeler JM (1968) Introduction of *C. esculenta* into the Pacific: the indications of the chromosome numbers. Ethnology 7:259–267. https://doi.org/10.2307/3772891

Chapter 17
Jackfruit (*Artocarpus heterophyllus*): An Overview of Nutritional and Functional Food Properties

Muhammad Ammar Khan, Md. Mokter Hossain, Zahida Qadeer, Saira Tanweer, Baila Ahmad, and Muhammad Waseem

17.1 Introduction: Jackfruit as the Fruit of Future

The scientific name of Jackfruit tree is *Artocarpus heterophyllus* Lam. It is generally cultivated in tropical areas (Saha et al. 2022). Jackfruit is the member of Moraceae family. It is a hardy crop (Singh and Baradevanal 2023; Rai et al. 2022) that has been grown in a wide range of agro-ecological zones (Nakintu et al. 2020; Weintraub et al. 2022). Jackfruit is a highly profitable tree, as it produces high fruit yield with low maintenance requirements (Das and Saha 2020). Its fruit has a distinct aroma and flavor. It is among one of the largest tree-borne fruits in the world, and is considered the fruit having the heaviest weight (Bensi et al. 2020).

Jackfruit is called a multiple fruit. Fruits may be classified as simple, aggregate, or multiple fruits. Simple fruits develop from one flower with a single or compound ovary and one pistil. It can be either fleshy or dry. Simple fruits can be classified into three types: berries, drupes, and pomes. A berry is a small fruit that is juicy, e.g., strawberries, raspberries, blueberries, and blackberries. A pepo is a type of berry with a hard rind, e.g., cucumbers, pumpkins, and watermelons. A drupe is a type of fleshy fruit containing a large seed, e.g., apricots, cherries, peaches, plums, and mangoes. Stone fruits are fruits that have a pits or stones in the center, e.g., peaches, plums, and cherries. Drupes and stone fruits are the same thing. A pome is a fruit that has a core with seeds surrounded by edible flesh, e.g., apples and pears. An aggregate fruit is formed from several ovaries of one flower that have merged

M. A. Khan (✉) · Z. Qadeer · S. Tanweer · B. Ahmad · M. Waseem
Department of Food Science & Technology, Faculty of Agriculture & Environment, The Islamia University of Bahawalpur, Bahawalpur, Pakistan
e-mail: ammar@iub.edu.pk

Md. M. Hossain
Department of Horticulture, Faculty of Agriculture, Bangladesh Agricultural University, Mymensingh, Bangladesh

T. Ismail et al. (eds.), *Neglected Plant Foods of South Asia*,
https://doi.org/10.1007/978-3-031-37077-9_17

together. On the other hand, multiple fruits are a result of the gynoecium of different flowers of a cluster of flowers (inflorescence), e.g., jackfruit, pineapple, fig, mulberry, etc. Multiple fruits like jackfruit and pineapple possess a rind that forms from their inflorescence. In the inflorescence of multiple fruits such as jackfruit, each flower produces a fruit that eventually matures and merges into one mass. Once the flowers have bloomed, the mass that forms is known as an infructescence.

Jackfruit has many nutrients (Bakewell-Stone 2023; Astuti et al. 2022; Islam et al. 2021a; Chhotaray and Priyadarshini 2022; Mijan et al. 2016), hence besides being a tasty fruit, it is also beneficial for health. Global food insecurity has hit millions of people, as almost 690 million people remained undernourished in 2019 as per a report published by the Food and Agriculture Organization (FAO) of the United Nations (UN) (Ayinde et al. 2020). The major contributors to global food insecurity were poverty, lack of nutritious food, climate change and natural disasters. Jackfruit can be eaten raw, as well as used in many food recipes including curries, stews, and soups (Ramli et al. 2017; Yudhistira 2022). Moreover, industrial products can be developed from by roasting, boiling, or grinding it into flour (Thatsanasuwan et al. 2023). It can also be used as a meat substitute, because of its unique meat-like texture and flavor (Keerthana Priya et al. 2022; Parihar et al. 2021). This makes it a valuable protein (Mijan et al. 2016) source of for those who cannot afford meat.

Beyond fruit production, jackfruit tree has multiple uses. The wood of jackfruit tree has been used for making furniture and building material (Gala et al. 2022). The bark and leaves of this important tree have also been traditionally used to treat many diseases (Jadhav et al. 2021; Dahanayake 2015). Although no religious importance of this tree is reported, a dye made from the wood chips of its tree is usually used to color the robes of Buddhist monks orange-red in the countries where jackfruit is cultivated in Buddhist communities (Prakash et al. 2009). Besides, timber purposes, jackfruit has environmental significance, as well. Jackfruit tree has been widely cultivated in agroforestry systems (Das and Saha 2020; Kader et al. 2022; Mondol 2021). Agroforestry systems integrate trees into crops and livestock systems. This not only leads to increased yields, but also increases monetary benefits for farmers and local communities. Agroforestry systems that incorporate jackfruit have been reported to improve soil quality, reduce soil erosion, and provide shade for other crops (Das and Saha 2020; Islam et al. 2021b; Riyadh et al. 2019).

Despite being grown in several countries of the world, jackfruit has been an underutilized fruit worldwide (Faujan et al. 2022). It is relatively unknown in several countries despite excellent potential for the development of nutritious products (Moreno Delgado 2021). Moreover, this fruit has great potential to contribute to the achievement of sustainable development goals (SDGs) of the UN. Recently, it has gained significant attention from the scientific community, as a crop and food source, mainly because of interest in its nutritional and medicinal properties (Khan et al. 2021), potential for sustainable agriculture, and economic development (Delfino et al. 2021). However, scientific community needs to know further about its origin, distribution, cultivation, nutrition, medicinal importance, culinary uses, value-addition, and economic importance.

17.2 Origin, Distribution, and Cultivation of Jackfruit

17.2.1 Origin and Distribution

The Jackfruit tree has been grown for almost thousands of years in Southeast Asia, where it is considered an important crop with cultural and religious significance (Rana et al. 2018). The Jackfruit tree is generally thought to be native to the Western Ghats in India. It has been mentioned in Hindu religious texts such as the Mahabharata and the Ramayana (Singh 2022), where it is described as a food fit for royalty and a symbol of abundance (Ravindran et al. 2007). Since it has high potential for adaptability to a variety of different farming and ecological environments (Das and Saha 2020), it is found in many tropical areas of many other parts of the world (Zhang et al. 2021). During the colonial period, jackfruit was introduced to many tropical areas of the world, including Caribbean, Brazil, and Africa. In the Caribbean, it was introduced by Indian indentured laborers, and it soon became a staple food in the region (Houston 2005). In Brazil, it was introduced during the nineteenth century, and at present it is cultivated in many parts of the country (Raíces et al. 2017). The economic and cultural significance of jackfruit in Southeast Asia and India has also led to the development of many traditional dishes and culinary practices. In Indonesia, for example, it is used in a variety of dishes, including gudeg, a sweet and spicy stew made with young jackfruit, coconut milk, and spices (Yudhistira 2022). In India, it has been utilized in a many recipes, including curries, stews, and desserts (Chhotaray and Priyadarshini 2022). In addition to its culinary and cultural significance, it has also been used to treat various diseases by traditional medicinal practitioners, usually known as Ayurveds, for thousands of years. The worldwide cultivation of jackfruit is more than 375,000 hectares (Fig. 17.1). The total area under jackfruit cultivation in India was 185,000 hectares in 2017–2018, Vietnam 72,200 hectares in 2021, Sri Lanka 44,730 hectares in 2018, Bangladesh 27,988

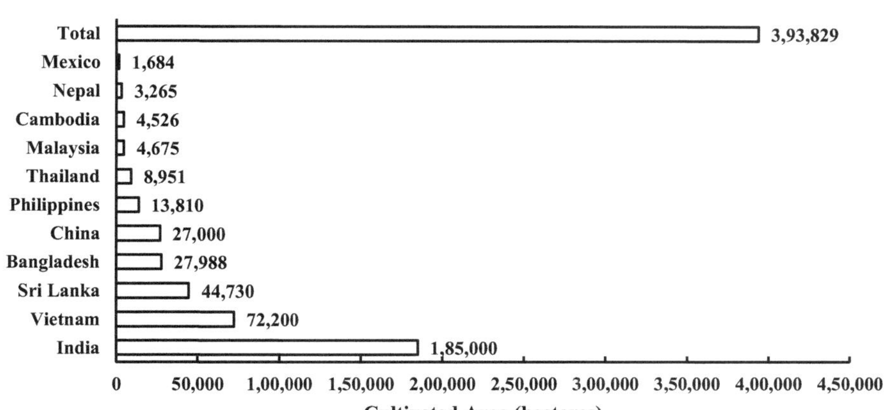

Fig. 17.1 Cultivated area under jackfruit in major jackfruit producing countries of the world

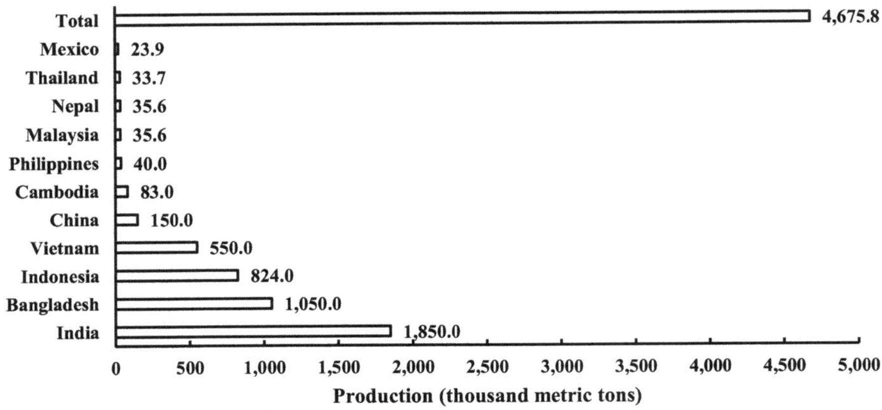

Fig. 17.2 Production of jackfruit in major jackfruit producing countries of the world

hectares in 2021–2022, China 27,000 hectares in 2019, Philippines 13,810 hectares in 2021, Thailand 8951 hectares, Malaysia 4675 hectares in 2020, Cambodia 4526 hectares in 2020, Nepal 3265 hectares in 2019–2020, and Mexico 1684 hectares in 2020 (Chauvet and Mitra 2023; Statistics BBO 2022).

Total worldwide jackfruit production is more than 4700 thousand tons (Fig. 17.2). The jackfruit production in India was 1850 thousand metric tons (MT) during 2018–2019 (Mithun and Kaleemullah 2019), Bangladesh 1050 thousand MT in 2021–2022 (Statistics BBO 2022), Indonesia 824 thousand MT in 2020, Vietnam 550 thousand MT in 2020, China 150 thousand MT, Cambodia 83 thousand MT, Philippines 40 thousand MT, Malaysia 35.6 thousand MT in 2020, Nepal 35.6 thousand MT, Thailand 33.7 thousand MT, and Mexico 23.9 thousand MT (Chauvet and Mitra 2023).

17.2.2 Cultivation

Jackfruit plant can be propagated by seeds (Witherup et al. 2019) as well as by vegetative means (Goswami et al. 2022), including grafting or budding. The trees developed from grafts start yielding from 5th year, while those developed from seedlings give fruit from 8th year (Ahmed et al. 2019). Jackfruit seedlings should be transplanted at a distance of 25 ft × 25 square feet if 69 trees are to be transplanted in an acre, or 30 × 30 square feet if 48 plants are to be transplanted in an acre to allow for proper growth and development, considering large tree size. The description from plant seedling, nursery, young plant, young tree, and fully-grown tree is given in Fig. 17.3.

Jackfruit tree requires a warm, humid climate with plenty of rainfall for optimum tree growth (Pathak et al. 2022). The optimum temperature required for the cultivation of jackfruit trees ranges from 20 to 35 °C, along with an annual rainfall is

Fig. 17.3 Life cycle of jackfruit from seedling, nursery, young plant, young tree, to fully-grown tree

1000–2400 mm (Haq 2006). The basic needs for efficient growth of jackfruit plant include a well-drained, fertile soil along with regular watering during the growing season (Jadhav et al. 2021). Moreover, fertilizer application is essential for optimum yield depending on soil analysis, as well as depending on the nutritional requirements of the plant (Briones and Lina 2020). Regular soil testing can help to identify nutrient deficiencies and guide fertilizer application. Irrigation should be done as per the requirement of the plant and the climatic conditions, avoiding water logging and over-irrigation. Mulching around the base of the plant can help to retain moisture, suppress weed growth, and improve soil fertility (Ghosh and Biswajit 2006).

Pests and diseases significantly reduce the productivity of agricultural crops. Jackfruit tree requires regular pest and disease management to prevent damage and ensure healthy growth (Kallekkattil et al. 2020). Jackfruit tree and leaves are susceptible to various pests and diseases, including fruit fly (Oraño et al. 2020), mealybugs (Kallekkattil and Mani 2022), anthracnose (Borges et al. 2022), and leaf spot (Habib et al. 2022). *Diaphania caesalis* is a pest that damages the shoots of jackfruit (Butani 1978). There are also several types of pests that bore into the bark of jackfruit trees, including *Indarbela quadrinotata*, *I. tetraonis*, *Batocera rufomaculata*, *Batocera rubus*, *Epepeotes luscus*, *Sthenias grisator*, and *Apriona germari*.

Ochyromera artocarpi is a type of pest that damages the buds of jackfruit and is a significant threat. *Cosmoscarta relata* is a type of spittle bug that can cause serious damage to jackfruit trees (Butani 1978). *Ferrisia virgata*, also known as the white-tailed mealybug, can be found in jackfruit trees as well (Butani 1978). Finally, there are two types of pests, *Perina nuda* and *Diaphania bivitralis*, that weave webs on the leaves of jackfruit trees, but they are not considered a major threats (Butani 1978). Regular monitoring of the plant for pest and disease incidence and timely action can help to prevent crop loss and ensure good yield. The use of Integrated Pest Management (IPM) practices may reduce the need for synthetic pesticides and promote sustainable production (Tait et al. 2021). In addition, these days, the trend for biopesticides is also growing (Srinivasan et al. 2019). Organic farming practices can be adopted for its cultivation to produce chemical-free and healthy fruit (Briones and Lina 2020; Su et al. 2021). Pruning of the plant should be done to remove diseased, damaged or unproductive branches, and to maintain the desired shape and size of the tree (Ahasan Ullah Khan et al. 2021). In addition, the use of genetically improved varieties that possess the potential to yield higher and exhibit higher resistance against pests and diseases can also increase it's the crop productivity of jackfruit (Ibrahim et al. 2020). Besides edible products for human, jackfruit leaves can also be used as a fodder for livestock, providing an additional source of income for farmers. Farmers have been provided with technical training, financial support, and market linkages to promote jackfruit cultivation and value addition (Mondol 2021). This has led to increased farmer incomes and improved food security in rural areas (Savita et al. 2017). In addition to its economic benefits, it has also been used for social development in some regions (Shams et al. 2015).

17.3 Nutritional Composition of Jackfruit

17.3.1 Nutritional Profile of Fruit

Although, jackfruit is a minor fruit, yet it has gained popularity recently in the scientific community mainly because of its nutritional composition that is useful for consumer health. The fruit contains all the major food components in abundant quantities (Fig. 17.4).

Jackfruit consumption has been associated with various health benefits (Kaur et al. 2023). One of the main nutritional benefits of jackfruit is its high fiber content. According to an earlier report, a 100 g serving of jackfruit contains about 1.6 g of fiber (Mijan et al. 2016), while RDI of total fiber has been 14 g per 1000 kcal per day for healthy adults (Trumbo et al. 2002). Jackfruit has low sucrose contents. The major carbohydrates found in jackfruit are shown in Fig. 17.5.

Jackfruit contains several minerals, which are essential for maintaining optimum growth and metabolism of humans. Although, the mineral contents of various varieties of jackfruit vary significantly, but overall jackfruit may be considered an

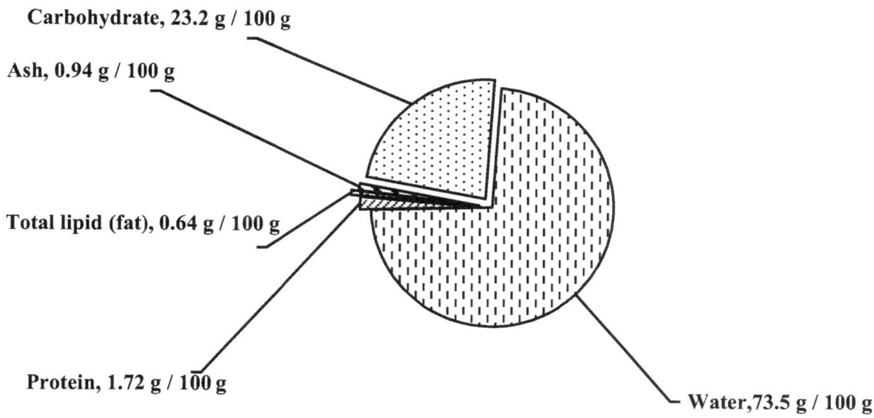

Fig. 17.4 Major food components of raw jackfruit. (Source: United States Department of Agriculture (2019))

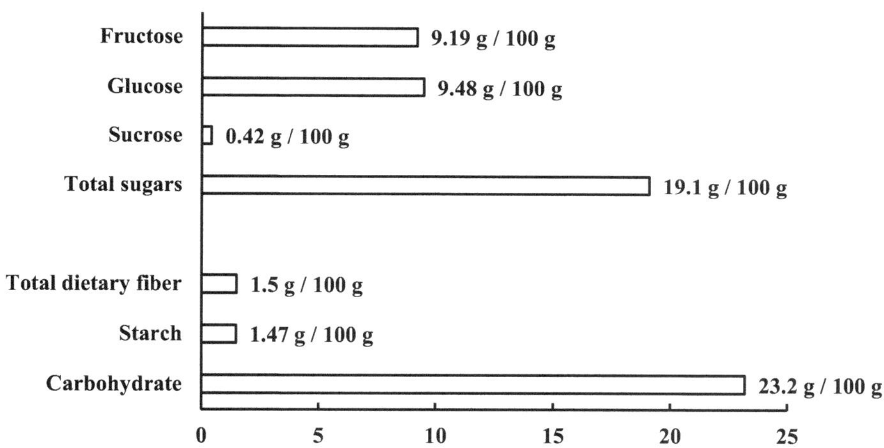

Fig. 17.5 Carbohydrates contents of raw jackfruit. (Source: United States Department of Agriculture (2019))

important source of potassium. One study reported 303 mg of potassium per 100 of fruit (Islam et al. 2021a). The United States Department of Agriculture (USDA) (2019) suggested even higher amount of potassium in raw jackfruit. Besides, the fruit also possess high amounts of magnesium, calcium, and phosphorus (Bakewell-Stone 2023; Islam et al. 2021a). This makes it a valuable source of nutrition for people who may not have access to other nutritious foods. The minerals profile according to USDA (Statistics A 2019) in raw jackfruit is shown hereunder (Fig. 17.6).

Vitamins are important food components that play role in maintaining human health. These are essential nutrients, which are usually divided into two

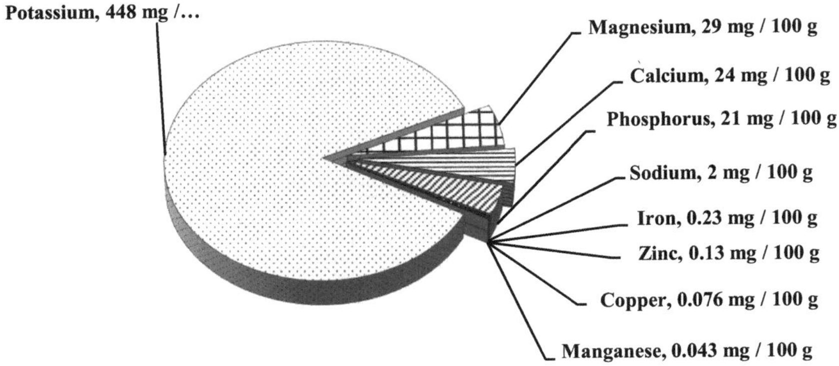

Fig. 17.6 Major minerals found in raw jackfruit. (Source: United States Department of Agriculture (2019))

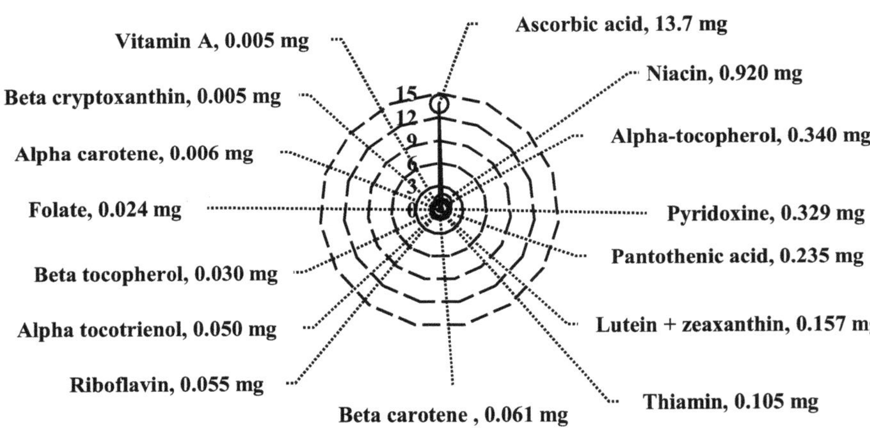

Fig. 17.7 Major vitamins found in raw jackfruit. Values are shown in milligrams per 100 g. (Source: United States Department of Agriculture (USDA) (2019))

categories: fat-soluble and water-soluble. Vitamins A, D, E and K belong the former category, whereas B-complex (niacin, thiamin, pyridoxine, folic acid, riboflavin, etc,) and vitamin C belong to the latter category. These play very essential role in maintaining human health and development. Jackfruit contains source of vitamins of the both categories, hence a very important source of nutrients. Some of the important vitamins as described by USDA (Statistics A 2019) in raw jackfruit are shown below (Fig. 17.7).

Fats and oils play important role in regulating human health. Fatty acids are building blocks of fats. Their two important types include saturated (SFA) and unsaturated fatty acids. The most important unsaturated fatty acids include

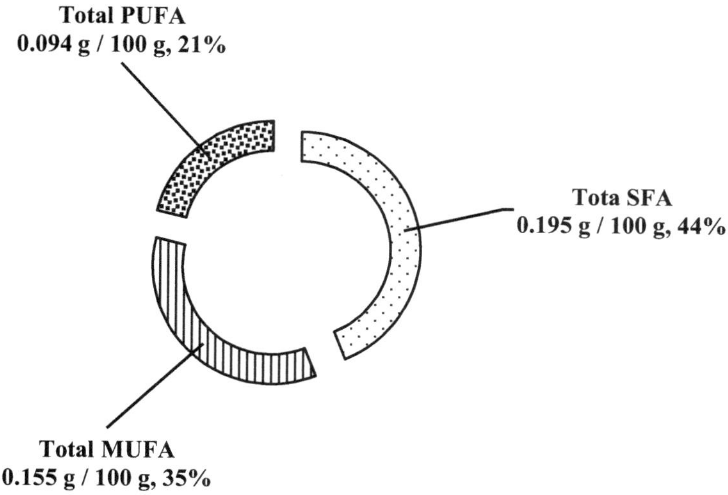

Fig. 17.8 Fatty acids composition of raw jackfruit. SFA stands for saturated fatty acids, MUFA for monounsaturated fatty acids, whereas PUFA for polyunsaturated fatty acids. Percentages shown describe percentage of total fatty acids. (Source: United States Department of Agriculture (2019))

monounsaturated (MOFA), and polyunsaturated fatty acids (PUFA). Jackfruit is low in fats (0.03 g/100 g of fruit) and calories (94 calories/100 g of fruit) (Abdullah 2017). The total SFA, MOFA, and PUFA contents of raw jackfruit according to USDA data (Statistics A 2019) are presented in Fig. 17.8.

Among fatty acids, PUFA are considered the most health friendly. Particularly, omega 3 (n-3) fatty acids are related with retina and brain development, hence important for vision and cognitive abilities. The detailed composition of fatty acids found in jackfruit is given in Fig. 17.9.

Jackfruit provides considerable amounts of plant-based proteins (Mijan et al. 2016), which make it a suitable option for vegetarians and vegans. Van et al. 2021 reported that a 100 g serving of jackfruit contains approximately 1.7 g of protein. The building blocks of proteins are amino acids, which are either essential or non-essential. Essential amino acids need to be consumed by dietary sources, because our bodies cannot synthesize them by their own. Protein percentage of jackfruit is about 1.7%, however, these proteins contain almost all essential amino acids, with the highest contents of leucine, followed by threonine and valine. The contents of essential amino acids found in raw jackfruit are shown below (Fig. 17.10).

In addition, the highest content of the non-essential amino acids found in jackfruit is for glutamic acid, followed by aspartic acid, and arginine, respectively. Besides, it contains almost all the non-essential amino acids, as well, which are presented below (Fig. 17.11).

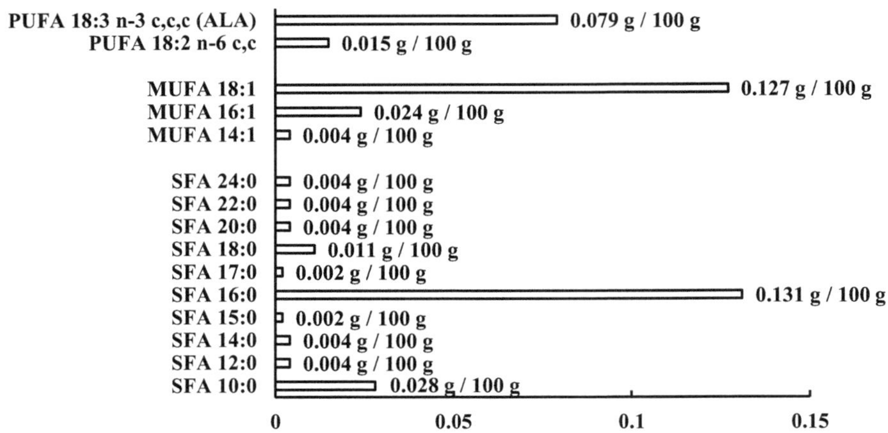

Fig. 17.9 The contents of saturated, monounsaturated and polyunsaturated fatty acids in raw jackfruit. SFA stands for saturated fatty acids, MUFA for monounsaturated fatty acids, whereas PUFA for polyunsaturated fatty acids. (Source: United States Department of Agriculture (2019))

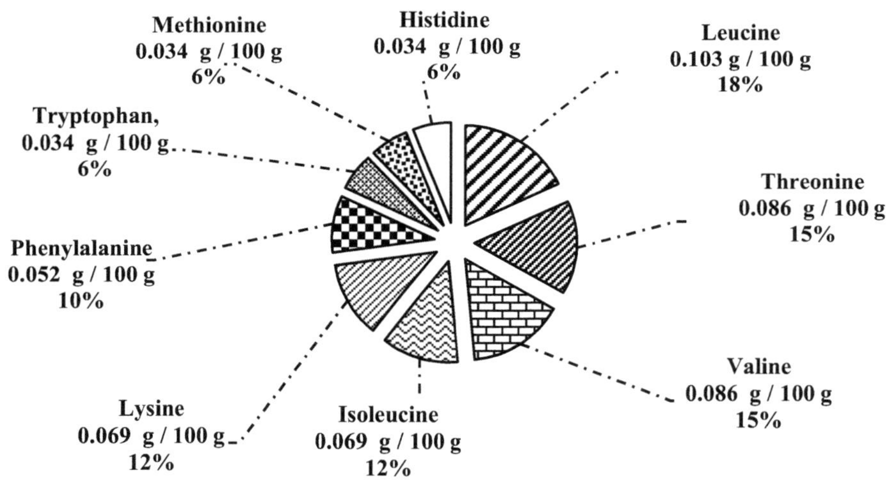

Fig. 17.10 Essential amino acid profile of raw jackfruit. (Source: United States Department of Agriculture (2019))

17.3.2 Nutritional Profile of Seed

Seed is present inside the flesh of jackfruit (Fig. 17.12). Jackfruit seed is also edible. It can be roasted as a snack, as well as it can be dried and converted into powder. The jackfruit seed powder can be processed as a supplement into a variety of products. Jackfruit seeds also provide considerable amounts of carbohydrates (Gupta et al. 2011), proteins (Gupta et al. 2011), starch (Shedge et al. 2022), fats (Fernandes

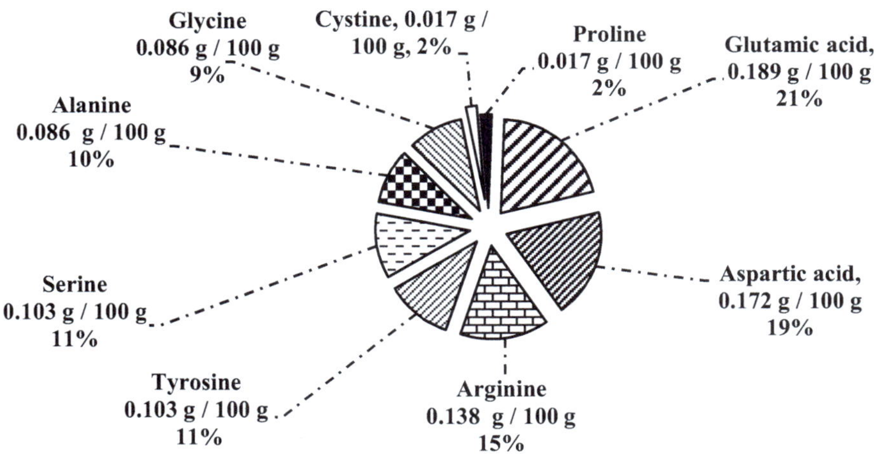

Fig. 17.11 Non-essential amino acid profile of raw jackfruit. (Source United States Department of Agriculture (2019))

Fig. 17.12 Unpeeled and peeled jackfruit seeds

et al. 2017), dietary fibers (Shedge et al. 2022), minerals (Akter and Haque 2018), vitamins, and various bioactive compounds. The carbohydrate content of jackfruit seeds is quite high, but usually ranges from 20% to 25%, however, Habibah et al. (2021) reported as high as 56.1 g/100 g. The seeds also contain significant amounts of dietary fibers; Shedge et al. (2022) reported about 3.19%, while Islam et al. (2015) reported about 2.49%. The protein content of jackfruit seeds is also high. Kushwaha et al. (2019) reported protein percent ranging from 13.3% to 16.6%,

Ranasinghe et al. (2019) reported 5.3% to 6.8%, Wu et al. (2022) reported up to 20%, while Palamthodi et al. (2021) reported up to 13.9%; these differences could be attributed to varietal differences or diverse agroclimatic conditions. The jackfruit seed proteins also contain significant amounts of essential amino acids (Wu et al. 2022) including leucine (Kumoro et al. 2020), alanine (Kumoro et al. 2020), threonine (Kumoro et al. 2020), and methionine (Kumoro et al. 2020), which are beneficial for maintaining human health. The seed may be converted into flour, which still contains high amounts of amino acid such as aspartic acid, serine, glutamic acid, glycine, histidine, arginine, threonine, alanine, proline, tyrosine, valine, lysine, isoleucine, leucine and phenylalanine upon thermal processing (Zuwariah et al. 2018). The fat content of jackfruit seeds ranges from 0.41% (Ndikuryayo et al. 2021). The seeds contain both saturated and unsaturated fatty acids, with linoleic acid and oleic acid being the major fatty acids. The major fatty acids in jackfruit seeds are linolenic and linoleic acids (Kumoro et al. 2020). Its seeds also contain 413.76 mg/100 g γ-aminobutyric acid (Shi et al. 2018). It contains 58.9% saturated fatty acids, 6.9% mono-unsaturated fatty acids, and 34.2% PUFA, which mainly included linoleic acid (C18:2) (Fernandes et al. 2017). The other important fatty acids include palmitic acid (C16:0) at 35.6%, stearic acid (C18:0) at 7.7%, behenic acid (C22:0) at 4.8%, oleic acid (C18:1) at 4.7%, arachidic acid (C20:0) at 3.2%, and lignoceric acid (C24:0) at 2.8%, each of heneicosanoic acid (C21:0), margaric acid (C17:0), and pentadecanoic acid (C15:0) at 1.0%, each of myristic acid (C14:0), palmitoleic acid (C16:1), and tricosanoic acid (C23:0) at 0.9%, margaroleic acid (C17:1) at 0.4%, as well as cis-10-Pentadecenoic (C15:1n-5c) at 0.1% (Fernandes et al. 2017).

In addition, its seeds contain considerable amounts of vitamins such as vitamin B (Prasad and Gupta 2018), vitamin A (Islam et al. 2015) and vitamin C (Islam et al. 2015). The important B vitamins in jackfruit seeds include thiamine, riboflavin, niacin, etc. There are many important minerals in jackfruit seeds. Akter and Haque (2018) reported calcium (308.7 mg), potassium (147.8 mg), iron (13.0 mg), sodium (6.066 mg), and copper (1.045 mg) per 100 g of jackfruit seed. The important bioactive compounds of seeds include flavonoids (Manurung et al. 2021), alkaloids (Manurung et al. 2021), terpenoids (Manurung et al. 2021), steroids (Manurung et al. 2021), saponins (Shedge et al. 2022), lignans (Shedge et al. 2022), lectins (Bakri et al. 2021), jacalin (Bakri et al. 2021), and isoflavones (Shedge et al. 2022). Its seed even contains more calcium than its pulp (Amadi et al. 2018).

17.4 Nutritional and Health Benefits

Almost all parts of jackfruit tree have been utilized for medicinal purposes. Ayurveda is the traditional Indian system of medicine, in which jackfruit is used to treat a variety of several health problems. Recently, several research articles have also demonstrated the medicinal properties of jackfruit, its seeds, leaves, bark, and roots. The most important treated diseases include asthma (Prakash et al. 2009), fever (Shedge et al. 2022), diarrhea (Shedge et al. 2022), and skin diseases (Prakash et al.

2009). Particularly, the fruit is believed to have anti-cancer (Saha et al. 2022; Khan et al. 2021; Vazhacharickal et al. 2015), anti-ulcer (Saha et al. 2022; Khan et al. 2021), anti-hypertensive (Saha et al. 2022), anti-aging (Saha et al. 2022), and anti-inflammatory properties (Vazhacharickal et al. 2015).

17.4.1 Health Benefits of Fruit

Fruit of the jackfruit tree, being delicious, is well liked among the local populations where it is being grown and marketed. It has been used in traditional ayurvedic medicine in Asian countries for centuries due to numerous health benefits. The nutritional benefits of the fruit are associated with the nutrients of which the fruit is composed of. Particularly, fiber content is important that improves the digestive health. In addition, the dietary fibers found in jackfruit also regulates blood sugar levels and reducing the risk of chronic diseases. Ghavami et al. (2023) suggested that the soluble fiber in jackfruit demonstrated cholesterol-lowering effects. Parihar et al. (2021) reported that the high fiber content of jackfruit improved the digestive health and reduced the risk of colon cancer. Shedge et al. (2022) found that eating jackfruit promoted regular bowel movements and prevented constipation mainly because of its high fiber content. Obesity is leading health issue globally, that is considered mother of many diseases. The low fats (Fernandes et al. 2017) and calories contents of jackfruit make it beneficial for people who are at risk of obesity or other diet-related diseases (Abdullah 2017). Moreover, healthy immune system is important for combating the effects of incoming pathogens and toxicants. Vitamins and minerals are also important for health. Particularly, vitamin C demonstrates wound healing properties.

Jackfruit contains significant amounts of vitamin C, vitamin A, and potassium which may contribute to maintenance of a healthy immune function, vision, and blood pressure regulation, respectively. The jackfruit is also rich in a number of bioactive components and phytochemicals. The medicinal properties of jackfruit are associated with the bioactive phytochemical constituents it contains. Cruz-Casillas et al. (2021) reported that the consumption of jackfruit demonstrated anti-inflammatory properties, which can helpful in reduction of the risk of many chronic diseases such as cardiovascular disease, diabetes, and cancer. Free radicals are leading cause of various chronic diseases such as cancer, heart disease, and Alzheimer's disease (Kiziltaş 2022). Jackfruit is a rich in antioxidants that help protect the body from the free radicals (Zheng et al. 2022). Moreover, Viresh et al. (2022) demonstrated that the antioxidants of jackfruit protected against cellular damage caused by oxidative stress. Chavez-Santiago et al. (2021) reported many important phenolic acids in jackfruit which included gallic acid, ferulic acid and tannic acid and flavonoids. Flavonoids are major bioactive components in jackfruit that possesses various pharmacological properties. Al-Khayri et al. (2022) demonstrated the antioxidant, anti-inflammatory, and anti-cancer properties of flavonoids. Moreover, flavonoids may exhibit significant anti-diabetic activity by lowering down the levels of blood glucose, by increasing insulin sensitivity, by exhibiting anti-inflammatory properties. Saponins are also another class of bioactive components found in jackfruit. They have also been found to exhibit various pharmacological properties such as antibacterial (Manivannan and Johnson 2020), anti-inflammatory (Dong et al. 2019), anti-cancer (Majnooni et al. 2023), and anti-diabetic (Uzayisenga et al. 2014) activities. Gupta et al. (2011) reported that jackfruit contains saponins. Shedge et al. (2022) demonstrated that saponin-rich extract exhibited significant anti-cancer activity. Lignans are also bioactive compounds that possess antioxidant (Jang et al. 2022), anti-cancer (Jang et al. 2022), anti-inflammatory (Jang et al. 2022), and anti-menopause (Jang et al. 2022) properties. Lectins have been reported to exhibit as anti-tumor (Wu and Bao 2013), anti-viral (Wu and Bao 2013), immunomodulatory (Wu and Bao 2013), anti-pathogenic properties (Lam and Ng 2011). Additionally, Pablo and Cimafranca (Pablo and Cimafranca 2022) documented the prebiotic properties of jackfruit and suggested beneficial role in supporting gut bacteria that may cause positive effects on immune function and overall health (Abasubong et al. 2022).

17.4.2 Health Benefits of Seeds

Jackfruit seeds are also edible. These have been used in traditional medicines. It contains a variety of phytochemicals, including lectins (Hashim et al. 1991), lignans (Shedge et al. 2022), and saponins (Gupta et al. 2011), which contribute to its medicinal properties. It can be converted into powder as well. Zuraidah et al. (2017) documented the presence of jacalin in its seed powder. Maurya (2016) suggested the ability of jackfruit seeds to treat digestive problems such as constipation. In addition, obesity (Goswami et al. 2021) reported the effects of its seed powder to mitigate obesity that may help reduce body weight and fat mass. Gunawan and Basoeki (2022) developed a drink from jackfruit seed powder mixed with watermelon rind was found to be useful for lactose intolerant people. Moreover, the seeds are also possess anti-diabetic properties (Ngwere and Mongi 2021) and may help regulate blood sugar levels (Li et al. 2019). Goswami et al. (2021) also reported antidiabetic properties of jackfruit seed powder mainly because it attenuated hyperphagia and hyperglycemia. Wang and Jiang (2022) reported the potential of jackfruit seed extract as a natural treatment for diabetes, as it showed increased glucose tolerance and insulin sensitivity.

Jackfruit seed extract has antioxidant and anti-inflammatory properties (Pu et al. 2022), which may be beneficial for the prevention of chronic diseases. Jackfruit seed extract can also be used for medicinal purposes. Its seed extract exhibits antioxidant properties, mainly by scavenging free radicals (Shanooba et al. 2020), antiviral, cytotoxic (Suzihaque et al. 2022), as well as antibacterial and antifungal activities against various pathogenic microorganisms (Suzihaque et al. 2022). Shanooba et al. (2020) reported flavonoids in its seed extract. Zuraidah et al. (2017) described that jacalin in jackfruit seed powder exhibited anti-cancer properties by suppressing the growth and spread of cancer cells. Furthermore, Mandhare et al. (2020), reported anti-arthritic properties jackfruit seed powder that helped reduce inflammation and pain associated with arthritis. Phukan et al. (2018) demonstrated hepatoprotective properties of the aqueous extract of seeds that might help protect against liver damage (Isinenyi and Andrew 2021). reported nephroprotective properties of boiled jackfruit seeds in rats, which protected against kidney damage. Sowmyashree and Devaraja (2022) noted that artin M, a d-mannose binding lectin, found in jackfruit seeds boosted the immune system by instigating the creation of proinflammatory cytokines. Besides many beneficial properties, Mahanta and Kalita (2015), however, demonstrated negative effects of lectins (named jacalin) on digestive system that suggests that further research is needed to determine the optimum dietary and pharmaceutical doses of all these chemicals.

17.4.3 Health Benefits of Other Parts

Bark, leaves, and roots of jackfruit tree have also been found to have potential medicinal uses (Ravindran et al. 2020). Jackfruit leaves have widely been utilized in traditional medicine in Asian countries to treat skin diseases such as eczema (Ningsih et al. 2018), leukoderma (Ahmed et al. 2019), and psoriasis (Ahmed et al. 2019). Moreover, jackfruit leaves are also believed to have antifungal (Vázquez-González et al. 2020) and antibacterial (Sousa et al. 2021) properties. In addition, jackfruit leaves are also rich in phytochemicals such as flavonoids (Rahma et al. 2020), saponins (Rahma et al. 2020), and tannins (Rahma et al. 2020), which contribute to its medicinal properties. Furthermore, Jackfruit leaves has been used to treat ulcer (Khan et al. 2021), and also found to possess anticonvulsant properties, which may help reduce the frequency and severity of seizures (Somashekhar et al. 2013). Particularly, Siregar, et al. (Siregar et al. 2018) reported antibacterial effect of jackfruit bark extract against *Propionibacterium acnes* and *Staphylococcus aureus*. Moreover, it has been used to treat skin diseases (Kemalasari et al. 2018). Furthermore, jackfruit bark has also been thought to possess anti-inflammatory properties (Ilmi et al. 2020) that may be helpful in reducing swelling. Jackfruit root has also been used to treat skin diseases, fever, diarrhea, and asthma (Maheswari and Valsan 2019). Finally, jackfruit peel has also found to have medicinal properties (Cruz-Casillas et al. 2021) reported that jackfruit peel extract demonstrated antidiabetic properties by regulating blood sugar levels.

17.4.4 Case Study of Diabetes

About 5% world population is victim of diabetes. Type 1 and Type 2 diabetes are the two important types of diabetes. Normally, the pancreas of human body produces a hormone named insulin which regulates the movement of sugar into cells. In Type

1 diabetes, a chronic disease, the ability of pancreas to produce insulin by itself is hindered or restricted. It is a genetic disorder, in which the cells in the pancreas that produce insulin get attacked by the immune system of body. Type 1 diabetes usually shows in early part of life; hence it is also known as juvenile diabetes or insulin-dependent diabetes. Type 2 diabetes is more prevalent in humans compared to Type 1 diabetes. On the other hand, Type 2 diabetes is a condition in which either the human body gets non-responsive to insulin or becomes unable to produce adequate amounts of insulin, hence cells do not get enough amount of sugar. Also, in this disease, cells may also respond poorly to insulin, hence do not get enough amount of sugar. Consequently, sugar level in bloodstream rises higher than normal. This condition is called hyperglycaemia. Hyperglycaemia is a condition in which the level of glucose rises high in the bloodstream. This condition is often an indicator of diabetes mellitus. In contrast, hypoglycaemia is a condition in which the level of glucose falls too low in the bloodstream. This condition arrives in diabetes mellitus patients when diabetes mellitus patients take insulin higher than required dose. Blood glucose is normally checked after overnight fasting. The normal blood glucose level is 99 mg/dL after overnight fasting. The people who have 100 to 125 mg/dL of blood glucose at fasting are considered prediabetics. The people who have 126 mg/dL or higher of blood glucose at fasting are considered diabetics. Interestingly, α-glucosidase is an enzyme that instigate carbohydrates digestion, hence increase the postprandial glucose level in diabetic patients. In the above study, Natta et al. (2023) demonstrated that jackfruit accessions with deep-yellow flakes inhibited α-glucosidase enzyme activity, hence it could be helpful in preventing the increase of postprandial glucose level in diabetic patients. Similarly, α-amylase is another enzyme that increases carbohydrates digestion. Moreover, β-glucosidase is an enzyme that converts cellobiose to glucose. Natta et al. (2023) also revealed that jackfruit accessions with orange-red flakes inhibited α-amylase and β-glucosidase activities, hence it could also be helpful in preventing the increase of postprandial glucose level in diabetic patients. Simple phenolic acids increase cellular uptake of glucose. Natta et al. (2023) found that jackfruit flakes had high levels of phenolic acids such as sinapic acid, ferulic acid, and quercetin, which were useful in management of hyperglycemic condition.

17.5 Culinary Uses of Jackfruit in Different Cultures

Jackfruit has been an underutilized fruit worldwide (Faujan et al. 2022). However, its culinary uses of jackfruit are not limited to household dishes, such as curry, biryani, smoothies, salads, chapatis and roti.

17.5.1 Jackfruit Curry

Curry is a regular Asian-style dish. It is usually prepared by frying the shredded onion, garlic, and ginger in ghee or oil, followed by partial cooking of the food stuff including meat, potatoes, or vegetables in this mixture. Subsequently, spices are added to adjust the taste. Afterwards, amount of water is adjusted in meat curries, and allowed the meat to be cooked on mild heat until meat becomes tender. Jackfruit can be used in Asian-style curries as a substitute for meat or poultry. This is a savory dish made with jackfruit as the main ingredient. Jackfruit curry is a dish made with ripe or unripe jackfruit cooked in a curry sauce. It has been a famous dish in Asian cuisine. It also has a unique fruity flavor that can add an interesting dimension to the overall taste of the curry. Jackfruit curry is a delicious and unique dish that offers a different flavor experience than regular curry. Lakshmana et al. (2013) developed a ready-to-eat (RTE) jackfruit curry from tender jackfruit (harvested at early stages of development of fruit) by cooking in a steam air retort at 15 PSI of pressure for 45 min, stored at it at 27–30 °C, and found reduction in its hardness from 39 N to 1 N after retort processing. The jackfruit curry demonstrated exceptional stability and acceptability, maintaining its optimal texture and sensory characteristics for up to 12 months under ambient conditions. The peroxide values and free fatty acids contents of the jackfruit-based RTE curry were found to exhibit non-significant changes (Lakshmana et al. 2013).

17.5.2 Jackfruit Biryani

Biryani is a famous rice-based dish of Indian subcontinent origin. Typically, boiled rice layered with well cooked meat prepared in spices. The dish is typically layered, with the rice and meat or vegetables cooked separately before being mixed together with spices and garnished with fried onions, nuts, and raisins. The exact ingredients and cooking method can vary depending on the region and cultural influences, but *biryani* is generally known for its fragrant aroma and rich, spicy flavor. Since tender jackfruit has meat-like texture and flavor, it can be used as meat replacement in *biryani* (Chougale et al. 2021). Jackfruit *biryani* can serve as a vegetarian version of the popular *biryani* dish.

17.5.3 Jackfruit Smoothies

Smoothies are thick, creamy and nutritious drinks made by blending fruits (Artés-Hernández et al. 2021), vegetables (Artés-Hernández et al. 2021), herbs (Victor-Aduloju et al. 2020), leaves (Victor-Aduloju et al. 2020), flowers (Gil et al. 2022), into some liquids such as milk, whey, etc. The resultant beverage contains nutritious contents of all the ingredients which they contain. Amador-Espejo et al. (2020) prepared delicious and nutritious jackfruit smoothies by blending with other fruits and vegetables, and subjected it to thermo-sonication. Thermo-sonication was found helpful in increasing the jackfruit shelf-life, mainly by reduction of microbial load and inactivation of enzymes. The process conditions reported for optimal shelf life were at an amplitude of 77.5% processed for 20 min at a temperature of 47.5 °C (Amador-Espejo et al. 2020).

17.5.4 Jackfruit Salad

Salad is a traditional side dish which is used with regular dishes during meals. It is usually prepared from sliced vegetables, fruits, legumes, etc. Sometimes, salads may also include seafoods and meat chunks. Nuts, seeds, herbs, and meat analogues are also used. Russian salad, kachumber salad, and fresh salad are famous types eaten in Asian countries. It is considered an important weight and hunger control tool for many health-conscious people. Moreover, the nutrients found in the ingredients that are used to prepare salads also contribute to nutritional profile of salads. Some obese people also skip a meal with salads, as it serves as a low-calorie dish. There are countless variations of salads, subject to ingredients, dressings, and flavors. Jackfruit salad is also a nutritious option for Asian people, particularly it is used in Thai salad called *Tam Kanun* (Islam et al. 2015).

17.5.5 Jackfruit Chapati

Chapati is the most important food of several countries of Asian subcontinent. It is a type of unleavened flatbread. It is made from a dough consisting of wheat flour, salt, and water. The dough is kneaded and then rolled out into thin, circular discs. The dough discs are cooked on a hot plate or tawa. It is eaten curries. *Roti* is also a flat bread famous in Asian culture, however, it is cooked in an open flame oven, usually called *Tandoor*. Since *roti* keeps attached to the wall of cooking medium on one side, so it is not turned during cooking, however, *chapati* is turned few times to ensure equal cooking on both sides. Inclusion of jackfruit in *chapati* or *roti* will increase the fiber content of these products. Sultana et al. (2014) reported that 5% white jackfruit seed flour (JSF) and 10% brown JSF along with 5% Bengal gram flour exhibited greater sensory acceptance than control. Moreover, the value-addition by using JSF resulted in higher nutraceutical value of *chapatis*.

17.6 Processing, Value-Addition and Commercial Significance

Primary processing represents to steps taken to prevent the post-harvest losses in the fruits, whereas secondary processing refers to value-addition that represents product development from jackfruit and its by-products.

17.6.1 Primary Processing

17.6.1.1 Harvesting

The quality at which a fruit and vegetable is harvested cannot be improved, however, all the post-harvest technologies and preservation technologies target to reduce the level of deterioration in it. So, it is utmost important that the harvest of a fruit is done at its peak quality. Proper harvesting at the correct maturity stage ensures higher fruit quality and long shelf life (Nidhina et al. 2022). Being a climacteric fruit (Morelos-Flores et al. 2022), it is prudent to harvest jackfruit at maturity prior to ripening. Jackfruit experiences significant post-harvest losses due to problems related to post-harvest technology and gaps of food supply chain (Vazhacharickal et al. 2015). The postharvest losses in fruits and vegetables occur due to microbial, physiological, physical, and environmental factors. So, proper postharvest management is necessary to ensure an optimum quality, as well as to enhance its shelf life. The postharvest activities include harvesting of jackfruit at maturity, preventing ripening at undesired time, prevention of microbial spoilage, and storage.

17.6.1.2 Maturity Indices

The maturity of jackfruit can be determined using several indices, including yellowish-green color, rough and bumpy skin, dull sound on tapping, and a fully developed brown colored seeds. Jackfruit is harvested when the fruit is fully mature and the rind turns from green to yellowish-brown (Sumari et al. 2022). Mature fruits also have a strong aroma. The spines on the fruit become more prominent. An immature fruit gives a metallic sound. Postharvest management practices such as washing, cleaning, grading, and storage should be followed to ensure quality and prevent spoilage (Ying et al. 2020). Furthermore, proper training and education of farmers on good cultivation practices can help to increase their knowledge and skills in jackfruit cultivation. In addition, adoption of Good Agricultural Practices (GAP) can improve the quality and safety of jackfruit and promote export opportunities.

17.6.1.3 Delay of Ripening

The primary purpose of postharvest management of climacteric fruits is to prevent their ripening, so that ripening may happen at a desired time and place. This is the primary method to improve the shelf life of climacteric fruits. Since jackfruit is also a climacteric fruit (Morelos-Flores et al. 2022), its ripening is delayed by various means. Ripening of climacteric fruit is generally influenced by high temperature, availability of oxygen, and exposure to ethylene. So, if maturity has to be delayed, then jackfruit need to be stored at low temperatures, such as 5 °C. Moreover, controlled atmospheric (CA) and modified atmospheric (MA) storage facilities ensure reduction of oxygen and increase of carbon dioxide. In MA storage, passive control of gases is achieved by wrapping the fruits in a semipermeable membrane, which allow the exchange of oxygen and carbon dioxide to desired levels. Controlled atmospheric storage, in contrast to MA storage also involves the control of gaseous environment around the food products. In CA storage, the control of gases is, however, active and without the use of any membrane. CA storage along with chitosan coating effectively increased the shelf storage life and sensory value of freshly cut jackfruit bulbs (Saxena et al. 2013). In addition, several chemical treatments such as exposure to 1-methylcyclopropene (1-MCP) can also delay or inhibit ripening of jackfruit (Vargas-Torres et al. 2017). Several chemical treatments can also be used as post-harvest treatments to reduce the spoilage of jackfruit.

17.6.1.4 Prevention of Post-harvest Losses

Microbial spoilage is an important mode of quality deterioration in fruits and vegetables. Many chemical treatments can be used to enhance the shelf life. Navindra et al. (2009) used anti-browning agents to enhance the shelf life of fresh-cut jackfruit. Fungal diseases deteriorate the quality of fruits at ambient temperature in the presence of high humidity. So, fungicide treatment can be used to prevent fungal growth. Enzymatic activities continue in fruits after harvest. Blanching has been an important heat-based method, in which, hot water treatment (various temperature-time combinations) can prevent fruit from enzymatic deterioration. Fruit surface is susceptible to the loss of moisture and attack of pests. It can be saved from these losses by coatings made from edible natural materials such as chitosan or carrageenan. These coating materials reduce the microbial spoilage by reducing water activity. In addition, Amulya et al. (2016) and Ahmed et al. (2020) described the importance of wax coating as an important post-harvest method to preserve shelf life of climacteric fruits. Wax coating can also be applied and tested to reduce post-harvest losses of jackfruit, as jackfruit is also a climacteric fruit [134]. High pressure processing is a novel processing method. It has also been researched to preserve the nutritional and sensory attributes of various food products. It can also be used for reducing post-harvest losses of jackfruit. Finally, control of temperature and humidity during storage and transportation of jackfruit may also reduce quality deterioration of jackfruit.

17.6.1.5 Storage at Low Temperature

After harvest, the fruits keep respiring. Respiration leads to fruit towards unplanned ripening and quick deterioration. Ripening of fruits can take place after 9 days of storage after harvest during storage at 27 °C and 70–80% RH (Rosnah et al. 2009). However, refrigeration temperature (less than 5 °C) can slow down the enzymatic activities responsible for respiration process. However, the fruit is not brought to the refrigeration temperature suddenly. If the fruits are suddenly brought to a temperature less than 12 °C directly from field temperature, there is a high chance that yellowing of the skin, softening of the flesh, and appearance of water-soaked spots appear on the skin. This condition is called chilling injury. This phenomenon is more prevalent in tropical fruits than temperate fruits. So, it is necessary to bring the fruit temperature down in a number of steps. First of all, field heat is removed by placing the fruit under shade. It is recommended to transfer the fruits at night time, so that the environmental temperature further drops. Once the fruit reaches the processing facility, then efforts to done to drop the fruit temperature to the refrigeration temperature. However, this is also done in steps. First pre-cooling is done to reduce the temperature. In this regard, a term ‘half cooling rate’ is important. It is the time at which the fruit temperature reaches half of the difference of fruit temperature with atmospheric temperature. There are several pre-cooling methods adopted in fruit processing industry. The most common methods include hydro-cooling, forced-air cooling, and vacuum cooling. In the next stage, the temperature of the fruits is brought down to required refrigeration temperature. Sometimes, the refrigeration temperature is not economically viable, then a temperature near 10 °C is considered appropriate for fruit storage. Moreover, CA storage and MA storage are also important types storage, which are used in combination with low-temperature storage. Jackfruit can also be frozen, either whole or in pieces, and used later for cooking. Freezing is the storage method with the highest shelf life.

17.6.1.6 Packing

After cooling, the fruits are sorted and graded based on a number of criteria including color, size, weight, etc. Further processing may be done in some industries, such as coating and waxing to prevent moisture losses and avoid microbial spoilage. Description of both the methods have been given above. The fruits are packed in suitable packaging material according to the consumer demand. The quality and nature of packaging material influences the quality of fruits. Primary packaging represents to the polyethylene bags materials which touch with the products. There is a rising trend these days that primary packaging materials are being made from biodegradable materials. Moreover, there is an increasing trend for vacuum packaging, MA packing, and active packaging, as well. In this regard, Saxena et al. (2008) used MA packaging (MAP) to enhance the shelf stability of minimally processed jackfruit bulbs. Moreover, Ng et al. (2020) tested vacuum packaging alng with high pressure processing (500 MPa for 5 min) to increase the color stability and shelf life of jackfruit during chilled storage at 5 °C. Primary packaged fruits and vegetables are then packed in secondary packing material, which are usually composed to hard material to give a proper shape to the food product. This is box packing, which is made of card or thick paper. These boxes are then packed in tertiary packaging material which are usually cartons. These are easy to stack in cold stores and easy for transportation. A packaging material must prevent contamination, protect packed food from the effect of environment, have appealing look, contain information about food it contains, packer industry, best before date, traceability, etc.

17.6.2 Value-Addition

If the fruits are not intended to be distributed in raw form, and further products are to be developed, then the processing steps are known as further processing or secondary processing. Food industry is continuously looking for innovative solutions. The global market for jackfruit and jackfruit-based products is growing rapidly due to its nutritional value, versatility, and sustainability (Shedge et al. 2022). Being rich in various nutrients that include dietary fiber, vitamins, minerals, and antioxidants (Wigati et al. 2023), it is considered a valuable ingredient in various food recipes. Moreover, it can be used as a meat substitute in vegan and vegetarian products. The unique texture and flavor of jackfruit have also led to the development of new products, such as jackfruit chips (Nuary and Firdaus 2022), snacks (Rajagopal 2023), and desserts (Mathew 2017). In addition to its use in food products, its seeds can be processed into flour, which can be used in baking and other food applications (Thatsanasuwan et al. 2023; Maradesha et al. 2022). Industrial byproducts of fruits are cause of significant environmental degradation. The use of jackfruit industrial by-products in various food recipes can contribute to the reduction of food waste (Pathak et al. 2022; Rengsutthi and Charoenrein 2011). Moreover, because of containing a number of bioactive components, such as flavonoids (Wigati et al. 2023) and phenolic acids (Pu et al. 2022); jackfruit wastes have potential to serve as an ingredient in various functional foods (Cruz-Casillas et al. 2021). Moreover, there is huge potential to use the development of value-added products from jackfruit waste that can contribute to rural development and poverty alleviation in developing countries (Islam et al. 2021a; Kushwaha et al. 2023).

17.6.2.1 Beverages and Fermented Products

Juice extraction is the most preferred mode of processing from the ripened jackfruit. After juice extraction, pulp and pomace are left, which are often discarded. However, recently, due to rising concern over environmental sustainability, research is in progress to develop valuable products from waste materials. Jackfruit juice can be extracted using simple pressing or squeezing apparatus. The fibrous, solid portion remained after the juice extraction is known as pulp. The skin, seeds, and other plant materials which are attached to the fruit but are not used for juice extraction, are known as pomace. Besides being a refreshing beverage, jackfruit juice has been tested for making into various products. Jackfruit juice may be drunk fresh, or may be dried into powder (Navarrete-Solis et al. 2020). Moreover, several fermented products may be developed from jackfruit juice. Yogurt is can be prepared by using up to 15% jackfruit juice with excellent taste and sensory values (Dey et al. 2014). Co-fermented collagen peptide-jackfruit juice has been found to be effective in regulating immune response (Ma et al. 2021). Moreover, a probiotic beverage can be developed from jackfruit juice (Muhialdin et al. 2021). Wine is another fermented product that can be prepared from jackfruit juice (Kumoro et al. 2012).

17.6.2.2 Jam and Ice Cream

Jam is a sweet fruit-based product in which fruit is cooked to a gel-consistency, sometimes using gelatin and pectin as additives. Jackfruit jam is often tested for its nutraceutical properties (Arshad et al. 2018). Jam is used as a spread on the bread, usually for breakfast. It is a profitable product that farmers can also prepare with minor investment, hence a great source of additional income for farmers. Ice cream is a frozen dessert prepared from milk, cream and sucrose. Up to 15% jackfruit pulp can be used in the manufacture of ice cream (Gaikwad et al. 2020). Addition of jackfruit pulp in the ice cream improves the nutritional profile of the ice cream. Moreover, it can be used as a natural flavoring agent for the ice cream. Puree is the fruit heated to cooking and then converted into a uniform paste like structure. Up to 15% jackfruit puree can be used for the manufacture of jackfruit ice cream (Thompson et al. 2009).

17.6.2.3 Chips

Chips are thinly sliced pieces of fruits and vegetables, which are deep-fried to create a crispy texture. The most renowned chips products available in market are potato chips. However, other fruits and vegetables can also be sliced and fried similar to potato chips. The countries where jackfruit cultivation is done, jackfruit chips are also known. Jackfruit chips is a novel product that has potential for export and generate valuable income. These are processed in frying oil until golden-brown color. Various novel processing technologies can be used for the manufacture of jackfruit chips. Chips may be pre-dried prior to frying. Yi et al. (2016) tested various drying methods, such as hot air-, freeze-, infrared-, microwave-, and vacuum-based drying, however, freeze-dried jackfruit chips gave the best sensory and nutrition results. Since, frying retains a large quantity of oil in the chips, therefore novel frying methods, such as vacuum frying have also been tested. Maity et al. (2015) demonstrated that other techniques such as coating with hydrocolloids such as pectin, carboxymethyl cellulose (CMC), gum arabic and sodium alginate can also be used to reduce the oil absorption in the vacuum-fried jackfruit chips without compromising the sensory attributes.

17.6.2.4 Extruded Products

Noodles and pasta can also be prepared from the jackfruit flour. Noodle and pasta are generally prepared from flour mixed with multiple ingredients such as egg, water, milk, etc., however, noodles are needle shaped long fibers, while pasta is cut into various shapes. Noodles are popular food worldwide, particularly in many Asian cuisines (Wang et al. 2021; Chung et al. 2010; Baviskar 2018). Addition of jackfruit flour or puree in general recipe of noodle (Hestiningtyas et al. 2022) and pasta (Swathi et al. 2019) can increase its nutritional value. Moreover, jackfruit flour is gluten-free (Nguyen Thi Kim 2017), and low in fat. In an earlier published study, Divakar (Divakar 2017) prepared noodles from 50% refined wheat flour, 20% JSF and 30% jackfruit bulb flour showed an improvement in the yield at reduced cooking time without affecting cooking and physical properties. The utilization of jackfruit for preparation of jackfruit-based noodles can be market as a health food, mainly because of gluten-free and high-fiber nature of jackfruit as compared to traditional wheat noodles.

17.6.2.5 Substitute of Meat

The meat mince or shreds combined to form a circular disc are known as burger patties. The burger patties are fried, and placed in a bun to make a hamburger. Due to rising concerns of environmental sustainability and religious concerns, there has been a rising trend for vegan burgers. Vegan/veggie burgers are popular in Europe and now new regulations have also been established by European parliament to allow promotion of this type of products (Bánáti 2020). These have been gaining tremendous popularity in US, as well (Kim and Chung 2022). Hence, vegan burgers made with jackfruit as a meat substitute have great scope to be marketed in in Europe and the United States, besides their native Asian markets. Jackfruit burgers can serve as a vegetarian or vegan version of burgers. The percentage of jackfruit used in the burger patty can vary from 25%, 50%, 75% and 100% (Abdullah 2017). This addition at these levels can improve consumer acceptability of patties (Abdullah 2017). Moreover, addition of these levels improves the nutritional properties of meat patties. Generally, beans, lentils, or mushrooms have been tried and tested to create a vegan burger patty that has a meaty texture (Pintado and Delgado-Pando 2020). Since, unripe jackfruit also possesses a structure similar to meat, it can be mixed with other ingredients such as beans, lentils, or mushrooms to create a vegan burger patty that has a meaty texture. Jackfruit provides meaty flavor (Faujan et al. 2022), which makes it useful in making it a popular meat substitute in vegetarian and vegan dishes.

Sausages are meat products usually made from meat mince packed into casings of various shapes. If the meat mince is simply made into sausage after mixing with required ingredients and giving a gel-like consistency upon cooking, it is known as a gel-type sausage. These have coarse texture. However, if the meat mince is converted into an emulsion prior to stuffing into casing, then it is known as emulsion-type sausage. These have a smooth texture. Within increasing consumer demand for vegan products, sausages with jackfruits have also been developed. Jackfruit gives similar texture and taste so meat, so it has great potential to be used for development of vegan sausages. Faujan et al. (2022) prepared beef-jackfruit (B:J) sausages using different B:J combinations such as 65:0, 48.75:16.26, 32.5:32.5, 16.25:48.75, and 0:65. The beef sausages substituted with unripe jackfruit exhibited higher carbohydrate, moisture and ash contents, but lower protein, fat, and fiber contents. Although, the consumers rated the control sausages as best, however, considered the one substituted with 25% unripe jackfruit with highest scores for texture, taste, and overall acceptance. Hence, incorporation of unripe jackfruit into meat products can potentially improve the nutritional value and sustainability of existing recipes, along with increasing the utilization this highly underutilized fruit.

17.6.2.6 Waste Valorization

Significant amounts of both edible and non-edible wastes are recovered after processing of jackfruit in the industry. The edible parts include the flesh, seeds, and arils. The edible waste generated after processing can be utilized for various purposes. The edible waste produced from jackfruit includes the outer rind, core, and seeds. These can be utilized for various purposes, including the production of food products and value-added products.

Roasted Jackfruit Seeds

Roasted seeds are traditional snack foods. Generally roasting of seeds is conducted in an iron pan filled with salt or sand (Kora 2019). Chickpeas (Coşkuner and Karababa 2004), pea (Bi et al. 2020), peanuts (Rossi-Márquez et al. 2021), corn (Lailiyah et al. 2020), pumpkin seeds (Akomolafe et al. 2021), etc. are liked and eaten worldwide. Roasted jackfruit seeds are eaten as a snack in many Southeast Asian countries (Spada et al. 2021). Moreover, these can be used in other beverages due to development of unique flavor. Spada et al. (2020) and Spada et al. (2022) reported that roasted jackfruit seeds possess chocolate aroma so they have been sometimes used in various food products to mimic the chocolate aroma. The jackfruit seeds are roasted until they have a crispy and crunchy texture. In some cultures, roasted jackfruit seeds are a popular snack food, as well as used as a coffee substitute (Spada et al. 2022). Odimegwu et al. (2019) suggested that roasted jackfruit seed contain proteins, dietary fibers, healthy fats, and minerals including iron, magnesium, and zinc, hence, not only serve as an important snack, but also as a source of nutrition, thus an important market commodity. Besides roasted jackfruit seeds, jackfruit seeds can be boiled and used as a vegetable in curries and stews, as well (Islam et al. 2015; Spada et al. 2021).

Jackfruit Seed Flour

Jackfruit seeds can also be processed into flour (JSF) and can be incorporated into various baked products as a substitute to wheat flour (Mhatre and Shekhar 2021). Incorporating JSF into food products has also been found to enhance their sensory properties, such as flavor and aroma of snack bar (Meethal et al. 2017), biscuits (Islam et al. 2015), bread (Hossain et al. 2014), buns (Ngwere and Mongi 2021), chocolate (Spada et al. 2020), and cappuccinos (Papa Spada et al. 2018). Moreover, JSF can serve as an ingredient in the development of nutraceutical food products. Gluten is an important protein that contributes to dough development properties, however, several consumers experience gluten intolerance, so there is high scope for gluten-free alternative to wheat flour. In this regard, Maskey, et al. (Maskey et al. 2020) prepare gluten-free cookies using JSF. JSF has been found to be a suitable replacement for wheat flour in the production of gluten-free laddus, biscuits, and crackers, due to its high starch content and low protein content (Mhatre and Shekhar 2021). Dietary fibers are useful for digestive health. Astuti et al. (2022) developed jackfruit seed crackers enriched with beetroot, in which highest soluble and insoluble dietary fibers were found in case of JSF. Ortega-González et al. (2022) prepared donuts from JSF supplemented wheat flour and reported improved scores for texture, hardness, adhesiveness and chewiness with supplementation of JSF in the recipe. The 5% replacement increased the consumer acceptability of donuts. Murali et al. (2017) assessed the gelling and thickening properties of jackfruit seed starch, and suggested its potential for the development of novel food products. Hence, incorporating JSF into food products can improve their textural properties, such as water-holding capacity and viscosity.

Jackfruit Seed Oil

Jackfruit by-products also possess antioxidant properties, which may help improve the shelf-life mainly by reducing the extent of lipid oxidation of food products. Jackfruit seed oil can be obtained by milling of seeds (Babu et al. 2017). Dias et al. (2022) demonstrated several functional properties of jackfruit seed oils including antibiotic properties. Fernandes et al. (2017) thought that jackfruit oil was nutritionally superior to other fatty acids, and had potential health benefits, mainly because it contained 34.2% PUFA, particularly due to presence of Linoleic acid (C18:2). Babu et al. (2017) tested jackfruit oil for frying potato chips, and noted similar attributes to that of sunflower oil. Ramli et al. (2021) utilized the essential oil and seed extracts of jackfruit on meat and reported higher shelf stability of meat mainly by decreasing TBARS.

Other Products

The outer rind of jackfruit is rich in bioactive compounds. The rind can also be utilized to develop into flour (Feili et al. 2013). The outer rind powder can be used to make bread (Felli et al. 2018), cookies (Ramya et al. 2020), and other food products. Moreover, research is ongoing to explore the potential of jackfruit non-edible waste for the development of value-added products such as biofuels (Pathak et al. 2022) and biodegradable plastics (Ooi et al. 2011; Razak et al. 2018). In addition, jackfruit peel, as well as its core contains pectin (Ahmmed et al. 2017). Pectin is a type of dietary fiber that possesses many functional properties. Dietary fibers been used traditionally as a natural laxative and to treat digestive disorders. These can be processed into a variety of food products.

17.6.3 Economic Significance

Jackfruit has high economic significance, particularly in rural areas where poverty and food insecurity are prevalent. In some regions, it has been promoted as a cash crop for small-scale farmers. Its fruit has a high market value, and its products such as canned fruit (Babu et al. 2022), chips (Nuary and Firdaus 2022), jam (Masuku 2021), roasted seeds (Spada et al. 2021), baked products made from its ground flour (Thatsanasuwan et al. 2023; Maradesha et al. 2022) have high market significance. In addition, jackfruit can promote sustainable economic development. It has been recognized as a miracle crop (Weintraub et al. 2022) due to its ability to thrive in diverse agro-ecological zones, high yield potential (Nansereko and Muyonga 2021), and low input requirements. This has made it an attractive option for small-scale farmers and agroforestry systems, leading to several benefits for both the environment and local communities. One of the key ways in which it has been used for sustainable agriculture is through agroforestry systems (Ahasan Ullah Khan et al. 2021). Agroforestry involves the integration of trees into crop and livestock systems to promote ecological and economic benefits. In many regions, it is planted as a multipurpose tree that provides shade, fruit, and timber, while also enhancing soil fertility and biodiversity (Islam et al. 2021b). Agroforestry systems that incorporate jackfruit can improve soil quality, increase water retention, and reduce soil erosion (Das and Saha 2020; Mondol 2021). Additionally, the shade provided by jackfruit trees (Riyadh et al. 2019) can help to reduce temperature stress on other crops, leading to increased yields and improved quality. Its traditional varieties can be highly variable in terms of fruit quality, yield, and susceptibility to pests and diseases. However, through selective breeding and genetic improvement, new cultivars have been developed that are more productive, disease-resistant, and have better fruit quality, particularly related to jackfruit bronzing disease (Ibrahim et al. 2020). These improved cultivars have the potential to increase farmer incomes, reduce post-harvest losses, and provide consumers with better quality fruit.

17.7 Challenges and Opportunities in Jackfruit-Based Agribusiness

Several challenges limit the growth of jackfruit agribusiness. However, challenges bring opportunities. Lack of initial investment may be a major constraint for small-scale farmers in jackfruit cultivation. This could be related to high labor requirements for planting and maintenance of crop (Deepa and Murthy 2019; Reddy et al. 2019). Moreover, wait for the first crop may also restrict the new farmer from its plantation, because its tree may take 5–6 years to reach maturity and bear fruit after planting (Ahmed et al. 2019). Moreover, the this crop has been limited to a few countries of the world despite huge potential, mainly because of lack of quality planting materials and seedling production systems (Rai et al. 2022). Furthermore, growing this cop requires costly inputs, including fertilizers, pesticides, and irrigation systems, which could be problem for new farmer without getting profit for initial years. So, it is recommended to initially start jackfruit cultivation in an agroforestry system to cope the above problems. Agroforestry system containing jackfruit may also serve as a tool for sustainable agriculture, as it can alleviate the effect of climate change (Das and Saha 2020; Kader et al. 2022; Mondol 2021). Despite high labor cost, jackfruit-based agroforestry systems are productive for small-scale farmers mainly due to additional source of income (Mondol 2021). Agroforestry systems may be complex, and there is lack of standardization, including planting density and spacing, which can limit the yield and quality of the fruit. Hence, training of farmers is essential for getting adequate cultivation of jackfruit and optimum yield (Mukta 2020). Training of low educated farmers belonging to far away locations may be another challenge, as outreach and lack of resources of trainers as well as farmer may hinder this process.

Crop yield and quality is subject to the attack of pests and diseases. Moreover, farmers may not possess adequate technical knowledge and skills for adequate pest and disease management (Vazhacharickal et al. 2015). The important pests of jackfruit include shoots borers, bark borer, bud pests, spittle bug, white-tailed mealybug, leaf webbers, particularly bud pests, spittle bugs, and mealybug are the major threats, which reduce jackfruit yield and productivity (Butani 1978). Another cause of crop susceptibility may be the lack of pest-resistant jackfruit varieties. In addition, there may be a lack of implementation of effective pest and disease management programs in the jackfruit cultivated areas (Johari et al. 2020; Borines et al. 2013). When old jackfruit trees are cut down for timber, these reduce the availability of quality planting material and loss of gene pool (Witherup et al. 2019). Moreover, the efforts regarding exploration of genetic diversity have not been sufficient, regarding traits of yield (Wangchu et al. 2013), pest resistance, and nutritional value (Ojwang et al. 2022).

Jackfruit yields fruit in various seasons; however, peak season may vary in various countries. In Sri Lanka (Medagoda 2011), there are two peak seasons i.e. March-June and a minor season from November-January. However, in Uganda it also has various fruiting seasons (Balamaze et al. 2019), but the peak season is in

December-January (Nakintu et al. 2019). This provides a great opportunity to the processors to have a predicted supply of jackfruit for development into various products. Another challenge may be the lack of availability of mechanical harvesting devices, which enforce the jackfruit farmers for manual harvesting by unskilled labor force, which may be time consuming and quality compromising, which results in significant quantities of fruit loss (Rana et al. 2018; Vazhacharickal et al. 2015). Although, deterioration of jackfruit is a serious concern that affects the income of farmers, yet wasted jackfruit can be converted in biochar at high temperature, which can serve to improve soil nutritional value for improving crop productivity (Nsubuga et al. 2020).

Since jackfruit is a climacteric fruit (Morelos-Flores et al. 2022), respiratory peak occurs in it. It means that once it is ripened, it has a very short shelf life. Postharvest management and marketing of jackfruit are, therefore, very challenging (Johari et al. 2020). It means that jackfruit farmers have a very short window of time to sell their crop. This leads to scarcity of the jackfruit throughout the year except peak fruiting seasons. Farmers experiences huge reduction in the income due to post-harvest losses during storage and supply chain that (Vazhacharickal et al. 2015), so development of adequate storage facilities and supply chain is another challenge. In addition, its inconsistent and inadequate market demand for jackfruit (Nakintu et al. 2020) and lack of proper marketing channels (Safari et al. 2019) for the fruit can be a significant challenge for farmers. Furthermore, inadequate infrastructure, including roads and transportation facilities, can limit the distribution and marketing of jackfruit, particularly in remote areas (Best 2015). Additionally, competition from other fruits, such as mango and banana in the jackfruit producing countries may also reduce the market demand for jackfruit.

Jackfruit has been a popular fruit in Asia and Africa (Corley 2007), but being a minor fruit, not much attention has been given on the development of its processing industry. So, with increasing research, more industry will be established in these regions, and they can generate significant revenue by not only developing various value-added jackfruit products selling locally but also to newly emerging jackfruit markets in other regions (Bánáti 2020; Kim and Chung 2022). In addition, the utilization of jackfruit as a vegan meat substitute has gained popularity in recent years, particularly in Western countries, due to its meat-like texture and ability to absorb flavors (Parihar et al. 2021). Although research has been in progress in development of various value-added products of jackfruit, yet their marketing and consumer acceptability is a challenge. Moreover, their industrial adaptation needs significant amount of work in dimensions of technology transfer, outreach, and investment. Finally, consumer campaigns are needed to increase the acceptability of these products among new consumers.

Finally, climate change is a looming concern worldwide. Jackfruit tree can help against the effect of climate change due to huge potential to be adopted in agroforestry systems, high yield, and huge structure (Weintraub et al. 2022). Although, it can help coping the effect of climate change, its flowering and fruit bearing capacity may itself be affected by unpredictable rainfalls and extreme weather, that can impact the growth and yield of jackfruit. With changing climatic pattern, there could be increased incidence of pests (Oraño et al. 2020; Kallekkattil and Mani 2022) and diseases (Borges et al. 2022; Habib et al. 2022; Hiremath et al. 2022). Rising temperatures and changing rainfall patterns can create favorable conditions for pests such as fruit flies (Oraño et al. 2020) and mealybugs (Kallekkattil and Mani 2022), and diseases such as anthracnose (Borges et al. 2022), and leaf spot (Habib et al. 2022), that may ultimately affect the crop growth rate and quality (Chowdhury and Hossain 2011). These pests and diseases can cause significant damage to jackfruit crops, leading to reduced yield and quality (Variasa 2018). Moreover, droughts and floods can damage to jackfruit trees as well, that may result in the loss of genetic diversity (Hassan et al. 2018). The extreme climatic change can cause stress on fruiting trees of jackfruit that may result in non-uniform fruit composition with lower nutritional value and flavor, hence postharvest losses are significant (Cruz-Casillas et al. 2021) and uniformity inn product quality is not achieved (Mortuza et al. 2014). Since it can be adopted in various agro-ecological zones (Mondol 2021), hence it has huge potential to promote food security and farmer income, hence can contribute to the achievement of several Sustainable Development Goals (SDGs), including zero hunger, responsible consumption and production, and climate action.

17.8 Conclusion and Future Prospects

The future prospects of jackfruit fruit are bright due to its versatility, unique flavor, and potential as a sustainable and healthy food source. Being an exotic fruit, it is gaining popularity in international markets, as well. There is significant potential for research and development in the areas of jackfruit breeding, production, and processing technologies. With more research, there is a possibility to enhance yield, quality, and nutritional attributes of jackfruit, leading to increased commercialization and market demand. Hence, there is a huge opportunity for jackfruit farmers and processors to expand their agribusinesses. Jackfruit can serve as a tool towards achievement of SDGs by contributing towards food security, zero hunger, and climate action by incorporating the concept of agroforestry. Due to unique nutritional composition, it has huge potential for development of nutraceuticals and value-added food products. Also, there is tremendous potential for industrial adaptation of this fruit and valorization of its waste; however, the challenges in the commercialization include the lack of standardization in processing and packaging, as well as the need for improved postharvest handling and storage practices. Furthermore, there is a rising global demand for plant-based foods, and jackfruit is a popular meat substitute for vegetarians and vegans due to its unique texture and flavor. Hence, it can significantly contribute the development agribusinesses in various regions of the world and create a positive impact on farmers and the economy.

Conflicts of Interest All authors have read and approved the final manuscript and declare that they have no conflict of interest.

References

Abasubong KP, Li XF, Adjoumani JJY, Jiang GZ, Desouky HE, Liu WB et al (2022) Effects of dietary xylooligosaccharide prebiotic supplementation on growth, antioxidant and intestinal immune-related genes expression in common carp Cyprinus carpio fed a high-fat diet. J Anim Physiol Anim Nutr 106:403–418

Abdullah I (2017) Effect of using unripe jackfruit as a meat substitute on nutrition composition and organoleptic characteristic of meat patty. J Eng Technol 2:96–106

Ahasan Ullah Khan M, Choudhury AR, Abdul M, Maleque CKD, Talucder MSA, Maukeeb ARM, Adnan M et al (2021) Management of insect pests and diseases of jackfruit (*Artocarpus heterophyllus* L.) in agroforestry system: a review. Acta Entomol Zool 2:37–46

Ahmed AA, Barua JP, Pathak S, Gogoi S et al (2019) Agroforestry model for Assam. India. https://krishi.icar.gov.in/jspui/bitstream/123456789/56315/1/Agroforestry%20Model%20for%20Assam.pdf

Ahmed A, Ali SW, Imran A, Afzaal M, Arshad MS, Nadeem M, Ikram A et al (2020) Formulation of date pit oil-based edible wax coating for extending the storage stability of guava fruit. J Food Process Preserv 44:1–10

Ahmmed R, Inam A, Alim M, Sobhan M, Haque M et al (2017) Extraction, characterization and utilization of pectin from jackfruit (*Artocarpus heterophyllus* L.) waste. J Environ Sci Nat Resour 32:57–66

Akomolafe SF, Olasehinde TA, Aluko BT et al (2021) Diets supplemented with raw and roasted pumpkin (Cucurbita pepo L) seeds improved some biochemical parameters associated with erectile function in rats. J Food Biochem 45:1–9

Akter B, Haque MA (2018) Utilization of Jackfruit (Artocarpus heterophyllus) seed's flour in food processing: a review. The Agriculturists 16:131–142

Al-Khayri JM, Sahana GR, Nagella P, Joseph BV, Alessa FM, Al-Mssallem MQ et al (2022) Flavonoids as potential anti-inflammatory molecules: a review. Molecules 27:1–24

Amadi JA, Ihemeje A, Afam-Anene O (2018) Nutrient and phytochemical composition of jackfruit (*Artocarpus heterophyllus*) pulp, seeds and leaves. Int J Innov Food Nutr Sustain Agric 6:27–32

Amador-Espejo GG, Chávez-Ocegueda J, Cruz-Cansino N, Suárez-Jacobo A, Gutiérrez-Martínez P, Valencia-Flores D, Velázquez Estrada R et al (2020) Thermosonication parameter effects on physicochemical changes, microbial and enzymatic inactivation of fruit smoothie. J Food Sci Technol 57:1680–1688

Amulya P, Sudheer K, Rifna E, Sreesha K, Nusaiba C et al (2016) Effect of edible wax coating and map on the quality of mango during storage. Int J Agric Sci Res 6:13–18

Arshad M, Vasudeva K, Krishna H, Shankarappa T, Halesh G et al (2018) Effect of Storage conditions on the nutritional quality of protein fortified jackfruit jam. Int J Curr Microbiol App Sci 7:1018–1024

Artés-Hernández F, Castillejo N, Martínez-Zamora L, Martínez-Hernández GB et al (2021) Phytochemical fortification in fruit and vegetable beverages with green technologies. Foods 10:1–32

Astuti P, Zumar HA, Fathonah S, Ansori M et al (2022) Differences in preferences, water content, and dietary fibre of jackfruit seed crackers enriched with red beetroot (Beta vulgaris). IOP Conf Ser: Earth Environ Sci, IOP Publishing 969. https://doi.org/10.1088/1755-1315/969/1/012009

Ayinde IA, Otekunrin OA, Akinbode SO, Otekunrin OA et al (2020) Food security in Nigeria: impetus for growth and development. J Agric Econ 6:808–820

Babu NG, Kumar S, Sundar S et al (2017) Extraction and comparison of properties of jackfruit seed oil and sunflower seed oil. Int J Sci Eng Res 8:635–639

Babu PS, Pulissery SK, Chandran SM, Mahanti NK, Pandiselvam R, Jaganath B, Kothakota A et al (2022) Non-invasive and rapid quality assessment of thermal processed and canned tender jackfruit: NIR spectroscopy and chemometric approach. Int J Food Sci Technol 57:6072–6081

Bakewell-Stone P (2023) Artocarpus heterophyllus (jackfruit). CABI Compendium. https://doi.org/10.1079/cabicompendium.1832

Bakri NFM, Ishak Z, Hassan H, Mohamad A, Razali R et al (2021) Effect of pretreatment on the level of antinutritional factors in jackfruit seeds. J Acad 9:11–20

Balamaze J, Muyonga J, Byaruhanga Y et al (2019) Production and utilization of jackfruit (Artocarpus heterophyllus) in Uganda. Afr J Food Agric Nutr Dev 19:14289–14302

Bánáti D (2020) Veggie burgers, vegan meats? The ruling of the European Parliament paved the way for meat substitutes with meat denominations. J Food Invest 66:3166–3174

Baviskar A (2018) Consumer citizenship: instant noodles in India. Gastronomica 18:1–10

Bensi PS, Divakar S, Kumari KS, Mini C, Joseph B, Krishnaja U et al (2020) Optimizing conditions for enzymatic extraction of juice from jackfruit (koozha) using response surface methodology. Eur J Nutr Food Saf 12:23–31

Best A (2015) Value-adding options for tropical fruit using jackfruit as a case study. RIRDC, Barton ACT

Bi S, Xu X, Luo D, Lao F, Pang X, Shen Q, Wu J et al (2020) Characterization of key aroma compounds in raw and roasted peas (Pisum sativum L.) by application of instrumental and sensory techniques. J Agric Food Chem 68:2718–2727

Borges R, Rossato M, Santos MDM, dos Santos Vieira WA, Câmara MP, Reis A et al (2022) Characterization of Colletotrichum isolates causing anthracnose on Artocarpus heterophyllus in Brazil. J Plant Pathol 105:299–305

Borines LM, Guadalquiver GA, Palermo VG, De la Cruz CS, Abuyabor J, Pedroso MA, Guest D et al (2013) Participatory action research (PAR) on Phytophthora management delivers positive outcomes for smallholder jackfruit farmers. Smallholder HOPES—horticulture, people and soil. In: Aciar proceedings, pp 31–44

Briones RL, Lina DP (2020) Effects of organic and inorganic fertilizer application on the postharvest quality of jackfruit (*Artocarpus heterophyllus* Lam.) Var. EVIARC sweet. Mindanao J Sci Technol 18:259–276

Butani DK (1978) Pests and diseases of jackfruit in India and their control. Fruits 33(5):351–357

Chauvet M, Mitra S (2023) Botany, origin, history and production. In: Jackfruit: botany, production and uses. GB, CABI, Wallingford, pp 1–19

Chavez-Santiago JO, Rodríguez-Castillejos GC, Montenegro G, Bridi R, Valdes-Gomez H, Alvarado-Reyna S, Santiago-Adame R et al (2021) Phenolic content, antioxidant and antifungal activity of jackfruit extracts (*Artocarpus heterophyllus* Lam.). Food Sci Technol 42:1–7

Chhotaray S, Priyadarshini B (2022) Nutritional composition and health benefits of jackfruit seed flour: a review. Variations 81:42–49

Chougale V, Shinde V, Khandekar R, Pawar C, Mahadik S et al (2021) Effect of stage of harvesting on various chemical properties of tender jack vegetable (*Artocarpus heterophyllus*). Pharm Innov J 10:1538–1540

Chowdhury M, Hossain I (2011) Effects of temperature, rainfall and relative humidity on leaf spot of jackfruit seedling and its eco-friendly management. The Agriculturists 9:126–136

Chung CE, Lee KW, Cho MS et al (2010) Effect of ramyen and noodles intake in diet & health status of Koreans. J Korean Soc Food Cult 25:109–116

Corley H (2007) Fruits for the future 10. Jackfruit (*Artocarpus heterophyllus*). By N. Haq. Southampton, UK: Southampton Centre for Underutilised Crops. Exp Agric Book Rev 43:407–408

Coşkuner Y, Karababa E (2004) Leblebi: a roasted chickpea product as a traditional Turkish snack food. Food Rev Int 20:257–274

Cruz-Casillas FC, García-Cayuela T, Rodriguez-Martinez V et al (2021) Application of conventional and non-conventional extraction methods to obtain functional ingredients from jackfruit (*Artocarpus heterophyllus* lam.) tissues and by-products. Appl Sci 11:1–14

Dahanayake N (2015) Some neglected and underutilized fruit-crops in Sri Lanka. Int J Sci Res Publ 5:1–7

Das K, Saha A (2020) Jackfruit (Artocarpus heterophyllus Lam.), a potential fruit crop of Tripura and exploring its nutritional benefits. J Med Plants 8:101–103

Deepa M, Murthy P (2019) Economic analysis of women empowerment through value addition of jackfruit in Bengaluru Rural District. Econ Aff 64:401–406

Delfino D, Octavian E, Sutandi J et al (2021) Utilization of jackfruit seed powder as a sustainable cocoa substitute: a review. In: Conference proceeding FOSTER (food science student conference) theme: pushing the boundaries: sustainable and responsible food innovation. SCU Knowledge Media

Dey KC, Begum R, Rahman MRT, Sultana A, Akter S, Janny RJ et al (2014) Development of fruit juice yogurt by utilization of jackfruit juice: a preliminary study on sensory evaluation, chemical composition and microbial analysis. Int J Eng Res 3:1074–1079

Dias CJF, Raposo A, de Sousa CDT, de Araújo-Neto JB, Tintino SR, Oliveira-Tintino CDDM, de Sousa EO et al (2022) Modification of antibiotic activity by fixed oil of the Artocarpus heterophyllus almond against standard and multidrug-resistant bacteria strains. Biology 11:1–10

Divakar S (2017) Quality analysis of raw jackfruit based noodles. Asian J Dairy Food Res 36:45–51

Dong J, Liang W, Wang T, Sui J, Wang J, Deng Z, Chen D et al (2019) Saponins regulate intestinal inflammation in colon cancer and IBD. Pharmacol Res 144:66–72

Faujan NH, Azizan N, Arifin N et al (2022) Proximate composition and sensory preference of beef sausages with unripe jackfruit. Malaysian J Med Health Sci 8:9–13

Feili R, Zzaman W, Abdullah WNW, Yang TA et al (2013) Physical and sensory analysis of high fiber bread incorporated with jackfruit rind flour. Food Sci Technol 1:30–36

Felli R, Yang TA, Abdullah WNW, Zzaman W et al (2018) Effects of incorporation of jackfruit rind powder on chemical and functional properties of bread. Trop Life Sci Res 29:113–126

Fernandes F, Ferreres F, Gil-Izquierdo A, Oliveira AP, Valentão P, Andrade PB et al (2017) Accumulation of primary and secondary metabolites in edible jackfruit seed tissues and scavenging of reactive nitrogen species. Food Chem 233:85–95

Gaikwad S, Kamble D, Jaybhay V et al (2020) Development of ice-cream by using jackfruit pulp. Pharma Innov J 9:533–539

Gala S, Sumarno S, Mahfud M et al (2022) Optimization of microwave-assisted extraction of natural dyes from jackfruit wood (Artocarpus heterophyllus Lamk) by response surface methodology. Eng Appl Sci Res 49:29–35

Ghavami A, Ziaei R, Talebi S, Barghchi H, Nattagh-Eshtivani E, Moradi S, Askari G et al (2023) Soluble fiber supplementation and serum lipid profile: a systematic review and dose-response meta-analysis of randomized controlled trials. Adv Nutr 1:1–10

Ghosh S, Biswajit B (2006) Proceedings of the national symposium on production, utilization and export of underutilized fruits with commercial potentialities. Bidhan Chandra Krishi Viswavidyalaya, Kalyani, Nadia, pp 150–153

Gil KA, Wojdyło A, Nowicka P, Montoro P, Tuberoso CIG et al (2022) Effect of apple juice enrichment with selected plant materials: focus on bioactive compounds and antioxidant activity. Foods 12:1–18

Goswami C, Kazal MKH, Alam O, Moon RJ, Khatun K, Hossan M, Chacrabati R et al (2021) Jackfruit seed powder supplementation attenuates high-sugar diet-induced hyperphagia and hyperglycemia in mice. Biol Life Sci Forum 6:92–105

Goswami RK, Neog P, Hazarika D, Baruah D et al (2022) Optimizing the time of grafting in jackfruit in the subtropical environment of Assam. Int J Environ Clim 12:3134–3145

Gunawan WB, Basoeki LEAS (2022) Incorporating watermelon rind into jackfruit seed drink: a functional drink rich in antioxidant and phenolic compounds. Edufortech 7:100–107

Gupta D, Mann S, Sood A, Gupta RK et al (2011) Phytochemical, nutritional and antioxidant activity evaluation of seeds of jackfruit (Artocarpus heterophyllus Lam.). Int J Pharma Bio Sci 2:336–345

Habib MT, Mia MJ, Uddin MS, Ahmed F et al (2022) An in-depth exploration of automated jackfruit disease recognition. J King Saud Univ – Comput Inf Sci 34:1200–1209

Habibah NU, Albaar N, Rasulu H et al (2021) The effect of substitution of seed flour of jackfruit (*Artocarpus heterophyllus* lam.) on the physicochemical and organoleptic characteristics of macrons. Int J Food Agric Nat Resour 2:19–24

Haq N (2006) Fruits for the future 10–Jackfruit (Artocarpus heterophyllus) monograph. Crops for the future. Southampton Centre for Underutilized Crops, University of Southampton, Southampton

Hashim OH, Ng CL, Gendeh GS, Jaafar MIN et al (1991) IgA binding lectins isolated from distinct Artocarpus species demonstrate differential specificity. Mol Immunol 28:393–398

Hassan S, Bhat KM, Jan A, Mehraj S, Wani SA, Khanday MUD, Bisati IA et al (2018) Managing genetic resources in temperate fruit crops. Econ Aff 63:987–996

Hestiningtyas KA, Wardani HE, Kurniawan A et al (2022) Proximate analysis of noodles with substitution of jackfruit seed flour and branch flour as alternative food for diabetes patients. In: 3rd International scientific meeting on public health and sports. Atlantis Press, pp 90–95

Hiremath SR, Santhoshkumar T, Prathapan K et al (2022) Outbreak of Glyphodes caesalis (Lepidoptera, Pyraloidea, Crambidae, Spilomelinae) on exotic varieties of jackfruit (*Artocarpus heterophyllus*) and a note on records of its occurrence in Kerala. J Trop Agric 60:125–128

Hossain MT, Hossain MM, Sarker M, Shuvo AN, Alam MM, Rahman MS et al (2014) Development and quality evaluation of bread supplemented with jackfruit seed flour. Int J Nutr Food Sci 3:484–487

Houston LM (2005) Food culture in the Caribbean. Greenwood Publishing Group, Westport

Ibrahim R, Ismail-Suhaimy NW, Shu-Qing T, Ismail SI, Ina-Salwany MY, Yusof MT, Zulperi D et al (2020) Draft genome sequencing data of a pathogenic Pantoea stewartii subspecies stewartii strain SQT1 causing bronzing disease of jackfruit in Malaysia. Data Brief 30:1–6

Ilmi HM, Elya B, Handayani R et al (2020) Association between total phenol and flavonoid contents in *Artocarpus heterophyllus* (jackfruit) bark and leaf extracts and lipoxygenase inhibition. Int J Appl Pharm 12:252–256

Isinenyi GO, Andrew N (2021) Study on the nutritional compositions of boiled jackfruit seed and its effect on the liver and kidney functions of Wistar albino rats. Int Netw Organ Sci Res 7:96–108

Islam MS, Begum R, Khatun M, Dey KC et al (2015) A study on nutritional and functional properties analysis of jackfruit seed flour and value addition to biscuits. Int J Eng Res Technol 4:139–147

Islam MJ, Sarker BC, Rahman MS, Hossen MS, Arthi MIJ et al (2021a) Preparation and preservation of value added products from jackfruit and evaluation of their biochemical properties. Int J Agric Med Plants 2:18–27

Islam MM, Miah MG, Saha SR, Rahman MA, Akanda MAM, Kamruzzaman M et al (2021b) Improvement of soil fertility through jackfruit-based multistoried agroforestry practices in terrace ecosystem. Ann Bangladesh Agric 25:105–115

Jadhav H, Mankar S, Bhosale M et al (2021) A review on jackfruit: it is profitable to human beings. Res J Pharmacogn Phytochem 13:51–54

Jang WY, Kim MY, Cho JY et al (2022) Antioxidant, anti-inflammatory, anti-menopausal, and anti-cancer effects of lignans and their metabolites. Int J Mol Sci 23:1–19

Johari AM, Rahman NAA, Baharuddin AS, Basha RK, Mohammed M, Parid D, Wakisaka M et al (2020) Effects of different low temperature storage conditions on the physico-chemical properties of Mastura (J37) jackfruit bulbs. J Agric Food Eng 1:1–6

Kader A, Sinha SN, Ghosh P et al (2022) Clonal fidelity investigation of micropropagated hardened plants of jackfruit tree (Artocarpus heterophyllus L.) with RAPD markers. J Genet Eng Biotechnol 20:1–14

Kallekkattil S, Mani M (2022) Pests and their management in jackfruit. In: Trends in horticultural entomology. Springer, Singapore, pp 701–717

Kallekkattil S, Krishnamoorthy A, Venkatesha MG et al (2020) Insect pest complex of jackfruit in South India. Int J Trop Insect Sci 40:513–525

Kaur J, Singh Z, Shah HMS, Mazhar MS, Hasan MU, Woodward A et al (2023) Insights into phytonutrient profile and postharvest quality management of jackfruit: a review. Crit Rev Food Sci Nutr 1:1–27

Keerthana Priya R, Rawson A, Vidhyalakshmi R, Jagan Mohan R et al (2022) Development of vegan sausage using banana floret (Musa paradisiaca) and jackfruit (Artocarpus heterophyllus Lam.) as a meat substitute: evaluation of textural, physico-chemical and sensory characteristics. J Food Process Preserv 46:1–9

Kemalasari N, Sumardi S, Febriani Y et al (2018) The effectiveness of total flavonoid effect of ethanol extracts of nangka stone (*Artocarpus heterophyllus* Lam) on healing of pass wounds in broiler chicken (*Gallus domesticus*). J Pharm Sci 1:1–6

Khan AU, Ema IJ, Faruk M, Tarapder SA, Khan AU, Noreen S, Adnan M et al (2021) A review on importance of Artocarpus heterophyllus L. (Jackfruit). J Multidiscip Appl Nat Sci 1:106

Kim S, Chung SJ (2022) Hedonic plasticity of vegan burger patty in implicit or explicit reference frame conditions. J Sci Food Agric 103:2641–2652

Kiziltaş H (2022) Comprehensive evaluation of Reseda lutea L.(Wild Mignonette) and 7 isolated flavonol glycosides: determination of antioxidant activity, anti-Alzheimer, antidiabetic and cytotoxic effects with in vitro and in silico methods. Turk J Chem 46:1185–1198

Kora AJ (2019) Applications of sand roasting and baking in the preparation of traditional Indian snacks: nutritional and antioxidant status. Bull Natl Res Cent 43:1–11

Kumoro AC, Sari DR, Pinandita APP, Retnowati DS, Budiyati CS et al (2012) Preparation of wine from jackfruit (*Artocarpus heterophyllus* lam) juice using baker yeast: effect of yeast and initial sugar concentrations. World Appl Sci J 16:1262–1268

Kumoro AC, Alhanif M, Wardhani DH et al (2020) A critical review on tropical fruits seeds as prospective sources of nutritional and bioactive compounds for functional foods development: a case of Indonesian exotic fruits. Int J Food Sci 2020:1–15

Kushwaha R, Singh V, Singh M, Kaur D et al (2019) Variation in physical and physicochemical properties of different cultivars of jackfruit seeds harvested from the northern region of India. Int J Adv Innov Res 6:43–48

Kushwaha R, Gupta A, Singh V, Kaur S, Puranik V, Kaur D et al (2023) Jackfruit seed flour-based waffle ice cream cone: optimization of ingredient levels using response surface methodology. Heliyon 9:1–11

Lailiyah M, Saputra SA, Sari F et al (2020) Antioxidant activity and sun protection factor evaluation for cream formulation of purified roasted corn silk extracts (*Zea Mays* L. Saccharata). Pharmaciana 10:371–380

Lakshmana JH, Jayaprahash C, Kumar R, Kumaraswamy MR, Kathiravan T, Nadanasabapathi S et al (2013) Development and evaluation of shelf stable retort pouch processed ready-to-eat tender jackfruit (*Artocarpus heterophyllus*) curry. J Food Proc Technol 4:274–281

Lam SK, Ng TB (2011) Lectins: production and practical applications. Appl Microbiol Biotechnol 89:45–55

Li B, Wang H, Wang X, Zhang Y, Tan Y, Zhang Y, Zhang Y et al (2019) Prediction of the postprandial blood sugar response estimated by enzymatic kinetics of in vitro digestive and fine molecular structure of artocarpus heterophyllus lam seed starch and several staple crop starches. Starch-Stärke 71:1–10

Ma T, Li C, Zhao F, Cao J, Zhang X, Shen X et al (2021) Effects of co-fermented collagen peptide-jackfruit juice on the immune response and gut microbiota in immunosuppressed mice. Food Chem 365:1–11

Mahanta CL, Kalita D (2015) Processing and utilization of jackfruit seeds. In: Processing and impact on active components in food. Academic Press, Amsterdam, pp 395–400

Maheswari TU, Valsan V (2019) Studies on the Nutritional and organoleptic characteristics of jackfruit halwa during storage. Res J Pharm Technol 12:4848–4850

Maity T, Bawa A, Raju P et al (2015) Use of hydrocolloids to improve the quality of vacuum fried jackfruit chips. Int Food Res J 22:1571–1577

Majnooni MB, Fakhri S, Ghanadian SM, Bahrami G, Mansouri K, Iranpanah A, Mojarrab M et al (2023) Inhibiting angiogenesis by anti-cancer saponins: from phytochemistry to cellular signaling pathways. Metabolites 13:1–28

Mandhare A, Banerjee P, Pande A, Gondkar A et al (2020) Jackfruit (Artocarpus heterophyllus): a comprehensive patent review. Curr Nutr Food Sci 16:644–665
Manivannan V, Johnson M (2020) Total phenolic, tannin, triterpenoid, flavonoid and sterol contents, anti-diabetic, anti-inflammatory and cytotoxic activities of Tectaria paradoxa (Fee.) Sledge. Toxicol Rep 7:1465–1468
Manurung B, Eyanoer PC, Fauzi TM et al (2021) Effect of ethanol extract jackfruit (Artocarpus heterophyllus Lam.) to decrease blood sugar levels in wistar female white rats with streptozotocin-induced gestational mellitus diabetes. Int J Biomed Herb Med 1:46–51
Maradesha T, Patil SM, Phanindra B, Achar RR, Silina E, Stupin V, Ramu R et al (2022) Multiprotein inhibitory effect of dietary polyphenol rutin from whole green jackfruit flour targeting different stages of diabetes mellitus: defining a bio-computational stratagem. Separations 9:1–26
Maskey B, Subedi S, Shrestha NK et al (2020) Effect of incorporation of jackfruit (Artocarpus heterophyllus) seed flour on the quality of cookies. Dristikon: Multidiscip J 10:60–72
Masuku MA (2021) The effect of sitrate acid addition towards organoleptic properties of jackfruit jam. Agrikan: Jurnal Agribisnis Perikanan 14:184–191
Mathew JJ (2017) Morphological diversity, nutritional quality and value addition of jackfruit (*Artocarpus heterophyllus*) in kerala. GRIN Publishing. Place of publication not identified
Maurya P (2016) Assessment of consumption practices of jackfruit (*Artocarpus heterophyllus* lam.) seeds in villages of Jalalpur block district Ambedarnagar (UP) India. Seeds 2:73–75
Medagoda I (2011) Jackfruit in Sri Lanka. The jackfruit, pp 369–392
Meethal SM, KAur N, Singh J, Gat Y et al (2017) Effect of addition of jackfruit seed flour on nutrimental, phytochemical and sensory properties of snack bar. Curr Res Nutr Food Sci J 5:154–158
Mhatre S, Shekhar A (2021) Shelf life study of gluten free and sugar free jackfruit seed laddus. Int J Creat Res Thoughts 9:820–829
Mijan NA, Lee HV, Alsultan AG, Taufiq-Yap YH et al (2016) Synthesis and characterization of silica-alumina supported Ca and Ni catalyst for deoxygenation of vegetable oil into diesel. Mater Sci Forum, Trans Tech Publications Ltd 840:353–358
Mithun K, Kaleemullah S (2019) Quality evaluation of osmosed jackfruit bulb slices. Int J Curr Microbiol App Sci 8:466–472
Mondol MA (2021) Role of jackfruit based traditional agroforestry in farmers income and livelihood improvements: evidence from Mymensingh district of Bangladesh. J Food Agric Environ 2(1):55–61
Morelos-Flores DA, Montalvo-González E, Chacón-López MA, Santacruz-Varela A, Zamora-Gasga VM, Torres-García G, de Lourdes García-Magaña M et al (2022) Comparative study of four jackfruit genotypes: morphology, physiology and physicochemical characterization. Horticulturae 8:1–15
Moreno Delgado KA (2021) Estudio de las propiedades nutricionales de la pulpa de jackfruit (Artocarpus heterophyllus lam) y su aplicación en la industria alimentaria
Mortuza M, Talukder S, Haque M et al (2014) Biochemical changes in jackfruit pulp as affected by cold temperature. J Environ Sci Nat Resour 7:93–97
Muhialdin BJ, Hussin ASM, Kadum H, Hamid AA, Jaafar AH et al (2021) Metabolomic changes and biological activities during the lacto-fermentation of jackfruit juice using Lactobacillus casei ATCC334. LWT 141:1–8
Mukta M (2020) Study on farmers'opinion about jackfruit tree cultivation and its importance in a selected upazila of Tangail District. Doctoral Dissertation, Department of Agroforestry and Environmental Science, Sher-e-Bangla Agricultural University, Dhaka
Murali S, Kar A, Patel AS, Mohapatra D, Krishnakumar P et al (2017) Optimization of rice bran oil encapsulation using jackfruit seed starch–whey protein isolate blend as wall material and its characterization. Int J Food Eng 13:1–11
Nakintu J, Olet A, Andama M, Lejju JB et al (2019) Ethno-varieties and distribution of jackfruit tree (Artocarpus heterophyllus Lam.) in Uganda: implications for trade, food security and germplasm conservation. East Afr J Sci Technol Innov 1:27–51

Nakintu J, Albrecht C, Müller CM, Kagoro-Rugunda G, Andama M, Olet EA, Gemeinholzer B et al (2020) Exploring the genetic diversity of jackfruit (Artocarpus heterophyllus Lam.) grown in Uganda based on SSR markers. Genet Resour Crop Evol 67:605–619

Nansereko S, Muyonga JH (2021) Exploring the potential of jackfruit (*Artocarpus heterophyllus* Lam). Asian Food Sci J 20:97–117

Natta S, Pal K, Alam BK, Mondal D, Dutta SK, Sahana N, Choudhury A et al (2023) In-depth evaluation of nutritive, chemical constituents and anti-glycemic properties of jackfruit (*Artocarpus heterophyllus* Lam) clonal accessions with flake colour diversity from Eastern Sub-Himalayan plains of India. Food Chem 407:1–12

Navarrete-Solis A, Hengl N, Ragazzo-Sánchez JA, Baup S, Calderón-Santoyo M, Pignon F, Ortiz-Basurto RI et al (2020) Rheological and physicochemical stability of hydrolyzed jackfruit juice (*Artocarpus heterophyllus* L.) processed by spray drying. J Food Sci Technol 57:663–672

Navindra B, Arvind R, Hassina BB et al (2009) Effects of antibrowning agents on the shelf life of fresh-cut green jackfruit (*Artocarpus heterophyllus* Lam.). J Appl Hortic 11:41–45

Ndikuryayo F, Nineza C, Bigumandondera P, Ndayizigiye S (2021) Nutritional composition of neglected seeds of jackfruit (Artocarpus heterophyllus) cultivated in Burundi. J Mater Environ Sci 12:766–774

Ng S, Tan T, Tan P, Chong G, Tan C et al (2020) Effect of selected high pressure processing parameters on the sensory attributes and shelf life of jackfruit (*Artocarpus heterophyllus* L.) bulb packed using different packaging materials. Int Food Res J 27:675–682

Nguyen Thi Kim N (2017) Application of jackfruit seed flour for non-fried instant noodle making: flour prooerties and noodle quality. Doctoral dissertation, International University-HCMC

Ngwere SS, Mongi RJ (2021) Nutritional composition, sensory profile and consumer acceptability of wheat-jackfruit seed composite buns. Tanz J Sci 47:1154–1164

Nidhina K, Abraham B, Fontes-Candia C, Martínez-Abad A, Martínez-Sanz M, Nisha P, Lopez-Rubio A et al (2022) Physicochemical and functional properties of pectin extracted from the edible portions of jackfruit at different stages of maturity. J Sci Food Agric 103:3194–3204

Ningsih I, Rosalinda DA, Kiranasari A, Dewi BE, Sjatha F et al (2018) In vitro antibacterial activity test of jackfruit (Artocarpus heterophyllus Lam.) leaf extract against methicillin-resistant Staphylococcus aureus (MRSA). In: BROMO Conference, symposium on natural product and biodiversity, pp 245–249

Nsubuga D, Banadda N, Kabenge I, Wydra KD et al (2020) Potential of jackfruit waste for biogas, briquettes and as a carbondioxide sink-a review. J Sustain Dev 13:60–75

Nuary AT, Firdaus MA (2022) The marketing strategy increasing the home of the crackup industry during the covid-19 pandemic in the village munjul, tanah sareal bogor. Abdi Dosen: Jurnal Pengabdian Pada Masyarakat 6:562–567

Odimegwu NE, Ofoedu CE, Omeire GC, Umelo MC, Eluchie CN, Alagbaoso SO, Ozoani PO et al (2019) Production and evaluation of breakfast cereals from flour blends of maize (*Zea mays*) and jackfruit (*Artocarpus heterophyllus* Lam.) seeds. Arch Curr Res Int 16:1–16

Ojwang RA, Muge EK, Nyaboga EN, Mbatia BN, Ogoyi DO et al (2022) Genetic diversity and relationships among populations of jackfruit, an underutilized nutrient-rich climate-smart fruit tree crop in Kenya and Uganda. J Crop Improv 36:619–637

Ooi Z, Ismail H, Abu Bakar A, Aziz N et al (2011) Effects of jackfruit waste flour on the properties of poly (vinyl alcohol) film. J Vinyl Addit Technol 17:198–208

Oraño JFV, Maravillas EA, Alia C et al (2020) Classification of jackfruit fruit damage using color texture features and backpropagation neural network. Int J Adv Sci Eng Inf Technol 10:1813–1820

Ortega-González L, Güemes-Vera N, Piloni-Martini J, Quintero-Lira A, Soto-Simental S et al (2022) Substitution of wheat flour by jackfruit (Artocarpus heterophyllus lam.) seed flour: effects on dough rheology and deep-frying doughnuts texture and sensory analysis. Int J Gastron Food Sci 30:1–6

Pablo AGP, Cimafranca LC (2022) Development of a water kefir beverage made from jackfruit processing by-product and its potential probiotic property. Philipp J Sci 151:1589–1598

Palamthodi S, Shimpi S, Tungare K et al (2021) A study on nutritional composition and functional properties of wheat, ragi and jackfruit seed composite flour. Food Sci Appl Biotechnol 4:63–75

Papa Spada F, da Silva PPM, Mandro GF, Margiotta GB, Spoto MHF, Canniatti-Brazaca SG et al (2018) Physicochemical characteristics and high sensory acceptability in cappuccinos made with jackfruit seeds replacing cocoa powder. PLoS One 13:1–12

Parihar PS, Brahma S, Shirole A et al (2021) Benefits of jackfruit and its natural properties to treat diabetes or colon cancer and weight loss related issues. Int Res J Eng Technol 8:810–815

Pathak N et al (2022) Valorization of jackfruit waste into value added products and their potential applications. Front Nutr 9:1–9

Phukan H, Singha LI, Mitra PK et al (2018) Hepato-protective effect of aqueous extract of seed, leaf and fruit of jackfruit (*Artocarpus heterophyllus* lam.) against CCl_4 induced hepatotoxicity on swiss albino mice. World J Pharm Res 7:768–779

Pintado T, Delgado-Pando G (2020) Towards more sustainable meat products: extenders as a way of reducing meat content. Foods 9:1–19

Prakash O, Kumar R, Mishra A, Gupta R et al (2009) Artocarpus heterophyllus (Jackfruit): an overview. Pharmacogn Rev 3:353–358

Prasad R, Gupta A (2018) Processing and utilization of jackfruit seeds, pearl millet and soybean flour for value addition. J Pharmacogn Phytochem 7:569–572

Pu SM, Chen WD, Zhang YJ, Li JH, Zhou W, Chen J, Liu CM et al (2022) Comparative investigation on the phytochemicals and biological activities of jackfruit (Artocarpus heterophyllus Lam.) pulp from five cultivars. Plant Foods Hum Nutr 1:1–10

Rahma FA, Mahati E, RMD RK, Halleyantoro R et al (2020) Effectivity of the jackfruit leaves infusion (Artocarpus heterophyllus l.) as a larvacide to the mortality rate of Aedes aegypti larva. Jurnal Profesi Medika: Jurnal Kedokteran dan Kesehatan 14:12–17

Rai R, Srivastava R, Kumar R, Singh V et al (2022) An efficient protocol for shoot organogenesis and plant regeneration in Jackfruit. Indian J Hortic 79:273–279

Raíces D, Ferreira PM, Mello JHF, de Godoy Bergallo H (2017) Smile, you are on camera or in a live trap! The role of mammals in dispersion of jackfruit and native seeds in Ilha Grande State Park, Brazil. Nat Conserv Res 2:78–89

Rajagopal K (2023) Effect of extrusion conditions on physical properties of an extruded snack incorporated with chicken meat, rice flour and jackfruit bulb and seed powders. J Pharm Innov 12:2058–2062

Ramli RA, Azmi A, Johari NR, Noor SM et al (2017) Minimally processed jackfruit: opportunity for the foodservice industry. J Food Res 6:1–6

Ramli ANM, Badrulzaman SZS, Hamid HA, Bhuyar P et al (2021) Antibacterial and antioxidative activity of the essential oil and seed extracts of Artocarpus heterophyllus for effective shelf-life enhancement of stored meat. J Food Process Preserv 45:1–10

Ramya H, Anitha S, Ashwini A et al (2020) Nutritional and sensory evaluation of jackfruit rind powder incorporated with cookies. Int J Curr Microbiol Appl Sci 9:3305–3312

Rana SS, Pradhan RC, Mishra S et al (2018) Variation in properties of tender jackfruit during different stages of maturity. J Food Sci Technol 55:2122–2129

Ranasinghe R, Maduwanthi S, Marapana R et al (2019) Nutritional and health benefits of jackfruit (Artocarpus heterophyllus Lam.): a review. Int J Food Sci 2019:1–13

Ravindran C, Kohli A, Murthy BS, Srinivasa A et al (2007) Fruit production in India. Chron Hortic 47:21–26

Ravindran A, Raman M, Babu N, Dinakaran A, Sankar TV, Srinivasa Gopal TK et al (2020) Diet chocolates and replacement of cocoa powder with jackfruit seed powder. Food Nutr Sci 11(3). https://doi.org/10.4236/fns.2020.113017

Razak SFA, Rahman WA, Majid NA et al (2018) Effect of jackfruit rind-based cellulose (JR-CEL.) on physical and mechanical properties of the biodegradable glycerol/gelatine matrix film. AIP Conf Proc 2020:1–9

Reddy IC, Prabakar C, Devi KS, Ponnarasi T, Peter YS et al (2019) An economic analysis on jackfruit production and marketing in Cuddalore District of Tamilnadu, India. Plant Arch 19:2801–2809

Rengsutthi K, Charoenrein S (2011) Physico-chemical properties of jackfruit seed starch (*Artocarpus heterophyllus*) and its application as a thickener and stabilizer in chilli sauce. LWT-Food Sci Technol 44:309–1313
Riyadh ZA, Rahman MA, Miah MG, Saha SR, Hoque MA, Saha S, Rahman MM et al (2019) Performance of aroid under jackfruit-based agroforestry system in terrace ecosystem of Bangladesh. Ann Bangladesh Agric 23:79–87
Rosnah S, Ling C, Ling C, Noraziah M, Osman H et al (2009) Chemical compositions of the jackfruit juice (Artocarpus) cultivar J33 during storage. J Appl Sci 9:3202–3204
Rossi-Márquez G, Helguera M, Briones M, Dávalos-Saucedo CA, Di Pierro P et al (2021) Edible coating from enzymatically reticulated whey protein-pectin to improve shelf life of roasted peanuts. Coatings 11:1–11
Safari S, Razali NA, Mustaffa R et al (2019) Distribution channel assessment: a case study in exporting Malaysian jackfruit to United Arab Emirates (UAE) by air shipment. Int J Agric For Plant 8:75–85
Saha S, Sarker M, Haque AR, Nayeem TA, Maukeeb ARM et al (2022) A review on tropical fruit: jackfruit (Artocarpus heterophyllus). Asian J Adv Res 5(1):298–307
Savita SM, Srinivasappa KN, Vasanthi BG, Manjunath B, Anandamanegar G, Padmavathi M et al (2017) Value addition and income generation using jack fruit-a scope for micro-enterprise. Trends Biosci 10:5541–5546
Saxena A, Bawa AS, Raju PS et al (2008) Use of modified atmosphere packaging to extend shelf-life of minimally processed jackfruit (*Artocarpus heterophyllus* L.) bulbs. J Food Eng 87:455–466
Saxena A, Saxena TM, Raju P, Bawa A et al (2013) Effect of controlled atmosphere storage and chitosan coating on quality of fresh-cut jackfruit bulbs. Food Bioprocess Technol 6:2182–2189
Shams R, Parkins JR, Swallow BM et al (2015) Socio-cultural dimensions of a Jackfruit-Burmese grape dominated agroforestry system in Bangladesh. Int J Dev Sustain 4:721–743
Shanooba PM, Tungare K, Sunariwal S, Sonawane S et al (2020) Extraction and characterization of polyphenols from artocarpus heterophyllus and its effect on oxidative stability of peanut oil. Int J Fruit Sci 20:S1134–S1155
Shedge MS, Haldankar PM, Ahammed Shabeer TP, Pawar CD, Kasture VV, Khandekar RG, Khapare LS et al (2022) Jackfruit: functional component related with human health and its application in food industry. Pharma Innov J 11:824–830
Shi J, Cao H, Guo F et al (2018) Nutrition analysis and security assessment of jackfruit seeds. Food Res Dev 39:193–196
Singh NK (2022) The sacred flora of India: a case for biodiversity conservation. Ecol Environ Conserv 28:87–92
Singh HS, Baradevanal G (2023) Jackfruit: botany, production and uses. CABI, GB, Wallingford, pp 217–243
Siregar AB, Bulan R, Yusak Y et al (2018) Antibacterial & antioxidant properties of leave & stem bark extract of *Artocarpus heterophyllus* as the component of peel-off mask. Int J Sci Technol Eng 5:101–106
Somashekhar M, Nayeem N, Sonnad B et al (2013) A review on family Moraceae (Mulberry) with a focus on Artocarpus species. World J Pharm Pharm Sci 2:2614–2626
Sousa DFD, Campos Filho PC, Conceição AOD et al (2021) Antibacterial activity of jackfruit leaves extracts and the interference on antimicrobial susceptibility of enteropathogen. Food Sci Technol 42:1–10
Sowmyashree G, Devaraja S (2022) Jackfruit and its beneficial effects in boosting digestion and immune-enhancing properties. In: Nutrition and functional foods in boosting digestion, metabolism and immune health. Academic Press, London, pp 267–287
Spada FP, Mandro GF, da Matta Junior MD, Canniatti-Brazaca SG et al (2020) Functional properties and sensory aroma of roasted jackfruit seed flours compared to cocoa and commercial chocolate powder. Food Biosci 37:1–9
Spada FP, Balagiannis DP, Purgatto E, do Alencar SM, Canniatti-Brazaca SG, Parker JK et al (2021) Characterisation of the chocolate aroma in roast jackfruit seeds. Food Chem 354:1–12

Spada FP, de Alencar SM, Purgatto E et al (2022) Comprehensive chocolate aroma characterization in beverages containing jackfruit seed flours and cocoa powder. Future Foods 6:1–13

Srinivasan R, Sevgan S, Ekesi S, Tamò M et al (2019) Biopesticide based sustainable pest management for safer production of vegetable legumes and brassicas in Asia and Africa. Pest Manag Sci 75:2446–2454

Statistics A (2019) United States Department of Agriculture, National Agricultural Statistical Service. United States government printing office, Washington, DC

Statistics BBO (2022) Summary crop statistics area, yield rates and productions of minor crops 2020–21 and 2021–2022 (update: 12-12-2022). http://bbs.portal.gov.bd/sites/default/files/files/bbs.portal.gov.bd/page/16d38ef2_2163_4252_a28b_e65f60dab8a9/2022-12-1-06-15-661c2e4e8ee906cbfb2b77bee3d74429.pdf

Su L, Bai T, Qin X, Yu H, Wu G, Zhao Q, Tan L et al (2021) Organic manure induced soil food web of microbes and nematodes drive soil organic matter under jackfruit planting. Appl Soil Ecol 166:1–12

Sultana A, Rahman MRT, Islam M, Rahman M, Alim MA et al (2014) Evaluation of quality of chapaties enriched with jackfruit seed flour and bengal gram flour. IOSR J Environ Sci Toxicol Food Technol 8:73–78

Sumari P, Kassim AM, Song-Quan O, Nair G, Aminuddin NF et al (2022) Classification of jackfruit and cempedak using convolutional neural network and transfer learning. IAES Int J Artif Intell 11:1353–1361

Suzihaque MUH, Zaki NAM, Alwi H, Ibrahim UK, Abd Karim SF, Anuar NK et al (2022) Jackfruit seed as an alternative replacement for starch flour. Mater Today: Proc 63:S451–S455

Swathi B, Lekshmi G, Sajeev M et al (2019) Cooking quality, nutritional composition and consumer acceptance of functional jackfruit pasta enriched with red amaranthus. Environ Conserv J 20:89–97

Tait G, Mermer S, Stockton D, Lee J, Avosani S, Abrieux A, Walton VM et al (2021) Drosophila suzukii (Diptera: Drosophilidae): a decade of research towards a sustainable integrated pest management program. J Econ Entomol 114:1950–1974

Thatsanasuwan N, Duangjai A, Suttirak P, Phanthurat N et al (2023) Proximate composition and sensory attributes of gluten-free pasta made from jackfruit seeds. Funct Foods Health Dis 13:11–21

Thompson RC, Swan SH, Moore CJ, Vom Saal FS et al (2009) Our plastic age. Philos Trans R Soc Lond B Biol Sci 364:1973–1976

Trumbo P, Schlicker S, Yates AA, Poos M et al (2002) Dietary reference intakes for energy, carbohydrate, fiber, fat, fatty acids, cholesterol, protein and amino acids.(Commentary). J Am Diet Assoc 102:1621–1631

Uzayisenga R, Ayeka PA, Wang Y et al (2014) Anti-diabetic potential of Panax notoginseng saponins (PNS): a review. Phytother Res 8:510–516

Van K, Pham VT, Phu ND, Nhan NPT, Nguyen HK, Du NH (2021) Effects of storage conditions on quality change of minimally dried jackfruit (Artocarpus heterophyllus Lam.) Bulbs. J Mater Sci Technol 1048. https://doi.org/10.4028/www.scientific.net/MSF.1048.531

Vargas-Torres A, Becerra-Loza AS, Sayago-Ayerdi SG, Palma-Rodríguez HM, de Lourdes García-Magaña M, Montalvo-González E et al (2017) Combined effect of the application of 1-MCP and different edible coatings on the fruit quality of jackfruit bulbs (Artocarpus heterophyllus Lam) during cold storage. Sci Hortic 214:221–227

Variasa TO (2018) Shade-grown coffee under fruit trees in highland forests as part of an environmental village restoration. E3S Web Conf 74:1–6

Vazhacharickal PJ, Sajeshkumar NK, Mathew JJ, Kuriakose AC, Abraham B, Mathew RJ, Jose S et al (2015) Chemistry and medicinal properties of jackfruit (Artocarpus heterophyllus): a review on current status of knowledge. Int J Innov Res Rev 3:83–95

Vázquez-González Y, Ragazzo-Sánchez JA, Calderón-Santoyo M et al (2020) Characterization and antifungal activity of jackfruit (*Artocarpus heterophyllus* Lam.) leaf extract obtained using conventional and emerging technologies. Food Chem 330:1–10

Victor-Aduloju A, Nwanja N, Ezegbe C, Okocha K, Aduloju T et al (2020) Phytochemicals and vitamin properties of smoothie flavoured with mint leaves extract. Int J Biochem Res 29:24–30
Viresh M, Mini C, Soni K et al (2022) Jackfruit (Artocarpus heterophyllus Lam.) non-edible parts: a potential source of antioxidants and anti-hyperglycemic agent. Pharm Innov J 11:1419–1422
Wang Y, Jiang Z (2022) Rapid large-scale preparation of polysaccharides from jackfruit peel waste by high-speed countercurrent chromatography and their antioxidant and hypoglycemic activities. J Sep Sci 45:771–779
Wang S, Gao L, Zheng M, Qiao L, Xu C, Wang K, Huang D et al (2021) Occurrences, congener group profiles, and risk assessment of short-and medium-chain chlorinated paraffins in cup instant noodles from China. Chemosphere 279:1–7
Wangchu L, Singh D, Mitra S et al (2013) Studies on the diversity and selection of superior types in jackfruit (*Artocarpus heterophyllus* Lam.). Genet Resour Crop Evol 60:1749–1762
Weintraub K, Rodrigues C, Tabai K et al (2022) Perspectives on sustainable management of jackfruit trees for food consumption in Rio de Janeiro, Brazil. Environ Sci Proc 15:1–8
Wigati D, Setyaningrum L, Pratoko DK et al (2023) The effect of extraction methods on the total phenols and total flavonoids content of jackfruit (*Artocarpus heterophyllus* Lamk) peels extract. EKSAKTA: Berkala Ilmiah Bidang MIPA 24:30–39
Witherup C, Zuberi MI, Hossain S, Zerega NJ et al (2019) Genetic diversity of Bangladeshi jackfruit (*Artocarpus heterophyllus*) over time and across seedling sources. Econ Bot 73:233–248
Wu L, Bao J (2013) Anti-tumor and anti-viral activities of Galanthus nivalis agglutinin (GNA)-related lectins. Glycoconj J 30:269–279
Wu J, Zhou X, Zhou L, Liu W, Zhong J, Zhang Y, Liu C et al (2022) Physicochemical, structural, and functional properties of protein fractions and protein isolate from jackfruit seeds. J Food Sci 87:1540–1551
Yi J, Zhou L, Bi J, Chen Q, Liu X, Wu X et al (2016) Influence of pre-drying treatments on physicochemical and organoleptic properties of explosion puff dried jackfruit chips. J Food Sci Technol 53:1120–1129
Ying JCL, Jaffar NS, Hussin WMRIW, Hamid MFA, Muhsin NI et al (2020) Maintaining postharvest quality of jackfruit during cold storage. Int J Agric For Plant 10:282–286
Yudhistira B (2022) The development and quality of jackfruit-based ethnic food, gudeg, from Indonesia. J Ethn Foods 9:1–10
Zhang Y, Li B, Xu F, He S, Zhang Y, Sun L, Tan L et al (2021) Jackfruit starch: composition, structure, functional properties, modifications and applications. Trends Food Sci Technol 107:268–283
Zheng YZ, Deng G, Zhang YC et al (2022) Multiple free radical scavenging reactions of flavonoids. Dyes Pigments 198:1–7
Zuraidah M, John BA, Kamaruzzaman Y et al (2017) Cytotoxicity on MCF7 cell lines exposed to an extract of the jacalin from jackfruit seed. Sci Herit J 1:16–18
Zuwariah I, Noor F, Hadijah M, Rodhiah R et al (2018) Comparison of amino acid and chemical composition of jackfruit seed flour treatment. Food Res 2:539–545

Chapter 18
Fenugreek (*Trigonella foenum-graecum*): An Overview of Food Uses and Health Benefits

Nazir Ahmad, Sakhawat Riaz, Anwar Ali, and Tariq Riaz

18.1 Introduction

Leguminous plant fenugreek (*Trigonella-Foenum Graecum*) is one of the oldest and most potent plants for medicinal purposes. *Methi* is a popular name for fenugreek in Hindi. Because of its dietary and therapeutic benefits as a herbal cure, this herb has been extensively employed for over 2500 years (Mukthamba and Srinivasan 2017). Approximately 70–97 species of fenugreek are cultivated worldwide. The species of the genus *foenum-graecum* signifies "Greek hay" since it comes from Greece (Zandi et al. 2017). Fenugreek is a herbaceous, leguminous, rain-fed crop that is used in the seed spices production. Fenugreek plant grows to a height in range of 30–60 cm. Due to its qualities, the leaves and seeds are utilized as unusual spices to add flavor and color to cuisines, and for changing foods consistency (Pal and Mukherjee 2020).

Fenugreek is known to be introduced for cultivation in the near East as food, feed and herb. According to Cato, fenugreek is classified as an herb, a leguminous plant and as a crop cultivated for animal feed. Wine and fenugreek were introduced to

N. Ahmad (✉)
Department of Nutritional Science, Government College University Faisalabad, Faisalabad, Pakistan
e-mail: drnazirahmad@gcuf.edu.pk

S. Riaz
Department of Home Economics, Government College University Faisalabad, Faisalabad, Pakistan

A. Ali
Department of Epidemiology and Health Statistics, Xiangya School of Public Health, Central South University, Changsha, China

T. Riaz
Office of the Human Resource Development Center Baltistan, Skardu, Pakistan

T. Ismail et al. (eds.), *Neglected Plant Foods of South Asia*,
https://doi.org/10.1007/978-3-031-37077-9_18

meals by the Romans. Fenugreek was farmed for the very first time as a staple food in Galilee in the first century (Nayak et al. 2018). This plant is extensively grown in India and Pakistan, as well as many other countries of Middle East Asia, West, United Kingdom, Australia, Mediterranean Europe, North America and Africa.

Fenugreek has considerable development potential in dry lands (rain fed) with little irrigation assistance. The crop, on the other hand, responds well under conventional irrigation system, making it excellent for sustainable and resilient farming due to its relatively low water requirements reducing farmers' cost of production. Fenugreek primarily being a dry land crop, is suitable for low-input agro-ecosystems in emerging and impoverished African, Latin American and Asian countries (Basu et al. 2018).

It is an annual diploid species ($2n = 16$) belonging to the Leguminosae family, fenugreek can be identified by a variety of names across the globe (Gołaszewska and Wierzbowska 2017). The plant has a long taproot and a primary stem that is sparsely branched and grows between 30 and 60 cm in height. The stem is cylindrical. Fenugreek has tall, pediculate compound leaves that are pinnately trifoliate and have toothed, triangular, lanceolate stipules that divide into simple leaflets. Bugs pollinate its hermaphroditic, white to yellow-white axillary and sessile blooms (Nayak and Pal 2018). Each flower is made up of five petals. The ovary is aldohexose and immature, with oval to spherical spore grains (Pal and Mukherjee 2020). Fenugreek is a hardy plant that thrives in dry grasslands, slopes, and planes, as well as on the field borders. It does, however, need a significant amount of daylight (Sharma et al. 1990). Fenugreek takes around 4–7 months to achieve full maturity. Flowers bloom from June to August, and seeds develop in the fall (August to September). The development of the flowers and leaves is dependent on temperature, although fenugreek is a drought-resistant plant that does well in tropical climates with moderate winters and summers that are hot (Basch et al. 2003).

The seeds are available as whole or powdered form, and are used as spices in a variety of meals, including teas, curry powders, and spice mixes. A firm, yellow embryo is enclosed in a corneous, thick layer of white, semi-transparent endosperm in fenugreek seeds (Wani and Kumar 2018). In different traditions such as those of Egyptian, *methi* or fenugreek has been used to help childbirth and boost milk production. Recent traditional knowledge on uses of fenugreek still identify Egyptian women to use fenugreek in easing period cramps, and to prepare tea (*hilba*) to relieve various sorts of stomach discomfort. In China, fenugreek tea is known as "*hu luba*", and is used to ease stomach pain (Wani and Kumar 2018; Khalid et al. 2022). *Methi* is used in the production of bread in Egypt and Ethiopia, while it is used as flavouring ingredient for cheese in Switzerland. In the United States, it is commonly used in the preparation of spice mixes for stews and soups (Passano 1995). In Europe, fenugreek was traditionally valued as a spice in cooking and it is still commonly used in Pakistani and other Asian nations like Bangladesh and India as principal ingredient curry powders, pickles, and spice blends. Irrespective of their abundance in certain ecological settings and its historical application as human food, fenugreek is currently considered a neglected or underutilized food plant (Ozturk et al. 2018). The Food & Drug Administration, USA has certified fenugreek

Table 18.1 Potential fenugreek production areas

Continental distribution	Countries
South and southeast Asia	Singapore, Bhutan, Bangladesh, Malaysia the Philippines, Maldives, Sri Lanka, Myanmar, Vietnam, Thailand, Laos PDR, Cambodia, New Guinea, Brunei-Darussalam, Indonesia, East Timor, and Papua
Central Asia	Mongolia, Uzbekistan, Kyrgyzstan, Armenia, Georgia, Azerbaijan, Kazakhstan, Tajikistan
Europe	Monaco, Bulgaria, Gibraltar, Ukraine, Malta, Bosnia-Herzegovina, Kosovo Montenegro, Cyprus, Albania, Serbia, Moldavia, Macedonia, Belorussia, and Georgia
Far East	South Korea and Japan
Africa	Western Sahara, Libya, Algeria, Sub-Saharan African nations, The Horn of Africa (Somalia and Djibouti)
America	Colombia, Brazil, Venezuela, Costa Rica, Suriname, Guyana, Ecuador, Uruguay
North America South	Costa, Mexico, Guatemala, Nicaragua, Belize, El Salvador, Honduras, Panama
Oceania	Pacific Islands

as safe for human ingestion, and has proven to be an ameliorative ingredient against many health challenges (Sureshkumar et al. 2018). A list of potential fenugreek producers across the globe is presented in Table 18.1 (Basu et al. 2018).

18.2 Nutritional Perspectives of Fenugreek

Fenugreek holds a variety of essential minerals and phytochemical of human health significance (Syed et al. 2020). Fenugreek seeds contain 5–10% lipids, 45–60% carbohydrates including mucilaginous fibers and galactomannan, and protein in a range between 20% and 30%. Saponins, free amino acids notably 4-hydroxyisoleucine, pyridine-type alkaloids (mainly trigonelline), and glycosides that form steroidal sapogenins, such as diosgenin, are other key components (Fatima et al. 2018). Fenugreek leaves are rich in several essential vitamins, minerals and choline. The seeds of fenugreek are astringent, aromatic, galactogogue, carminative, and hold antibacterial properties. The leaves are a valuable source of dietary fiber since the foliage of fenugreek contains 50% carbohydrates as dietary fiber. The fiber part is of the leaves is constituted of 20% soluble and 30% insoluble fractions, whose main component is galactomannan (Fatima et al. 2018). Additionally, fenugreek includes a variety of micronutrients such as vitamin C (220 mg/100 g), calcium (75 mg/100 g), zinc (2.4 mg/100 g), manganese, copper (0.9 mg/100 g), potassium (603 mg/100 g), iron (25.8 mg/100 g), β-carotene (19 mg/100 g), and magnesium (42 mg/100 g) (Wani and Kumar 2018; Al-Jasass and Al-Jasser 2012). Phenylalanine tyrosine, glutamic acid, leucine, and aspartic acid, are just a few of the vital amino acids present in fenugreek. It also contains trace levels of methionine and cysteine which seem to be crucial in the body's physiological functions. 4-hydroxyisoleucine is among the most abundant free amino acid in fenugreek.

The non-protein amino acid i.e., 4-hydroxyisoleusine accounts for approximately 80% of the amino acid contents of the dried fenugreek seeds. Available data on nutritional quality of fenugreek suggest fenugreek proteins as more promising on quality index as compared to the protein of many other edible plants (Syed et al. 2020). Feyzi and his team conducted research in which they evaluated the protein quality of fenugreek matches that of soybean protein isolate suggesting fenugreek seeds to hold better protein profile and a superior amino acid makeup than soy protein isolates (Feyzi et al. 2015).

18.3 Primary and Secondary Bioactive Metabolites of Fenugreek

Fenugreek contains several alkaloids, including trigonella, trimethyl coumarin, trigocoumarin, and nicotinic acid. It also contains polyphenols and flavonoids, which are significant and beneficial elements (Ahmad et al. 2023). Alcoholic extracts of fenugreek plant have been reported to comprise a variety of flavonoids such as luteolin, vitexin, 7,4-dimethoxy flavanonest, and quercetin (Syed et al. 2020). Other studies have shown aglycones, naringenin tricin, kaempferol, and quercetin as potential bioactive compounds of fenugreek (Petropoulos 2002). Mature seeds also contain gentianine, choline, argenine, lysine, trigonelline, histidine, triethylamine, leucine, neurin, isoleucine, 4-hydroxyisoleucine, carpine, betaine, L-tryptophan, (Dixit et al. 2005). Table 18.2 highlights the nutritional makeup of fenugreek. Fenugreek seeds contain 0.1–0.9% diosgenin. Multiple coumarin compounds alongside alkaloids have been isolated from the fenugreek seeds such as gentianine, carpaine, trigonelline (Acharya et al. 2006).

Polyphenol like isovitexin and rhaponticin are also believed to be a few among the primary bioactive ingredients of seeds (He et al. 2015). A list of bioactive compounds identified from the various fractions is given in Table 18.3.

18.4 Health Perspectives of Fenugreek and Its Metabolites

The value of integrating medicinal herbs in cuisines is widely established to maintain a healthy lifestyle and avoiding risks of food-related illnesses such as inflammation, hypertension, cardiovascular disease, diabetes and cancers (Thorat and Gaikwad 2019; Khalid et al. 2022). The fenugreek seed and leaves are used as traditional therapeutics as antioxidant (Mukthamba and Srinivasan 2017) anti-diabetic

Table 18.2 Fenugreek seed and leaf composition

Component	Seeds	Leaves
Fiber	50 g	1.0 g
Protein	30 g	4.4 g
Moisture	–	86.0 g
Trigonelline	380 mg	–
Sapogenins	2 g	–
Vitamin C	43 mg	52 mg
Thiamine	340 μg	40 μg
Riboflavin	290 μg	310 μg
Folic acid	84	–
Nicotinic acid	1.1 μg	800 μg
Fe	14 mg	16.5 mg
P	370 mg	51 mg
Ca	160 mg	395 mg
Na	19 mg	76 mg
Mg	160 mg	67 mg
Cl	165 mg	165 mg
S	16 mg	167 mg

Source: Srinivasan et al. (2007)

Table 18.3 The bioactive substances present in fenugreek

Chemical group	Compounds	References
Amino acids	4-Hydroxyisoleucine, arginine, lysine, tyrosine, histidine, cystine, tryptophan	Ruby et al. (2005); Gupta et al. (2001); El Nasri and El Tinay (2007)
Coumarins	Trigocumarin, methyl coumarin, trimethyl coumarin	Raju et al. (2001)
Alkaloids	Trigonelline, choline, carpaine	Kaviarasan et al. (2007); Rababah et al. (2011)
Flavonoids	Quercetin, tricin 7-O-D glucopyranoside, vitexin, isoorientin, saponaretin, kaempferol, orientin, saponaretin, lilyn, naringenin, luteolin	Ahmad et al. (2016)
Saponins	Foenugracin, Fenugrin, Diosgenin, Neotigogenin, Sarsasapogenin, Yamogenin, Yuccagenin, Glycoside, Gitogenin, Hederagin, Tigonenin, Yuccagenin, Smilagenin	Yoshinari and Igarashi (2010); Gupta et al. (2001)
Others	Vitamin A, Thiamin, Nicotinic Acid, Biotin, Ascorbic Acid, Riboflavin, Gum	Hamden et al. (2010); Chatterjee et al. (2010)

(Bahmani et al. 2016; Swaroop et al. 2018), anti-inflammatory (Tavakoly et al. 2018), anti-cancer (El Bairi et al. 2017), and anti-microbial (Vickers et al. 2017) referred to the Fig. 18.1. Furthermore, fenugreek seeds have been reported as potent radical-scavenging agent (Baba et al. 2018).

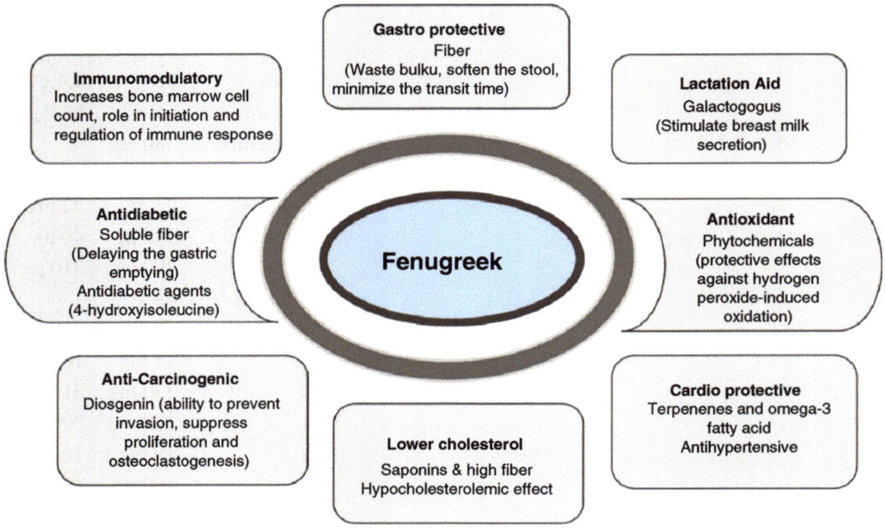

Fig. 18.1 Health benefits of using fenugreek

18.4.1 Cholesterol-Lowering Effect

Dyslipidemia, namely hypercholesterolemia, is one of the most serious risk variables for the onset of cardiovascular diseases. A plethora of research in animal models and those of the clinical studies validate hypocholesterolemic effects of fenugreek seeds. Feeding 15–60% fenugreek seed based cholesterol enriched diet to the rats had anticipated a significant decline in the blood cholesterol levels of the experimental model (Sharma 1984). In Wistar rats, feeding fenugreek seeds 10% of the feed contents had a hypocholesterolemic outcome coupled with a high-cholesterol diet (Mukthamba and Srinivasan 2015). Atherosclerosis or development of plaques in arteries is strongly associated with the increased circulating levels of low-density lipoprotein (LDL) and very-low-density lipoprotein (VLDL) cholesterol. Lowering the LDL-cholesterol concentrations within the blood is a critical technique for lowering the possibility of a heart attack. Research has looked into a variety of spices of fenugreek for their advantageous lipid-lowering properties (Srinivasan 2005). It has been demonstrated to lower cholesterol levels in both animal studies and a few clinical trials (Srinivasan et al. 2007). High fiber contents of the fenugreek and substantial fraction of saponins are considered principle ingredients responsible for reduction in cholesterol levels (Sharma and Raghuram 1990). The protein in diet have also been found impactful in maintaining normal cholesterol levels. Plant proteins tend to reduce cholesterol levels in general (O'Keefe et al. 2004). Cholesterol lowering effect of fenugreek at hepatic level is presented in Fig. 18.2. Hypolipidemic effect of fenugreek derived galactomannans is predominately linked to enhanced HMGA-CoA reductase activity, as a consequence, bile acid and neutral sterol

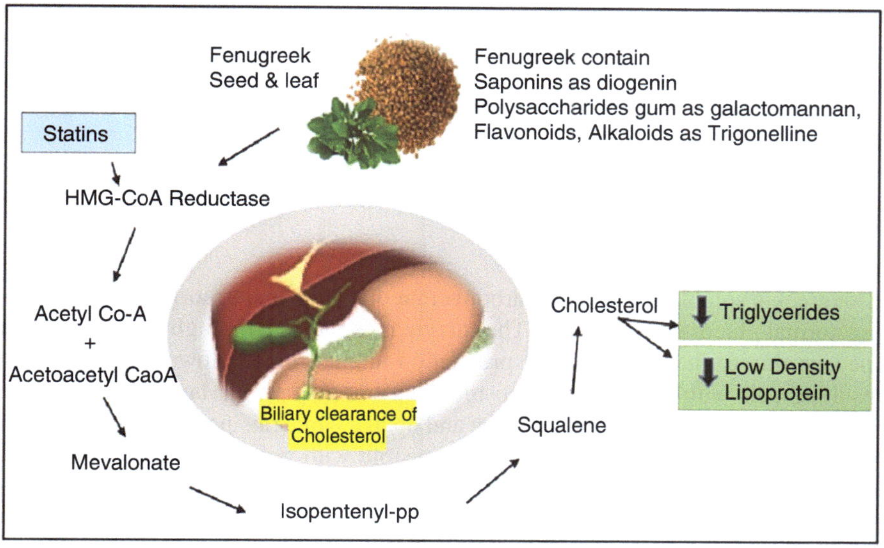

Fig. 18.2 Cholesterol lowering effect of fenugreek

excretion in the faeces increases (Ramulu et al. 2011). Cholesterol absorption is decreased by Diosgenin, a furostanol saponin derived from fenugreek, also decreasing hepatic cholesterol concentrations while increasing biliary cholesterol excretion, lowering serum cholesterol concentrations (Sfar et al. 2018).

18.4.2 Anti-carcinogenic Activities

Cancer is a lethal and challenging disease caused by uncontrolled and unpredictable cell development, and its prevalence rate is steadily growing (Khorshidian et al. 2016). Multiple research studies indicate that fenugreek has anticarcinogenic properties (Amin et al. 2005; Al-Oqail et al. 2013; Manzoor et al. 2022). Diosgenin (C27H42O3), in fenugreek, is regarded as an important ingredient in the manufacture of steroid hormones like progesterone & cortisol Listed steroid hormones have been indicated in cancer therapy due to their ability to suppress necrosis factor and hence decrease invasion, proliferation, and osteoclastogenesis. In response to cytokines and chemotherapeutic drugs, NF-kappa B regulates gene expression and promotes apoptosis (Fatima et al. 2018). In the H-29 colorectal cancer cell line, Fenugreek diosgenin has also been demonstrated to inhibit cell growth and promote apoptosis (Raju and Bird 2006). It had been reported that a fenugreek-derived compound namely protodioscincan suppress HL60 cell proliferation by inducing apoptotic changes (Hibasami et al. 2003). Another research discovered that fenugreek seed extract can help inhibit the occurrence of mammary hyperplasia produced by dimethylbenz-anthracene in rats. It can increase apoptosis on regular consumption of fenugreek and seems to have an effect on breast cancer. The extract of the whole fenugreek plant demonstrates *in vitro* cytotoxicity against the most of human cancer cell lines, notably IMR-32, HT29, and neuroblastoma (Rambhade et al. 2010).

18.4.3 Antidiabetic Activities

Fenugreek's soluble fiber slows carbohydrate digestion and absorption thus ensuring normal glucose homeostasis. The food mass's passage through the intestines is sped up due to the extensive gel production along with the low viscosity of the resulting gels within the gut that postpone gastric emptying (Hannan et al. 2007). Fenugreek reduced glucose breakdown and uptake while raising or imitating insulin activity to enhance glucose regulation in people with type 1 and type 2 diabetes. Fenugreek is easily available, and the seeds are regularly consumed, signifying low toxicity of the seeds, legume, and leaves of this plant. Fenugreek is therefore reflected as a potentially effective dietary remedy for diabetes therapy, as well as a prospective foundation for the progress of novel orally active anti-diabetic treatments (Hannan et al. 2007). Fenugreek seeds contain rhaponticin, polyphenols & stilbenes, which can increase sensitivity of insulin (Li et al. 2018). Fenugreek seed consumption significantly decreased glycosylated hemoglobin concentration and fasting blood sugar levels in type – 2 diabetic patients (Ranade and Mudgalkar 2017).

18.4.4 Antioxidant Activity

Polyphenols in fenugreek seeds have been linked to many health effects (Rayyan et al. 2010). Antioxidant action of plants including the edible plants is generally attributed to their bioactive phytochemicals composition (Ali et al. 2022; Riaz et al. 2022). Antioxidant quality of the extracts are attributed to flavonoids, phenolic compounds bearing high antioxidant capacity (Maqbool et al. 2022). The capacity of antioxidant the seeds is due to the flavonoidsm, polyphenols and many other bioactive soluble fractions (Raju and Bird 2006). Polyphenols of the fenugreek have been shown to protect erythrocytes from hemolysis and lipid peroxidation when subjected to hydrogen peroxide-induced oxidation (Kaviarasan et al. 2004). The seed extracts of fenugreek shows radical scavenging action when combined with acetone, ethanol, methanol, dichloromethane, ethyl acetate, and hexane (Bukhari et al. 2008). When tested vide scavenging technique of free-radical, the fenugreek seed husk extracts demonstrated 72% antioxidant activity, fenugreek seed showed 64% radical scavenging activity that was higher than one recorded for the fenugreek endosperm i.e., 56% (Madhava Naidu et al. 2011). Existing literature on antioxidant activity of fenugreek seeds indicate flavonoids namely tricin, quercetin, vitexin, naringenin, tricin-7-O-beta-D-glucopyranoside, and the seed crude extract to show dose dependent antiradical and antioxidant activities (Shang 1998).

18.4.5 Immunomodulatory

Natural immunomodulators work effectively and seldom cause harm. The immune-stimulatory activities of fenugreek are well documented (Guardiola et al. 2018). Various fenugreek extracts revealed to have immunomodulatory effects on immunological functioning in Swiss albino mice (Bin-Hafeez et al. 2003) and rats (Anarthe et al. 2014). Fenugreek extracts triggers both specific and non-specific immune mechanisms. At 100 mg/kg dosage by body weight, the extracts of fenugreek stimulated the immune mechanism in Swiss albino mice. It stimulates bone marrow hematopoietic stem cells by increasing bone marrow cell count. *T. foenum graecum* induced macrophage stimulation (Rehman et al. 2018). Some of the elements may have mitogenic actions, which in turn cause the immune-competent cells to have immunomodulatory effect. Its influence on macrophages is handled by the high concentration of dietary fiber and mucilage. The study's overall findings show that it has an immunomodulatory function (Bin-Hafeez et al. 2003).

18.4.6 Lactation Aid

Herbs and organic products have been traditionally used in many cultures to boost milk production since ancient times. Fenugreek is well-known galactagogues that boost production of breast milk (Khorshidian et al. 2016). Fenugreek is thought to increase sweat production, which increases breast milk output because the breast is a modified sweat gland (Khorshidian et al. 2016). Estrogen-like compounds are found in fenugreek seeds that in MCF-7 cell lines promote the production of pS2 (Sreeja and Anju 2010). Fenugreek diosgenin and phytoestrogen appear to be accountable for the rise in milk flow (Wiley et al. 2003). It has also been shown to anticipate estrogenic action that promotes breast milk production (Khorshidian et al. 2016).

18.4.7 Other Health Benefits

Research on rats model had suggested hepatoprotective effects of fenugreek seeds by minimizing lipid peroxidation liver and brain cells after ethanol toxicity (Wiley et al. 2003). The seeds of fenugreek have been shown to have immunomodulatory action in mice, as demonstrated by their stimulation of immunological processes, cellularity of the lymphoid organs, delayed hypersensitive response and relative thymus weight (Bin-Hafeez et al. 2003). In experimental rats, dietary fenugreek dramatically boosted the actions of pancreatic lipase and chymotrypsin (Platel and Srinivasan 2000). Additionally, studies have shown that fenugreek seeds have gastro-protective effects on stomach ulcers brought on by ethanol (Pandian et al. 2002). Fenugreek seeds can lengthen the transit duration of waste through the colon, soften the stool by holding more water, and increase the weight of gut waste (Johanson 2007). The majority of anti-inflammatory drugs lower cytokine concentrations associated with inflammatory diseases. Blocking cytokines and myeloperoxidase may alleviate symptoms of rheumatoid arthritis. In cultivated THP-1 cells, fenugreek seed extract inhibited the generation of inflammatory cytokines (Sindhu et al. 2018). In experimentally generated arthritis, fenugreek mucilage significantly reduced the levels of myeloperoxidase and cytokines (TNF-alpha and IL-6) activity (Yacoubi et al. 2011; Suresh et al. 2012).

The anti-nociceptive properties of fenugreek leaf, seed, and stem extracts were examined by a number of studies. In animal models, fenugreek extracts have demonstrated antipyretic and anti-nociceptive characteristics comparable to non-steroidal anti-inflammatory drugs (NSAIDs) (Ahmadiani et al. 2001; Abbas et al. 2016). Fenugreek seeds' anti-nociceptive properties may be related to the alkaloid compounds found in them (Kaviarasan et al. 2007). The extracts by ethanolic of fenugreek seeds displayed antinociceptive properties in rats model, perhaps owing to the combination of endogenous opioids (Biswal et al. 2003). The aqueous and ethanolic extracts of fenugreek seed have been shown to have antibacterial action against a variety of pathogenic microorganisms, including *Staphylococcus aureus* and *Pseudomonas aeruginosa* (Latif 2015; Elnour et al. 2015; Al-Timimi 2019). Similarly, methanolic extracts of the stem and leaves of fenugreek were found effective against *S. aureus* and *E. coli* (Sharma et al. 2016). One of the leading avoidable contributors to mortality in the globe is high blood pressure or hypertension (Kumar et al. 2021). Diuretics, beta blockers, angiotensin-II receptor blockers, calcium channel blockers, angiotensin-converting enzyme (ACE) inhibitors, and aldosterone antagonists are all employed as conventional drugs to treat hypertension (Thomopoulos et al. 2018). Methanolic extracts of fenugreek seeds were reported antihypertensive by 5-HT2 serotonergic antagonistic effect (Balaraman et al. 2006). Fenugreek seeds essential oils carrying terpenes and omega-3 fatty acids have also been found antihypertensive by lowering ACE activity (Wang et al. 2022; Hamden et al. 2011). In a different research, feeding fenugreek extracts increased sodium and potassium ion excretion, resulting in a hypocalciuric effect (AL-atwi 2010). Oxidative stress, inflammation, diabetes, and elevated cholesterol levels, are implicated in various CNS illnesses, resulting in alterations in brain integrity and functioning (Zameer et al. 2018). Fenugreek has been shown to have a neuroprotective effect due to its efficiency in avoiding Parkinson's disease caused by 6-hydroxy dopamine (6-OHDA) and restoring regular brain functioning (Mirzaie et al. 2016; Maqbool et al. 2022). Fenugreek seeds have also been explored for remediating renal stones formation in rat's model. The results suggest fenugreek as a potential traditional remedy against calcic urolithiasis (Laroubi et al. 2007).

18.5 Fenugreek Uses as Food Additive

The increased consumer demand for nutritious diets has driven the food industry to create products with additional health-promoting qualities for customers. Foods that are sufficiently healthy, nutritious, disease-preventive, or health-supportive have received commercial recognition (Khorshidian et al. 2016). Food hydrocolloids are a broad family of long-chain polymers that create viscous dispersions and/or gels when dispersed in water. The presence of several hydroxyl (−OH) groups increases their propensity to bind more water molecules, making them hydrophilic (Burey et al. 2008). Food hydrocolloids have been used in food systems for several decades as emulsifiers, gelling agents, thickeners, bulking agents, stabilizers, and filmmakers (Giosafatto et al. 2007). These biopolymers, if correctly explored as additive, might be used as emulsifiers, thickeners, restorative agents, stabilizers, excipients, bio-based encapsulates, texture modifiers, and coating agents (Guo et al. 2016; Timilsena et al. 2016). Galactomannan of fenugreek origin has been demonstrated as an outstanding natural additive bearing functional properties such as foaming stability, foaming capacity, oil-holding capacity, and better swelling index (Rashid et al. 2018). Incorporating fenugreek fiber to refined flours, the fenugreek fibers help strengthen dough structure by providing a balance of soluble and insoluble fiber. Extruded snacks have been developed using fenugreek seed flour and fenugreek leaf powder (Wani and Kumar 2018). Fenugreek seed husk is high in dietary fiber and contains a variety of essential elements. This fiber-rich functional component is suitable for use in the production of high-fiber muffins. Fenugreek fiber supplemented muffins had enough volume, a soft texture, and a medium-fine grain (Srivastava et al. 2012). In addition to generating feed and seed as direct agronomic products, fenugreek is a useful plant for the nutraceutical, phytochemical, and food industries.

18.6 Fenugreek-Based Food Products and Its Commercial Significance

The demand for processed food is quickly growing in emerging nations due to increased urbanization. Individuals require cheaper but protein-rich diets that take into account of their age, physical activity, gender, and physiological demands (Tamiru Kasaye 2015). The appropriate amino acid composition of the staple diet must thus be strategically supplemented with low-cost, high-protein resources to increase its nutritional significance and alleviate the problem of malnutrition in developing countries (Addisu 2011; Ahmed et al. 2021). By increasing the amino acid profiles, adding non-wheat flour to wheat flour increases the nutritious content of wheat products. To achieve many advantages, fenugreek seed can be used to add value to cereal-based food items. It can be added in crackers, cereals, sauces, cookies, bagels, pizza dough, bread, biscuits, salads, doughnuts, pasta, and chocolates.

Its leaves are abundant in beta-carotene, calcium, iron, and other vitamins and are frequently consumed as a green leafy vegetable in India (Sharma et al. 1990; Ali et al. 2021). Studies revealed that supplementing wheat flour with fenugreek resulted in superior flour for bread making (Tamiru Kasaye 2015). However, the gluten level decreased when the amount of fenugreek flour was increased. The inclusion of fermented fenugreek flour boosted the fiber, protein, ash, and mineral content of goods (Tamiru Kasaye 2015). Some essential minerals are needed as food additives, and gum proteins are frequently used for their emulsifying properties. Biscuits have been made with up to 10% fenugreek flour without any negative effects on performance. The mechanical, sensory, and nutritional qualities of biscuits including 10% germinated fenugreek flour showed that they were the best combo of fenugreek flour biscuits. Therefore, the development and consumption of such multipurpose meals would advantage those with neurodegenerative diseases as well as improve the nutritional status of the community (Hooda and Jood 2005; Khalid et al. 2022).

When mixed with wheat flour and germinated fenugreek powder, fenugreek leaves, seeds, and wheat flour increased the amount of total fibres, iron, proteins, calcium, zinc, vitamin B2, vitamin C, β-carotene, and vitamin E in meals by 5–10% (Wani and Kumar 2018). From basic components, the extrusion process creates a variety of food items with varying qualities, fragrances, flavors, colors, and forms. Extruded snacks have been developed using seed flour and leaf powder. In one experiment, a blend of about 1.78% fenugreek seed flour and 0.66% leaf powder combined with the raw product was utilised to show high levels of demand for functionality, physiological, and colour requirements and may be extruded while keeping adequate quality standards (Wani and Kumar 2016). When combined with rice and chickpeas, fenugreek, in the form of debittered polysaccharide, can be utilized up to 15% to produce snacks with exceptional physical and sensory qualities and a low GI (Gamlath and Ravindran 2009).

18.7 Strategies to Improve the Bioavailability of Fenugreek

Bioactivity is the amount of a chemical that is liberated from its matrix in the gastrointestinal tract and therefore made available for intestinal absorption (Fernández-García et al. 2009). Solid-state fermentation (SSF) is being employed to improve the nutritional content and antioxidant profile of several grains and legumes (Dhull et al. 2020). Fermentation is chosen over alternative procedures for the generation of polyphenolic compounds, possibly owing to the higher growth rate of microorganisms, simple downstream processing, low cost, and environmentally favorable nature of the system. Fermenting bacteria create a variety of enzymes (phytase, amylase, lipase, protease, and so on) that may hydrolyze carbohydrate, protein, and lipid components into easily digestible constituents with a pleasing texture and flavor. Meanwhile, many anti-nutritional substances like protease, phytate, and tannins inhibitors are substantially destroyed by these enzymes (Soetan and Oyewole 2009). This heightened interest in SSF as a means of improving bioavailability of certain minerals found in seeds (Dhull et al. 2021). According to research, the antioxidant profile of fenugreek improves dramatically following SSF. According to research, the bioavailability of trace minerals such as Zn, Ca, and Fe improved significantly when compared to their unfermented equivalents (Dhull et al. 2021). Sapogenins from hydrolyzed fenugreek extracts (HFE) showed high bio-accessibility after in vitro digestion (76%). The digestion of sapogenin & saponin standards revealed that additional extract chemicals were improving bioaccessibility. Other minor bio-active chemicals (tocols, certain phenolics, alkylresorcinols, and phytosterols,) have high bioaccessibility (70–100%) (Navarro del Hierro et al. 2020). Galactomannans are heterogeneous macromolecular polysaccharides with a wide range of activities and use in oil production, food and drilling, textile manufacturing, cosmetics, and other sectors (Wielinga 2009). Fenugreek seed is a source that is used in a variety of ways (Hamdani et al. 2019). Since boosting solubility is one of the most important aspects of bioactive compound dissolution, there are two primary ways to take into account: Increasing surface area by reducing particle size and promoting curcumin dispensability with chemicals. Three of the most often used and effective techniques include the use of hydrophobic carriers, surfactants, and inclusion complexes (Jäger et al. 2014; Jamwal 2018). The most cutting-edge methods for improving the bio-availability of nutraceuticals like curcumin include colloidal delivery technology and excipient foods. Excipient foods don't have any bioactivity on their own, but once they're in the gut, they create a situation that makes whatever substances they brought with them more bio-accessible, stable, or absorbable (McClements 2019; Kharat and McClements 2019; Zheng and McClements 2020).

18.8 Conclusion

Fenugreek is high in protein, fiber, and phytochemical constituents, making it a promising nutritional and health aid. Fenugreek is utilized as an important culinary ingredient in a variety of recipes, as well as to treat a variety of ailments. Additionally, the dietary herb with all its fractions is a well-known conventional ameliorative herb because of its emollient, gastro-protective effects, as a demulcent, wound healing ingredient, and source of bioactive compounds bearing anti-inflammatory, antibacterial, expectorant, aphrodisiac, anthelmintic, and anticancer properties. Fenugreek has been used in the food sector as a source of gum, food emulsifier, food adhesive, food stabilizer, and in the production of a wide range of baked and extruded items. Given all of these health benefits, fenugreek may be recommended and incorporated into our daily diet, as well as mixed into meals to produce functional foods.

Acknowledgments We are thankful to the digital library of GCUF for providing us access to the Publication.

Conflicts of Interest All authors read and approved the final manuscript and declare that they have no conflict of interest.

References

Abbas N, Al Rubaish AR, Ali S (2016) Study of anti-inflammatory analgesic and antipyretic antimicrobial effect of aqueous extract of Trigonella foenum-graecum sprouts and effect of glacial acetic acid on blood cells of mice. Int J Pharm Pharm Sci 1(1):67–73

Acharya S, Srichamroen A, Basu S et al (2006) Improvement in the nutraceutical properties of fenugreek (Trigonella foenum-graecum L.). J Sci Technol 28(1):1–9

Addisu K (2011) Optimization of extrusion conditions for developing ready-to-eat defatted soy-rice and chocolate blend extrudate. Dissertation, Addis Ababa University

Ahmad A, Alghamdi SS, Mahmood K, Afzal M (2016) Fenugreek a multipurpose crop: potentialities and improvements. Saudi J Biol Sci 23(2):300–310

Ahmad N, Riaz S, Ali A (2023) Ingredients for food products. In: Jeguirim M, Khiari B, Jellali S (eds) Palm trees and fruits residues, 1st edn. Elsevier, Amsterdam, pp 115–153

Ahmadiani A, Javan M, Semnanian S et al (2001) Anti-inflammatory and antipyretic effects of Trigonella foenum-graecum leaves extract in the rat. J Ethnopharmacol 75:283–286

Ahmed N, Ali A, Riaz S, Ahmad A, Aqib M (2021) Vegetable proteins: nutritional value, sustainability, and future perspectives. In: Yildirim E, Ekinci M (eds) Vegetable crops-health benefits and cultivation, 1st edn. IntechOpen, London, pp 42–56

Al-atwi LF (2010) Clinical evaluation for the diuretic effect of the alcoholic extract of Trigonella faenum-gracum seeds (fenugreek) on rabbits. Kufa J Vet Med Sci 1(1):116–121

Ali A, Ain Q, Saeed A et al (2021) Bio-molecular characteristics of whey proteins with relation to inflammation. In: Qureshi MS (ed) New advances in the dairy insdustry. IntechOpen, London, pp 3–10

Ali A, Riaz S, Sameen A et al (2022) The disposition of bioactive compounds from fruit waste, their extraction, and analysis using novel technologies: a review. Processes 10(10):2014

Al-Jasass FM, Al-Jasser MS (2012) Chemical composition and fatty acid content of some spices and herbs under Saudi Arabia conditions. Sci World J 859892:1–5. https://doi.org/10.1100/2012/859892

Al-Oqail MM, Farshori NN, Al-Sheddi ES et al (2013) In vitro cytotoxic activity of seed oil of fenugreek against various cancer cell lines. Asian Pac J Cancer Prev 14(3):1829–1832

Al-Timimi LAN (2019) Antibacterial and anticancer activities of fenugreek seed extract. Asian Pac J Cancer Prev 20(12):3771–3776

Amin A, Alkaabi A, Al-Falasi S, Daoud SA (2005) Chemopreventive activities of Trigonella foenum graecum (fenugreek) against breast cancer. Cell Biol Int 29(8):687–694

Anarthe SJ, Sunitha D, Raju MG (2014) Immunomodulatory activity for methanolic extract of Trigonella foenum graecum whole plant in wistar albino rats. Am J Phytomed Clin Ther 2(9):1081–1092

Baba WN, Tabasum Q, Muzzaffar S et al (2018) Some nutraceutical properties of fenugreek seeds and shoots (Trigonella foenum-graecum L.) from the high Himalayan region. Food Biosci 23:31–37

Bahmani M, Shirzad H, Mirhosseini M et al (2016) A review on ethnobotanical and therapeutic uses of fenugreek (Trigonella foenum-graceum L). J Evid Based Complementary Altern Med 21(1):53–62

Balaraman R, Dangwal S, Mohan M (2006) Antihypertensive effect of Trigonella foenum-greacum. Seeds in experimentally induced hypertension in rats. Pharm Biol 44(8):568–575

Basu SK, Zandi P, Cetzal-Ix W (2018) Fenugreek (Trigonella foenum-graecum L.): distribution, genetic diversity, and potential to serve as an industrial crop for the global pharmaceutical, nutraceutical, and functional food industries. In: Watson RR, Singh RB, Takahashi T (eds) The role of functional food security in global health, 1st edn. Elsevier Inc, London, pp 471–497

Basch E, Ulbricht C, Kuo G, Szapary P, Smith M (2003) Therapeutic applications of fenugreek. Altern Med Rev. 8(1):20–27. PMID: 12611558.

Bin-Hafeez B, Haque R, Parvez S et al (2003) Immunomodulatory effects of fenugreek (Trigonella foenum graecum L.) extract in mice. Int Immunopharmacol 3(2):257–265

Biswal S, Das MC, Nayak P (2003) Antinociceptive activity of seeds of Trigonella foenum gracecum in rats. Indian J Physiol Pharmacol 47(4):479–480
Bukhari SB, Bhanger MI, Memon S (2008) Antioxidative activity of extracts from fenugreek. Pak J Anal Environ Chem 9(2):78–83
Burey P, Bhandari BR, Howes T, Gidley MJ (2008) Hydrocolloid gel particles: formation, characterization, and application. Crit Rev Food Sci Nutr 48(5):361–377
Chatterjee S, Variyar PS, Sharma A (2010) Bioactive lipid constituents of fenugreek. Food Chem 119(1):349–353
Dhull SB, Punia S, Kidwai MK et al (2020) Solid-state fermentation of lentil (*Lens culinaris* L.) with *Aspergillus awamori*: effect on phenolic compounds, mineral content, and their bioavailability. Legum Sci 2(3):e37
Dhull SB, Punia S, Kumar R et al (2021) Solid state fermentation of fenugreek (Trigonella foenum-graecum): implications on bioactive compounds, mineral content and in vitro bioavailability. J Food Sci Technol 58(5):1927–1936
Dixit P, Ghaskadbi S, Mohan H, Devasagayam TPA (2005) Antioxidant properties of germinated fenugreek seeds. Phytother Res 19(11):977–983
El Bairi K, Ouzir M, Agnieszka N, Khalki L (2017) Anticancer potential of Trigonella foenum graecum: cellular and molecular targets. Biomed Pharmacother 9:479–491
El Nasri NA, El Tinay AH (2007) Functional properties of fenugreek (Trigonella foenum graecum) protein concentrate. Food Chem 103(2):582–589
Elnour ME, Ali AM, Saeed BE (2015) Antimicrobial activities and phytochemical screening of callus and seeds extracts of fenugreek (Trigonella foenum-graecum). Int J Curr Microbiol Appl Sci 4(2):147–157
Fatima T, Maqbool K, Scholar P et al (2018) Potential health benefits of fenugreek. J Med Plants Stud 6(2):166–169
Fernández-García E, Carvajal-Lérida I, Pérez-Gálvez A (2009) In vitro bioaccessibility assessment as a prediction tool of nutritional efficiency. Nutr Res 29(11):751–760
Feyzi S, Varidi M, Zare F, Varidi MJ (2015) Fenugreek (Trigonella foenum graecum) seed protein isolate: extraction optimization, amino acid composition, thermo and functional properties. J Sci Food Agric 95(15):3165–3176
Gamlath S, Ravindran G (2009) Extruded products with fenugreek (Trigonella foenum-graecium) chickpea and rice: physical properties, sensory acceptability and glycaemic index. J Food Eng 90(1):44–52
Giosafatto CVL, Mariniello L, Ring S (2007) Extraction and characterization of Foeniculum vulgare pectins and their use for preparing biopolymer films in the presence of phaseolin protein. J Agric Food Chem 55(4):1237–1240
Gołaszewska KŻ, Wierzbowska J (2017) Fenugreek: productivity, nutritional value and uses. J Elem 22(3):1067–1080
Guardiola FA, Bahi A, Esteban MA (2018) Effects of dietary administration of fenugreek seeds on metabolic parameters and immune status of gilthead seabream (Sparus aurata L.). Fish Shellfish Immunol 74:372–379
Guo R, Ai L, Cao N et al (2016) Physicochemical properties and structural characterization of a galactomannan from Sophora alopecuroides L. seeds. Carbohydr Polym 140:451–460
Gupta A, Gupta R, Lal B (2001) Effect of Trigonella foenum-graecum (fenugreek) seeds on glycaemic control and insulin resistance in type 2 diabetes mellitus: a double blind placebo controlled study. J Assoc Physicians India 49:1057–1061
Hamdani AM, Wani IA, Bhat NA (2019) Sources, structure, properties and health benefits of plant gums: a review. Int J Biol Macromol 135:46–61
Hamden K, Masmoudi H, Carreau S, Elfeki A (2010) Immunomodulatory, β-cell, and neuroprotective actions of fenugreek oil from alloxan-induced diabetes. Immunopharmacol Immunotoxicol 32(3):437–445
Hamden K, Keskes H, Belhaj S et al (2011) Inhibitory potential of omega-3 fatty and fenugreek essential oil on key enzymes of carbohydrate-digestion and hypertension in diabetes rats. Lipids Health Dis 10(1):226

Hannan JMA, Ali L, Rokeya B et al (2007) Soluble dietary fibre fraction of Trigonella foenum-graecum (fenugreek) seed improves glucose homeostasis in animal models of type 1 and type 2 diabetes by delaying carbohydrate digestion and absorption, and enhancing insulin action. Br J Nutr 97(3):514–521

He Y, Ding C, Wang X et al (2015) Using response surface methodology to optimize counter-current chromatographic separation of polyphenol compounds from fenugreek (Trigonella foenum-graecum L.) seeds. J Liq Chromatogr Relat Technol 38(1):29–35

Hibasami H, Moteki H, Ishikawa K et al (2003) Protodioscin isolated from fenugreek (Trigonella foenumgraecum L.) induces cell death and morphological change indicative of apoptosis in leukemic cell line H-60, but not in gastric cancer cell line KATO III. Int J Mol Med 11:23–26

Hooda S, Jood S (2005) Organoleptic and nutritional evaluation of wheat biscuits supplemented with untreated and treated fenugreek flour. Food Chem 90(3):427–435

Jäger R, Lowery RP, Calvanese AV (2014) Comparative absorption of curcumin formulations. Nutr J 13(1):11

Jamwal R (2018) Bioavailable curcumin formulations: a review of pharmacokinetic studies in healthy volunteers. J Integr Med 16(6):367–374

Johanson JF (2007) Review of the treatment options for chronic constipation. MedGenMed 9(2):25

Kaviarasan S, Vijayalakshmi K, Anuradha CV (2004) Polyphenol-rich extract of fenugreek seeds protect erythrocytes from oxidative damage. Plant Foods Hum Nutr 59(4):143–147

Kaviarasan S, Sundarapandiyan R, Anuradha CV (2007) Protective action of fenugreek (Trigonella foenum graecum) seed polyphenols against alcohol-induced protein and lipid damage in rat liver. Paper presented at VIIIth conference of the International Society for Trace Element Research in Humans (ISTERH), the IXth conference of the Nordic Trace Element Society (NTES), and the VIth conference of the Hellenic Trace Element Society (HTES). Springer, Dordrecht, pp 111–120

Khalid W, Gill P, Arshad MS et al (2022) Functional behavior of DHA and EPA in the formation of babies brain at different stages of age, and protect from different brain-related diseases. Int J Food Prop 25(1):1021–1044

Kharat M, McClements DJ (2019) Recent advances in colloidal delivery systems for nutraceuticals: a case study – delivery by design of curcumin. J Colloid Interface Sci 557:506–518

Khorshidian N, Yousefi Asli M, Arab M et al (2016) Fenugreek: potential applications as a functional food and nutraceutical. Nutr Food Sci Res 3(1):5–16

Kumar N, Kumar M, Verma MK et al (2021) Bioactive effects and safety profiles of fenugreek (Trigonella foenum-graecum L.) for pharmaceutical and medicinal applications. Pharma Innov 10(12):912–919

Laroubi A, Touhami M, Farouk L et al (2007) Prophylaxis effect of Trigonella foenum graecum L. seeds on renal stone formation in rats. Phytother Res 21(10):921–925

Latif AD (2015) Evaluation of the antimicrobial (antibacterial and antifungal) activity of ethanolic extracts of some medical plants. J Nat Sci Res 5(22):87–95

Li G, Luan G, He Y et al (2018) Polyphenol stilbenes from fenugreek (Trigonella foenum-graecum L.) seeds improve insulin sensitivity and mitochondrial function in 3T3-L1 adipocytes. Oxidative Med Cell Longev 2018:1–9

Madhava Naidu M, Shyamala BN, Pura Naik J et al (2011) Chemical composition and antioxidant activity of the husk and endosperm of fenugreek seeds. LWT – Food Sci Technol 44(2):451–456

Manzoor MF, Hussain A, Tazeddinova D, Abylgazinova A, Xu B (2022) Assessing the nutritional-valuebased therapeutic potentials and non-destructive approaches for mulberry fruit assessment: An overview. Computational intelligence and neuroscience, 2022.

Maqbool Z, Arshad MS, Ali A et al (2022) Potential role of phytochemical extract from saffron in development of functional foods and protection of brain-related disorders. Oxidative Med Cell Longev 2022:6480590

McClements DJ (2019) Food nanotechnology: harnessing the power of the miniature world inside our foods. In: McClements DJ (ed) Future foods, 1st edn. Springer Nature, Cham

Mirzaie M, Khalili M, Kiasalari Z, Roghani M (2016) Neuroprotective and antiapoptotic potential of trigonelline in a striatal 6-hydroxydopamine rat model of Parkinson's disease. Neurophysiology 48(3):176–183

Mukthamba P, Srinivasan K (2015) Beneficial hypolipidemic influence of a combination of dietary fenugreek (Trigonella foenum-graecum) seeds and garlic (Allium sativum) in induced hypercholesterolemic rats. Eur Food Res Technol 240(5):1049–1058

Mukthamba P, Srinivasan K (2017) Dietary fenugreek (Trigonella foenum-graecum) seeds and garlic (Allium sativum) alleviates oxidative stress in experimental myocardial infarction. Food Sci Hum Wellness 6(2):77–87

Navarro del Hierro J, Reglero G, Martin D (2020) Chemical characterization and bioaccessibility of bioactive compounds from saponin-rich extracts and their acid-hydrolysates obtained from fenugreek and quinoa. Foods 9(9):1159

Nayak AK, Pal D (2018) Functionalization of tamarind gum for drug delivery. In: Thakur VK (ed) Functional bioplomers, 1st edn. Springer Nature, Cham, pp 25–56

Nayak AK, Bera H, Hasnain M, Pal D (2018) Synthesis and characterization of graft copolymers of plant polysaccharides. In: Thakur VK (ed) Biopolymer grafting, 1st edn. Elsevier, Amsterdam, pp 1–62

O'Keefe JH, Captain BK, Jones PG, Harris WS (2004) Atorvastatin reduces remnant lipoproteins and small, dense low-density lipoproteins regardless of the baseline lipid pattern. Prev Cardiol 7(4):154–160

Ozturk M, Hakeem KR, Ashraf M, Ahmad MSA (eds) (2018) Global perspectives on underutilized crops. Springers Nature, Cham

Pal D, Mukherjee S (2020) Fenugreek (Trigonella foenum) seeds in health and nutrition. In: Preedy VR, Watson RR (eds) Nuts and seeds in health and disease prevention, 1st edn. Elsevier, Amsterdam, pp 161–170

Pandian RS, Anuradha CV, Viswanathan P (2002) Gastroprotective effect of fenugreek seeds (Trigonella foenum graecum) on experimental gastric ulcer in rats. J Ethnopharmacol 81(3):393–397

Passano P (1995) The many uses of methi. Manushi 2(1):31–34

Petropoulos GA (ed) (2002) Fenugreek: the genus Trigonella. CRC Press, Florida

Platel K, Srinivasan K (2000) Influence of dietary spices and their active principles on pancreatic digestive enzymes in albino rats. Nahrung – Food 44(1):42–46

Rababah TM, Ereifej KI, Esoh RB et al (2011) Antioxidant activities, total phenolics and HPLC analyses of the phenolic compounds of extracts from common Mediterranean plants. Nat Prod Res 25(6):596–605

Raju J, Bird RP (2006) Alleviation of hepatic steatosis accompanied by modulation of plasma and liver TNF-α levels by Trigonella foenum graecum (fenugreek) seeds in Zucker obese (fa/fa) rats. Int J Obes 30(8):1298–1307

Raju J, Gupta D, Rao AR et al (2001) Trigonellafoenum graecum (fenugreek) seed powder improves glucose homeostasis in alloxan diabetic rat tissues by reversing the altered glycolytic, gluconeogenic and lipogenic enzymes. Mol Cell Biochem 224:45–51

Rambhade S, Chakraborty A, Patil U, Rambhade A (2010) Diabetes mellitus – its complications, factors influencing complications and prevention: an overview. J Chem Pharm Res 2(6):7–25

Ramulu P, Giridharan NV, Udayasekhararao P (2011) Hypolipidemic effect of soluble dietary fiber (galactomannan) isolated from fenugreek seeds in WNIN (GR-Ob) obese rats. J Med Plant Res 5(19):4804–4813

Ranade M, Mudgalkar N (2017) A simple dietary addition of fenugreek seed leads to the reduction in blood glucose levels: a parallel group, randomized single-blind trial. Int Q J Res Ayurveda 38(1):24–27

Rashid F, Hussain S, Ahmed Z (2018) Extraction purification and characterization of galactomannan from fenugreek for industrial utilization. Carbohydr Polym 180:88–95

Rayyan S, Fossen T, Andersen ØM (2010) Flavone c-glycosides from seeds of fenugreek, Trigonella foenum-graecum L. J Agric Food Chem 58(12):7211–7217

Rehman T, Owais Ghauri A, Tayyeba Rehman C (2018) Nutritional and health benefits of fenugreek: a short review. Int J Home Sci 2(2):28–30
Riaz S, Ahmad A, Farooq R et al (2022) Edible mushrooms, a sustainable source of nutrition, biochemically active compounds and its effect on human health. In: Shiomi N, Savitskaya A (eds) Current topics in functional food, 1st edn. IntechOpen, London
Ruby BC, Gaskill SE, Slivka D, Harger SG (2005) The addition of fenugreek extract (Trigonella foenum-graecum) to glucose feeding increases muscle glycogen resynthesis after exercise. Amino Acids 28(1):71–76
Sfar M, Jaouani A, Ben-Mahrez K et al (2018) The effect of fenugreek on blood glucose, lipid parameters, insulin and resistin in obese women with type 2 diabetes. Int J Pharm Pharm Res 11(3):108–123
Shang M (1998) Studies of flavonoids from fenugreek (Trigonella foenumgraecum L.). Zhongguo Zhong Yao Za Zhi 23:614–616
Sharma RD (1984) Hypocholesterolemic activity of fenugreek (T. foenum graecum). An experimental study in rats. Nutr Rep Int 30(1):221–231
Sharma RD, Raghuram TC (1990) Hypoglycaemic effect of fenugreek seeds in non-insulin dependent diabetic subjects. Nutr Res 10(7):731–739
Sharma RD, Raghuram TC, Rao NS (1990) Effect of fenugreek seeds on blood glucose and serum lipids in type I diabetes. Eur J Clin Nutr 44(4):301–306
Sharma V, Singh P, Rani A (2016) Antimicrobial activity of Trigonella foenum-graecum L. (fenugreek). Eur Exp Biol 7(1):1–4
Sindhu G, Shyni GL, Pushpan CK et al (2018) Evaluation of anti-arthritic potential of Trigonella foenum graecum L. (fenugreek) mucilage against rheumatoid arthritis. Prostaglandins Other Lipid Mediat 138:48–53
Soetan KO, Oyewole OE (2009) The need for adequate processing to reduce the anti-nutritional factors inplants used as human foods and animal feeds: a review. Afr J Food Sci 3(9):223–232
Sreeja S, Anju VS (2010) In vitro estrogenic activities of fenugreek Trigonella foenum graecum seeds. Indian J Med Res 131(6):814–819
Srinivasan, K. (2005). Plant foods in the management of diabetes mellitus: spices as beneficial antidiabetic food adjuncts. Int J Food Sci Nutr 56(6), 399–414
Srinivasan M, Sudheer AR, Menon VP (2007) Ferulic acid: therapeutic potential through its antioxidant property. J Clin Biochem Nutr 40(2):92–100
Srivastava D, Rajiv J, Naidu MM et al (2012) Effect of fenugreek seed husk on the rheology and quality characteristics of muffins. Food Nutr Sci 03(11):1473–1479
Suresh P, Kavitha CN, Babu SM et al (2012) Effect of ethanol extract of Trigonella foenum graecum (fenugreek) seeds on Freund's adjuvant-induced arthritis in albino rats. Inflammation 35(4):1314–1321
Sureshkumar D, Begum S, Johannah NM et al (2018) Toxicological evaluation of a saponin-rich standardized extract of fenugreek seeds (FenuSMART®): acute, sub-chronic and genotoxicity studies. Toxicol Rep 5:1060–1068
Swaroop A, Bagchi M, Preuss HG, Bagchi D (2018) Safety and antidiabetic efficacy of a novel Trigonella foenum-graecum seed extract. In: Bagchi D, Nair S (eds) Nutritional and therapeutic interventions for diabetes and metabolic syndrome, 1st edn. Elsevier, Amsterdam, pp 357–364
Syed QA, Rashid Z, Ahmad MH et al (2020) Nutritional and therapeutic properties of fenugreek (Trigonella foenum-graecum): a review. Int J Food Prop 23(1):1777–1791
Tamiru Kasaye A (2015) Evaluation of composite blends of fermented fenugreek and wheat flour to assess its suitability for bread and biscuit. Int J Nutr Food Sci 4(1):29
Tavakoly R, Maracy MR, Karimifar M, Entezari MH (2018) Does fenugreek (Trigonella foenum-graecum) seed improve inflammation, and oxidative stress in patients with type 2 diabetes mellitus? A parallel group randomized clinical trial. Eur J Integr Med 18:13–17
Thomopoulos C, Parati G, Zanchetti A (2018) Effects of blood pressure-lowering treatment on cardiovascular outcomes and mortality. J Hypertens 36(8):1637–1647

Thorat R, Gaikwad D (2019) Pharmacognostical and phytophysicochemical investigations of Trigonella foenum – graecum Linn. J Drug Deliv Ther 9:138–145
Timilsena YP, Adhikari R, Kasapis S, Adhikari B (2016) Molecular and functional characteristics of purified gum from Australian chia seeds. Carbohydr Polym 136:128–136
Vickers CE, Williams TC, Peng B, Cherry J (2017) Recent advances in synthetic biology for engineering isoprenoid production in yeast. Curr Opin Chem Biol 40:47–56
Wang L, Zheng W, Yang J et al (2022) Mechanism of Astragalus membranaceus alleviating acquired hyperlipidemia induced by high-fat diet through regulating lipid metabolism. Nutrients 14(5):954
Wani SA, Kumar P (2016) Fenugreek enriched extruded product: optimization of ingredients using response surface methodology. Int Food Res J 23(2):630–637
Wani SA, Kumar P (2018) Fenugreek: a review on its nutraceutical properties and utilization in various food products. J Saudi Soc Agric Sci 17(2):97–106
Wielinga WC (2009) Galactomannans. In: Phillips GO, Williams PA (eds) Handbook of hydrocolloids, 1st edn. Elsevier, Amsterdam, pp 228–251
Wiley J, Thirunavukkarasu V, Anuradha CV, Viswanathan P (2003) Protective effect of fenugreek (Trigonella foenum graecum) seeds in experimental ethanol toxicity. Phytother Res 17(7):737–743
Yacoubi L, Rabaoui L, Hamdaoui MH et al (2011) Anti-oxidative and anti-inflammatory effects of Trigonella foenum-graecum Linnaeus, 1753 (fenugreek) seed extract in experimental pulmonary fibrosis. J Med Plant Res 5(17):4315–4325
Yoshinari O, Igarashi K (2010) Anti-diabetic effect of trigonelline and nicotinic acid, on KK-Ay mice. Curr Med Chem 17(20):2196–2202
Zameer S, Najmi AK, Vohora D, Akhtar M (2018) A review on therapeutic potentials of Trigonella foenum graecum (fenugreek) and its chemical constituents in neurological disorders: complementary roles to its hypolipidemic, hypoglycemic, and antioxidant potential. Nutr Neurosci 21(8):539–545
Zandi P, Basu SK, Cetzal-Ix W et al (2017) Fenugreek (Trigonella foenum-graecum L.): an important medicinal and aromatic crop. In: El-Shemy H (ed) Active ingredients from aromatic and medicinal plants, 1st edn. IntechOpen, London, pp 207–224
Zheng B, McClements DJ (2020) Formulation of more efficacious curcumin delivery systems using colloid science: enhanced solubility, stability, and bioavailability. Molecules 25(12):2791

Index

T. Ismail et al. (eds.), *Neglected Plant Foods Of South Asia*,
https://doi.org/10.1007/978-3-031-37077-9

M

N

O

P

Q

R

MIX
Papier aus verantwortungsvollen Quellen
Paper from responsible sources
FSC® C105338

If you have any concerns about our products,
you can contact us on
ProductSafety@springernature.com

In case Publisher is established outside the EU,
the EU authorized representative is:
Springer Nature Customer Service Center GmbH
Europaplatz 3, 69115 Heidelberg, Germany

Printed by Libri Plureos GmbH
in Hamburg, Germany